CHEMISTRY AND BIOLOGY OF SERPINS

ADVANCES IN EXPERIMENTAL MEDICINE AND BIOLOGY

CHEMISTRY AND BIOLOGY OF SERPINS

Edited by

Frank C. Church
University of North Carolina School of Medicine
Chapel Hill, North Carolina

Dennis D. Cunningham
University of California
Irvine, California

David Ginsburg
Howard Hughes Medical Institute
University of Michigan Medical Center
Ann Arbor, Michigan

Maureane Hoffman
Duke University and
Durham VA Medical Centers
Durham, North Carolina

Stuart R. Stone
Late of Monash University
Clayton, Victoria, Australia

and

Douglas M. Tollefsen
Washington University School of Medicine
St. Louis, Missouri

SPRINGER SCIENCE+BUSINESS MEDIA, LLC

Library of Congress Cataloging-in-Publication Data

Chemistry and biology of serpins / edited by Frank C. Church, ... [et
al.].
       p.   cm. -- (Advances in experimental medicine and biology ; v.
425)
    "Proceedings of the International Symposium on the Chemistry and
Biology of Serpins, held April 13-16, 1996, in Chapel Hill, North
Carolina"--T.p. verso.
    Includes bibliographical references and index.
    ISBN 978-0-306-45698-5   ISBN 978-1-4615-5391-5 (eBook)
    DOI 10.1007/978-1-4615-5391-5
    1. Serpins--Physiological effect--Congresses. 2. Serpins-
-Pathophysiology--Congresses.   I. Church, Frank C.
II. International Symposium on the Chemistry and Biology of Serpins
(1996 : Chapel Hill, N.C.) III. Series.
QP607.S47C48 1997
612'.01575--dc21   97-31140
                                                                  CIP

Proceedings of the International Symposium on the Chemistry and Biology of Serpins,
held April 13 – 16, 1996, in Chapel Hill, North Carolina

ISBN 978-0-306-45698-5

© 1997 Springer Science+Business Media New York
Originally published by Plenum Press, New York in 1997

http://www.plenum.com

10 9 8 7 6 5 4 3 2 1

STUART STONE 1951–1996

The inherent tragedy in the loss of a family man in the prime of life is such that it seems almost trivial to speak of the effect of the loss on his profession. Yet for Stuart Stone's colleagues there is a special sense of grief, not only for someone whose contributions helped establish the field we all work in, but also for someone whose career was just on the threshold of opening new understandings in our science.

Stuart Stone was born in Sydney on 1st November 1951 and he died, suddenly whilst swimming, on 16th December of 1996, shortly after his return to Australia to take up a senior Professorship at Monash University in Melbourne. His training had commenced as a student in Agricultural Science at the University of Sydney where he obtained his degree in 1973 with 1st class honours and won the University Medal for the outstanding graduate of the year. He stayed in Sydney to complete his PhD and was awarded a series of prizes and fellowships that took him to post-doctoral posts in Iowa State University from 1978–80 and as a Research Fellow to the John Curtin School of Medical Research in Canberra from 1980–83. During this time he carried out the studies that elucidated the catalytic mechanism of dihydrofolate reductase.

Stuart Stone's first substantive post, and the one in which he built the foundation for his career research, was in 1984 as a research scientist and then group leader in the Friedrich-Miescher Institut in Basel. Here over the next seven years, with Jan Hofsteenge, he carried out a series of key studies on the function of thrombin and the mechanism of its inhibition by hirudin. Their work, together with the structural group at Martinsried, showed that hirudin interacts with regions of thrombin distinct from the active site. The conclusions from these studies provided the basis for the subsequent development of the recombinant analogues of hirudin.

In 1991 Stuart joined me in the Department of Haematology in the University of Cambridge as a University Lecturer. This tenured post provided him with the chance to undertake full-time independent research. He rapidly built up a strong team with a core of post-doctorals but with its main strength in a lively and ever-changing group of graduate students. All of his graduates were trained to meet the standards he expected of himself — there were no compromises. I would sometimes hear him shouting from the floor below, "This is just not good enough!" to be followed seconds later by an angry and flustered graduate student coming up the stairs and striding out of the door. Next day all would be well and Stuart would have grown in the respect and affection of his research students.

A reason for Stuart's success was his approach to science. The alternative styles of research can be compared to that of artists, some fill their canvas with broad strokes and later add detail, others start with fine detail in one corner and work out from that to cover the canvas. The latter was Stuart's style. His constant careful attention to detail made him an excellent research supervisor and a sought-after participant in Gordon Conferences and the like. Here he was at his best — equally able to exchange technical data and to debate broad concepts. His knowledge, his open manner, good humour and optimism all added together to inspire others and to provide motivation to the field as a whole.

In Cambridge he commenced a series of kinetic studies on thrombin that, coupled with *in vivo* studies of inhibitors in animal models of thrombosis, have provided guidelines for the development of thrombin inhibitors as antithrombotic agents. In particular Stuart's later work, on the detailed interactions of peptides, receptors and serpins with the active site of serine proteases, formed the basis for the planned work in his new post, as Professor of Biochemistry and Molecular Biology in Monash. It is a great sadness that his sudden death came at a time when he was just completing the organization of the research that he realistically believed would give international leadership in the field by the turn of the millennium.

For those who aspire to success in research and career it is worth noting that Stuart Stone achieved this by placing his family first. His children: Brangwen, Meredith and Simon, and his wife Eleanor (herself a scientist of distinction) were at the centre of his life and a constant support to him in his work. Stuart had a quiet but deep religious faith and the family came to be much loved in the village of Newnham in Cambridge where Stuart was a member of the vestry and the family were loyal supporters of the parish church. Newnham, Monash, Basel grieve for him, as do all his colleagues in our small but interdependent field of research.

Robin Carrell

REMEMBERING STUART R. STONE (NOVEMBER 1, 1951–DECEMBER 16, 1996)

Dedication of the Serpin Symposium Proceedings

It seems like such a short time ago that I had the opportunity to be friends with Stuart Stone. He was very intelligent, hardworking, friendly, funny, personable, a family man who loved his wife and children very much, loyal to his lab, his University, true to his science, his family, and always available for advice, wisdom and words. He was a remarkable person. Susie Bauman, one of my graduate students, said that Stuart was her role model as a scientist, she admired his work very much. She had the opportunity, as we all did, to enjoy Stuart's company at the Serpin Symposium in Chapel Hill in April 1996. We then realized that all the things said about Stuart were true: he was a scholar, he was fun, he was caring, and he was an amazing person.

Recently, my wife, Gwyn Cutsforth, and I had dinner with Stuart and his wife, Eleanor Mackie, in London. My wife instantly liked both Stuart and Eleanor, and said that they were both so friendly, caring and very much in love.

Stuart touched everyone in his field of science. When I phoned, E-mailed or visited him he was always ready to talk science, to talk life experiences, or to talk about family and friends. I always wanted to have completed my best and most clever experiment when thinking about an upcoming chat with Stuart, I wanted to impress him and hoped that my science was as significant as his. But he was never disappointed in anyone's work, always supportive, always offering advice.

A saying by the great American football coach Vince Lombardi comes to mind when I think about the kind of person Stuart was: "The difference between men is in energy, in singleness of purpose, and in invincible determination. But the great difference between men is in sacrifice, in self-denial, in fearlessness and humility, in love and loyalty, and the perfectly disciplined will. This is not the only difference between men, but this is the difference between great men and little men." Stuart Stone was a giant among people, a leader among scientists. He has already been missed by all, he will continued to be missed by all. His legacy of excellence is established and his colleagues and friends will maintain and perpetuate his work as a scientist and as a person.

The Organizers of the Serpin Symposium unanimously agreed that it would be a fitting affirmation to dedicate this book in memory of Stuart R. Stone.

Frank C. Church, Ph. D.
School of Medicine
The University of North Carolina at Chapel Hill
March 1997

PREFACE

Serpins (*ser*ine *p*rotease *in*hibitors) are a superfamily of proteins whose physiological action is primarily targeted to inhibiting serine proteases. There are instances where serpins are not inhibitors (and can carry steroid hormones for instance), yet key structural and functional elements found in all serpins are maintained in these 'non-inhibitor' serpins. Many serpins have well-described biological properties which influence pathophysiological events, including: antithrombin (historically called antithrombin III), α_1-protease inhibitor (historically called α_1-antitrypsin), and plasminogen activator inhibitor-1, just to mention a few. A deficiency or defect in antithrombin leads to venous thromboembolic disease, while a deficiency or defect in α_1-protease inhibitor is associated with chronic obstructive pulmonary emphysema. In contrast, it has been suggested that increased levels of plasminogen activator inhibitor-1 may be a predisposition to myocardial infarction. The list goes on for each of our own "favorite" serpin. The biological roles found for serpins are key participants in almost every physiological event. In other words, serine proteases are needed for many events in biology and the role of serpins to down regulate these proteases is essential. Thus, just using these three examples above for serpins and their pathophysiological roles reminds us that the medical costs to control such events is significant worldwide.

From a biochemical perspective, serpins have become one of the model systems to study regarding protein structure and function relationships, and the docking of proteases to the reactive site loop of a serpin is also a model of protein:protein interactions. In fact, the "Holy Grail" of serpin protein biochemists must be the crystallization of a bimolecular complex between a serpin and its cognate protease. The general structural features of a serpin contain three β-sheets and nine α-helices. Serpins that are active as protease inhibitors have an exposed reactive site loop (to serve as a 'pseudosubstrate'), which resides between the A- and C-sheets. Serpins that are not active protease inhibitors have variations on this theme. There are other structural elements of importance in serpins outside of the reactive site loop, one notable example is the "heparin-binding" serpins (with the A- and D-helices) such as antithrombin and heparin cofactor II whose inhibitory action is enhanced by glycosaminoglycans like heparin and dermatan sulfate. Again, for every one's own "favorite" serpin there is a key structural element that points to an important biological function.

When I think about papers that have influenced me over the years in my research with serpins, I can think of many citations. I would like to mention just a few which were "milestone" papers to the serpin field, (I realize that each of us would have a different list of most important papers, so I apologize if your paper is not included or my list is too limited). Egeberg showed that a deficiency in antithrombin resulted in thrombophilia (1); Rosenberg and Damus described the mechanism of action of antithrombin in the presence of heparin (2); Hunt and Dayhoff reported the existence of a new superfamily with a surprising group of proteins including antithrombin, α_1-protease inhibitor and ovalbumin (3); Kurachi et al. cloned and sequenced α_1-protease inhibitor (4); Owen et al. described α_1-protease inhibitor Pittsburgh which occurred in a young boy that led to a fatal bleeding diathesis (5); Loebermann et al. presented the crystal structure of reactive site looped-cleaved α_1-protease inhibitor (6); and Carrell and Travis coined the term 'serpin' and reviewed their general features (7).

This brief historical and personal introduction reminds us that serpins are intimately involved in regulating serine proteases important in most physiological processes ranging from blood coagulation to fibrinolysis to inflammation to reproduction to tumor cell invasion. Further, this is the "golden-age" of protein biochemistry and molecular medicine and those that work in the field of serpins are poised to contribute lasting information about biochemical processes, pathological functions, and medical treatment for many of these essential pathways and processes.

Considering the advances made in our understanding of the chemistry, biology and physiology of serpins over the past decade, it seemed timely and apropos to have a meeting focused on serpins. The genesis of this symposium started at another meeting, the "Heparin and Related Polysaccharides" symposium held in Uppsala, Sweden (September 1991). I can distinctly remember talking to Doug Tollefsen, Ingemar Björk, Steve Olson, Ulrich Abildgaard, and Dennis Cunningham during this magnificent symposium about whether anyone remembered a meeting focused on serpins, and whether someone should organize or was planning such a meeting. Over the next two years, at every meeting I attended I asked the very same questions, and to my excitement everyone said unanimously, "let's do it!" It is somewhat foggy about what happened next, but it seems that I starting organizing the event during 1994. For many different reasons, it was decided that mid-April 1996 would be the best time.

I got the Program Committee organized and it consisted of Dennis Cunningham, David Ginsburg, Stuart Stone, and Doug Tollefsen. Also, a local advisory panel consisting of Don Gabriel, Maureane Hoffman, and Harold Roberts was formed. The next event which solidified the time was a trip I took in September 1994, to visit Stuart Stone and Robin Carrell in Cambridge, England. I spent some time talking with Robin and I mentioned the idea for the serpin meeting and he immediately embraced the concept and his secretary reserved this time in April 1996. Upon returning to Chapel Hill, I immediately contacted Ingemar Björk and Jim Travis to check their availability to participate in the symposium. Once done, I had gotten the "Holy Triumvirate" of serpin scientists (in my opinion) scheduled for the meeting! It was all downhill from there. In selecting speakers for the six different sessions planned for the symposium, my School of Medicine colleagues asked that we have many extra names selected because it was their experience that you had to go three-deep sometimes just to get one speaker to commit. I am proud to say that everyone instantly accepted the invitation to speak at the serpin meeting except for two cases: one declination was for surgery scheduled immediately prior to the Symposium, and one speaker canceled at the last minute because of a clinical scheduling conflict. It was very gratifying that everyone was excited about the serpin meeting.

For those of you who were able to attend the Serpin Symposium, the meeting was almost "magical": the science significant, the weather perfect, the Carolina Inn accommodations outstanding, and the social events were a great way to end a day of "serpin-oriented" science (many science and life stories were exchanged during the evening spectacles). The keynote address at the banquet by Jim Travis was one for the memories; I am sorry that it was not videotaped for posterity purposes. It was especially noteworthy to view the posters (which numbered close to 100 total) each day and to see such wonderful science and that every poster was dedicated to serpins. Finally, it is important to state that there were over 200 scientists, clinicians and students (graduate, medical and undergraduate) in attendance who represented 13 different countries (photo page 339).

I wish to thank my fellow Program Committee Members who took the time to develop and plan the agenda for the symposium and who enthusiastically endorsed the purpose of this meeting: Dennis Cunningham, Ph.D.; Don Gabriel, M.D./Ph.D.; David Ginsburg, M.D.; Maureane Hoffman, M.D./Ph.D.; Harold Roberts, M.D.; Stuart Stone, Ph.D.; and Doug Tollefsen, M.D./Ph.D. The collective input from each member contributed directly to the foundation on which this meeting was based, and their contributions insured for a successful meeting. It was an absolute delight to work with this group.

I also wish to thank the numerous financial grants and contributions by federal, state and by numerous industries that allowed us to have this symposium, this volume, and a permanent memory and record of the symposium. Without these generous contributions, the symposium would have been mediocre instead of the reality, a significant event. A list of contributors is included in this volume. I encourage all readers of this book to note each contributor on the list.

This meeting could not have been planned or even existed without the help of the Office of Continuing Medical Education at The University of North Carolina at Chapel Hill School of Medicine. This office provided constant guidance and logistical support for each phase of the meeting. In particular, two individuals should be acknowledged, Ms. Jane Radford and Dr. William Easterling. Jane was instrumental in making many of the arrangements for the meeting and she provided a constant source of encouragement during all of the planning stages and she was present during the symposium 'working tirelessly' and 'putting out fires.'

My most profound thanks go to Ms. Jaime Welch-Donahue of the Center for Thrombosis and Hemostasis of The University of North Carolina at Chapel Hill School of Medicine. Jaime was the mainstay of the entire mission for this meeting. Without her constant efforts to organize the symposium, to stay in contact with speakers, to keep us informed on fund raising, to serve as the liaison with CME, and to keep me focused on the calendar for when things "had to get done," this meeting would have never happened, it would have been a disaster. Her persistent efforts resulted in the success that became the Serpin Symposium. And her job did not end there, it is her vigilance that has gotten this volume to the publishers within a year of the actual meeting.

Finally, I would like to acknowledge my wife, Gwyn Cutsforth, Ph.D., and my laboratory group. Without their constant support and encouragement to organize this meeting, and their tolerance of my absence during its planning stages, this meeting would have never materialized. Their collective enthusiasm for the Serpin Symposium fueled my fire to go through with the necessary steps to host and plan this symposium. Without the help and support of everyone mentioned here (and many others not even noted), there would have been no symposium and no proceedings!

Frank C. Church, Ph.D.
School of Medicine
The University of North Carolina at Chapel Hill

Note Added in Proof: I am sorry to note that we learned of the death of Professor Ruth Sager, a contributor to this volume (and speaker at the Symposium) and a senior figure in the field, as this book entered production. Ruth was a wonderful person and a very successful scientist who late in her scientific career got involved in serpin science by discovering maspin. Her research in this area provided new emphasis and importance to both the basic and clinical sciences regarding the role of serpins in tumor cell biology.

REFERENCES

1. Egeberg, O. (1965) Inherited antithrombin deficiency causing thrombophilia. *Thromb. Daith. Haemorrh. (Stuttg.)* **13:** 516–530.
2. Rosenberg, R.D. and P.S. Damus (1973) The purification and mechanism of action of human antithrombin-heparin cofactor. *J. Biol. Chem.* **248:** 6490–6505.
3. Hunt, L.T. and M.O. Dayhoff (1980) A surprising new protein superfamily containing ovalbumin, antithrombin-III, and alpha 1-proteinase inhibitor. *Biochem. Biophys. Res. Commun.* **95:** 864–871.
4. Kurachi, K., T. Chandra, S.J. Friezner Degen, T.T. White, T.L. Marchioro, S.L.C. Woo, and E.W. Davie (1981) Cloning and sequence of cDNA coding for α_1-antitrypsin. *Proc. Natl. Acad. Sci. (U.S.A.)* **78:** 6826–6830.
5. Owen, M.C., S.O. Brennan, J.H. Lewis, and R.W. Carrell (1983) Mutation of antitrypsin to antithrombin. Alpha 1-antitrypsin Pittsburgh (358 Met leads to Arg), a fatal bleeding disorder. *N. Engl. J. Med.* **309:** 694–698.
6. Loebermann, H., R. Tokuoka, J. Deisenhofer, and R. Huber (1984) Human α_1-protease inhibitor. Crystal structure analysis of two crystal modifications, molecular model and preliminary analysis of the implication for function. *J. Mol. Biol.* **177:** 531–556.
7. Carrell, R.W. and J. Travis (1985) α_1-Antitrypsin and the serpins: variations and countervariation. *Trends Biochem. Sci.* **10:** 20–24.

ACKNOWLEDGMENTS

The editors wish to thank the following sources for their generous educational grants in support of the symposium.

Bayer Corp.
Baxter Healthcare Corporation
Biogen
Centeon
Chromogenix
Immuno–U.S., Inc.
LXR Biotechnology Inc.
Merck Research Laboratories
National Heart, Lung, and Blood Institute[*]
North Carolina Biotechnology Center[†]
Novo Nordisk A/S
N.V. Organon
Organon Teknika
Ortho Diagnostic Systems Inc.
Sanofi Winthrop Pharmaceuticals
Searle
SmithKline Beecham
Zeneca Pharmaceuticals
ZymoGenetics

---

[*] The Public Health Service participated in the support of this meeting under grant #1 R13 HL/HD56220-01, from (awarding units) NHLBI, NICHD.

[†] This material is based upon work supported in whole or part by the North Carolina Biotechnology Center. Any opinions, findings, conclusions or recommendations expressed in this publication are those of the author(s) and do not necessarily reflect the views and policies of the North Carolina Biotechnology Center.

SPONSORS

The International Symposium on the Chemistry and Biology of Serpins was presented by the Center for Thrombosis and Hemostasis and sponsored by the Office of Continuing Medical Education and Alumni Affairs of the University of North Carolina at Chapel Hill School of Medicine.

CONTENTS

IV. FIBRINOLYSIS

V. DEVELOPMENT AND REPRODUCTION

VI. INFLAMMATION

VI. NONINHIBITOR SERPINS

VII. ABSTRACTS

CHEMISTRY AND BIOLOGY
OF SERPINS

SERPINS

From the Way It Was to the Way It Is

James Travis

Department of Biochemistry and Molecular Biology
University of Georgia
Athens, Georgia 30602

During the period of 1950–1970 an immense interest developed in the field of proteinase inhibitors. This was primarily due to the work of Mike Laskowski, Jr. and Bob Feeney, each of whom had his own ideas on how such molecules functioned. Indeed, there was considerable controversy over whether tetrahedral or acyl intermediate complexes were the forms found when specific inhibitors interacted with their target proteinases. Significantly, none of the inhibitors investigated at that time were from blood, most being from either plants or eggs. It had already been known since the late 19th century that there was inhibitory activity towards proteinases in plasma, but it wasn't until the early 50s that a serious attempt was made to isolate these molecules. However, the real breakthrough did not occur until 1962 when Carl Laurell and Sten Eriksson found a deficiency in a plasma protein referred to as alpha-1-antitrypsin (alpha-1-AT) in individuals prone to developing emphysema at an early age, and this was the impetus needed for intense efforts to begin on the investigation of plasma proteinase inhibitors, particularly alpha-1-AT.

My introduction to this field also started in the late 60s but not on inhibitors. Rather, it began with an attempt to isolate and characterize human pancreatic proteinases. Together with Ron Roberts, Peter Mallory, and Michael Coan we succeeded in purify two forms of trypsin, two forms of chymotrypsin, and an enzyme referred to as proteinase E from this tissue. It was the isolation of human trypsin which brought me face to face with the inhibitor field, as I was invited in early 1970 to discuss the properties of this enzyme at the first meeting on Pulmonary Emphysema and Proteinase held in Duarte, CA. Quite frankly, I had never heard of alpha-1-AT, but by its very name I was assured that it would inhibit my enzyme and that probably human trypsin must be involved in the development of lung disease. What a leap of faith! I was astonished that clinicians could think that there was some way for trypsin to get into the lungs. Perhaps it was some new duct from the pancreas which had never been seen before! In any event, at this meeting it became clear to me that a biochemist could have a scientifically lucrative career in trying to unravel the role of alpha-1-AT in the protection of the lung against emphysema.

Chemistry and Biology of Serpins, edited by Church *et al.*
Plenum Press, New York, 1997

My first trials at obtaining pure alpha-1-AT (Don't worry, I'll get around to the name change soon.) were fraught with disaster. I had an excellent student named Ralph Pannell who tried several ways to isolate the inhibitor, the most ingenious I thought being the use of sepharose-trypsin. We could easily bind all of the activity, but we never recovered it from the column, even after treatment with acid pH, and this was the first evidence that we might be dealing with an inhibitor which had significantly different properties from the well characterized plant and egg white proteins. A breakthrough for us was reached through pure serendipity when I visited a colleague, Uly Seal, at the VA hospital in Minneapolis. Uly was also working on the isolation of proteins from plasma and was using a large sephadex sizing column in one of his purification steps. He had added blue dextran to his sample so that he could easily detect the void volume on his column, a technique commonly utilized at the time. However, what I noted was the fact that some of the blue color was retarded, obviously binding to albumin. The latter had been a terrible contamination problem for anyone trying to isolate virtually any plasma protein since it represented approximately 50% of the protein in this fluid. Back in Georgia we quickly seized upon this binding property of albumin and obtained several Cibacron dyes from Ciba-Geigy which we linked to sepharose. It was a colorful time, to say the least, and we soon found that Cibacron Blue (the dye which is the blue in blue dextran) was the best choice for binding albumin, while either only weakly binding other proteins. In particular, alpha-1-AT did not bind at all to this dye-matrix, and we were quickly able to purify the inhibitor. Of course, today a lot of inhibitors and other proteins are being routinely isolated by including a dye-matrix sepharose column as one of the purification steps, usually to get rid of the albumin contaminant.

The term alpha-1-antitrypsin and alpha-1-antichymotrypsin (alpha-1-Achy) had been coined earlier by Norbert Heimburger and his group at Behringwerke, in Germany. It still isn't clear to me how they came upon those names since both can inhibit chymotrypsin. Perhaps, it was the fact that alpha-1-Achy could only inhibit this enzyme and not trypsin which resulted in its name. Thankfully, the true function of alpha-1-AT became clearer because of the work of the late Aaron Janoff who pioneered in determining the role of neutrophil enzymes in phagocytosis and in disease. His was a labor of love, and we not only became friends but also fierce competitors in trying to understand the relevance of both inhibitors and proteinases in connective tissue disease. From his data it was clear that there was a major proteinase burden from neutrophils (now known to be due to elastase, cathepsin G, and proteinase III) which was placed on inflamed and/or infected tissues and that alpha-1-Proteinase Inhibitor (alpha-1-PI) must be involved in regulating at least part of this activity. We had renamed this protein primarily because it was a far better inhibitor of neutrophil elastase than trypsin, although, in retrospect, a better name might have been alpha-1-elastase inhibitor. However, we didn't want to cause any more controversy at that time in trying to describe its function and, therefore, utilized a more generic term.

It was not until the late 70s that the mystery as to how alpha-1-PI, neutrophil elastase, and cigarette smoke were involved in the development of pulmonary emphysema became untangled. One of the first clues came while I was on sabbatical with Alan Barrett. Although my primary goal at that time was to investigate macrophage proteinases, I soon realized that a) this was not going to be an easy task and b) I was in a laboratory where cysteine proteinases were the hot item. I, therefore, came back to my first love, the plasma inhibitors, and began to investigate whether they could inactivate papain and cathepsin B. There was no evidence for inhibition by alpha-1-PI, but I did the unusual experiment to determine whether the inhibitor/cysteine proteinase mixture was still functional against trypsin. Surprisingly, it was not so I decided to run some gels and see what had occurred.

This intrigued Barrett, in fact so much so that when I came in one morning I found that the gel stainer was missing. Alan had taken it home to see the results for himself! What we both noted was that the inhibitor had been cleaved by both papain and cathepsin B (much later we found that it had really been contaminating cathepsin L which had been the culprit), and this was the first evidence of the instability of a serpin to proteinases with which it could not complex.

Eventually, Dave Johnson and I took advantage of this specific cleavage and inactivation of alpha-1-PI to determine the reactive site sequence of the inhibitor, and when we noted a methionine residue at the reactive site, suggested at about the same time but not actually proven by Janoff, it didn't take us long to come to the conclusion that oxidation of this residue by cigarette smoke and/or endogenous oxidants was responsible for the reduction of inhibitory inactivity, increased neutrophil proteinase burden, and uncontrolled connective tissue disease associated with smokers emphysema. This was neatly confirmed by the first really good kinetic data to indicate the biological properties of serpins which was performed by Jo Bieth while visiting in my laboratory, working with Keith Beatty, and by Jim Powers at Georgia Tech. Individually, each showed that if you used either oxidized alpha-1-PI (our laboratory) or oxidized synthetic substrates containing methionine in the P_1-position (Powers group) there was a marked reduction in the ability of the oxidized serpin to complex neutrophil elastase or for the latter to hydrolyze the oxidized substrates. It was at this time that Bieth also showed that cathepsin G and mast cell chymase were regulated by alpha-1-Achy, finally giving a putative function for this primary acute phase protein.

The mechanism by which alpha-1-PI inhibited serine proteinases was totally misunderstood by all of us in the field, primarily because of the dogged work of Mike Laskowski, Jr. who was convincing in his hypothesis that there was a standard mechanism for proteinase inhibition. Unfortunately, that didn't appear to apply to the plasma-derived inhibitors since, unlike the well characterized plant, egg white, or tissue inhibitors, a) no one could easily demonstrate reversible complex formation, b) there were few if any disulfide bonds to hold together the reactive site loop, c) complexes with target proteinases appeared to be stable in SDS, and d) inhibitory activity was rapidly lost in the presence of non-target proteinases. It became even clearer that these were very different types of proteinase inhibitors when Huber and his colleagues crystallized a post-complex form of alpha-1-PI in which the reactive site P_1 and P_1'-residues had moved apart by more than 70 Å. Nevertheless, I believe that at least part of the standard mechanism for proteinase: proteinase inhibitor complex formation must be observed in the case of the plasma inhibitors. It is the details of this mechanism which seem to be so controversial at this time, including our own suggestion that a reversible tetrahedral complex is formed with functional plasma inhibitors. Obviously, this will all play out if and when someone crystallizes and resolves the structure of a native plasma inhibitor:native proteinase complex.

The story of how Serpins got their name is somewhat unusual. While I was on sabbatical during 1975–76 with Barrett at the Strangeways Laboratories on the outskirts of Cambridge, Robin Carrell was in the middle of the city trying to determine the structure of alpha-1-PI (He still calls it alpha-1-AT, but clinicians are always resistant to change!). The day after I returned to the United States he came to Alan's laboratory to visit me, but it was too late. It wasn't until the first Gordon Conference on Proteinases and Inhibitors in 1980 that we met. After that, we had a lot of correspondence and we both agreed that we should begin a collaboration which was initiated during a sabbatical which I took in 1980 to Robin's laboratory in Christchurch, New Zealand. It wasn't long after I arrived that he proposed the term Serpin to cover a new superfamily of proteins which inhibit serine proteinases. I knew that this might meet with controversy, particularly from Mike Laskowski, but I agreed to rework the paper, adding a few suggestions of my own. However, it should be made quite clear that the term Serpin was far more an idea of Robin's than mine.

Quite obviously, we were not the first to see a relationship between alpha-1-PI and other proteins. This had been first observed by the late Margaret Dayhoff who was truly the primary mover in developing systems for following protein alignments and, therefore, evolutionary relationships. However, as other sequences were elucidated, either through protein or gene analysis, it soon became clear that there was a family of proteins emerging, most of which had inhibitory activity, some of which needed cofactors for activity, and a few of which were not inhibitors at all. Indeed, there are now over 60 different serpins which have been identified, either as purified proteins or through gene sequence analysis, and it is likely that many more are yet to come.

So, where do we go from here? Obviously, someone is going to have to crystallize and resolve the structure of a native serpin:native proteinase complex, in order to finally put to rest the controversy as to how complexes are formed and/or stabilized? Frankly speaking, while understanding the mechanism of action of serpins as proteinase inhibitors is certainly of fundamental interest, it appears to me that more research could be performed to determine potential therapeutic uses for serpins, not only in controlling host proteinases but also in regulating those from parasites. In addition, there is now ample evidence to indicate that serpins can also regulate cysteine proteinases, as confirmed by the recent data indicating that a viral serpin (CrmA) can regulate the cysteine proteinase ICE. Thus, it is surprising that there is not more interest in trying to develop mutant serpins which might be useful in reducing tissue damage during infection and/or inflammation, as well as pathogen growth and proliferation. Surely, with the overwhelming evidence indicating a major upshift in the occurrence and resistance of pathogens to antibiotics, this latter area should be a ripe area for future research.

In looking back, it is apparent to those of us who have worked with serpins for most of our career that interest in this group of proteins has increased dramatically over the past twenty five years. However, as Sinatra puts it in a well known song of the 50s and 60s, "The best is yet to come and won't it be fine." As an eternal optimist, I'm counting on it.

2

SERPINS

A Mechanistic Class of Their Own

Stuart R. Stone, James C. Whisstock, Stephen P. Bottomley, and
Paul C. R. Hopkins[*]

Department of Haematology
University of Cambridge
MRC Centre
Hills Road, Cambridge CB2, 2QH
United Kingdom

There has been great progress over the last five years in our understanding of serpin structure. Despite this progress, the inhibitory mechanism of serpins is not well understood and much remains to be elucidated. It is clear, however, that the inhibitory mechanism of serpins is more complex than that of other inhibitors of serine proteases. In this article, we will contrast the inhibitory reaction of the so-called "standard-mechanism" inhibitors with that of serpins, highlighting important differences. In particular, we will focus on the role of conformational changes in the inhibitory mechanism.

1. STANDARD-MECHANISM INHIBITORS

Standard-mechanism inhibitors of serine proteases are relatively small proteins; they have usually less than 100 residues compared with the 400 residues of serpins. The standard mechanism inhibitors can be divided into a number of families, with the Kunitz and Kazal families being among the best known (1). They have been extensively studied by X-ray crystallography and NMR; numerous structures of both free inhibitor and protease-inhibitor complexes are available (2). Like serpins, standard-mechanism inhibitors bind to the active site of their target protease by a surface loop known as the reactive-site loop. This reactive-site loop serves as a pseudo-substrate for the protease and docks to its active site in a substrate-like manner. The structure of the reactive-site loop of standard mechanism inhibitors is the same in both the unbound and bound forms of the inhibitor. More-

[*] Current addresses: Department of Biochemistry and Molecular Biology, Monash University, Clayton, Victoria 3168, Australia (JCW, SPB); Gladstone Institute of Cardiovascular Disease, University of California, San Francisco 94141-9100(PCRH), deceased (SRS).

Chemistry and Biology of Serpins, edited by Church *et al.*
Plenum Press, New York, 1997

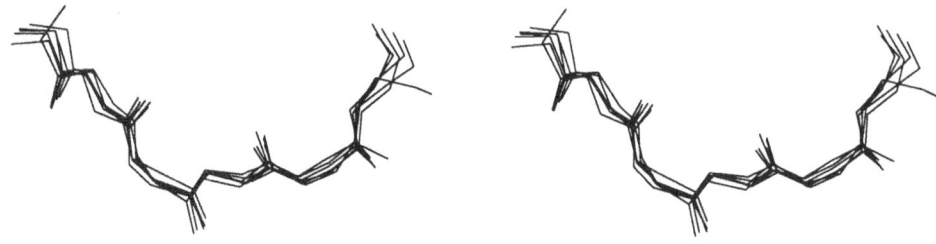

Inhibitor Loops   Inhibitor Loops

Figure 1. Comparison of the structure of the reactive-site loops of different standard-mechanism inhibitors. A superimposition of the reactive-site loops of six different standard-mechanism inhibitors is shown. The inhibitors shown are bovine pancreatic trypsin inhibitor, pancreatic secretory trypsin inhibitor, eglin c, barley chymotrypsin inhibitor 2, turkey ovomucoid inhibitor domain 3, and streptomyces subtilisin inhibitor. The co-ordinates of these and other structures were taken from the Brookhaven Protein Databank. The structures of the peptide backbones of the inhibitors differ only slightly in spite of large differences in other structural elements between the different inhibitors (2). The conserved conformation of the reactive-site loop is termed the "canonical conformation".

over, the same structure is exhibited by different inhibitors families which have no sequence homology (Figure 1). This canonical structure is complementary to the protease's active site; it permits an antiparallel β-strand interaction with the active-site and orients the side chains of the inhibitor towards the appropriate binding pockets in the protease (Figure 2). The residues within the reactive-site loop are numbered from N- to C-terminus as follows: $P_n-...-P_3-P_2-P_1-P_1'-P_2'-P_3'...-P_n'$, where cleavage would occur at the P_1-P_1' bond in substrates (3). The binding sites on the protease for these residues are similarly denoted $S_n-...-S_3-S_2-S_1-S_1'-S_2'-S_3'...-S_n'$.

The canonical conformation of the reactive-site loop is stabilized by intramolecular bonds (disulphide bridges or hydrogen bonds) between the loop and the body of the inhibitor. The tightness of the complex between standard-mechanism inhibitors and their

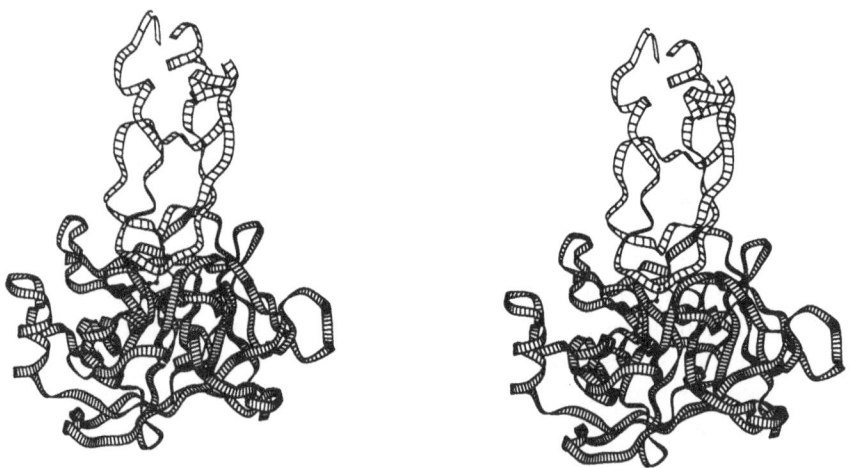

Figure 2. Complex between trypsin and bovine pancreatic trypsin inhibitor. The α-carbon trace of trypsin (dark) and bovine pancreatic trypsin inhibitor (light) are shown. The reactive-site loop of the inhibitor docks to the active site of trypsin; the P_1 residue of the inhibitor is shown in ball-and-stick form. The conformation of the reactive-site loop is complementary to that of the active site of trypsin.

target protease is due to the complementarity of the inhibitor's reactive-site loop and the active site of the protease. Moreover, because the canonical conformation is stabilized in a form appropriate for binding to the protease's active site, very little conformational entropy is lost upon binding. The stabilization of the canonical conformation by intra-molecular interactions also has important implications for the inhibitory mechanism which is summarized by the following scheme:

$$E + I \longleftrightarrow EI \longleftrightarrow EI^* \longleftrightarrow EI' \longleftrightarrow E + I'$$

In this scheme, the inhibitor (I) binds to the protease (E) to form an initial complex (EI) which is equivalent to the Michaelis complex in substrate reactions. The reaction of the inhibitor with the protease then proceeds in a similar manner to substrate cleavage with the final step being the release of cleaved inhibitor (I'). The intermediate EI* complex actually represents a number of distinct species on the reaction pathway including tetrahe-dral and acyl intermediates, while the EI' complex is a Michaelis complex with cleaved inhibitor. The tetrahedral and acyl intermediates involve a covalent bond with the active-site serine. For standard-mechanism inhibitors, all the steps in the reaction pathway are reversible. It is possible to obtain intact inhibitor after incubating cleaved inhibitor with the protease (1). This reversibility of the reaction is due to the efficient stabilization of the canonical conformation in which the ends of the cleaved bond are held in close proximity and, thus, the resynthesis of the bond is favoured.

Although there is kinetic evidence for all the intermediates in the above scheme, all the crystal structures of complexes exist in the Michaelis (EI) form (2). This predomi-nance of the Michaelis intermediate is consistent with the observed small contribution to the stability of the complex made by interactions with the active-site serine. Relatively tight complexes are formed between standard-mechanism inhibitors and proteases in which the active-site serine has been chemical modified to yield anhydroalanine (1). Such proteases are catalytically inactive and, thus, incapable of forming either tetrahedral or acyl intermediates. Consequently, it can be concluded that these covalent complexes are of relatively minor importance.

2. SERPINS ARE NOT STANDARD-MECHANISM INHIBITORS

Serpins differ from standard-mechanism inhibitors in a number of important aspects: 1. the reactive-site loop of serpins is flexible and adopts a number of different conforma-tions; 2. rearrangement of the reactive-site loop is an integral part of the reaction mecha-nism; i.e., unlike standard-mechanism inhibitors, the structure of the reactive-site loop changes upon complex formation; 3. release of the cleaved serpin from the protease-serpin complex is irreversible; 4. the active-site serine of the protease plays a critical role in the formation of the stable complex. These points will be discussed in detail in the following sections.

2.1. Flexibility of the Serpin Reactive-Site Loop

The principal difference between serpins and standard-mechanism inhibitors is that the reactive-site loop of serpins is flexible and can assume a number of different confor-mations. Three of the observed conformations are shown in Figures 3–5. The first serpin structure to be determined was that of cleaved α_1-antitrypsin (4). In this structure, the

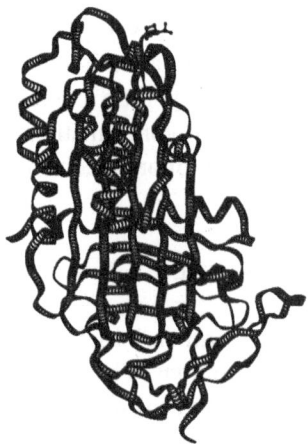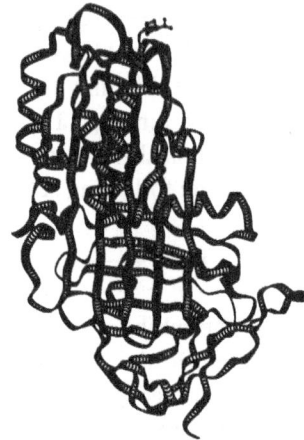

Figure 3. Structure of cleaved α₁-antitrypsin. The α-carbon trace is shown together with the P_1 methionine in ball-and-stick form (at the bottom of the model). The reactive-site loop is inserted as a central strand of the A β-sheet which is at the front of the model. The B β-sheet runs perpendicular to the A β-sheet at the back of the model. The C β-sheet is located at the top. The F helix runs obliquely to the A β-sheet at the bottom of the model.

reactive-site loop from P_3–P_{15} was incorporated as the central strand in the A β-sheet in the centre of the molecule. It has subsequently been shown that an intact reactive-site loop can also be incorporated into the A β-sheet in the latent form of serpins (5,6); serpins are termed latent if they possess no inhibitory activity, but can be reactivated by denaturation and refolding. In contrast to cleaved α₁-antitrypsin, the reactive-site loop of active antithrombin and other active serpins is not incorporated into the A β-sheet. The conformation of the reactive-site loop has, however, varied greatly ranging from the partial β-strand character seen for antithrombin in Figure 4 to a distorted helix for a variant of α₁-antichymotrypsin (7). Most recently, a structure very similar to the canonical conformation has been observed for an uncleaved variant of α₁-antitrypsin (8). In view of the

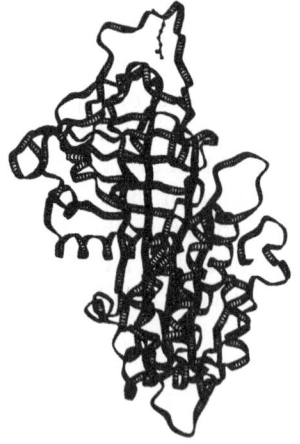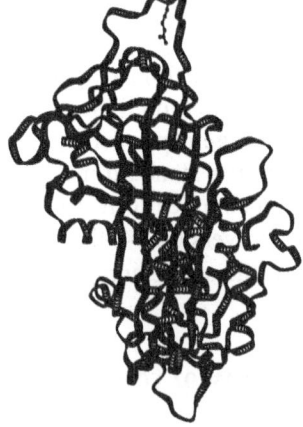

Figure 4. Structure of active antithrombin. The orientation of the model is the same as that shown for cleaved α₁-antitrypsin in Figure 3. The reactive-site loop is intact with the P_1 arginine pointing towards the body of the molecule. There is partial insertion of the reactive-site loop into the top of the A β-sheet.

Figure 5. Structure of active α_1-antitrypsin. The orientation of the model is the same as that shown for cleaved α_1-antitrypsin in Figure 3. However, the reactive-site loop is intact with the P_1 methionine in a conformation suitable for docking with the target protease. The reactive-site loop is in a canonical conformation without insertion of residues into the top of the A β-sheet.

diverse conformations observed for the reactive-site loop of serpins, it can be concluded that unlike the standard-mechanism inhibitors, the reactive-site loop is not stabilized into a canonical conformation by intramolecular bonds. Furthermore, the tight protease-serpin complex is not the result of the binding of a preformed reactive-site loop. Moreover, examination of the structure of cleaved α_1-antitrypsin indicates that the release of cleaved serpin will be irreversible. The N- and C-terminal ends of the cleaved bond are separated by 70 Å and the reactive-site loop is buried in the A β-sheet. Consequently, the protease will not be able to bind to this form of the serpin and resynthesize the cleaved bond in contrast to standard-mechanism inhibitors.

2.2. Mobility of the Reactive-Site Loop and the Inhibitory Mechanism

The ability of the serpin reactive-site loop to assume a number of different conformations appears to be central to the inhibitory mechanism. In order to bind to the active site of the target protease, the reactive-site loop of the serpin must adopt a canonical conformation similar to that shown in Figure 1; other conformations that have been observed for the reactive-site loop are not suitable for binding to the active site. A highly conserved region at the N-terminal end of the reactive-site loop appears to be very important for productive binding of the reactive-site loop to the active site of its target protease. There is evidence from the behaviour of natural and recombinant mutants that insertion of one or more residues of this region into the A β-sheet is essential for the inhibitory activity of serpins (9–13). The region of the reactive-site loop involved in this insertion consists of residues P_{10}–P_{15} and has been termed the hinge region. Although the insertion of all these residues is not required for the formation of the canonical conformation (see Figure 5), the residues also participate in a turn out of the A β-sheet (see Figures 4 and 5). As the name of the region implies, it is involved in hinging the rest of the reactive-site loop to the A β-sheet. The sequence of the hinge region is highly conserved in inhibitory serpins; the residues in this region have small side chains and are well adapted to forming turns. Point mutations in the hinge region which would either restrict the ability of the region to insert

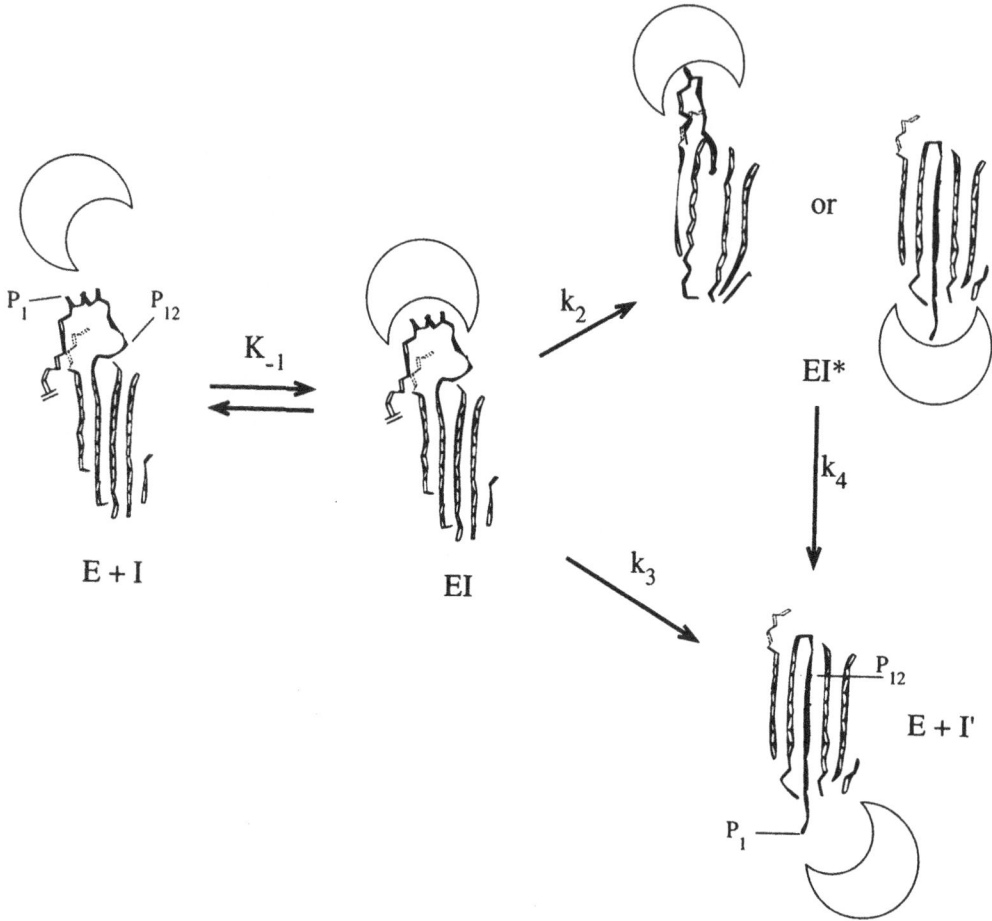

Figure 6. Suicide-substrate mechanism. Only the reactive-site loop, strand 1C and the A β-sheet of the serpin are shown. The protease is represented by the half-moon symbol. The structures of the intact and cleaved serpins are based on ovalbumin and α_1-antitrypsin, respectively. The structures of both protease-serpin complexes (EI and EI*) are not known. Two possible structures are shown for the final complex, with the protease at the top or bottom of the A β-sheet.

into the A β-sheet, or to make a turn, decrease the rate of formation of the stable protease-serpin complex and also increase the fraction of the serpin that is inactivated by cleavage (see Figure 6).

The observation that under some conditions, cleaved (inactive) serpin is released during the formation of the inhibited complex led to the proposal that serpins act as suicide-substrate inhibitors (14). The term suicide substrate was first used for synthetic inhibitors; the enzyme metabolized these compounds like substrates to generate a reactive species that was capable inactivating the enzyme. However, inactivation of the enzyme was not observed for every substrate-like reaction and a fraction of the suicide substrate was metabolized to an inactive form before the enzyme was fully inhibited. This behaviour is similar to that observed for serpins: a fraction of the serpin is released in an inactive form during the course of the inhibition reaction. The reaction scheme presented in Figure 6 represents a minimal reaction pathway for the serpin-protease interaction. After the formation of an initial (Michaelis) complex (EI), the reaction partitions between an inhibitory path leading to the formation of the stable complex (EI*) and a cleavage route resulting in the release of cleaved serpin (I') and free protease (E). The stable complex may also break down slowly to yield cleaved serpin (15,16). The two most important constants associated with the suicide-substrate mechanism of Figure 6 are the rate constant for the formation of stable complex (k_{on}) and the stoichiometry of inhibition (SI). Most serpin-protease reactions are studied under conditions where an insignificant concentration of the initial complex forms, and in this case, the rate constant for the formation of the stable complex will be k_2/K_{-1}. The stoichiometry of inhibition (SI) is the number of molecules of serpin that are required to inhibit one molecule of protease; this ratio is determined by the relative rates of the two pathways such that SI equals $1 + k_3/k_2$. Mutations in the hinge region can lead to marked decreases in the value of k_{ass} with concomitant increases in the SI value (10–12). These results are consistent with mutations in the hinge region affecting the value of the rate constant for the formation of the stable complex from the initial one (k_2); i.e., the mobility of the hinge region is important for the formation of the stable complex.

2.3. Interactions with the Active-Site Serine of the Protease Are Important in the Formation of the Serpin–Protease Complex

In contrast to standard-mechanism inhibitors for which interactions with the active site of the protease play a relatively minor role in stabilizing the complex, such interactions are crucially important for the stability of the protease-serpin complex. Serpins form extremely tight complexes with their target proteases; the dissociation constants of these complexes determined in kinetic experiments is usually less than 1 nM. Modification of the active-site serine of the protease results in a dramatic increase in the dissociation constant (K_i) for protease-serpin complex. For instance, chemical modification of the active-site serine of trypsin to yield anhydroalanine markedly increased the dissociation constant of trypsin with α_1-antitrypsin, antithrombin and plasminogen activator inhibitor-1 (PAI-1). The K_i value for the inactive trypsin with PAI-1 was 0.23 μM, whereas the values with antithrombin and α_1-antitrypsin were greater than 10 μM (17). Other studies have found somewhat tighter binding of serpins to proteases with a modified active-site serine. Lijnen et al. (18) estimated a K_i value of 3–5 nM for the complex of PAI-1 with the Ser[195]→Ala mutant of tissue-plasminogen activator (tPA); anhydro-urokinase, in which the active-site serine had been chemically modified to anhydroalanine, exhibited a similar K_i with PAI-1. The K_i value for serpins with the Ser[195]→Ala mutant of thrombin varied from 1 nM for protease nexin-1 in the presence of heparin to 3 μM for antithrombin in the absence of heparin (19). Although the importance of the interactions with the active-site serine varies from one protease-serpin pair to another, it can be calculated that such interactions contribute roughly 50% of the binding energy of the complex. However, the wide range of K_i values observed for the complexes of serpins with catalytically inactivated proteases suggests that the relative importance of the active-site serine varies between protease-serpin interactions; other contacts stabilize the complex to varying extents. Thus, complementarity between the reactive-site loop of the serpin and the active site of the protease may play a significant role in stabilizing some complexes.

Although the structure of the protease-serpin complex is not known, there are three plausible models (20). The reaction mechanism for cleavage of a peptide bond consists of the formation of an initial (Michaelis) complex followed by a tetrahedral intermediate involving the active-site serine and carbonyl group of the scissile bond. This tetrahedral intermediate breaks down to yield an acyl-enzyme in which the peptide on the N-terminal side of the scissile bond is covalently bound to the protease while the C-terminal fragment is released. The N-terminal fragment of the peptide is released after the formation of another tetrahedral intermediate (20). It is thought that the protease-serpin complex takes the form of one of these intermediates on the catalytic pathway; i.e., the Michaelis, tetrahedral or acyl intermediates. As noted above, the stable complex between serine proteases and standard-mechanism inhibitors takes the form of a Michaelis complex (2). The large effects of modification of the active-site serine on the stability of the protease-serpin complexes indicate that these complexes do not exist in the Michaelis form. The large effects of the mutation are consistent with the complex taking the form of either a tetrahedral or acyl-enzyme intermediate. There is experimental evidence for both forms of the complex. NMR studies of the complex between pancreatic elastase and α_1-antitrypsin suggest that this complex exists as a tetrahedral intermediate (21). In other studies, the α-amino groups of the P_1' residues of PAI-1 and α_1-antitrypsin were found to be accessible to chemical modification in complexes with proteases, suggesting that the P_1-P_1' bond is cleaved and that the complex exists as an acyl-intermediate (22,23). The effects seen when Ser^{195} is mutated or modified are consistent with both the tetrahedral intermediate and acyl-enzyme models. The catalytic activity of Ser^{195} is central to both models; both involve a covalent bond with the hydroxyl of Ser^{195}. Moreover, the decreases in binding energy for serpin-protease complexes caused by the modification of $Ser^{195} \rightarrow Ala$ (20–30 kJ mol^{-1}; refs. 17,19) are of the same order as the contribution of interactions with Ser^{195} to stabilization of the transition state in hydrolysis (24,25).

2.4. Extent of Insertion of the Reactive-Site Loop into the A β-Sheet during Complex Formation

Although it is apparent that some insertion of the reactive-site loop is essential for the formation of the serpin-protease complexes (see section 2.2), the extent of insertion of the reactive-site loop in the stable complex is unknown; Figure 6 indicates schematically two possibilities. Wright and Scarsdale (26) have proposed that the reactive-site loop of

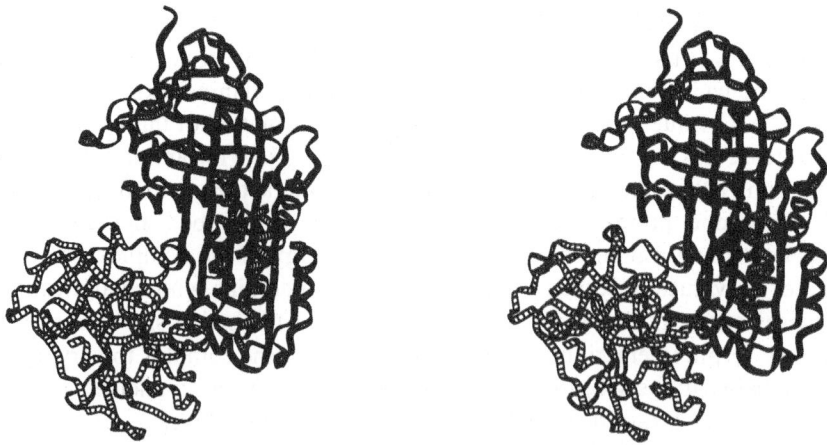

Figure 7. Model of the protease-serpin complex with the reactive-site loop fully inserted into the A β-sheet. The model is based on the structures of human neutrophil elastase (light) and α_1-antitrypsin (dark). The complex was modelled assuming that the side chain of the P_1 residue was bound in the primary specificity pocket of elastase and the orientation of elastase adjusted to avoid steric clashes. Details of the model will be published elsewhere.

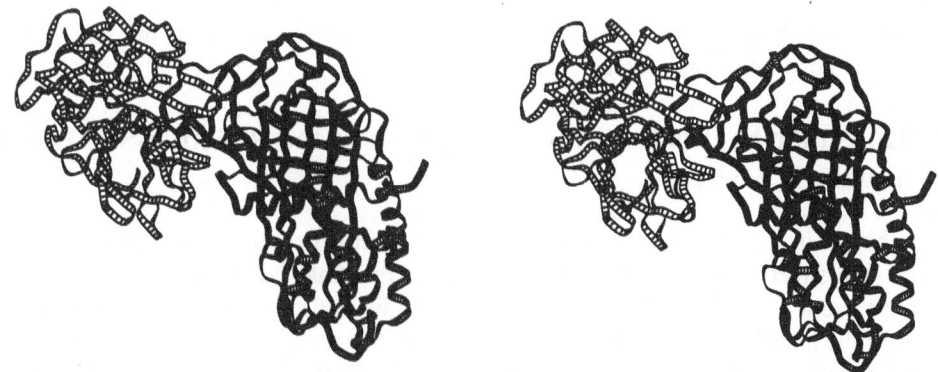

Figure 8. Model of the protease-serpin complex with the intact reactive-site loop only inserted into the A β-sheet as far as the P_{12} residue. The model is based on the structures of human neutrophil elastase (light) and α_1-antitrypsin (dark). The complex was modelled as described in Whisstock et al. (28).

the serpin is fully inserted into the A β-sheet in the stable complex as shown in Figure 7. The full insertion results in the translocation of the protease from the top to the bottom of the serpin during formation of the complex. Experimental evidence for full insertion of the reactive-site loop comes from results obtained with a variant of PAI-1 labelled with a fluorescent probe on the P_9 residue (27). In the structure of latent PAI-1, the reactive-site loop is fully inserted into the A β-sheet and the P_9 residue packs underneath the F helix which runs across the top of the A β-sheet (see Figure 3). The fluorescent label on the P_9 residue moves to a more hydrophobic environment in latent and cleaved PAI-1 consistent with the packing of this residue under the F helix. In the complex between PAI-1 and tPA, the fluorescent probe on P_9 also moves to a similar hydrophobic environment, suggesting that the P_9 residue is located under the F helix in the complex. Such a position for the P_9 residue would be consistent with full insertion of the reactive-site loop and translocation of the protease in the stable complex. Molecular modelling, however, suggests that the P_9 residue can move to a hydrophobic environment without full insertion of the reactive-site loop. Figure 8 shows a model of the complex between elastase and α_1-antitrypsin in which the reactive-site loop is uncleaved and insertion into the A β-sheet only occurs as far as P_{12} (28). Even with this small insertion, the P_9 residue packs against the F helix. Data obtained with a variant of α_1-antitrypsin containing a P_{10} glutamate also argue against full insertion of the reactive-site loop in the stable complex. Although the kinetic parameters for the formation of the complex between elastase and the P_{10}Glu variant of α_1-antitrypsin were indistinguishable from those of wild-type α_1-antitrypsin, the complex with P_{10}Glu-antitrypsin was not stable; it broke down to yield cleaved serpin (12). The P_{10}Glu mutation did not affect the stability of insertion of the reactive-site loop from P_{15}-P_6 into the A β-sheet of α_1-antitrypsin cleaved by papain (12). Consequently, the P_{10}Glu mutation would not be expected to alter the stability of the complex if the reactive-site loop is fully inserted into the A β-sheet. It may, however, affect the stability of a small insertion where the contribution of a single residue would be larger. The P_{10}Glu mutation may also affect the stability of a turn such as that shown in Figure 8. Thus, experimental data in support of both full and partial insertion of the reactive-site loop in the final complex have been obtained and further experiments are required to resolve this problem. The most satisfactory resolution would be a crystal structure of the protease-serpin complex. However, the complex gradually breaks to yield cleaved serpin (Figure 6) and the half-life of the complex may not be long enough to permit its crystallization. The use of other biophysical techniques will thus be required to elucidate the extent of insertion of the reactive-site loop and the position of the protease in the final complex.

2.5. Molecular Events Involved in Insertion of the Reactive-Site Loop into the A β-Sheet

Whereas it is generally accepted that insertion (partial or full) of the reactive-site loop into the A β-sheet occurs during the formation of the final complex, the nature of the molecular event triggering this insertion remains to be definitively established. Does the binding of the protease to the reactive-site loop induce insertion, or is cleavage of the P_1–P_1' bond necessary? Results obtained with fluorescently labelled variants of PAI-1 indicate that cleavage and insertion occur simultaneously (22,27). For the reaction between PAI-1 and tPA, formation of an acyl intermediate resulted in a greater exposure of the P_1' residue to solvent. This movement of the P_1' residue coincided with movement of the P_9 residue to a more hydrophobic environment. These data indicate that cleavage and insertion are coupled, but it is not possible to decide whether cleavage precedes insertion or *vice versa*. Results obtained with the $Ser^{195} \rightarrow Ala$ mutant of thrombin suggest that insertion precedes the formation of the acyl intermediate and is rate-limiting in the formation of the final complex. For the mechanism represented in Figure 6, the initial complex (EI) is converted to the final complex (EI*) with a rate constant of k_2. The value of k_2 for interaction of both protease nexin-1 and antithrombin with thrombin in the presence of heparin was not affected by the $Ser^{195} \rightarrow Ala$ mutation (19). These results indicate that the rate-limiting process in the conversion of the initial to the final complex does not involve the catalytic serine. Other data suggest that the rate-limiting step is probably partial insertion of the hinge region into the A β-sheet; mutations in the hinge region which would impair its insertion decrease the rate of the formation of the stable complex (10–12). Since this step occurs with the $Ser^{195} \rightarrow Ala$ mutant of thrombin, binding rather than catalysis must induce partial insertion.

However, the extent of insertion induced by protease binding cannot be deduced from the available data. It seems possible that binding may only induce the insertion of a few residues and that catalysis in a subsequent step may allow further insertion of the reactive-site loop into the A β-sheet. As noted above, the results of Shore et al. (27) indicate that the P_9 residue moves to a more hydrophobic environment upon interaction with tPA which suggests that a significant portion of the reactive-site loop is inserted in the tPA-PAI-1 complex. The binding of anhydrotrypsin to PAI-1 did not, however, induce the same change in the fluorescent properties of the probe attached to the P_9 residue of PAI-1 (17), suggesting that interactions with Ser^{195} may be essential for a fuller insertion of the reactive-site loop. Taken together, the data favour a model in which protease binding initiates insertion of the reactive-site loop into the A β-sheet in a step that is rate-limiting in the formation of the stable complex. In a subsequent step, interactions with Ser^{195} promote a fuller insertion of the reactive-site loop which prevents hydrolysis and stabilizes the serpin-protease complex.

REFERENCES

1. Laskowski, M. J. and Kato, I. (1980) Protein inhibitors of proteinases. *Annu. Rev. Biochem.* 49: 593–626.
2. Bode, W. and Huber, R. (1992) Natural protein proteinase-inhibitors and their interaction with proteinases. *Eur. J. Biochem.* 204: 433–451.
3. Schechter, I. and Berger, A. (1967) On the size of the active site in proteases. I. Papain. *Biochem. Biophys. Res. Commun.* 27: 157–162.

4. Loebermann, H., Tokuoka, R., Deisenhofer, J. and Huber, R. (1984) Human α_1-antiproteinase inhibitor: Crystal structure analysis of two crystal modifications, molecular model and preliminary analysis of function. *J. Mol. Biol.* 177: 531–556.

5. Mottonen, J., Strand, A., Symersky, J., Sweet, R. M., Danley, D. E., Geoghegan, K. F., Gerard, R. D. and Goldsmith, E. J. (1992) Structural basis of latency in plasminogen activator inhibitor-1. *Nature* 355: 270–273.

6. Carrell, R. W., Stein, P. E., Fermi, G. and Wardell, M. R. (1994) Biological implications of a 3 Å structure of dimeric antithrombin. *Structure* 2: 257–270.

7. Wei, A., Rubin, H., Cooperman, B. S. and Christianson, D. W. (1994) Crystal structure of an uncleaved serpin reveals the conformation of the inhibitory reactive loop. *Nature Struct. Biol.* 1: 251–258.

8. Elliott, P. R., Lomas, D. A., Carrell, R. W. and Abrahams, J. P. (1996) Inhibitory conformation of the reactive-site loop of α_1-antitrypsin. *Nature Struct. Biol.*, in press.

9. Skriver, K., Wikoff, W. R., Patston, P. A., Tausk, F., Schapira, M., Kaplan, A. P. and Bock, S. C. (1991) Substrate properties of C1 inhibitor Ma (alanine 434→glutamic acid). Genetic and structural evidence suggesting that the P12 region contains critical determinants of serine protease inhibitor/substrate status. *J. Biol. Chem.* 266: 9216–9221.

10. Hood, D. B., Huntington, J. A. and Gettins, P. G. W. (1994) α_1-Antiproteinase inhibitor variant T345R. Influence of P14 residue on substrate and inhibitory pathways. *Biochemistry* 33: 8533–8547.

11. Hopkins, P. C. R., Carrell, R. W. and Stone, S. R. (1993) Effects of mutations in the hinge region of serpins. *Biochemistry* 32: 7650–7657.

12. Hopkins, P. C. R., and Stone, S. R. (1995) The contribution of the conserved hinge region residues of α_1-antitrypsin to its reaction with elastase. *Biochemistry* 34: 15872–15879.

13. Stein, P. E. and Carrell, R. W. (1995) What do dysfunctional serpins tell us about molecular mobility and disease? *Nature Struct. Biol.* 2: 96–113.

14. Fish, W. W. and Björk, I. (1979) Release of two-chain antithrombin from the antithrombin-thrombin complex. *Eur. J. Biochem.* 101: 31–38.

15. Cooperman, B. S., Stavridi, E., Nickbarg, E., Rescorla, E., Schechter, N. M. and Rubin, H. (1993) Antichymotrypsin interaction with chymotrypsin — partitioning of the complex. *J. Biol. Chem.* 268: 23616–23625.

16. Patston, P. A., Gettins, P., Beechem J. and Schapira, M. (1991) Mechanism of serpin action - evidence that C1 inhibitor functions as a suicide substrate. *Biochemistry* 30: 8876–8882.

17. Olson, S. T., Bock, P. E., Kvassman, J., Shore, J. D., Lawrence, D. A., Ginsburg, D. and Björk, I. (1995) Role of the catalytic serine in the interactions of serine proteinases with protein inhibitors of the serpin family. *J. Biol. Chem.* 270: 30007–30017.

18. Lijnen, H. R., Van Hoef, B. and Collen, D. (1991) On the reversible interaction of plasminogen-activator inhibitor-1 with tissue-type plasminogen activator and with urokinase-type plasminogen-activator. *J. Biol. Chem.* 266: 4041–4044.

19. Stone, S. R. and Le Bonniec, B. F. (1996) Inhibitory mechanism of serpins. Identification of steps involving the active-site serine of the protease. *Submitted for publication*.

20. Engh, R. A., Huber, R., Bode, W. and Schulze, A. J. (1995) Divining the serpin inhibition mechanism: A suicide substrate 'springe'? *Trends Biotechnol.* 13: 503–510.

21. Matheson, N. R., van Halbeek, H. and Travis, J. (1991) Evidence of a tetrahedral intermediate complex during serpin-proteinase interactions. *J. Biol. Chem.* 266: 13489–13491.

22. Lawrence, D. A., Ginsburg, D., Day, D. E. , Berkenpas, M. B., Verhamme, I. M, Kvassman., J.-O. and Shore , J. D. (1995) Serpin-protease complexes are trapped as stable acyl-enzyme intermediates. *J. Biol. Chem.* 270: 25309–25312.

23. Wilczynska, M., Fa, M., Ohlsson, P.-I. and Ny, T. (1995) The inhibition mechanism of serpins. Evidence that the mobile reactive center loop is cleaved in the native protease-inhibitor complex. *J. Biol. Chem.* 270: 29652–29655.

24. Carter, P. and Wells, J. A. (1988) Dissecting the catalytic triad of a serine protease. *Nature*. 332: 564–568.

25. Corey, D. R. and Craik, C. S. (1992) An investigation into the minimum requirements for peptide hydrolysis by mutation of the catalytic triad of trypsin. *J. Am. Chem. Soc.* 114: 1784–1790.

26. Wright, H. T. and Scarsdale, J. N. (1995) Structural basis for serpin inhibitor activity. *Proteins Struct. Funct. Genet.* 22: 210–225.

27. Shore, J. D., Day, D. E., Francis-Chmura, A. M., Verhamme, I., Kvassman, J., Lawrence, D. A. and Ginsburg, D. (1995) A fluorescent probe study of plasminogen activator inhibitor-1. Evidence for reactive center loop insertion and its role in the inhibitory mechanism. *J. Biol. Chem.* 270: 5395–5398.

28. Whisstock, J. C., Lesk, A. M. and Carrell, R. W. (1996) Modelling of serpin-protease complexes: Antithrombin-thrombin, antitrypsin-thrombin, antitrypsin-trypsin and antitrypsin-elastase. *Proteins Struct. Funct. Genet.*, in press.

3

ANTITHROMBIN

A Bloody Important Serpin

Ingemar Björk[1] and Steven T. Olson[2]

[1]Department of Veterinary Medical Chemistry
Swedish University of Agricultural Sciences
Uppsala Biomedical Center, S-751 23 Uppsala, Sweden
[2]Center for Molecular Biology of Oral Diseases
University of Illinois
Chicago, Illinois 60612-7213

1. INTRODUCTION

Antithrombin is a major regulator of blood clotting. It is a plasma proteinase inhibitor that inactivates a number of coagulation proteinases, although its physiologically most important target enzymes are Factor Xa and thrombin. The enzymes are inactivated by being trapped in tight, equimolar complexes with the inhibitor. The rates of these reactions are moderate under normal conditions but are greatly accelerated by the sulfated mast-cell glycosaminoglycan, heparin, and certain species of the related cell-surface molecule, heparan sulfate. The latter polysaccharide presumably functions as an activator of antithrombin on the vessel wall. The importance of antithrombin in normal hemostasis is borne out by the tendency of individuals with antithrombin deficiencies to develop venous thromboses.

2. STRUCTURE

Antithrombin is an inhibitor of the serpin type[1,2] and is thus homologous with most other plasma proteinase inhibitors, such as α_1-proteinase inhibitor, α_1-antichymotrypsin, α_2-antiplasmin, plasminogen activator inhibitor-1 and heparin cofactor II[3]. It is a glycoprotein with 432 amino acid residues, three disulfide bonds and four glycosylation sites[4-7]. All these sites carry oligosaccharide chains in the dominating form in plasma, α-antithrombin, whereas the site at Asn-135 is not glycosylated in the minor form, β-antithrombin[8-10]. The incomplete glycosylation is due to the presence of Ser instead of Thr in the recognition sequence for core oligosaccharide addition at this site[11]. All four carbohydrate side chains of human α-antithrombin have identical biantennary structures, except in their terminal sialic

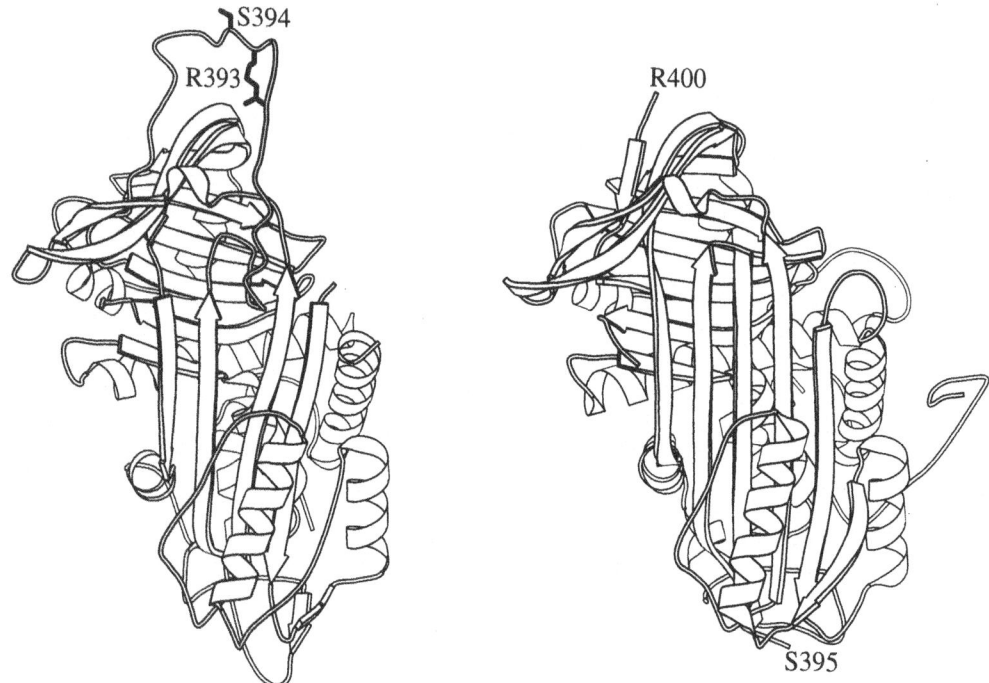

Figure 1. The three-dimensional structures of intact and cleaved antithrombin. (Left) Intact human antithrombin. The reactive-bond loop, with the Arg-393–Ser-394 reactive bond, is at the top and helix F with the underlying A β-sheet is at the front of the molecule. (Right) Cleaved bovine antithrombin, in a similar orientation as the intact human inhibitor. The cleaved bond is Ser-395–Leu-396, one bond carboxy-terminal of the reactive bond (which is Arg-394–Ser-395 in the bovine inhibitor). The carboxy-terminal segment of the cleaved reactive-bond-loop is at the top of the molecule (although no electron density is apparent between residues 396 and 399), whereas the amino-terminal segment of the loop is inserted as a third strand in the A β-sheet. The figures were drawn by Dr. Mathias Eriksson (Dept of Molecular Biology, The Biomedical Center, Uppsala, Sweden) based on the structures determined by Mourey et al.[25] and Schreuder et al.[19] with Bobscript v. 2.0 (Robert Esnouf's extensions to Mol-script v. 1.4[185]).

acids[12,13], giving a molecular weight of 58,200 for this form. Recombinant antithrombins with properties similar to those of the plasma protein, although generally with different glycosylation patterns, have been expressed in several systems, including baculovirus-infected insect cells and mammalian cell lines[14–18].

Two X-ray structures of intact antithrombin and one of an inactive, cleaved form of the inhibitor have been determined. The structures of intact antithrombin[19,20] show two prominent features, a dominating five-stranded β-sheet, the A sheet, and a large exposed loop containing the reactive bond, Arg-393–Ser-394 (Fig. 1). These features are similar to those of other intact inhibitory serpins for which structures have been reported, viz. α₁-proteinase inhibitor and α₁-antichymotrypsin[21,22]. However, the antithrombin structures are unique in that the N-terminal two residues of the reactive-bond loop are inserted between the third and fourth strands of the A sheet (Fig. 1). Moreover, the loop shows no apparent regular structure in antithrombin, but assumes a helix-like conformation in the other two serpins, akin to that in the noninhibitory serpin, ovalbumin[23]. Whether these characteristics are similar in antithrombin in solution[24] or are a result of the close interaction of the reactive-bond loop with the neighboring molecule in the unit cell is uncertain.

The overall structure of inactive antithrombin, cleaved near the reactive bond, is similar to that of the intact inhibitor[25]. However, the N-terminal part of the reactive-bond loop is completely inserted as a middle strand in the A β-sheet, which thereby becomes six-stranded (Fig. 1). The resulting conformational change leads to an appreciable stabilization of the molecule against unfolding[26,27]. The loop is similarly inserted in a "latent", but intact, form of antithrombin, analogous to latent plasminogen activator inhibitor-1[28], which constitutes the second molecule of the unit cell in antithrombin crystals[19,20]. Synthetic peptides having the sequence of the reactive-bond-loop can experimentally be inserted as a middle strand into the A-sheet of intact antithrombin, as well as into that of intact α_1-proteinase inhibitor, inducing the stabilizing effect[29-32].

3. REACTION WITH PROTEINASES

Antithrombin inactivates its physiologic target enzymes, thrombin and Factor Xa, with bimolecular rate constants of $7-11 \times 10^{-3}$ and 2.5×10^{-3} $M^{-1} \cdot s^{-1}$, respectively, at 25°C[33-37]. Rapid-kinetics studies have shown that the inactivation is a two-step process[33,38]. An initial encounter complex, with a dissociation equilibrium constant of $\sim 1.4 \times 10^{-3}$ M, is converted by a conformational change to a stable complex at a rate constant of ~ 10 s^{-1} [33]. This conformational change is reflected in altered antigenic, spectroscopic and heparin-binding properties of antithrombin[37,39-41].

The inactivation of a target proteinase by antithrombin is initiated by the enzyme recognizing the reactive bond of the inhibitor[42-44]. The P1 residue of this bond, Arg-393, is essential for the recognition, as shown by congenital antithrombin variants with mutations of this residue being inactive[45-47]. The P1′ residue, Ser 394, is not of similar cardinal importance, although the size of side chains in this position in mutant recombinant antithrombins affects the rate of inhibition, primarily of thrombin[48-50]. Similarly, hydrophobic or negatively charged side chains at the P2 residue, Gly-392 in wild-type antithrombin, impair the reactivity with both thrombin and Factor Xa[51,52], whereas a proline residue moderately increases the rate of thrombin inhibition[53]. In thrombin, the active site serine[54] (see further below) and Trp-60D in the B-insertion loop[55,56] have been shown to be essential for interaction with antithrombin.

It has been proposed that recognition of proteinases by antithrombin requires that the N-terminal two to three residues of the reactive-bond loop is inserted into the A β-sheet of the inhibitor, like in the X-ray structures of antithrombin. As a consequence, the rest of the loop would assume a "canonical" conformation complementary to the active-site cleft of the enzyme and be able to form a highly stable, lock-and-key complex with the latter[30,57]. The interaction would thus be analogous to that between low-molecular-weight, non-serpin proteinase inhibitors and their target enzymes[58]. This proposal is supported by NMR experiments indicating that the reactive bond of serpins forms a tetrahedral intermediate in complexes with enzymes[59] and is also consistent with the reported reversibility of certain serpin–proteinase complexes[60]. It is more difficult to reconcile, however, with the change of antithrombin conformation shown to accompany complex formation.

Evidence is increasing, however, suggesting that the target enzyme instead attacks the reactive bond in the fully exposed loop as in a regular substrate. The reaction initially proceeds through the intermediates of normal proteolysis, but the enzyme is trapped by a conformational change, most likely induced at the acyl-intermediate stage. Antithrombin thus essentially acts as a suicide inhibitor[54,61-64]. In favor of this mechanism, antithrombin and other serpins bind the inactive, serine-modified trypsin derivative, anhydrotrypsin, with insignificant affinity, in contrast with the tight binding by low-molecular-weight, non-serpin proteinase inhibitors. Moreover, a catalytically active enzyme is required for induction of the conformational change accompanying complex formation[54]. That catalytic activity and cleavage of the reactive bond are involved in proteinase binding is further suggested by the demonstration of a new amino-terminal Ser residue in the antithrombin-thrombin complex[65]. Such cleavage has also been indicated for other serpin-proteinase complexes, even under conditions which would not be expected to perturb a possible tetra-hedral intermediate towards the acyl-intermediate state[66,67]. The stability of antithrombin-proteinase complexes, as well as other serpin-proteinase complexes, under denaturing conditions and the ability of nucleophiles to dissociate these complexes[44,61,68–70] are also compatible with an acyl bond stabilizing the complexes.

Considerable evidence indicates that the conformational change trapping the proteinase involves extensive insertion of the reactive-bond loop into the A β-sheet, in a similar manner as in cleaved antithrombin. Several natural or engineered antithrombin mutants, in which Ala residues in the N-terminal portion of the loop (the P10–P12 region) have been replaced by residues with larger side chains, are thus ineffective inhibitors of thrombin and act primarily as substrates of the enzyme[71–74]. This behavior is compatible with the bulkier residues preventing insertion of the loop into the A sheet[75,76]. Moreover, complexes between intact antithrombin and synthetic peptides having the sequence of the reactive-bond-loop are inactive, presumably because the synthetic peptide incorporated into the A sheet blocks insertion of the inhibitor reactive-bond loop[29,31]. Further support is provided by the demonstration that the same epitopes, not present in intact antithrombin, are exposed in antithrombin-proteinase complexes as in several forms of antithrombin in which loop insertion has occurred[77]. Also, fluorescence labelling of the P9 residue of the related serpin, plasminogen activator inhibitor-1, has shown that identical fluorescence changes are induced on complex formation with target proteinases as on loop insertion following cleavage of the reactive bond[78].

The reactive-bond loop may be partially or fully inserted into the A sheet in the complexes with proteinases. However, it is presumably inserted at least to the P10 residue[73,74], i.e. considerably farther than in the X-ray structures of antithrombin. An attractive hypothesis is that severance of the reactive-bond loop at the acyl-intermediate stage of proteolysis induces the loop to be fully inserted in the same manner as in cleaved antithrombin[79]. The proteinase is concomitantly dragged along the surface of the inhibitor to the pole of the inhibitor opposite from the reactive bond (Fig. 2). During this migration, and in the final complex, the acyl bond is protected from hydrolysis by tight inhibitor-proteinase interactions, excluding water, or by distortion of the catalytic machinery of the proteinase, induced by interaction with the inhibitor[50,54,79,80]. This mechanism is consistent with the observed slow ($t_{1/2}$ 0.5–3 days) dissociation of the final antithrombin-proteinase complex to yield cleaved inhibitor and active enzyme[54,81,82]. Although the model is in apparent disagreement with the reported reversibility of certain serpin-proteinase complexes[60], it should be noted that such reversibility has not been convincingly demonstrated for complexes of antithrombin with proteinases[54,81,82].

A notable feature of serpin-proteinase reactions, first demonstrated for antithrombin, is that a proportion of the inhibitor is cleaved at the reactive bond as a normal substrate, concomitant with complex formation[62,83,84]. This substrate pathway (Fig. 2) presumably arises from the reactive bond being hydrolysed and the enzyme released before the final complex has been stabilized by loop insertion. The acyl-intermediate thus may not be fully protected from hydrolysis during this time. The low proportion of cleaved antithrombin formed normally ($\leq$ 5%) must be due to stabilization being considerably faster than cleavage[62,84]. Reduction of the rate of loop insertion, however, by mutations of the P10–P12 region ([71–76]; see also above) results in increasing amounts of cleaved inhibitor. Binding of

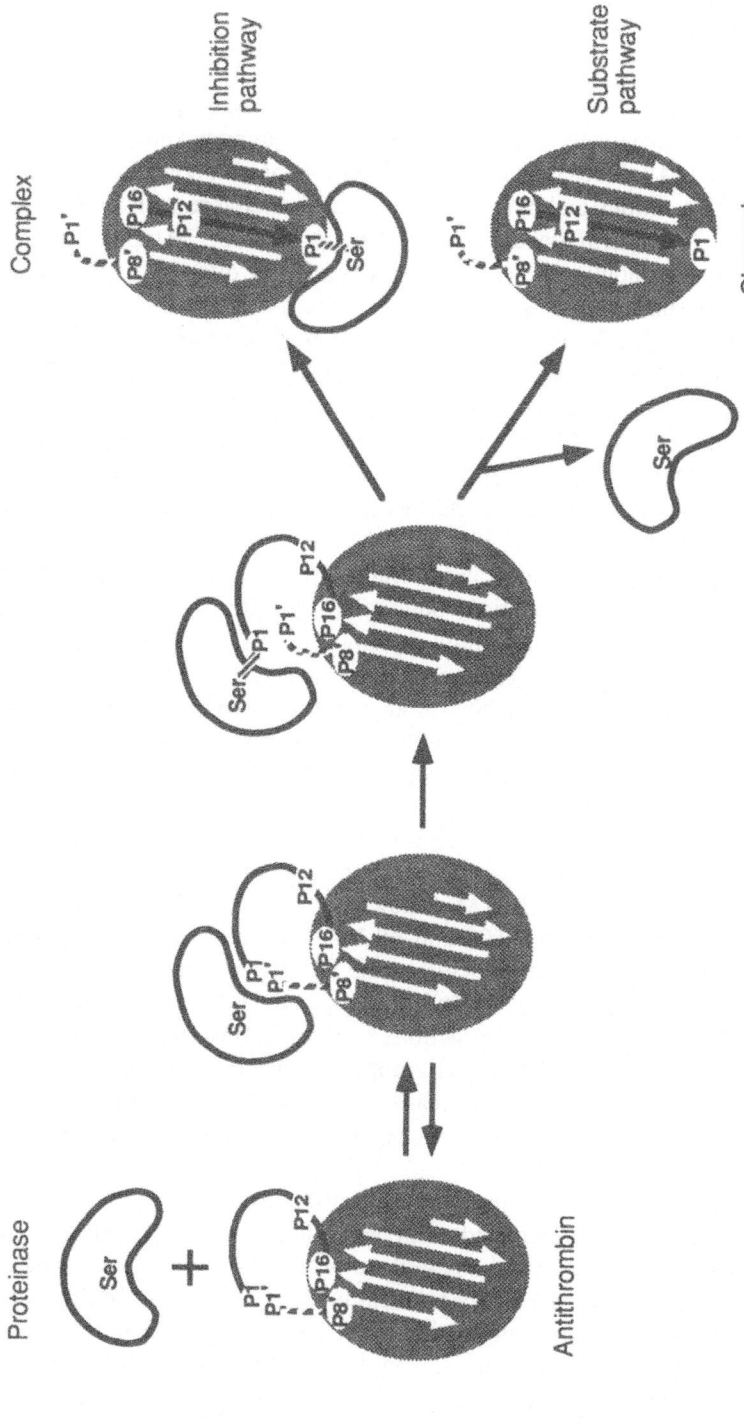

Figure 2. Model for antithrombin inhibition of target proteinases. The loop with the reactive bond (P1–P1′) of antithrombin is assumed to be fully exposed and accessible to the proteinase. The enzyme binds to the reactive bond as in a regular substrate reaction, but the scission of the bond at the acyl-intermediate stage triggers insertion of the loop into the A sheet. In the inhibition pathway, full insertion of the loop traps the proteinase in a stable complex, in which the acyl bond linking the active-site serine of the enzyme and the P1 Arg residue of the inhibitor is protected from hydrolysis and thus stabilizes the complex. In the substrate pathway, however, some proteinase molecules escape the trapping by completing the cleavage of the bond before the loop has been fully inserted, leading to production of a cleaved inactive inhibitor.

heparin to antithrombin, particularly at low ionic strength, has a comparable effect[62], although it is not clear which rate of the two pathways is affected. Reactive-bond-cleaved antithrombin can be identified in plasma under conditions involving activation of the coagulation system[85].

4. HEPARIN EFFECT

Heparin is a strongly negatively charged polymer of alternating hexuronic (D-glucuronic or L-iduronic) acid and D-glucosamine residues. This polymer is highly sulfated, having sulfate groups in various positions on both types of residues, which results in a considerable structural heterogeneity[86,87]. Heparin is exclusively synthesized by a special type of mast cells. However, a related glycosaminoglycan, heparan sulfate, primarily differing in the extent of sulfation, is present on the surface of most cells[86,87].

Heparin dramatically accelerates the rate of inhibition of clotting proteinases by antithrombin. The resulting anticoagulant activity is the basis for the extensive use of heparin for prophylaxis and treatment of venous thrombosis. The rate of the antithrombin-thrombin reaction is $1.5–4 \times 10^7$ $M^{-1} \cdot s^{-1}$ in the presence of optimal amounts of heparin, i.e. the reaction is accelerated 2000–4000-fold[88–91]. Factor Xa inhibition is accelerated to a somewhat lower extent, 500- to 1000-fold, and the maximal rate constant is almost tenfold lower than that of thrombin inhibition[88,91].

Heparin exerts its effect by binding to antithrombin. However, only about one third of all heparin chains in commercial preparations bind with high affinity and have high anticoagulant activity[92–94]. The antithrombin-binding region of these chains has been identified as a pentasaccharide sequence with a specific composition and sulfation pattern. In particular, this pentasaccharide contains a unique 3-O-sulfate substituent on a glucosamine residue that is not found elsewhere in the high-affinity molecules or in the low-affinity species[95–99]. The pentasaccharide, which has been synthesized chemically[100–102], accounts for ~95% of the free energy of binding of the high-affinity heparin species to antithrombin[91].

High-affinity heparin binds to antithrombin with a K_d of $1–2 \times 10^{-8}$ M at pH 7.4, I 0.15[91,103–105]. The binding is a two-step process, involving formation of an initial complex in rapid equilibrium with a K_d of $~4 \times 10^{-5}$ M, followed by a conformational change with forward and reverse rate constants of ~500 and ~0.2 s^{-1}, respectively[91,105]. The altered conformation induced in the second step increases the affinity ~2000-fold. Low-affinity heparin binds about 1000-fold more weakly and does not induce the full conformational change of high-affinity heparin[106]. Ionic interactions between sulfate and carboxyl groups on high-affinity heparin and positively charged groups on antithrombin are important for the interaction, as reflected in a marked decrease of the affinity with increasing ionic strength[90,103–105,107]. However, nonionic interactions also contribute substantially to the binding energy[90,91].

Considerable evidence indicates that the heparin binding site in antithrombin involves a number of positively charged residues in the region of the A and D helices (Fig. 3). Congenital antithrombin variants with substitutions of Arg-47 on the A helix and Arg-129 on the D-helix thus have decreased heparin affinity[108–112]. Chemical modification of Lys-125[113] on the D-helix and substitution of this residue in a recombinant antithrombin variant[114] also markedly decreased the heparin affinity. Congenital variants and chemical modification have further implicated Arg-24, Lys-107, Lys-114, Lys-136 and Arg-145, all located in the A/D helix region, as being important for the interaction[115–117]. This localization is also consistent with the finding that the fully glycosylated α-form of antithrombin

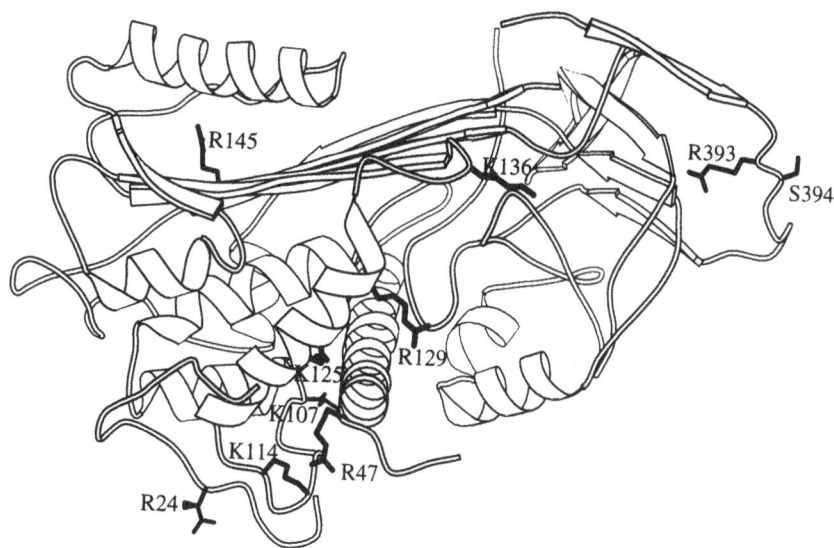

Figure 3. The proposed heparin binding site of human antithrombin. The loop with the Arg-393–Ser-394 reactive bond is to the right and helix F with the underlying A β-sheet is at the top of the molecule. The side chains of basic residues proposed to be involved in heparin binding are shown and labelled. The figure was drawn based on the structure determined by Carrell et al.[20] in the same manner as Fig. 1.

binds heparin more weakly than the β-form, lacking a carbohydrate chain at Asn-135, presumably due to interference by the oligosaccharide at this position[9,10]. The pentasaccharide binding region of heparin has been docked to the proposed binding site in antithrombin by computer modelling[25,118,119]. Congenital mutations of several uncharged residues of antithrombin, such as Pro-41, Leu-99, Ser-116, Gln-118 and certain residues in the C-terminal region, e.g. Phe-402, Ala-404 and Pro-407, have also been shown to be associated with reduced heparin binding[120–125]. Most of these substitutions presumably affect the conformation of the heparin binding site, although some may affect nonionic interactions involved in heparin binding.

The nature of the conformational change of antithrombin induced by heparin is still uncertain. The change is reflected in appreciably altered spectroscopic properties of the protein, indicative of perturbations of aromatic and His residues[126–130]. In particular, it affects the environment of a buried Trp residue, resulting in a fluorescence enhancement[128]. Substitution of Arg-393 of the reactive bond by Cys and labelling this residue with a reporter group have shown that the conformational change is transmitted to the reactive bond[131]. It has been suggested that the change involves partial insertion of the reactive-bond loop into the A β-sheet, thereby bringing the loop into an active conformation[30]. A contrasting proposal is that the reactive-bond loop is partially inserted in the A sheet in the absence of heparin and that heparin binding induces an elongation of the D-helix by one turn, leading to expulsion of the loop from the sheet and increasing its reactivity with proteinases[119].

Heparin interacts also with the target enzymes of antithrombin, thrombin and factor Xa, although appreciably more weakly than with antithrombin. In contrast with the specific binding of antitrombin to the pentasaccharide region, thrombin binds through non-specific electrostatic interactions to overlapping binding sites on the heparin chain, each the size of a hexasaccharide, with an intrinsic site K_d of 6–10 μM[132]. Thrombin thus binds about 1000-fold

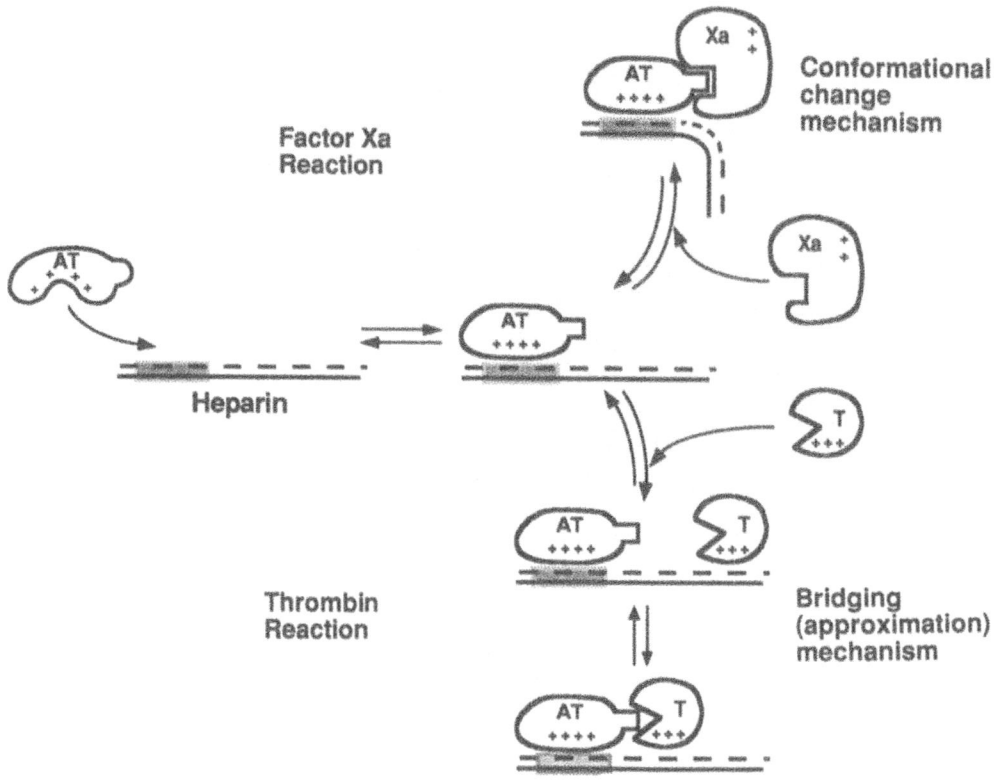

Figure 4. Mechanism of heparin acceleration of antithrombin inhibition of Factor Xa and thrombin. Antithrombin (AT) first binds to the pentasaccharide region of heparin (shaded). This binding induces a conformational change of antithrombin that increases the affinity for heparin and alters the structure of the reactive-bond loop. The latter change enhances Factor Xa recognition of the loop, leading to rate enhancement, but minimally affects thrombin (T) recognition. The rate of thrombin inactivation is instead primarily enhanced by the enzyme binding to the same heparin chain as antithrombin, so that the interaction between the two reactants is greatly facilitated through a ternary-complex bridging or approximation effect.

more weakly to heparin than antithrombin. Site-directed mutagenesis has shown that exosite II in thrombin, a groove comprising Lys-252, Lys-248, Lys-245, Arg-89, Arg-98 and Arg-180, is the heparin binding site[133,134]. The binding of Factor Xa to heparin apparently is also nonspecific and weaker than thrombin binding[132,135].

Heparin accelerates the rates at which antithrombin inactivates thrombin and Factor Xa by partly different mechanisms (Fig. 4). The tight binding of high-affinity heparin to antithrombin is a prerequisite for the inhibition of both enzymes[88,136]. The resulting conformational change in the reactive-bond loop of antithrombin increases the rate of reaction with thrombin, although only by an insignificant two-fold[91]. The predominant effect of heparin in accelerating thrombin inhibition instead arises from a surface approximation or bridging mechanism, due to antithrombin and thrombin binding to the same heparin chain[89–91,136–143]. Antithrombin binds to the polysaccharide before thrombin, because of its high affinity for the pentasaccharide sequence, and binds specifically to this sequence. Thrombin then binds nonspecifically at any number of sites on the same heparin chain and diffuses along the polysaccharide to encounter the inhibitor[90,132]. In contrast with this mechanism, the heparin-induced conformational change of the antithrombin reactive-bond loop is responsible for most of the accelerating effect on Factor Xa inhibition by increasing the rate by about 300-fold[91]. Surface approximation plays a minor role for this enzyme[91,139,143–145] but can contribute in the case of larger heparin chains[143,146]. In both mechanisms, the conformational change of antithrombin that traps the proteinase results in a large decrease in the heparin affinity of the inhibitor[39,136]. Heparin thus rapidly dissociates from the antithrombin-proteinase complex and can then bind to another antithrombin molecule, i.e. heparin accelerates antithrombin-proteinase reactions in a catalytic manner[104,136,147–150].

5. PHYSIOLOGICAL FUNCTION

Antithrombin is the major inactivator of thrombin, factor Xa and factor IXa in plasma[151–154]. Antithrombin is also the primary inhibitor of factor VIIa, factor XIa and kallikrein when these enzymes are associated with cofactor proteins[155–157]. The physiological role of antithrombin in regulating clotting enzymes is further borne out by the predisposition to thrombosis of individuals with inherited or acquired deficiencies of the inhibitor[158,159]. Inherited deficiencies may arise from mutations in the antithrombin gene that yield low circulating levels of the inhibitor or an abnormal protein. Low antithrombin levels result from mutations which introduce a frameshift or an aberrant stop codon or which lead to reduced synthesis[160,161]. Abnormal antithrombins, which have single amino acid substitutions as a result of single base substitutions, may be defective in proteinase binding, heparin binding or both[122,160,161], as discussed above. Heterozygous individuals with a proteinase binding defect are at risk for developing thrombosis. No individuals homozygous for such defects are known, suggesting that complete loss of antithrombin may not be compatible with life. In contrast, heterozygotes with heparin binding defects are normal and only homozygotes with such defects are predisposed to thrombosis, implying that heparin activation of antithrombin is physiologically important but not essential for survival[160,161].

Mast cell heparin is not present in blood, but the related glycosaminoglycan, heparan sulfate, is found in the subluminal and luminal surface of blood vessels[162,163]. A small fraction of these heparan sulfate chains contains the antithrombin-binding pentasaccharide sequence and presumably functions to localize circulating antithrombin and accelerate its inhibition of target proteinases[163–165]. Evidence has been presented that the β-form of antithrombin binds to these endothelial cell heparan sulfate species preferentially to the α-form, due to its higher affinity for the glycosaminoglycan[166]. In the case of thrombin, binding to the endothelial cell receptor, thrombomodulin, also results in accelerated inhibition by antithrombin, due to an associated chondrotin sulfate glycosaminoglycan[167–169]. Mast cell heparin may additionally function to activate extravascular antithrombin at the site of an injured vessel and thereby to regulate extravascular fibrin deposition. Clotting enzymes bound to platelets or endothelial cells at sites of vascular injury appear to be protected from antithrombin inhibition, suggesting that circulating antithrombin has an important function as a scavenger of clotting enzymes which escape from their sites of action[170–173]. The predominant plasma α-form may function mainly as such a scavenger, while the β-form may be primarily responsible for the acceleration of clotting enzyme inhibition at the vessel wall and thereby mediate the nonthrombogenic properties of this wall.

Antithrombin activity may be regulated by inactivation of the inhibitor by nontarget enzymes. Elastase released by neutrophils at inflammatory sites inactivates antithrombin by cleavage in the reactive-bond loop, although not at the reactive bond[174,175]. This inactivation is accelerated by heparin, apparently by a bridging mechanism[176]. In this manner, pathologic conditions which lead to systemic activation of neutrophils may reduce the levels of circulating antithrombin and predispose affected individuals to thrombosis[177]. Matrix metalloproteinases have also been suggested to regulate antithrombin activity in a similar way[178].

Heparin activation of antithrombin may be similarly regulated by plasma heparin binding proteins, the most predominant being histidine-rich glycoprotein, fibronectin, platelet factor 4, and high molecular weight kininogen[179–182]. These proteins, which can be elevated in individuals with thromboembolic disease[183], bind nonspecifically to heparin and limit the amount of anticoagulantly active polysaccharide available for binding to antithrombin. The presence of anticoagulantly inactive heparin or heparan sulfate species with low antithrombin affinity therefore appears to be important to reduce the ability of such proteins to compete with antithrombin[184].

REFERENCES

1. Hunt, L.T. and Dayhoff, M.O. (1980) A surprising new protein superfamily containing ovalbumin, antithrombin III, and alpha₁-proteinase inhibitor. Biochem. Biophys. Res. Commun. 95: 864–871.

2. Carrell, R. and Travis, J. (1985) α_1-Antitrypsin and the serpins: variation and countervariation. Trends Biochem. Sci. 10: 20–24.

3. Huber, R. and Carrell, R.W. (1989) Implications of the three-dimensional structure of alpha 1-antitrypsin for structure and function of serpins. Biochemistry 28: 8951–8966.

4. Petersen, T.E., Dudek-Wojciechowska, G., Sottrup-Jensen, L. and Magnusson, S. (1979) Primary structure of antithrombin III (heparin cofactor). Partial homology between α_1-antitrypsin and antithrombin III. pp. 43–54, In "The Physiological Inhibitors of Blood Coagulation and Fibrinolysis" (Eds., Collen, D., Wiman, B. and Verstraete, M.), Elsevier/North-Holland, Amsterdam.

5. Bock, S.C., Wion, K.L., Vehar, G.A. and Lawn, R.M. (1982) Cloning and expression of the cDNA for human antithrombin III. Nucleic Acids Res. 10: 8113–8125.

6. Prochownik, E.V., Markham, A.F. and Orkin, S.H. (1983) Isolation of a cDNA clone for human antithrombin III. J. Biol. Chem. 258: 8389–8394.

7. Chandra, T., Stackhouse, R., Kidd, V.J. and Woo, S.L. (1983) Isolation and sequence characterization of a cDNA clone of human antithrombin III. Proc. Natl. Acad. Sci. U. S. A. 80: 1845–1848.

8. Carlson, T.H. and Atencio, A.C. (1982) Isolation and partial characterization of two distinct types of antithrombin III from rabbit. Thromb. Res. 27: 23–34.

9. Peterson, C.B. and Blackburn, M.N. (1985) Isolation and characterization of an antithrombin III variant with reduced carbohydrate content and enhanced heparin binding. J. Biol. Chem. 260: 610–615.

10. Brennan, S.O., George, P.M. and Jordan, R.E. (1987) Physiological variant of antithrombin-III lacks carbohydrate sidechain at Asn 135. FEBS Lett. 219: 431–436.

11. Picard, V., Ersdal-Badju, E. and Bock, S.C. (1995) Partial glycosylation of antithrombin III asparagine-135 is caused by the serine in the third position of its N-glycosylation consensus sequence and is responsible for production of the β-antithrombin III isoform with enhanced heparin affinity. Biochemistry 34: 8433–8440.

12. Franzén, L.E., Svensson, S. and Larm, O. (1980) Structural studies on the carbohydrate portion of human antithrombin III. J. Biol. Chem. 255: 5090–5093.

13. Mizuochi, T., Fujii, J., Kurachi, K. and Kobata, A. (1980) Structural studies of the carbohydrate moiety of human antithrombin III. Arch. Biochem. Biophys. 203: 458–465.

14. Stephens, A.W., Siddiqui, A. and Hirs, C.H. (1987) Expression of functionally active human antithrombin III. Proc. Natl. Acad. Sci. U. S. A. 84: 3886–3890.

15. Zettlmeissl, G., Conradt, H.S., Nimtz, M. and Karges, H.E. (1989) Characterization of recombinant human antithrombin III synthesized in Chinese hamster ovary cells. J. Biol. Chem. 264: 21153–21159.

16. Gillespie, L.S., Hillesland, K.K. and Knauer, D.J. (1991) Expression of biologically active human antithrombin III by recombinant baculovirus in Spodoptera frugiperda cells. J. Biol. Chem. 266: 3995–4001.

17. Ersdal-Badju, E., Lu, A., Peng, X., Picard, V., Zendehrouh, P., Turk, B., Björk, I., Olson, S.T. and Bock, S.C. (1995) Elimination of glycosylation heterogeneity affecting heparin affinity of recombinant human antithrombin III by expression of a β-like variant in baculovirus-infected insect cells. Biochem. J. 310: 323–330.

18. Fan, B., Crews, B.C., Turko, I.V., Choay, J., Zettlmeissl, G. and Gettins, P. (1993) Heterogeneity of recombinant human antithrombin III expressed in baby hamster kidney cells. Effect of glycosylation differences on heparin binding and structure. J. Biol. Chem. 268: 17588–17596.

19. Schreuder, H.A., De Boer, B., Dijkema, R., Mulders, J., Theunissen, H.J.M., Grootenhuis, P.D.J. and Hol, W.G.J. (1994) The intact and cleaved human antithrombin III complex as a model for serpin-proteinase interactions. Nature Struct. Biol. 1: 48–54.

20. Carrell, R.W., Stein, P.E., Fermi, G. and Wardell, M.R. (1994) Biological implications of a 3 Å structure of dimeric antithrombin. Structure 2: 257–270.

21. Wei, A., Rubin, H., Cooperman, B.S. and Christianson, D.W. (1994) Crystal structure of an uncleaved serpin reveals the conformation of an inhibitory reactive loop. Nature Struct. Biol. 1: 251–258.

22. Song, H.K., Lee, K.N., Kwon, K.S., Yu, M.H. and Suh, S.W. (1995) Crystal structure of an uncleaved α_1-antitrypsin reveals the conformation of its inhibitory reactive loop. FEBS Lett. 377: 150–154.

23. Stein, P.E., Leslie, A.G., Finch, J.T. and Carrell, R.W. (1991) Crystal structure of uncleaved ovalbumin at 1.95 Å resolution. J. Mol. Biol. 221: 941–959.

24. Chang, W.S.W., Wardell, M.R., Lomas, D.A. and Carrell, R.W. (1996) Probing serpin reactive-loop conformations by proteolytic cleavage. Biochem. J. 314: 647–653.

25. Mourey, L., Samama, J.-P., Delarue, M., Petitou, M., Choay, J. and Moras, D. (1993) Crystal structure of cleaved bovine antithrombin III at 3.2 Å resolution. J. Mol. Biol. 232: 223–241.

26. Bruch, M., Weiss, V. and Engel, J. (1988) Plasma serine proteinase inhibitors (serpins) exhibit major conformational changes and a large increase in conformational stability upon cleavage at their reactive sites. J. Biol. Chem. 263: 16626–16630.

27. Mast, A.E., Enghild, J.J., Pizzo, S.V. and Salvesen, G. (1991) Analysis of the plasma elimination kinetics and conformational stabilities of native, proteinase-complexed, and reactive site cleaved serpins: comparison of alpha 1-proteinase inhibitor, alpha 1-antichymotrypsin, antithrombin III, alpha 2-antiplasmin, angiotensinogen, and ovalbumin. Biochemistry 30: 1723–1730.

28. Mottonen, J., Strand, A., Symersky, J., Sweet, R.M., Danley, D.E., Geoghegan, K.F., Gerard, R.D. and Goldsmith, E.J. (1992) Structural basis of latency in plasminogen activator inhibitor-1. Nature 355: 270–273.

29. Schulze, A.J., Baumann, U., Knof, S., Jaeger, E., Huber, R. and Laurell, C.B. (1990) Structural transition of alpha 1-antitrypsin by a peptide sequentially similar to beta-strand s4A. Eur. J. Biochem. 194: 51–56.

30. Carrell, R.W., Evans, D.L. and Stein, P.E. (1991) Mobile reactive centre of serpins and the control of thrombosis. Nature 353: 576–578.

31. Björk, I., Ylinenjärvi, K., Olson, S.T. and Bock, P.E. (1992) Conversion of antithrombin from an inhibitor of thrombin to a substrate with reduced heparin affinity and enhanced conformational stability by binding of a tetradecapeptide corresponding to the P1 to P14 region of the putative reactive bond loop of the inhibitor. J. Biol. Chem. 267: 1976–1982.

32. Schulze, A.J., Frohnert, P.W., Engh, R.A. and Huber, R. (1992) Evidence for the extent of insertion of the active site loop of intact α_1 proteinase inhibitor in β-sheet A. Biochemistry 31: 7560–7565.

33. Olson, S.T. and Shore, J.D. (1982) Demonstration of a two-step reaction mechanism for inhibition of alpha-thrombin by antithrombin III and identification of the step affected by heparin. J. Biol. Chem. 257: 14891–14895.

34. Danielsson, A. and Björk, I. (1982) Mechanism of inactivation of trypsin by antithrombin. Biochem. J. 207: 21–28.

35. Latallo, Z.S. and Jackson, C.M. (1986) Reaction of thrombins with human antithrombin III: II. Dependence of rate of inhibition on molecular form and origin of thrombin. Thromb. Res. 43: 523–537.

36. Craig, P.A., Olson, S.T. and Shore, J.D. (1989) Transient kinetics of heparin-catalyzed protease inactivation by antithrombin III. Characterization of assembly, product formation, and heparin dissociation steps in the factor Xa reaction. J. Biol. Chem. 264: 5452–5461.

37. Wong, R.F., Windwer, S.R. and Feinman, R.D. (1983) Interaction of thrombin and antithrombin. Reaction observed by intrinsic fluorescence measurements. Biochemistry 22: 3994–3999.

38. Stone, S.R. and Hermans, J.M. (1995) Inhibitory mechanism of serpins. Interaction of thrombin with antithrombin and protease nexin 1. Biochemistry 34: 5164–5172.

39. Carlström, A.S., Liedén, K. and Björk, I. (1977) Decreased binding of heparin to antithrombin following the interaction between antithrombin and thrombin. Thromb. Res. 11: 785–797.

40. Wallgren, P., Nordling, K. and Björk, I. (1981) Immunological evidence for a proteolytic cleavage at the active site of antithrombin in the mechanism of inhibition of coagulation serine proteases. Eur. J. Biochem. 116: 493–496.

41. Peterson, C.B. and Blackburn, M.N. (1987) Antithrombin conformation and the catalytic role of heparin. I. Does cleavage by thrombin induce structural changes in the heparin-binding region of antithrombin? J. Biol. Chem. 262: 7552–7558.

42. Jörnvall, H., Fish, W.W. and Björk, I. (1979) The thrombin cleavage site in bovine antithrombin. FEBS Lett. 106: 358–362.

43. Björk, I., Danielsson, Å., Fenton, J.W.,II and Jörnvall, H. (1981) The site in human antithrombin for functional proteolytic cleavage by human thrombin. FEBS Lett. 126: 257–260.

44. Björk, I., Jackson, C.M., Jörnvall, H., Lavine, K.K., Nordling, K. and Salsgiver, W.J. (1982) The active site of antithrombin. Release of the same proteolytically cleaved form of the inhibitor from complexes with factor IXa, factor Xa, and thrombin. J. Biol. Chem. 257: 2406–2411.

45. Erdjument, H., Lane, D.A., Panico, M., Di Marzo, V. and Morris, H.R. (1988) Single amino acid substitutions in the reactive site of antithrombin leading to thrombosis. Congenital substitution of arginine 393 to cysteine in antithrombin Northwick Park and to histidine in antithrombin Glasgow. J. Biol. Chem. 263: 5589–5593.

46. Owen, M.C., Beresford, C.H. and Carrell, R.W. (1988) Antithrombin Glasgow, 393 Arg to His: a P1 reactive site variant with increased heparin affinity but no thrombin inhibitory activity. FEBS Lett. 231: 317–320.

47. Lane, D.A., Erdjument, H., Thompson, E., Panico, M., Di Marzo, V., Morris, H.R., Leone, G., De Stefano, V. and Thein, S.L. (1989) A novel amino acid substitution in the reactive site of a congenital variant antithrombin. Antithrombin pescara, Arg393 to Pro, caused by a CGT to CCT mutation. J. Biol. Chem. 264: 10200–10204.

48. Stephens, A.W., Siddiqui, A. and Hirs, C.H. (1988) Site-directed mutagenesis of the reactive center (serine 394) of antithrombin III. J. Biol. Chem. 263: 15849–15852.

49. Theunissen, H.J.M., Dijkema, R., Grootenhuis, P.D.J., Swinkels, J.C., De Poorter, T.L., Carati, P. and Visser, A. (1993) Dissociation of heparin-dependent thrombin and factor Xa inhibitory activities of antithrombin-III by mutations in the reactive site. J. Biol. Chem. 268: 9035–9040.

50. Olson, S.T., Stephens, A.W., Hirs, C.H.W., Bock, P.E. and Björk, I. (1995) Kinetic characterization of the proteinase binding defect in a reactive site variant of the serpin, antithrombin. Role of the P1′ residue in transition-state stabilization of antithrombin-proteinase complex formation. J. Biol. Chem. 270: 9717–9724.

51. Blajchman, M.A., Fernandez-Rachubinski, F., Sheffield, W.P., Austin, R.C. and Schulman, S. (1992) Antithrombin-III-Stockholm: a codon 392 (Gly—Asp) mutation with normal heparin binding and impaired serine protease reactivity. Blood 79: 1428–1434.

52. Sheffield, W.P. and Blajchman, M.A. (1994) Site-directed mutagenesis of the P2 residue of human antithrombin. FEBS Lett. 339: 147–150.

53. Sheffield, W.P. and Blajchman, M.A. (1994) Amino acid substitutions of the P2 residue of human antithrombin that either enhance or impair function. Thromb. Res. 75: 293–305.

54. Olson, S.T., Bock, P.E., Kvassman, J., Shore, J.D., Lawrence, D.A., Ginsburg, D. and Björk, I. (1995) Role of the catalytic serine in the interactions of serine proteinases with protein inhibitors of the serpin family — Contribution of a covalent interaction to the binding energy of serpin-proteinase complexes. J. Biol. Chem. 270: 30007–30017.

55. Le Bonniec, B.F., Guinto, E.R. and Stone, S.R. (1995) Identification of thrombin residues that modulate its interactions with antithrombin III and α1-antitrypsin. Biochemistry 34: 12241–12248.

56. Rezaie, A.R. (1996) Tryptophan 60-D in the B-insertion loop of thrombin modulates the thrombin antithrombin reaction. Biochemistry 35: 1918–1924.

57. Perry, D.J., Daly, M., Harper, P.L., Tait, R.C., Price, J., Walker, I.D. and Carrell, R.W. (1991) Antithrombin Cambridge II, 384 Ala to Ser. Further evidence of the role of the reactive centre loop in the inhibitory function of the serpins. FEBS Lett. 285: 248–250.

58. Laskowski, M.,Jr. and Kato, I. (1980) Protein inhibitors of proteinases. Annu. Rev. Biochem. 49: 593–626.

59. Matheson, N.R., van Halbeek, H. and Travis, J. (1991) Evidence for a tetrahedral intermediate complex during serpin-proteinase interactions. J. Biol. Chem. 266: 13489–13491.

60. Shieh, B.H., Potempa, J. and Travis, J. (1989) The use of alpha 2-antiplasmin as a model for the demonstration of complex reversibility in serpins. J. Biol. Chem. 264: 13420–13423.

61. Fish, W.W. and Björk, I. (1979) Release of a two-chain form of antithrombin from the antithrombin-thrombin complex. Eur. J. Biochem. 101: 31–38.

62. Olson, S.T. (1985) Heparin and ionic strength-dependent conversion of antithrombin III from an inhibitor to a substrate of alpha-thrombin. J. Biol. Chem. 260: 10153–10160.

63. Patston, P.A., Gettins, P., Beechem, J. and Schapira, M. (1991) Mechanism of serpin action: evidence that C1 inhibitor functions as a suicide substrate. Biochemistry 30: 8876–8882.

64. Cooperman, B.S., Stavridi, E., Nickbarg, E., Rescorla, E., Schechter, N.M. and Rubin, H. (1993) Antichymotrypsin interaction with chymotrypsin. Partitioning of the complex. J. Biol. Chem. 268: 23616–23625.

65. Ferguson, W.S. and Finlay, T.H. (1983) Formation and stability of the complex formed between human antithrombin-III and thrombin. Arch. Biochem. Biophys. 220: 301–308.

66. Lawrence, D.A., Ginsburg, D., Day, D.E., Berkenpas, M.B., Verhamme, I.M., Kvassman, J.O. and Shore, J.D. (1995) Serpin-protease complexes are trapped as stable acyl-enzyme intermediates. J. Biol. Chem. 270: 25309–25312.

67. Wilczynska, M., Fa, M., Ohlsson, P.I. and Ny, T. (1995) The inhibition mechanism of serpins — Evidence that the mobile reactive center loop is cleaved in the native protease inhibitor complex. J. Biol. Chem. 270: 29652–29655.

68. Rosenberg, R.D. and Damus, P.S. (1973) The purification and mechanism of action of human antithrombin-heparin cofactor. J. Biol. Chem. 248: 6490–6505.

69. Owen, W.G. (1975) Evidence for the formation of an ester between thrombin and heparin cofactor. Biochim. Biophys. Acta 405: 380–387.

70. Jesty, J. (1979) Dissociation of complexes and their derivatives formed during inhibition of bovine thrombin and activated factor X by antithrombin III. J. Biol. Chem. 254: 1044–1049.

71. Molho-Sabatier, P., Aiach, M., Gaillard, I., Fiessinger, J.N., Fischer, A.M., Chadeuf, G. and Clauser, E. (1989) Molecular characterization of antithrombin III (ATIII) variants using polymerase chain reaction. Identification of the ATIII Charleville as an Ala 384 Pro mutation. J. Clin. Invest. 84: 1236–1242.

72. Caso, R., Lane, D.A., Thompson, E.A., Olds, R.J., Thein, S.L., Panico, M., Blench, I., Morris, H.R., Freyssinet, J.M. and Aiach, M. (1991) Antithrombin Vicenza, Ala 384 to Pro (GCA to CCA) mutation, transforming the inhibitor into a substrate. Br. J. Haematol. 77: 87–92.

73. Austin, R.C., Rachubinski, R.A., Ofosu, F.A. and Blajchman, M.A. (1991) Antithrombin-III-Hamilton, Ala 382 to Thr: an antithrombin-III variant that acts as a substrate but not an inhibitor of alpha-thrombin and factor Xa. Blood 77: 2185–2189.

74. Austin, R.C., Rachubinski, R.A. and Blajchman, M.A. (1991) Site-directed mutagenesis of alanine-382 of human antithrombin III. FEBS Lett. 280: 254–258.

75. Skriver, K., Wikoff, W.R., Patston, P.A., Tausk, F., Schapira, M., Kaplan, A.P. and Bock, S.C. (1991) Substrate properties of Cl inhibitor Ma (alanine 434–glutamic acid). Genetic and structural evidence suggesting that the P12-region contains critical determinants of serine protease inhibitor/substrate status. J. Biol. Chem. 266: 9216–9221.

76. Hopkins, P.C.R., Carrell, R.W. and Stone, S.R. (1993) Effects of mutations in the hinge region of serpins. Biochemistry 32: 7650–7657.

77. Björk, I., Nordling, K. and Olson, S.T. (1993) Immunologic evidence for insertion of the reactive-bond loop of antithrombin into the A β-sheet of the inhibitor during trapping of target proteinases. Biochemistry 32: 6501–6505.

78. Shore, J.D., Day, D.E., Francis-Chmura, A.M., Verhamme, I., Kvassman, J., Lawrence, D.A. and Ginsburg, D. (1995) A fluorescent probe study of plasminogen activator inhibitor-1. Evidence for reactive center loop insertion and its role in the inhibitory mechanism. J. Biol. Chem. 270: 5395–5398.

79. Wright, H.T. and Scarsdale, J.N. (1995) Structural basis for serpin inhibitor activity. Proteins 22: 210–225.

80. Lawrence, D.A., Strandberg, L., Ericson, J. and Ny, T. (1990) Structure-function studies of the SERPIN plasminogen activator inhibitor type 1. Analysis of chimeric strained loop mutants. J. Biol. Chem. 265: 20293–20301.

81. Danielsson, Å. and Björk, I. (1980) Slow, spontaneous dissociation of the antithrombin-thrombin complex produces a proteolytically modified form of the inhibitor. FEBS Lett. 119: 241–244.

82. Danielsson, A. and Björk, I. (1983) Properties of antithrombin-thrombin complex formed in the presence and in the absence of heparin. Biochem. J. 213: 345–353.

83. Fish, W.W., Orre, K. and Björk, I. (1979) The production of an inactive form of antithrombin through limited proteolysis by thrombin. FEBS Lett. 98: 103–106.

84. Björk, I. and Fish, W.W. (1982) Production in vitro and properties of a modified form of bovine antithrombin, cleaved at the active site by thrombin. J. Biol. Chem. 257: 9487–9493.

85. Lindo, V.S., Kakkar, V.V. and Melissari, E. (1995) Cleaved antithrombin (ATc): A new marker for thrombin generation and activation of the coagulation system. Br. J. Haematol. 89: 157–162.

86. Lane, D.A. and Lindahl, U. (1989) "Heparin. Chemical and Biological Properties. Clinical Applications", Edward Arnold, London.

87. Lane, D.A., Björk, I. and Lindahl, U. (1992) "Heparin and Related Polysaccharides", Plenum Press, New York.

88. Jordan, R.E., Oosta, G.M., Gardner, W.T. and Rosenberg, R.D. (1980) The kinetics of haemostatic enzyme-antithrombin interactions in the presence of low molecular weight heparin. J. Biol. Chem. 255: 10081–10090.

89. Griffith, M.J. (1982) Kinetics of the heparin-enhanced antithrombin III/thrombin reaction. Evidence for a template model for the mechanism of action of heparin. J. Biol. Chem. 257: 7360–7365.

90. Olson, S.T. and Björk, I. (1991) Predominant contribution of surface approximation to the mechanism of heparin acceleration of the antithrombin-thrombin reaction. Elucidation from salt concentration effects. J. Biol. Chem. 266: 6353–6364.

91. Olson, S.T., Björk, I., Sheffer, R., Craig, P.A., Shore, J.D. and Choay, J. (1992) Role of the antithrombin-binding pentasaccharide in heparin acceleration of antithrombin-proteinase reactions. Resolution of the antithrombin conformational change contribution to heparin rate enhancement. J. Biol. Chem. 267: 12528–12538.

92. Lam, L.H., Silbert, J.E. and Rosenberg, R.D. (1976) The separation of active and inactive forms of heparin. Biochem. Biophys. Res. Commun. 69: 570–577.

93. Höök, M., Björk, I., Hopwood, J. and Lindahl, U. (1976) Anticoagulant activity of heparin: Separation of high-activity and low-activity heparin species by affinity chromatography on immobilized antithrombin. FEBS Lett. 66: 90–93.

94. Andersson, L.O., Barrowcliffe, T.W., Holmer, E., Johnson, E.A. and Sims, G.E.C. (1976) Anticoagulant properties of heparin fractionated by affinity chromatography on matrix-bound antithrombin III and by gel filtration. Thromb. Res. 9: 575–583.

95. Lindahl, U., Bäckström, G., Thunberg, L. and Leder, I.G. (1980) Evidence for a 3-O-sulfated D-glucosamine residue in the antithrombin-binding sequence of heparin. Proc. Natl. Acad. Sci. U.S.A. 77: 6551–6555.

96. Casu, B., Oreste, P., Torri, G., Zoppetti, G., Choay, J., Lormeau, J.C., Petitou, M. and Sinay, P. (1981) The structure of heparin oligosaccharide fragments with high anti-(factor Xa) activity containing the minimal antithrombin III-binding sequence. Chemical and 13C nuclear-magnetic-resonance studies. Biochem. J. 197: 599–609.

97. Thunberg, L., Bäckström, G. and Lindahl, U. (1982) Further characterization of the antithrombin-binding sequence in heparin. Carbohydr. Res. 100: 393–410.

98. Atha, D.H., Stephens, A.W. and Rosenberg, R.D. (1984) Evaluation of critical groups required for the binding of heparin to antithrombin. Proc. Natl. Acad. Sci. U. S. A. 81: 1030–1034.

99. Atha, D.H., Lormeau, J.C., Petitou, M., Rosenberg, R.D. and Choay, J. (1985) Contribution of monosaccharide residues in heparin binding to antithrombin III. Biochemistry 24: 6723–6729.

100. Choay, J., Petitou, M., Lormeau, J.C., Sinay, P., Casu, B. and Gatti, G. (1983) Structure-activity relationship in heparin: a synthetic pentasaccharide with high affinity for antithrombin III and eliciting high anti-factor Xa activity. Biochem. Biophys. Res. Commun. 116: 492–499.

101. Sinay, P., Jacquinet, J.C., Petitou, M., Duchaussoy, P., Lederman, I., Choay, J. and Torri, G. (1984) Total synthesis of a heparin pentasaccharide fragment having high affinity for antithrombin III. Carbohydr. Res. 132: C5–C9.

102. van Boeckel, C.A.A. and Petitou, M. (1993) The unique antithrombin III binding domain of heparin: A lead to new synthetic antithrombotics. Angew. Chem. 32: 1671–1690.

103. Nordenman, B., Danielsson, Å. and Björk, I. (1978) The binding of low-affinity and high-affinity heparin to antithrombin. Fluorescence studies. Eur. J. Biochem. 90: 1–6.

104. Jordan, R.E., Beeler, D.L. and Rosenberg, R.D. (1979) Fractionation of low molecular weight heparin species and their interaction with antithrombin. J. Biol. Chem. 254: 2902–2913.

105. Olson, S.T., Srinivasan, K.R., Björk, I. and Shore, J.D. (1981) Binding of high affinity heparin to antithrombin III. Stopped flow kinetic studies of the binding interaction. J. Biol. Chem. 256: 11073–11079.

106. Streusand, V.J., Björk, I., Gettins, P.G.W., Petitou, M. and Olson, S.T. (1995) Mechanism of acceleration of antithrombin-proteinase reactions by low affinity heparin. Role of the antithrombin binding pentasaccharide in heparin rate enhancement. J. Biol. Chem. 270: 9043–9051.

107. Nordenman, B. and Björk, I. (1981) Influence of ionic strength and pH on the interaction between high-affinity heparin and antithrombin. Biochim. Biophys. Acta 672: 227–238.

108. Koide, T., Odani, S., Takahashi, K., Ono, T. and Sakuragawa, N. (1984) Antithrombin III Toyama: replacement of arginine-47 by cysteine in hereditary abnormal antithrombin III that lacks heparin-binding ability. Proc. Natl. Acad. Sci. U. S. A. 81: 289–293.

109. Borg, J.Y., Owen, M.C., Soria, C., Soria, J., Caen, J. and Carrell, R.W. (1988) Proposed heparin binding site in antithrombin based on arginine 47. A new variant Rouen-II, 47 Arg to Ser. J. Clin. Invest. 81: 1292–1296.

110. Borg, J.Y., Brennan, S.O., Carrell, R.W., George, P., Perry, D.J. and Shaw, J. (1990) Antithrombin Rouen-IV 24 Arg–Cys. The amino-terminal contribution to heparin binding. FEBS Lett. 266: 163–166.

111. Gandrille, S., Aiach, M., Lane, D.A., Vidaud, D., Molho-Sabatier, P., Caso, R., de Moerloose, P., Fiessinger, J.N. and Clauser, E. (1990) Important role of arginine 129 in heparin-binding site of antithrombin III. Identification of a novel mutation arginine 129 to glutamine. J. Biol. Chem. 265: 18997–19001.

112. Najjam, S., Chadeuf, G., Gandrille, S. and Aiach, M. (1994) Arg-129 plays a specific role in the conformation of antithrombin and in the enhancement of factor Xa inhibition by the pentasaccharide sequence of heparin. Biochim. Biophys. Acta 1225: 135–143.

113. Peterson, C.B., Noyes, C.M., Pecon, J.M., Church, F.C. and Blackburn, M.N. (1987) Identification of a lysyl residue in antithrombin which is essential for heparin binding. J. Biol. Chem. 262: 8061–8065.

114. Fan, B., Turko, I.V. and Gettins, P.G.W. (1994) Lysine-heparin interactions in antithrombin. Properties of K125M and K290M,K294M,K297M variants. Biochemistry 33: 14156–14161.

115. Liu, C.S. and Chang, J.Y. (1987) The heparin binding site of human antithrombin III. Selective chemical modification at Lys114, Lys125, and Lys287 impairs its heparin cofactor activity. J. Biol. Chem. 262: 17356–17361.

116. Chang, J.Y. (1989) Binding of heparin to human antithrombin III activates selective chemical modification at lysine 236. Lys-107, Lys-125, and Lys-136 are situated within the heparin-binding site of antithrombin III. J. Biol. Chem. 264: 3111–3115.

117. Sun, X.J. and Chang, J.Y. (1990) Evidence that arginine-129 and arginine-145 are located within the heparin binding site of human antithrombin III. Biochemistry 29: 8957–8962.

118. Grootenhuis, P.D.J. and van Boeckel, C.A.A. (1991) Constructing a molecular model of the interaction between antithrombin III and a potent heparin analogue. J. Am. Chem. Soc. 113: 2743–2747.

119. van Boeckel, C.A.A., Grootenhuis, P.D.J. and Visser, A. (1994) A mechanism for heparin-induced potentiation of antithrombin III. Nature Struct. Biol. 1: 423–425.

120. Chang, J.Y. and Tran, T.H. (1986) Antithrombin III Basel. Identification of a Pro-Leu substitution in a hereditary abnormal antithrombin with impaired heparin cofactor activity. J. Biol. Chem. 261: 1174–1176.

121. Olds, R.J., Lane, D.A., Boisclair, M., Sas, G., Bock, S.C. and Thein, S.L. (1992) Antithrombin Budapest 3. An antithrombin variant with reduced heparin affinity resulting from the substitution L99F. FEBS Lett. 300: 241–246.

122. Lane, D.A., Olds, R.J., Conard, J., Boisclair, M., Bock, S.C., Hultin, M., Abildgaard, U., Ireland, H., Thompson, E., Sas, G., Horellou, M.H., Tamponi, G. and Thein, S.-L. (1992) Pleiotropic effects of antithrombin strand 1C substitution mutations. J. Clin. Invest. 90: 2422–2433.

123. Okajima, K., Abe, H., Maeda, S., Motomura, M., Tsujihata, M., Nagataki, S., Okabe, H. and Takatsuki, K. (1993) Antithrombin III Nagasaki (Ser116-Pro): A heterozygous variant with defective heparin binding associated with thrombosis. Blood 81: 1300–1305.

124. Mille, B., Watton, J., Barrowcliffe, T.W., Mani, J.-C. and Lane, D.A. (1994) Role of N- and C-terminal amino acids in antithrombin binding to pentasaccharide. J. Biol. Chem. 269: 29435–29443.

125. Chowdhury, V., Mille, B., Olds, R.J., Lane, D.A., Watton, J., Barrowcliffe, T.W., Pabinger, I., Woodcock, B.E. and Thein, S.L. (1995) Antithrombins Southport (Leu 99 to Val) and Vienna (Gln 118 to Pro): Two novel antithrombin variants with abnormal heparin binding. Br. J. Haematol. 89: 602–609.

126. Villanueva, G.B. and Danishefsky, I. (1977) Evidence for a heparin-induced conformational change on antithrombin III. Biochem. Biophys. Res. Commun. 74: 803–809.

127. Nordenman, B. and Björk, I. (1978) Binding of low-affinity and high-affinity heparin to antithrombin. Ultraviolet different spectroscopy and circular dichroism studies. Biochemistry 17: 3339–3344.

128. Olson, S.T. and Shore, J.D. (1981) Binding of high affinity heparin to antithrombin III. Characterization of the protein fluorescence enhancement. J. Biol. Chem. 256: 11065–11072.

129. Gettins, P. (1987) Antithrombin III and its interaction with heparin. Comparison of the human, bovine, and porcine proteins by 1H NMR spectroscopy. Biochemistry 26: 1391–1398.

130. Fan, B., Turko, I.V. and Gettins, P.G.W. (1994) Antithrombin histidine variants. ^{1}H NMR resonance assignments and functional properties. FEBS Lett. 354: 84–88.

131. Gettins, P.G.W., Fan, B., Crews, B.C., Turko, I.V., Olson, S.T. and Streusand, V.J. (1993) Transmission of conformational change from the heparin binding site to the reactive center of antithrombin. Biochemistry 32: 8385–8389.

132. Olson, S.T., Halvorson, H.R. and Björk, I. (1991) Quantitative characterization of the thrombin-heparin interaction. Discrimination between specific and nonspecific binding models. J. Biol. Chem. 266: 6342–6352.

133. Gan, Z.-R., Li, Y., Chen, Z., Lewis, S.D. and Shafer, J.A. (1994) Identification of basic amino acid residues in thrombin essential for heparin-catalyzed inactivation by antithrombin III. J. Biol. Chem. 269: 1301–1305.

134. Sheehan, J.P. and Sadler, J.E. (1994) Molecular mapping of the heparin-binding exosite of thrombin. Proc. Natl. Acad. Sci. U. S. A. 91: 5518–5522.

135. Jordan, R.E., Oosta, G.M., Gardner, W.T. and Rosenberg, R.D. (1980) The binding of low molecular weight heparin to hemostatic enzymes. J. Biol. Chem. 255: 10073–10080.

136. Olson, S.T. and Shore, J.D. (1986) Transient kinetics of heparin-catalyzed protease inactivation by antithrombin III. The reaction step limiting heparin turnover in thrombin neutralization. J. Biol. Chem. 261: 13151–13159.

137. Laurent, T.C., Tengblad, A., Thunberg, L., Höök, M. and Lindahl, U. (1978) The molecular-weight-dependence of the anti-coagulant activity of heparin. Biochem. J. 175: 691–701.

138. Pomerantz, M.W. and Owen, W.G. (1978) A catalytic role for heparin. Evidence for a ternary complex of heparin cofactor, thrombin, and heparin. Biochim. Biophys. Acta 535: 66–77.

139. Oosta, G.M., Gardner, W.T., Beeler, D.L. and Rosenberg, R.D. (1981) Multiple functional domains of the heparin molecule. Proc. Natl. Acad. Sci. U. S. A. 78: 829–833.

140. Holmer, E., Kurachi, K. and Söderstrom, G. (1981) The molecular-weight dependence of the rate-enhancing effect of heparin on the inhibition of thrombin, factor Xa, factor IXa, factor XIa, factor XIIa and kallikrein by antithrombin. Biochem. J. 193: 395–400.

141. Nesheim, M.E. (1983) A simple rate law that describes the kinetics of the heparin-catalyzed reaction between antithrombin III and thrombin. J. Biol. Chem. 258: 14708–14717.

142. Hoylaerts, M., Owen, W.G. and Collen, D. (1984) Involvement of heparin chain length in the heparin-catalyzed inhibition of thrombin by antithrombin III. J. Biol. Chem. 259: 5670–5677.

143. Danielsson, Å., Raub, E., Lindahl, U. and Björk, I. (1986) Role of ternary complexes, in which heparin binds both antithrombin and proteinase, in the acceleration of the reactions between antithrombin and thrombin or factor Xa. J. Biol. Chem. 261: 15467–15473.

144. Thunberg, L., Lindahl, U., Tengblad, A., Laurent, T.C. and Jackson, C.M. (1979) On the molecular-weight dependence of the anticoagulant activity of heparin. Biochem. J. 181: 241–243.

145. Lane, D.A., Denton, J., Flynn, A.M., Thunberg, L. and Lindahl, U. (1984) Anticoagulant activities of heparin oligosaccharides and their neutralization by platelet factor 4. Biochem. J. 218: 725–732.

146. Ellis, V., Scully, M.F. and Kakkar, V.V. (1986) The relative molecular mass dependence of the anti-factor Xa properties of heparin. Biochem. J. 238: 329–333.

147. Björk, I. and Nordenman, B. (1976) Acceleration of the reaction between thrombin and antithrombin III by non-stoichiometric amounts of heparin. Eur. J. Biochem. 68: 507–511.

148. Griffith, M.J. (1982) The heparin-enhanced antithrombin III/thrombin reaction is saturable with respect to both thrombin and antithrombin III. J. Biol. Chem. 257: 13899–13302.

149. Pletcher, C.H. and Nelsestuen, G.L. (1983) Two-substrate reaction model for the heparin-catalyzed bovine antithrombin/protease reaction. J. Biol. Chem. 258: 1086–1091.

150. Evington, J.R., Feldman, P.A., Luscombe, M. and Holbrook, J.J. (1986) The catalysis by heparin of the reaction between thrombin and antithrombin. Biochim. Biophys. Acta 870: 92–101.

151. Fuchs, H.E. and Pizzo, S.V. (1983) Regulation of factor Xa in vitro in human and mouse plasma and in vivo in mouse. Role of the endothelium and plasma proteinase inhibitors. J. Clin. Invest. 72: 2041–2049.

152. Fuchs, H.E., Trapp, H.G., Griffith, M.J., Roberts, H.R. and Pizzo, S.V. (1984) Regulation of factor IXa in vitro in human and mouse plasma and in vivo in the mouse. Role of the endothelium and the plasma proteinase inhibitors. J. Clin. Invest. 73: 1696–1703.

153. Gitel, S.N., Medina, V.M. and Wessler, S. (1984) Inhibition of human activated Factor X by antithrombin III and alpha 1-proteinase inhibitor in human plasma. J. Biol. Chem. 259: 6890–6895.

154. Jesty, J. (1986) The kinetics of inhibition of alpha-thrombin in human plasma. J. Biol. Chem. 261: 10313–10318.

155. Lawson, J.H., Butenas, S., Ribarik, N. and Mann, K.G. (1993) Complex-dependent inhibition of factor VIIa by antithrombin III and heparin. J. Biol. Chem. 268: 767–770.

156. Olson, S.T., Sheffer, R. and Francis, A.M. (1993) High molecular weight kininogen potentiates the heparin-accelerated inhibition of plasma kallikrein by antithrombin: Role for antithrombin in the regulation of kallikrein. Biochemistry 32: 12136–12147.

157. Olson, S.T., Sheffer, R. and Shore, J.D. (1994) Parallel procoagulant and anticoagulant pathways for high molecular weight kininogen coagulant function. Agents Actions Suppl. 38: 241–248.

158. Egeberg, O. (1965) Inherited antithrombin deficiency causing thrombophilia. Thromb. Diath. Haemorrh. 13: 516–530.

159. Abildgaard, U. (1981) Antithrombin and related inhibitors of blood coagulation. Recent Adv. Blood Coag. 3: 151–173.

160. Lane, D.A., Ireland, H., Olds, R.J., Thein, S.L., Perry, D.J. and Aiach, M. (1991) Antithrombin III: a database of mutations. Thromb. Haemost. 66: 657–661.

161. Lane, D.A., Olds, R.J., Boisclair, M., Chowdhury, V., Thein, S.L., Cooper, D.N., Blajchman, M., Perry, D., Emmerich, J. and Aiach, M. (1993) Antithrombin III mutation database: First Update. Thromb. Haemost. 70: 361–369.

162. De Agostini, A.I., Watkins, S.C., Slayter, H.S., Youssoufian, H. and Rosenberg, R.D. (1990) Localization of anticoagulantly active heparan sulfate proteoglycans in vascular endothelium: antithrombin binding on cultured endothelial cells and perfused rat aorta. J. Cell Biol. 111: 1293–1304.

163. Felsch, J.S. and Owen, W.G. (1994) Endogenous antithrombin associated with microvascular endothelium. Quantitative analysis in perfused rat hearts. Biochemistry 33: 818–822.

164. Marcum, J.A. and Rosenberg, R.D. (1984) Anticoagulantly active heparin-like molecules from vascular tissue. Biochemistry 23: 1730–1737.

165. Marcum, J.A., Atha, D.H., Fritze, L.M., Nawroth, P., Stern, D. and Rosenberg, R.D. (1986) Cloned bovine aortic endothelial cells synthesize anticoagulantly active heparan sulfate proteoglycan. J. Biol. Chem. 261: 7507–7517.

166. Witmer, M.R. and Hatton, M.W. (1991) Antithrombin III-beta associates more readily than antithrombin III-alpha with uninjured and de-endothelialized aortic wall in vitro and in vivo. Arterioscler. Thromb. 11: 530–539.

167. Preissner, K.T., Delvos, U. and Muller-Berghaus, G. (1987) Binding of thrombin to thrombomodulin accelerates inhibition of the enzyme by antithrombin III. Evidence for a heparin-independent mechanism. Biochemistry 26: 2521–2528.

168. Bourin, M.C., Ohlin, A.K., Lane, D.A., Stenflo, J. and Lindahl, U. (1988) Relationship between anticoagulant activities and polyanionic properties of rabbit thrombomodulin. J. Biol. Chem. 263: 8044–8052.

169. Bourin, M.C. and Lindahl, U. (1990) Functional role of the polysaccharide component of rabbit thrombomodulin proteoglycan. Effects on inactivation of thrombin by antithrombin, cleavage of fibrinogen by thrombin and thrombin-catalysed activation of factor V. Biochem. J. 270: 419–425.

170. Miletich, J.P., Jackson, C.M. and Majerus, P.W. (1978) Properties of the factor Xa binding site on human platelets. J. Biol. Chem. 253: 6908–6916.

171. Schoen, P. and Lindhout, T. (1987) The in situ inhibition of prothrombinase-formed human alpha-thrombin and meizothrombin(des F1) by antithrombin III and heparin. J. Biol. Chem. 262: 11268–11274.

172. Hogg, P.J. and Jackson, C.M. (1989) Fibrin monomer protects thrombin from inactivation by heparin-antithrombin III: implications for heparin efficacy. Proc. Natl. Acad. Sci. U. S. A. 86: 3619–3623.

173. Eisenberg, P.R., Siegel, J.R., Abendschein, D.R. and Miletich, J.P. (1993) Importance of Factor Xa in determining the procoagulant activity of whole-blood clots. J. Clin. Invest. 91: 1877–1883.

174. Jochum, M., Lander, S., Heimburger, N. and Fritz, H. (1981) Effect of human granulocytic elastase on isolated human antithrombin III. Hoppe-Seyler's Z. Physiol. Chem. 362: 103–112.

175. Jordan, R.E., Kilpatrick, J. and Nelson, R.M. (1987) Heparin promotes the inactivation of antithrombin by neutrophil elastase. Science 237: 777–779.

176. Jordan, R.E., Nelson, R.M., Kilpatrick, J., Newgren, J.O., Esmon, P.C. and Fournel, M.A. (1989) Inactivation of human antithrombin by neutrophil elastase. Kinetics of the heparin-dependent reaction. J. Biol. Chem. 264: 10493–10500.

177. Carrell, R.W. and Owen, M.C. (1985) Plakalbumin, alpha 1-antitrypsin, antithrombin and the mechanism of inflammatory thrombosis. Nature 317: 730–732.

178. Mast, A.E., Enghild, J.J., Nagase, H., Suzuki, K., Pizzo, S.V. and Salvesen, G. (1991) Kinetics and physiologic relevance of the inactivation of alpha 1-proteinase inhibitor, alpha 1-antichymotrypsin, and antithrombin III by matrix metalloproteinases-1 (tissue collagenase), -2 (72-kDa gelatinase/type IV collagenase), and -3 (stromelysin). J. Biol. Chem. 266: 15810–15816.

179. Hayaishi, M. and Yamada, K.M. (1982) Divalent cation modulation of fibronectin binding to heparin and to DNA. J. Biol. Chem. 257: 5263–5267.

180. Lijnen, H.R., Hoylaerts, M. and Collen, D. (1983) Heparin binding properties of human histidine-rich glycoprotein. Mechanism and role in the neutralization of heparin in plasma. J. Biol. Chem. 258: 3803–3808.

181. Lane, D.A., Pejler, G., Flynn, A.M., Thompson, E.A. and Lindahl, U. (1986) Neutralization of heparin-related saccharides by histidine-rich glycoprotein and platelet factor 4. J. Biol. Chem. 261: 3980–3986.

182. Björk, I., Olson, S.T., Sheffer, R.G. and Shore, J.D. (1989) Binding of heparin to human high molecular weight kininogen. Biochemistry 28: 1213–1221.

183. Young, E., Prins, M., Levine, M.N. and Hirsh, J. (1992) Heparin binding to plasma proteins, an important mechanism for heparin resistance. Thromb. Haemost. 67: 639–643.

184. Barrowcliffe, T.W., Merton, R.E., Havercroft, S.J., Thunberg, L., Lindahl, U. and Thomas, D.P. (1984) Low-affinity heparin potentiates the action of high-affinity heparin oligosaccharides. Thromb. Res. 34: 125–133.

185. Kraulis, P.J. (1991) MOLSCRIPT: a program to produce both detailed and schematic plots of protein structures. J. Appl. Cryst. 24: 946–950.

4

HEPARIN COFACTOR II

Douglas M. Tollefsen[*]

Division of Hematology
Department of Internal Medicine
Washington University School of Medicine
St. Louis, Missouri

Heparin cofactor II (HCII) is a serpin that inhibits thrombin rapidly in the presence of dermatan sulfate or heparin. Both of these glycosaminoglycans bind to HCII and increase the rate of inhibition of thrombin >1000-fold. This review will focus on the biochemistry of HCII and the mechanism by which glycosaminoglycans stimulate its activity.

STRUCTURE, BIOSYNTHESIS, AND METABOLISM OF HCII

HCII is a single-chain glycoprotein with a molecular weight of ~66,500 (1). The cDNA for human HCII encodes a polypeptide 480 amino acids in length preceded by a 19-residue signal peptide (2,3). HCII is ~30% identical in sequence to antithrombin and other serpins. The greatest similarity is in the C-terminal two-thirds of the protein. By contrast, the N-terminal 80 amino acid residues of HCII share no homology with other serpins (Fig. 1). The N-terminal region includes a tandem repeat of two highly acidic sequences, each of which contains a tyrosine residue that becomes O-sulfated during biosynthesis (4). This region of HCII is required for rapid inhibition of thrombin in the presence of a glycosaminoglycan (see below). In addition, peptides cleaved from the N-terminal region of HCII by neutrophil proteases have potent chemotactic activity for neutrophils and monocytes (5). HCII has three potential Asn-linked glycosylation sites and contains ~10% carbohydrate by weight (1,6). Three cysteine residues with free sulfhydryl groups are also present in the polypeptide chain (7).

The reactive site peptide bond in HCII is Leu^{444}–Ser^{445} (8). The presence of leucine at the P1 position of the reactive site is unusual for a thrombin substrate and explains the fact that HCII inhibits chymotrypsin more rapidly than thrombin in the absence of a gly-

---

[*] Correspondence to: Dr. Douglas M. Tollefsen, Division of Hematology — Box 8125, Washington University School of Medicine, 660 South Euclid Avenue, St. Louis, MO 63110. Phone: (314) 362-8830; Fax: (314) 362-8826; E-mail: tollefsen@visar.wustl.edu.

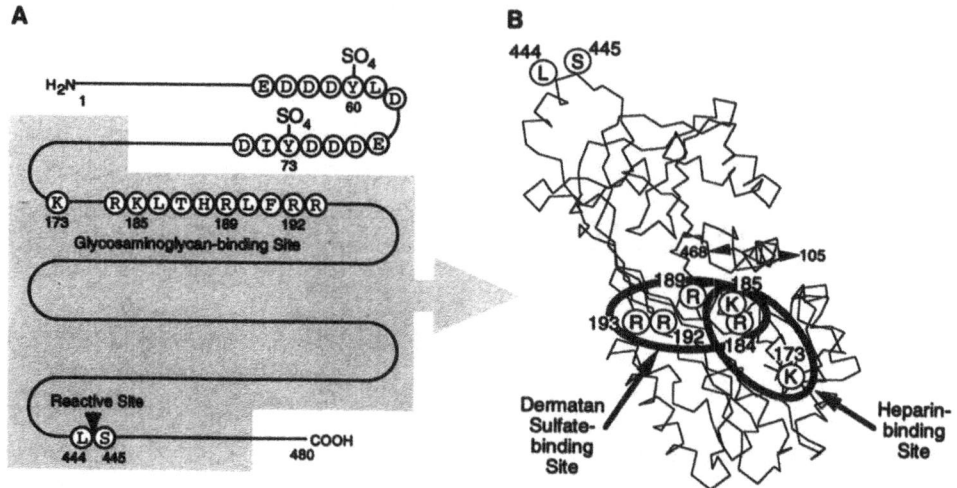

Figure 1. Schematic diagram of HCII. The shaded portion of the protein shown in *panel A* is projected onto the α-carbon backbone of ovalbumin determined by X-ray diffraction (*panel B*). Arginine (*R*) and lysine (*K*) residues implicated in binding to dermatan sulfate and heparin are indicated. The arrows in *panel B* that extend from residues 105 and 468 indicate the portions of HCII that are not represented in the ovalbumin crystal structure.

cosaminoglycan (9). Moreover, HCII does not inhibit other trypsin-like proteases of the coagulation cascade (10). Mutation of Leu[444] to arginine increases the rate of inhibition of thrombin in the absence of a glycosaminoglycan ~100-fold and eliminates the ability of HCII to inhibit chymotrypsin (11).

The gene for human HCII is located on chromosomal band 22q11 (12). A detailed physical map of the region of chromosome 22 encompassing the HCII gene has been published recently (13). The HCII gene contains five exons distributed over ~16 kb of DNA (12,14). The positions of the introns are similar to those in the genes for α1-antitrypsin, α1-antichymotrypsin, and angiotensinogen but are different from those in the genes for most other serpins, including antithrombin. Human liver contains a 2.3-kb mRNA for HCII (2,15), and biosynthesis of the protein has been demonstrated in cultured human hepatoma cells (4,16). HCII mRNA has not been detected by Northern blot analysis in other tissues, including heart, brain, placenta, lung, muscle, kidney, or pancreas.

HCII is secreted by the liver into the bloodstream, where it is present at a concentration of ~1.2 μM in human plasma (17). The half-life of HCII in the circulation is ~2.5 days (18), whereas thrombin-HCII complexes are cleared much more rapidly by the low density lipoprotein receptor-related protein on hepatocytes (19,20). Turnover studies of labeled HCII in man suggest that ~40% of the protein equilibrates into an extravascular compartment (18), but the tissue distribution of HCII has not yet been determined. Normal HCII levels are present during oral anticoagulant use, in the vast majority of patients with venous thrombosis, and in most patients with hereditary antithrombin deficiency (21,22). The plasma concentration of HCII is low in neonates and in some adult patients with liver disease or disseminated intravascular coagulation; in these situations, the HCII and antithrombin levels are usually reduced from the normal adult levels to a similar degree (17,22,23). The HCII concentration may be elevated during acute inflammatory reactions (24). The mechanism of this elevation is unknown, since the production of HCII in hepatoma cells does not appear to be regulated by the inflammatory cytokines interleukin-6, interleukin-1β, or tumor necrosis factor-α (25).

CONSERVATION OF HCII IN MAMMALS

HCII has been identified in the mouse, rabbit, rat, and dog (15,26,27), but it has not yet been detected in species other than mammals. Murine plasma contains two electrophoretically distinct forms of HCII that appear to differ only in the composition of their N-linked oligosaccharides (15). Human, murine, rat, and rabbit HCII are 84–87% identical to one another in amino acid sequence. The sequences of the reactive site, the glycosaminoglycan-binding site, and the N-terminal acidic domain are even more highly conserved. Some other human serpins, such as α1-antitrypsin and α1-antichymotrypsin, are only ~60% identical to their murine homologues. Analysis of murine genomic DNA by Southern hybridization indicates the presence of a single HCII gene on chromosome 16 (15). By contrast, the genes at the murine *Spi-1* and *Spi-2* loci (homologous to human α1-antitrypsin and α1-antichymotrypsin, respectively) have been duplicated in certain strains of mice to form clusters of 5–10 copies of each gene (28,29). The genes within each locus have diverged from one another primarily in their reactive site coding sequences, a process that appears to have altered the protease specificites of the duplicated gene products. The high degree of conservation of HCII suggests that it serves the same physiologic function in different species.

ACTIVATION OF HCII BY GLYCOSAMINOGLYCANS

The rate of inhibition of thrombin by HCII is increased more than 1000-fold by heparin, heparan sulfate, or dermatan sulfate (30). A variety of other natural and synthetic polyanions also activate HCII to various degrees (31–35). Heparin binds to HCII with a lower affinity than to antithrombin; a 10-fold higher concentration of heparin is required to accelerate thrombin inhibition by HCII (1,30). Antithrombin binds to a specific pentasaccharide structure in heparin that includes a 3-O-sulfated glucosamine residue (36). By contrast, HCII binds to most heparin oligosaccharides ≥4 monosaccharide units in length regardless of composition (37). Heparin chains that contain at least 20 monosaccharide units are required for maximal stimulation of the thrombin-HCII reaction (38–40).

HCII is unique among serpins in its ability to be activated by dermatan sulfate. Dermatan sulfate is a repeating polymer of D-glucuronic or L-iduronic acid and N-acetyl-D-galactosamine (41). O-Sulfation of iduronic acid residues at the C2 position and of N-acetylgalactosamine residues at the C4 and C6 positions occurs to a variable extent, yielding heterogeneous structures within the polymer. In contrast to the non-specific binding of HCII to heparin oligosaccharides, HCII binds to a minority of dermatan sulfate oligosaccharides ≥6 monosaccharide units in length. The high-affinity binding site for HCII in porcine skin dermatan sulfate is a tandem repeat of three iduronic acid 2-sulfate→N-acetylgalactosamine 4-sulfate disaccharide subunits (42). The binding site for HCII in dermatan sulfate from other tissues may also contain iduronic acid→N-acetylgalactosamine 4,6-disulfate subunits (43). It has been unclear whether activation of HCII by dermatan sulfate depends only on the charge density of the polymer or requires sulfate groups at precise locations in the polysaccharide. A repeating polymer of iduronic acid 2-sulfate→N-acetylgalactosamine 6-sulfate isolated from the tunicate *Ascidia nigra* activates HCII poorly (44), suggesting that sulfation of C4 in N-acetylgalactosamine is required for the interaction of dermatan sulfate with HCII.

Analysis of the natural variant HCII Oslo (Arg[189]→His) established that heparin and dermatan sulfate interact with different amino acid residues on the surface of the inhibitor

(45,46). This mutation causes a large decrease (~60-fold) in the affinity of HCII for dermatan sulfate but does not affect the affinity of the inhibitor for heparin. Arg[189] occurs in a cluster of basic amino acid residues in helix D that can be aligned with the proposed heparin-binding site of antithrombin but are poorly conserved in other serpins. Mutations of Lys[173], Arg[184], and Arg[185] in recombinant HCII diminish the binding of heparin and its ability to stimulate the thrombin-HCII reaction, whereas mutations of Arg[184], Arg[185], Arg[189], Arg[192], and Arg[193] affect the interaction with dermatan sulfate (47–50). Thus, the binding sites for heparin and dermatan sulfate appear to be overlapping but not identical.

MECHANISM OF ACTIVATION OF HCII

The N-terminal acidic domain of HCII resembles the C-terminal portion of the leech anticoagulant hirudin, which interacts with anion-binding exosite I of thrombin (51). A synthetic peptide corresponding to residues 54–75 of HCII competitively inhibits binding of hirudin to thrombin but does not interfere with the ability of thrombin to hydrolyze tripeptide *p*-nitroanilide substrates (52). Studies with mutant forms of recombinant HCII suggest that thrombin interacts with the N-terminal acidic domain in the intact inhibitor (48,49,53,54). For example, deletion of residues 1–67, which include the first acidic repeat, reduces the rate of thrombin inhibition ~100-fold in the presence of heparin and ~1000-fold in the presence of dermatan sulfate (53). Thus, the first acidic repeat is essential for rapid inhibition of thrombin in the presence of a glycosaminoglycan. In the absence of a glycosaminoglycan, the native and truncated forms of HCII inhibit thrombin at essentially the same (slow) rate. Deletions or point mutations of the acidic repeats increase the affinity of HCII for heparin-agarose (48,49,53), suggesting that the acidic domain occupies the glycosaminoglycan-binding site.

The experiments described above support a model in which binding of a glycosaminoglycan to HCII displaces the N-terminal acidic domain from the glycosaminoglycan-binding site, thereby allowing the acidic domain to interact with exosite I of thrombin

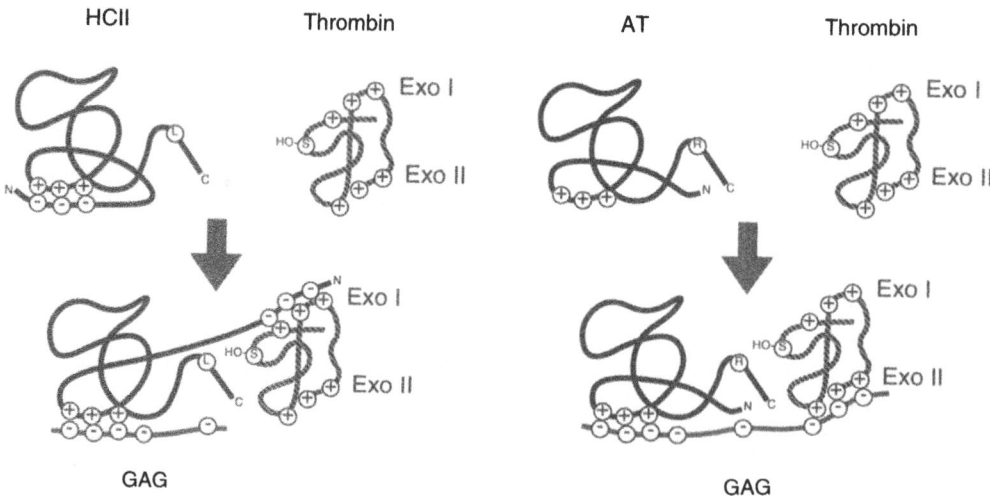

Figure 2. Comparison of the proposed mechanisms of inhibition of thrombin by HCII and antithrombin in the presence of a glycosaminoglycan. AT, antithrombin. GAG, glycosaminoglycan. Exo I, thrombin exosite I (hirudin-binding site). Exo II, thrombin exosite II (glycosaminoglycan-binding site).

(Fig. 2). This interaction could facilitate formation of a stable complex by bringing the active site of thrombin into approximation with the reactive site of HCII. Additional support for this model has been obtained from experiments with γ-thrombin (53,55), thrombin Quick I (56), and the thrombin mutant Arg^{68}→Glu (57), each of which is defective in exosite I. In the absence of a glycosaminoglycan, each of these variants is inhibited by HCII at a rate similar to that of native thrombin, but the maximal rate of inhibition in the presence of a glycosaminoglycan is decreased at least 100-fold. By contrast, mutations of exosite I have little or no effect on the rate of inhibition of thrombin by antithrombin in the presence of heparin (57).

The rate of inhibition of thrombin by HCII decreases in the presence of high concentrations of heparin or dermatan sulfate (30). This phenomenon could be explained by a template mechanism, which requires that the protease and inhibitor bind simultaneously to a single glycosaminoglycan chain for rapid inhibition to occur (58). The high-affinity dermatan sulfate hexasaccharide causes a 50- to 100-fold increase in the rate of inhibition of thrombin by HCII, and this effect is completely eliminated by deletion of the N-terminal acidic domain (53). Since the hexasaccharide is probably too small to bind HCII and thrombin simultaneously, activation of HCII apparently can occur independent of a template mechanism. Nevertheless, the maximal rate of inhibition is obtained only with longer dermatan sulfate oligosaccharides (59).

Experiments with exosite II variants of thrombin suggest that binding of a glycosaminoglycan to the protease plays only a minor role in inhibition by HCII (60). Mutations of Arg^{89}, Arg^{245}, Lys^{248}, and Lys^{252} in thrombin eliminate binding to dermatan sulfate-agarose and greatly reduce the affinity for heparin-agarose. Exosite II mutations do not significantly affect the rate of inhibition of thrombin by HCII in the absence of a glycosaminoglycan. More importantly, these mutations have no effect on the rate of inhibition in the presence of dermatan sulfate and relatively little effect ($\leq$ 7-fold reduction) in the presence of heparin. By contrast, exosite II mutations decrease the rate of inhibition of thrombin by antithrombin ~100-fold in the presence of heparin (61). These results suggest that when glycosaminoglycans bind to HCII or antithrombin, the resulting complexes interact with thrombin differently. Thus, the heparin/antithrombin complex interacts with thrombin exosite II, whereas the dermatan sulfate/heparin cofactor II complex interacts with thrombin exosite I.

ANTITHROMBOTIC ACTIVITY OF DERMATAN SULFATE

The ability of dermatan sulfate to prolong the partial thromboplastin time of human plasma suggests that non-heparin glycosaminoglycans might be useful anticoagulant agents (62). A variety of experiments, including absorption of plasma with anti-HCII antibodies, indicate that the anticoagulant activity of dermatan sulfate *in vitro* is mediated by HCII (10,63,64). Initially, dermatan sulfate was shown to have antithrombotic activity in the rabbit jugular venous stasis model (65). Interestingly, oral administration of the β-D-xyloside naroparcil to rabbits causes the release of dermatan sulfate-like material into the circulation and produces a systemic antithrombotic effect (66). Recent studies suggest that dermatan sulfate is a more potent inhibitor than heparin of clot-bound thrombin (67,68) and that dermatan sulfate in combination with low-molecular-weight heparin may produce a synergistic antithrombotic effect (69). Human clinical trials have shown that dermatan sulfate prevents thrombosis during hemodialysis (70) and in the post-operative setting (71,72).

PHYSIOLOGIC FUNCTION OF HCII

The presence of thrombin-HCII complexes in normal human plasma indicates that HCII inhibits thrombin *in vivo* (73,74). Cultured fibroblasts and vascular smooth muscle cells accelerate inhibition of thrombin by HCII, but endothelial cells do not (75,76). In the case of fibroblasts, which synthesize both heparan sulfate and dermatan sulfate, a small dermatan sulfate proteoglycan is responsible for the stimulatory effect. Two isolated dermatan sulfate-containing proteoglycans, biglycan and decorin, are capable of stimulating HCII activity *in vitro* (77). These results suggest that HCII may inhibit thrombin at extravascular sites where dermatan sulfate is present.

Heterozygous deficiency of HCII is found in ~1% of apparently healthy individuals (78) and in ~1% of patients with venous thrombosis (21,79). Thus, reported associations between HCII deficiency and thrombosis may be coincidental (80–83). The failure to identify a homozygous individual with HCII deficiency could be explained by (i) the low expected prevalence of homozygosity in the general population (1 per 40,000 if the rate of heterozygosity is 1%), (ii) the assumption that HCII deficiency would be associated with thrombosis (which may be incorrect), or (iii) the possibility that homozygous HCII deficiency is lethal *in utero*.

Circumstantial evidence suggests that the activity of HCII is increased during pregnancy, when both the maternal and fetal plasma contain trace amounts of a dermatan sulfate proteoglycan that stimulates inhibition of thrombin by HCII (84). The placenta is rich in dermatan sulfate and may be the source of this proteoglycan (85). Elevated concentrations of HCII have been reported in women who are pregnant or who use oral contraceptives (86), and thrombin-HCII complexes are elevated ~3- to 6-fold over baseline at term and immediately post-partum (74,87). Conversely, decreased HCII levels have been reported in patients with pre-eclampsia (88). Thus, HCII could be activated locally to inhibit coagulation of maternal blood in the placenta.

Thrombin has a variety of activities unrelated to blood coagulation or platelet activation (89). For example, it causes proliferation of fibroblasts and other cells, induces monocyte chemotaxis, promotes adhesion of neutrophils to endothelial cells, stimulates production of prostacyclin and other mediators by endothelial cells, and inhibits neurite outgrowth. The ability of HCII to block one or more of these activities may be important in the regulation of wound healing, inflammation, or neural development.

REFERENCES

1. Tollefsen, D.M., Majerus, D.W. and Blank, M.K. (1982) Heparin cofactor II. Purification and properties of a heparin-dependent inhibitor of thrombin in human plasma. J. Biol. Chem. 257: 2162–2169.
2. Ragg, H. (1986) A new member of the plasma protease inhibitor gene family. Nucleic Acids Res. 14: 1073–1088.
3. Blinder, M.A., Marasa, J.C., Reynolds, C.H., Deaven, L.L. and Tollefsen, D.M. (1988) Heparin cofactor II: cDNA sequence, chromosome localization, restriction fragment length polymorphism, and expression in Escherichia coli. Biochemistry 27: 752–759.
4. Hortin, G., Tollefsen, D.M. and Strauss, A.W. (1986) Identification of two sites of sulfation of human heparin cofactor II. J. Biol. Chem. 261: 15827–15830.
5. Church, F.C., Pratt, C.W. and Hoffman, M. (1991) Leukocyte chemoattractant peptides from the serpin heparin cofactor II. J. Biol. Chem. 266: 704–709.
6. Kim, Y.-S., Lee, K.-B. and Linhardt, R.J. (1988) Microheterogeneity of plasma glycoproteins heparin cofactor II and antithrombin III and their carbohydrate analysis. Thromb. Res. 51: 97–104.

7. Church, F.C., Meade, J.B. and Pratt, C.W. (1987) Structure-function relationships in heparin cofactor II: spectral analysis of aromatic residues and absence of a role for sulfhydryl groups in thrombin inhibition. Arch. Biochem. Biophys. 259: 331–340.

8. Griffith, M.J., Noyes, C.M., Tyndall, J.A. and Church, F.C. (1985) Structural evidence for leucine at the reactive site of heparin cofactor II. Biochemistry 24: 6777–6782.

9. Church, F.C., Noyes, C.M. and Griffith, M.J. (1985) Inhibition of chymotrypsin by heparin cofactor II. Proc. Natl. Acad. Sci. U. S. A. 82: 6431–6434.

10. Parker, K.A. and Tollefsen, D.M. (1985) The protease specificity of heparin cofactor II. Inhibition of thrombin generated during coagulation. J. Biol. Chem. 260: 3501–3505.

11. Derechin, V.M., Blinder, M.A. and Tollefsen, D.M. (1990) Substitution of arginine for Leu444 in the reactive site of heparin cofactor II enhances the rate of thrombin inhibition. J. Biol. Chem. 265: 5623–5628.

12. Herzog, R., Lutz, S., Blin, N., Marasa, J.C., Blinder, M.A. and Tollefsen, D.M. (1991) Complete nucleotide sequence of the gene for human heparin cofactor II and mapping to chromosomal band 22q11. Biochemistry 30: 1350–1357.

13. Kim, U.-J., Shizuya, H., Kang, H.-L., Choi, S.-S., Garrett, C.L., Smink, L.J., Birren, B.W., Korenberg, J.R., Dunham, I. and Simon, M.I. (1996) A bacterial artificial chromosome-based framework contig map of human chromosome 22q. Proc. Natl. Acad. Sci. U. S. A. 93: 6297–6301.

14. Ragg, H. and Preibisch, G. (1988) Structure and expression of the gene coding for the human serpin hLS2. J. Biol. Chem. 263: 12129–12134.

15. Zhang, G.S., Mehringer, J.H., Van Deerlin, V.M., Kozak, C.A. and Tollefsen, D.M. (1994) Murine heparin cofactor II: purification, cDNA sequence, expression, and gene structure. Biochemistry 33: 3632–3642.

16. Jaffe, E.A., Armellino, D. and Tollefsen, D.M. (1985) Biosynthesis of functionally active heparin cofactor II by a human hepatoma-derived cell line. Biochem. Biophys. Res. Commun. 132: 368–374.

17. Tollefsen, D.M. and Pestka, C.A. (1985) Heparin cofactor II activity in patients with disseminated intravascular coagulation and hepatic failure. Blood 66: 769–774.

18. Sié, P., Dupouy, D., Pichon, J. and Boneu, B. (1985) Turnover study of heparin cofactor II in healthy man. Thromb. Haemost. 54: 635–638.

19. Pratt, C.W., Church, F.C. and Pizzo, S.V. (1988) *In vivo* catabolism of heparin cofactor II and its complex with thrombin: evidence for a common receptor-mediated clearance pathway for three serine proteinase inhibitors. Arch. Biochem. Biophys. 262: 111–117.

20. Kounnas, M.Z., Church, F.C., Argraves, W.S. and Strickland, D.K. (1996) Cellular internalization and degradation of antithrombin III-thrombin, heparin cofactor II-thrombin, and α1-antitrypsin-trypsin complexes is mediated by the low density lipoprotein receptor-related protein. J. Biol. Chem. 271: 6523–6529.

21. Bertina, R.M., van der Linden, I.K., Engesser, L., Muller, H.P. and Brommer, E.J.P. (1987) Hereditary heparin cofactor II deficiency and the risk of development of thrombosis. Thromb. Haemost. 57: 196–200.

22. Ezenagu, L.C. and Brandt, J.T. (1986) Laboratory determination of heparin cofactor II. Arch. Pathol. Lab. Med. 110: 1149–1151.

23. Andrew, M., Paes, B., Milner, R., Johnston, M., Mitchell, L., Tollefsen, D.M. and Powers, P. (1987) Development of the human coagulation system in the full-term infant. Blood 70: 165–172.

24. Sandset, P.M. and Andersson, T.R. (1989) Coagulation inhibitor levels in pneumonia and stroke: changes due to consumption and acute phase reaction. J. Intern. Med. 225: 311–316.

25. Koike, C., Hayakawa, Y., Niiya, K., Sakuragawa, N. and Sasaki, H. (1996) The production of heparin cofactor II is not regulated by inflammatory cytokines in human hepatoma cells: comparison with plasminogen activator inhibitor type-1. Thromb. Haemost. 75: 298–302.

26. Westrup, D. and Ragg, H. (1994) Secondary thrombin-binding site, glycosaminoglycan binding domain and reactive center region of leuserpin-2 are strongly conserved in mammalian species. Biochim. Biophys. Acta 1217: 93–96.

27. Sheffield, W.P., Schuyler, P.D. and Blajchman, M.A. (1994) Molecular cloning and expression of rabbit heparin cofactor II: a plasma thrombin inhibitor highly conserved between species. Thromb. Haemost. 71: 778–782.

28. Borriello, F. and Krauter, K.S. (1991) Multiple murine α1-protease inhibitor genes show unusual evolutionary divergence. Proc. Natl. Acad. Sci. U. S. A. 88: 9417–9421.

29. Inglis, J.D. and Hill, R.E. (1991) The murine Spi-2 proteinase inhibitor locus: a multigene family with a hypervariable reactive site domain. EMBO J. 10: 255–261.

30. Tollefsen, D.M., Pestka, C.A. and Monafo, W.J. (1983) Activation of heparin cofactor II by dermatan sulfate. J. Biol. Chem. 258: 6713–6716.

31. Pratt, C.W., Whinna, H.C., Meade, J.B., Treanor, R.E. and Church, F.C. (1989) Physicochemical aspects of heparin cofactor II. Ann. N. Y. Acad. Sci. 556: 104–115.

32. Klauser, R.J. (1991) Interaction of the sulfated lactobionic acid amide LW 10082 with thrombin and its endogenous inhibitors. Thromb. Res. 62: 557–565.

33. Hayakawa, Y., Hayashi, T., Hayashi, T., Niiya, K. and Sakuragawa, N. (1995) Selective activation of heparin cofactor II by a sulfated polysaccharide isolated from the leaves of *Artemisia princeps*. Blood Coagul. Fibrinolysis 6: 643–649.

34. Nagase, H., Enjyoji, K., Minamiguchi, K., Kitazato, K.T., Kitazato, K., Saito, H. and Kato, H. (1995) Depolymerized holothurian glycosaminoglycan with novel anticoagulant actions: antithrombin III- and heparin cofactor II-independent inhibition of factor X activation by factor IXa-factor VIIIa complex and heparin cofactor II-dependent inhibition of thrombin. Blood 85: 1527–1534.

35. Melton, L.G., Church, F.C. and Erickson, B.W. (1995) Designed polyanionic coiled-coil proteins: acceleration of heparin cofactor II inhibition of thrombin. Int. J. Pept. Protein Res. 45: 44–52.

36. Bourin, M.-C. and Lindahl, U. (1993) Glycosaminoglycans and the regulation of blood coagulation. Biochem. J. 289: 313–330.

37. Maimone, M.M. (1990) Characterization of heparin and dermatan sulfate molecules that bind and activate heparin cofactor II. Ph. D. thesis. Washington University, St. Louis.

38. Maimone, M.M. and Tollefsen, D.M. (1988) Activation of heparin cofactor II by heparin oligosaccharides. Biochem. Biophys. Res. Commun. 152: 1056–1061.

39. Sié, P., Petitou, M., Lormeau, J.-C., Dupouy, D., Boneu, B. and Choay, J. (1988) Studies on the structural requirements of heparin for the catalysis of thrombin inhibition by heparin cofactor II. Biochim. Biophys. Acta 966: 188–195.

40. Bray, B., Lane, D.A., Freyssinet, J.-M., Pejler, G. and Lindahl, U. (1989) Anti-thrombin activities of heparin. Effect of saccharide chain length on thrombin inhibition by heparin cofactor II and by antithrombin. Biochem. J. 262: 225–232.

41. Conrad, H.E. (1989) Structure of heparan sulfate and dermatan sulfate. Ann. N. Y. Acad. Sci. 556: 18–28.

42. Maimone, M.M. and Tollefsen, D.M. (1990) Structure of a dermatan sulfate hexasaccharide that binds to heparin cofactor II with high affinity. J. Biol. Chem. 265: 18263–18271.

43. Mascellani, G., Liverani, L., Prete, A., Guppola, P.A., Bergonzini, G. and Bianchini, P. (1994) Relative influence of different disulphate disaccharide clusters on the HCII-mediated inhibition of thrombin by dermatan sulphates of different origins. Thromb. Res. 74: 605–615.

44. Pavão, M.S.G., Mourão, P.A.S., Mulloy, B. and Tollefsen, D.M. (1995) A unique dermatan sulfate-like glycosaminoglycan from ascidian. Its structure and the effect of its unusual sulfation pattern on anticoagulant activity. J. Biol. Chem. 270: 31027–31036.

45. Andersson, T.R., Larsen, M.L. and Abildgaard, U. (1987) Low heparin cofactor II associated with abnormal crossed immunoelectrophoresis pattern in two Norwegian families. Thromb. Res. 47: 243–248.

46. Blinder, M.A., Andersson, T.R., Abildgaard, U. and Tollefsen, D.M. (1989) Heparin cofactor II Oslo. Mutation of Arg-189 to His decreases the affinity for dermatan sulfate. J. Biol. Chem. 264: 5128–5133.

47. Blinder, M.A. and Tollefsen, D.M. (1990) Site-directed mutagenesis of arginine 103 and lysine 185 in the proposed glycosaminoglycan-binding site of heparin cofactor II. J. Biol. Chem. 265: 286–291.

48. Ragg, H., Ulshöfer, T. and Gerewitz, J. (1990) On the activation of human leuserpin-2, a thrombin inhibitor, by glycosaminoglycans. J. Biol. Chem. 265: 5211–5218.

49. Ragg, H., Ulshöfer, T. and Gerewitz, J. (1990) Glycosaminoglycan-mediated leuserpin-2/thrombin interaction. Structure-function relationships. J. Biol. Chem. 265: 22386–22391.

50. Whinna, H.C., Blinder, M.A., Szewczyk, M., Tollefsen, D.M. and Church, F.C. (1991) Role of lysine 173 in heparin binding to heparin cofactor II. J. Biol. Chem. 266: 8129–8135.

51. Rydel, T.J., Ravichandran, K.G., Tulinsky, A., Bode, W., Huber, R., Roitsch, C. and Fenton, J.W., II (1990) The structure of a complex of recombinant hirudin and human α-thrombin. Science 249: 277–280.

52. Hortin, G.L., Tollefsen, D.M. and Benutto, B.M. (1989) Antithrombin activity of a peptide corresponding to residues 54–75 of heparin cofactor II. J. Biol. Chem. 264: 13979–13982.

53. Van Deerlin, V.M.D. and Tollefsen, D.M. (1991) The N-terminal acidic domain of heparin cofactor II mediates the inhibition of α-thrombin in the presence of glycosaminoglycans. J. Biol. Chem. 266: 20223–20231.

54. Sheffield, W.P. and Blajchman, M.A. (1995) Deletion mutagenesis of heparin cofactor II: defining the minimum size of a thrombin inhibiting serpin. FEBS Lett. 365: 189–192.

55. Rogers, S.J., Pratt, C.W., Whinna, H.C. and Church, F.C. (1992) Role of thrombin exosites in inhibition by heparin cofactor II. J. Biol. Chem. 267: 3613–3617.

56. Phillips, J.E., Shirk, R.A., Whinna, H.C., Henriksen, R.A. and Church, F.C. (1993) Inhibition of dysthrombins Quick I and II by heparin cofactor II and antithrombin. J. Biol. Chem. 268: 3321–3327.

57. Sheehan, J.P., Wu, Q., Tollefsen, D.M. and Sadler, J.E. (1993) Mutagenesis of thrombin selectively modulates inhibition by serpins heparin cofactor II and antithrombin III. Interaction with the anion-binding exosite determines heparin cofactor II specificity. J. Biol. Chem. 268: 3639–3645.

58. Griffith, M.J. (1983) Heparin-catalyzed inhibitor/protease reactions: kinetic evidence for a common mechanism of action of heparin. Proc. Natl. Acad. Sci. U. S. A. 80: 5460–5464.

59. Tollefsen, D.M., Peacock, M.E. and Monafo, W.J. (1986) Molecular size of dermatan sulfate oligosaccharides required to bind and activate heparin cofactor II. J. Biol. Chem. 261: 8854–8858.

60. Sheehan, J.P., Tollefsen, D.M. and Sadler, J.E. (1994) Heparin cofactor II is regulated allosterically and not primarily by template effects. Studies with mutant thrombins and glycosaminoglycans. J. Biol. Chem. 269: 32747–32751.

61. Sheehan, J.P. and Sadler, J.E. (1994) Molecular mapping of the heparin-binding exosite of thrombin. Proc. Natl. Acad. Sci. U. S. A. 91: 5518–5522.

62. Teien, A.N., Abildgaard, U. and Höök, M. (1976) The anticoagulant effect of heparan sulfate and dermatan sulfate. Thromb. Res. 8: 859–863.

63. Tollefsen, D.M. and Blank, M.K. (1981) Detection of a new heparin-dependent inhibitor of thrombin in human plasma. J. Clin. Invest. 68: 589–596.

64. Sié, P., Ofosu, F., Fernandez, F., Buchanan, M.R., Petitou, M. and Boneu, B. (1986) Respective role of antithrombin III and heparin cofactor II in the in vitro anticoagulant effect of heparin and of various sulphated polysaccharides. Br. J. Haematol. 64: 707–714.

65. Fernandez, F., van Ryn, J., Ofosu, F.A., Hirsh, J. and Buchanan, M.R. (1986) The haemorrhagic and antithrombotic effects of dermatan sulfate. Br. J. Haematol. 64: 309–317.

66. Masson, P.J., Coup, D., Millet, J. and Brown, N.L. (1995) The effect of the β-D-xyloside naroparcil on circulating plasma glycosaminoglycans. An explanation for its known antithrombotic activity in the rabbit. J. Biol. Chem. 270: 2662–2668.

67. Buchanan, M.R., Liao, P., Smith, L.J. and Ofosu, F.A. (1994) Prevention of thrombus formation and growth by antithrombin III and heparin cofactor II-dependent thrombin inhibitors: importance of heparin cofactor II. Thromb. Res. 74: 463–475.

68. Bendayan, P., Boccalon, H., Dupouy, D. and Boneu, B. (1994) Dermatan sulfate is a more potent inhibitor of clot-bound thrombin than unfractionated and low molecular weight heparins. Thromb. Haemost. 71: 576–580.

69. Cosmi, B., Agnelli, G., Young, E., Hirsh, J. and Weitz, J. (1993) The additive effect of low molecular weight heparins on thrombin inhibition by dermatan sulfate. Thromb. Haemost. 70: 443–447.

70. Lane, D.A., Ryan, K., Ireland, H., Curtis, J.R., Nurmohamed, M.T., Krediet, R.T., Roggekamp, M.C., Stevens, P. and ten Cate, J.W. (1992) Dermatan sulphate in haemodialysis. Lancet 339: 334–335.

71. Prandoni, P., Meduri, F., Cuppini, S., Toniato, A., Zangrandi, F., Polistena, P., Gianese, F. and Maffei-Faccioli, A. (1992) Dermatan sulphate: a safe approach to prevention of postoperative deep vein thrombosis. Br. J. Surg. 79: 505–509.

72. Agnelli, G., Cosmi, B., Di Filippo, P., Ranucci, V., Veschi, F., Longetti, M., Renga, C., Barzi, F., Gianese, F., Lupattelli, L., Rinonapoli, E. and Nenci, G.G. (1992) A randomised, double-blind, placebo-controlled trial of dermatan sulphate for prevention of deep vein thrombosis in hip fracture. Thromb. Haemost. 67: 203–208.

73. Andersson, T.R., Sié, P., Pelzer, H., Aamodt, L.-M., Nustad, K. and Abildgaard, U. (1992) Elevated levels of thrombin-heparin cofactor II complex in plasma from patients with disseminated intravascular coagulation. Thromb. Res. 66: 591–598.

74. Liu, L., Dewar, L., Song, Y., Kulczycky, M., Blajchman, M.A., Fenton, J.W., II, Andrew, M., Delorme, M., Ginsberg, J., Preissner, K.T. and Ofosu, F.A. (1995) Inhibition of thrombin by antithrombin III and heparin cofactor II in vivo. Thromb. Haemost. 73: 405–412.

75. McGuire, E.A. and Tollefsen, D.M. (1987) Activation of heparin cofactor II by fibroblasts and vascular smooth muscle cells. J. Biol. Chem. 262: 169–175.

76. Hiramoto, S.A. and Cunningham, D.D. (1988) Effects of fibroblasts and endothelial cells on inactivation of target proteases by protease nexin-1, heparin cofactor II, and C1-inhibitor. J. Cell. Biochem. 36: 199–207.

77. Whinna, H.C., Choi, H.U., Rosenberg, L.C. and Church, F.C. (1993) Interaction of heparin cofactor II with biglycan and decorin. J. Biol. Chem. 268: 3920–3924.

78. Andersson, T.R., Larsen, M.L., Handeland, G.F. and Abildgaard, U. (1986) Heparin cofactor II activity in plasma: application of an automated assay method to the study of a normal adult population. Scand. J. Haematol. 36: 96–102.

79. Awidi, A.S., Abu-Khalaf, M., Herzallah, U., Abu-Rajab, A., Shannak, M.M., Abu-Obeid, T., Al-Taher, I. and Anshasi, B. (1993) Hereditary thrombophilia among 217 consecutive patients with thromboembolic disease in Jordan. Am. J. Hematol. 44: 95–100.

80. Sié, P., Dupouy, D., Pichon, J. and Boneu, B. (1985) Constitutional heparin co-factor II deficiency associated with recurrent thrombosis. Lancet 2: 414–416.

81. Tran, T.H., Marbet, G.A. and Duckert, F. (1985) Association of hereditary heparin co-factor II deficiency with thrombosis. Lancet 2: 413–414.

82. Weisdorf, D.J. and Edson, J.R. (1991) Recurrent venous thrombosis associated with inherited deficiency of heparin cofactor II. Br. J. Haematol. 77: 125–126.

83. Matsuo, T., Kario, K., Sakamoto, S., Yamada, T., Miki, T., Hirase, T. and Kobayashi, H. (1992) Hereditary heparin cofactor II deficiency and coronary artery disease. Thromb. Res. 65: 495–505.

84. Andrew, M., Mitchell, L., Berry, L., Paes, B., Delorme, M., Ofosu, F., Burrows, R. and Khambalia, B. (1992) An anticoagulant dermatan sulfate proteoglycan circulates in the pregnant woman and her fetus. J. Clin. Invest. 89: 321–326.

85. Brennan, M.J., Oldberg, A., Pierschbacher, M.D. and Ruoslahti, E. (1984) Chondroitin/dermatan sulfate proteoglycan in human fetal membranes: demonstration of an antigenically similar proteoglycan in fibroblasts. J. Biol. Chem. 259: 13742–13750.

86. Massouh, M., Jatoi, A., Gordon, E.M. and Ratnoff, O.D. (1989) Heparin cofactor II activity in plasma during pregnancy and oral contraceptive use. J. Lab. Clin. Med. 114: 697–699.

87. Andersson, T., Lorentzen, B., Høgdahl, H., Clausen, T., Mowinckel, M.-C. and Abildgaard, U. (1996) Thrombin-inhibitor complexes in the blood during and after delivery. Thromb. Res. 82: 109–117.

88. Sandset, P.M., Hellgren, M., Uvebrandt, M. and Bergström, H. (1989) Extrinsic coagulation pathway inhibitor and heparin cofactor II during normal and hypertensive pregnancy. Thromb. Res. 55: 665–670.

89. Coughlin, S.R., Vu, T.-K.H., Hung, D.T. and Wheaton, V.I. (1992) Characterization of a functional thrombin receptor. Issues and opportunities. J. Clin. Invest. 89: 351–355.

5

PCI: PROTEIN C INHIBITOR?

Scott T. Cooper[1] and Frank C. Church[2]

[1]Department of Biology and Microbiology
University of Wisconsin-La Crosse
La Crosse, Wisconsin 54601
[2]Departments of Pathology and Lab Medicine and Medicine,
    and Center for Thrombosis and Hemostasis, School of Medicine
The University of North Carolina at Chapel Hill
Chapel Hill, North Carolina, 27599-7035

1. HISTORY

1.1. Coagulation

Blood coagulation is a complex chain reaction of enzyme activations in which inactive zymogens are activated by proteolytic cleavage. Each newly activated enzyme in turn cleaves the next zymogen in the chain, resulting in its activation. The final enzyme in the cascade, thrombin, then cleaves soluble fibrinogen into insoluble fibrin which forms a thrombus. A single enzyme can cleave and activate multiple zymogens, thus a small initial stimulus can amplify into a massive wave of coagulation. In addition, thrombin can feedback and stimulate its own synthesis by proteolytically activating two necessary cofactors in thrombin generation, Factor Va and Factor VIIIa. Thus, the coagulation cascade proceeds by both amplification and a feedback positive mechanism, a potentially deadly combination if not closely regulated.

Hemostasis requires a balance of procoagulant and anticoagulant controls. The anticoagulant control mechanisms are both robust and elegant. The first line of defense against blood clotting is the inhibition of thrombin by the serine protease inhibitors (serpins) antithrombin (AT) and heparin cofactor II (HCII) found circulating in the blood. Serpins have reactive site loops that insert into the active site cleft of a serine protease. The protease then cleaves the serpin between the P1–P1' bond, resulting in the formation of a stable covalent intermediate (Travis and Salvesen, 1983). Both AT and HCII act through this common mechanism, blocking thrombin activity by inserting their reactive site loop into the active site of thrombin. The drug heparin stimulates the reaction between thrombin and AT and HCII several thousand-fold. The mechanism is thought to involve the formation of a ternary complex in which heparin binds to heparin binding sites on both thrombin and the serpin, thus forming a "bridge" between the two molecules (Björk and Lindahl, 1982; Rosenberg,

1977). The formation of a ternary complex alone does not fully account for the dramatic increase in activity seen in the presence of heparin. AT goes through a conformational change upon binding heparin which makes it more reactive with thrombin (Olson and Björk, 1992; Rosenberg and Damus, 1973; Shore et al., 1989). Heparin is thought to displace an acidic domain of HCII which can bind to thrombin, thus increasing the rate of inhibition in the presence of heparin (van Deerlin and Tollefsen, 1991). While AT and HCII are important in inhibiting thrombin activity it soon became apparent that some other process was responsible for shutting off thrombin production. This lead to the discovery of the protein C pathway (Esmon, 1992; Kisiel, 1979; Stenflo, 1976).

1.2. Protein C

Once activated, thrombin can proceed down one of two pathways. It can continue to cleave fibrinogen and activate Factor Va and Factor VIIIa, thus stimulating clot formation, or it can bind to thrombomodulin. Thrombomodulin is a membrane bound protein on the surface of endothelial cells. Once thrombin is bound to thrombomodulin it no longer recognizes fibrinogen as a substrate, thus blocking clot formation. Instead thrombin recognizes another inactive zymogen of a serine protease, protein C, as a substrate (Esmon, 1989). Proteolytically activated protein C (APC) requires two cofactors, Factor V and protein S, for full activity (Suzuki et al., 1983). APC specifically cleaves the two cofactors involved in thrombin generation, Factor Va and Factor VIIIa (Kane and Davie, 1988). The destruction of these necessary cofactors results in cessation of thrombin synthesis, and thus stops clot formation. The importance of this pathway in the regulation of coagulation is apparent in the recent discovery that familial thrombophilia is due to a mutation in Factor V which destroys its activity as a cofactor for APC (Dahlback, 1994; Dahlback et al., 1993; Dahlback and Hildebrand, 1994). In this elegant control system thrombin activates a pathway which feeds back and inhibits its own synthesis.

1.3. Protein C Inhibitor

There are several known inhibitors of APC in plasma. α_1-Protease inhibitor and α_2-macroglobulin are both present in high concentrations in plasma and inhibit APC, albeit at a fairly low rate (Heeb and Griffin, 1988; Hoogendoorn et al., 1991; van der Meer et al., 1989). The presence of a heparin-accelerated inhibitor of APC in plasma was first described by Marlar and Griffith in 1980 (Marlar and Griffin, 1980). Three years later Suzuki et al. purified protein C inhibitor (PCI) from plasma using heparin affinity chromatography (Suzuki et al., 1983). PCI was shown to be a glycoprotein with a molecular weight of 57,000 D. The cDNA for PCI was subsequently obtained in 1987 by screening a human liver library (Suzuki et al., 1987). Sequence analysis showed that PCI belonged to the serpin family, the highest homology being with α1-antichymotrypsin. PCI is a very good inhibitor of APC in vitro. However, the importance of PCI in inhibiting APC was best demonstrated in vivo.

APC was injected into chimpanzees and pre- and post-infusion plasma samples were examined by immunoblotting with anti-protein C antibodies. Complexes between APC and PCI were observed first, and only after PCI had been depleted were complexes between APC and α_1-protease inhibitor or α_2-macroglobulin observed (Hoogendoorn et al., 1990). In the presence of heparin, complexes between PCI and APC are also formed most readily in human plasma, suggesting that PCI is the primary inhibitor of APC in plasma (Heeb et al., 1989). Complexes between PCI and APC have also been detected in humans following pulmonary embolism, disseminated intravascular coagulation (DIC), and myocardial infarction (Espana et al., 1990; Marlar et al., 1985; Minamikawa et al., 1994; Tanigawa et al., 1995). Patients suffering from DIC undergo a dramatic, often fatal, stimulation of coagulation. In the course of this dramatic increase in thrombin production, APC is generated and forms complexes with PCI. The role of PCI, if any, in the control of DIC is unknown.

2. STRUCTURE/FUNCTION STUDIES ON PCI

2.1. Protease Inhibition

PCI inhibits many serine proteases in addition to APC, including: thrombin, Factor Xa, Factor XIa, kallikrein, tissue-type plasminogen activator, urokinase, prostate specific antigen (PSA), acrosin, chymotrypsin and trypsin (Espana et al., 1991; Ecke et al., 1992; Espana et al., 1989; Espana et al., 1991; Meijers et al., 1988; Phillips et al., 1994; Pratt et al., 1989; Suzuki et al., 1989; Suzuki et al., 1983). The rates of inhibition of these serine proteases by PCI vary several-fold. This broad range of protease inhibition makes PCI an ideal inhibitor to use in studying differences in specificity among many serine proteases. While such comparisons have been made using small chromogenic substrates, using a macromolecule such as PCI allows for more realistic physiological comparisons, especially as most chromogenic substrates do not bind to the enzyme on both sides of the scissile P1–P1' bond.

Comparing the rates of inhibition of different serine proteases by PCI reveals some interesting trends. PCI inhibits thrombin with the highest rate of inhibition, followed by APC, Factor Xa, urokinase and then tissue-type plasminogen activator. Comparison of the structures of these proteins, based upon X-ray crystal structures and molecular models suggests that serine proteases with more restrictive active sites have the highest rates of inhibition by PCI (Cooper et al., 1995). Thrombin has a very restricted active site and shows the highest rates of inhibition by PCI, this ability of PCI to inhibit the procoagulant enzyme thrombin better than it inhibits the anticoagulant enzyme APC is an apparent paradox that will be addressed in detail later. APC has a slightly less restricted active site and has a lower rate of inhibition by PCI. Proteases with exposed active sites have much lower rates of inhibition by PCI. The accessibility of the active site thus appears to dramatically influence the rate of inhibition, perhaps by preventing water from penetrating and hydrolyzing the covalent bond between the protease and PCI. This is also supported by looking at the PCI inhibition of two thrombin mutants in which portions of the loops that extend over the active site were deleted, des-ETW and des-PPW (LeBonniec et al., 1992; LeBonniec et al., 1993). In both cases, inhibition rates by PCI were decreased, suggesting that these regions are important in the formation of complexes between PCI and thrombin (Cooper and Rezaie, unpublished results).

2.2. Expression of Recombinant PCI

Recombinant PCI has been expressed in several cell lines including, COS-1 cells, baby hamster kidney cells and in insect cells using baculovirus (Phillips et al., 1994; Suzuki et al., 1989). The ability to express recombinant PCI allows for the introduction of specific mutations into PCI. Mutations in the reactive site loop of PCI have also shed light on the mechanism by which different proteases recognize this serpin. Twenty different mutations were introduced into the reactive site loop of PCI from P3 to P3′ (Cooper and Church, 1995; Cooper et al., 1995; Phillips et al., 1994). These mutant proteins were then assayed with thrombin, APC, Factor Xa, urokinase and tissue-type plasminogen activator. The results indicated that each serpin has different regions of the inhibitor that influence activity the most. Thrombin is very sensitive to changes in the P2 residue, while APC and Factor Xa are affected most by changing the P3 residue of PCI to an arginine. Thrombin contains a unique 60 insertion loop that appears to interact with the P2 position of PCI in molecular models of the complex between thrombin and PCI. Inhibition of urokinase was increased by any changes in PCI that made its reactive site loop sequence more like that of plasminogen activator inhibitor (PAI-1), a potent natural inhibitor of urokinase. Mutations in the P2′ and P3′ had little effect on the interaction between PCI and any of the serine proteases. By comparing the effects of mutations engineered into a single serpin on multiple proteases much can be learned about the nature of the interactions between serpins and different proteases.

2.3. Molecular Modeling of PCI–Protease Complexes

When these kinetic data are coupled with molecular models, predictions of the individual amino acids that interact between the protease and PCI can be made. At least three molecular models have been prepared for PCI (Cooper et al., 1995; Kuhn et al., 1990; Suzuki et al., 1989). These have been based upon the crystal structure of ovalbumin. One region that cannot be fully resolved at this time is the amino terminus. This region is predicted to form a positively charged α-helix, the A+ helix, which has been proposed to form part of the heparin binding domain of PCI. A model of the complex between PCI and thrombin has been generated using the crystal structure of thrombin and a model of PCI. This model is shows contacts between PCI and thrombin consistent with the mutagenesis data described above (Cooper et al., 1995). These models have also been used to clearly define the heparin binding site of PCI, and its orientation relative to the heparin binding sites of thrombin and APC.

3. GLYCOSAMINOGLYCAN INTERACTIONS

3.1. Heparin-Binding Domain of PCI

The glycosaminoglycan heparin stimulates the inhibition of both thrombin and APC by PCI. The reaction with APC is stimulated 50-fold with an optimum heparin concentration of 100 μg heparin/ml. PCI inhibition of thrombin is stimulated 15-fold with an optimum heparin concentration of 10 μg heparin/ml (Pratt and Church, 1992; Pratt et al., 1992). This stimulation does not appear to involve a conformational change in the structure of PCI, as is seen with AT, or displacement of an additional binding domain, as is proposed with HCII. Three regions have been proposed as the heparin binding domain of PCI. The A+ helix near the amino terminus, the D-helix and the H-helix (Kuhn et al., 1990; Pratt et al., 1992; Shirk et al., 1994). The D-helicies of AT and HCII contain multiple positively charged residues and make up part of the heparin binding domain of these serpins. PCI does not have a positively charged D-helix, making this an unlikely heparin binding domain. Changing residues Arg_{269} and Lys_{270} in the H-helix of PCI to Ala decreases both heparin affinity and heparin acceleration of inhibition of thrombin and APC (Shirk et al., 1994). In addition, H-helix synthetic peptides blocked PCI binding to heparin, whereas the A+ synthetic peptides did not (Pratt and Church, 1992). Further, deleting the A+ helix had little effect on PCI heparin binding ability (Shirk et al., 1994). Thus, it appears that the heparin binding site in PCI is distinct from that of either AT or HCII. The heparin binding sites of thrombin and APC are thought to lie on different sides of the respective molecules. This suggests that heparin may be able to bind to PCI in different orientations to align it with the different proteases. Comparing molecular models of PCI complexed with APC or thrombin, the heparin binding sites are closer together in the PCI–APC complex relative to the PCI–thrombin complex. This is consistent with the kinetic data showing stronger heparin stimulation of the PCI–APC reaction than the PCI–thrombin reaction.

3.2. *In Vivo* Binding to Glycosaminoglycans

As PCI can bind to heparin, it is plausible that some PCI may be bound to the capillary endothelium. It has been shown that PCI activity with urokinase is stimulated by glycosaminoglycans bound to the kidney cell line TCL-598 (Geiger et al., 1991). Given that the plasma concentration of PCI is so low (70–90 nM, compared with 2–5 μM for AT or 1 μM for HCII) PCI bound to the endothelium could represent a large pool of available PCI. Following heparin injection into humans, circulating PCI concentrations were unchanged, indicating that a heparin-releasable pool of PCI was not bound to glycosaminoglycans lining the capillary walls (Cooper et al., 1996). This was supported immunohistochemically, showing PCI localization just beneath the endothelium. This evidence may support an extravascular function for PCI, perhaps in inflammatory regulation. PCI levels in the plasma do drop following heparin therapy, probably through formation of complexes with plasma proteases (Tabernero et al., 1990). The PCI observed just below the endothelium could either have entered the extravascular space by diffusion, or could be produced directly by endothelial cells.

4. WHAT IS THE ROLE OF PCI IN REGULATIN COAGULATION?

The sub-endothelial localization of PCI is also interesting because PCI has recently been shown to be a very potent inhibitor of thrombin when bound to thrombomodulin (Rezaie et al., 1995). In this capacity PCI is a better inhibitor of thrombin bound to thrombomodulin than is AT. Thrombomodulin contains a chondroitin sulfate chain which stimulates the AT-thrombin reaction through a mechanism akin to heparin stimulation of this reaction (Bourin et al., 1986; Hofsteenge et al., 1986; Preissner et al., 1990). In contrast, thrombomodulin lacking chondroitin sulfate is still capable of stimulating the inhibition of thrombin by PCI, suggesting that a conformational change occurs in thrombin upon binding to thrombomodulin which increases its reactivity with PCI (Rezaie et al., 1995). Further, thrombomodulin is found on endothelial cells, in much the same location as PCI (Cooper et al., 1996). This recent discovery that PCI is a potent inhibitor of thrombin bound to thrombomodulin partially explains the paradox noted earlier, that PCI inhibits thrombin at a ten-fold greater rate than it inhibits APC. One could then envision a model where PCI not only inhibits APC, but also inhibits the generation of APC by inhibiting thrombin bound to thrombomodulin. Thus, PCI would control both the activity and production of APC.

Why would we want to have a protein around that inhibits this important safeguard pathway? One possibility is that inhibiting APC is not the primary role of PCI. Before the discovery that PCI was such a good inhibitor of thrombin when bound to thrombomodulin this would have been a difficult argument to dispel. While it is possible to imagine a rogue protein escaping from the testis and happening to inhibit a plasma protein such as APC, it is more difficult to imagine how this same rogue protein could also just happen to inhibit the generation of APC as well. Such a scenario would be too much of a coincidence. A second possibility is that PCI inhibits APC to prevent the small amounts APC present in plasma from degrading Factor V or Factor VIII. Thus, a small amount of PCI is kept around to regulate the "leaky" production of APC. When the coagulation cascade is triggered a huge wave of thrombin is generated. This local increase in thrombin would deplete an area of PCI, both because thrombin concentrations would be much higher than APC, and because thrombin has a higher activity with PCI than does APC. Once thrombin had removed PCI from the scene, APC activation and subsequent inhibition of thrombin generation could proceed. This model also accounts for the apparently low concentrations of PCI found in plasma, relative to the other serpins that regulate coagulation. If PCI concentrations were higher, PCI would not be depleted by thrombin, and thus APC generation would be blocked. This could also explain why no patients with PCI deficiency have been discovered, they would probably manifest as a mild to moderate hemophilia, due to unchecked APC mediated degradation of Factor V or Factor VIII. A similar model has been proposed to explain the presence of relatively low concentrations of PCI in the acrosome of spermatozoa. Unfortunately, these models will be very difficult to test definitively.

5. OTHER POTENTIAL ROLES OF PCI

5.1. Tissue Distribution

In addition to being found in the blood, PCI is also found in urine, tears, saliva, breast milk, cerebral spinal fluid and amniotic fluid (Laurell et al., 1992). PCI has been detected in seminal fluid at concentrations 40-fold higher than seen in plasma (Laurell et al., 1992). Using northern blots PCI mRNA has been seen in liver, spleen, prostate, kidney, testis, ovary and pancreas. PCI has also been detected immunohistochemically in testis, ovary, and in endothelial cells lining vessel walls. This tissue distribution has been substantiated in tissue culture. PCI secretion has been observed in cultured PC3 cells from the prostate (Neese and Church, unpublished results), HepG2 cells from the liver (Morito et al., 1985) and CCL7.1 rhesus monkey kidney cells (Radtke et al., 1994). PCI mRNA has been detected in kidney cells by in situ hybridization (Radtke et al., 1994). While liver may be the primary source of plasma PCI, the relatively high PCI mRNA levels in testis, and the high concentrations of PCI in seminal fluid suggest that a large amount of PCI is being made in the male reproductive system.

5.2. Role in Male Reproductive System

What function could PCI be serving the reproductive system? Complexes between PCI and other proteases have been detected. PCI complexes with both urokinase and tissue type plasminogen activator have been detected in plasma and semen (Espana et al., 1993; Geiger et al., 1989). In human seminal fluid complexes with PSA have been detected (Espana et al., 1991). Complexes between tissue kallikrein with protein C inhibitor have also been detected in human semen and urine. (Espana et al., 1995). This in vivo evidence suggests that PCI may be inhibiting more than APC in different parts of the body. PCI is known to inhibit acrosin in vivo, and has been localized to the acrosome in spermatozoa (Zheng et al., 1994; Zheng et al., 1993). This suggests that PCI may play a role in preventing the premature activation of acrosin in spermatozoa. Once the sperm mix with PSA from the prostate during the formation of an ejaculate the massive excess of PSA to PCI would prevent PCI from inhibiting acrosin. This may be important in allowing acrosin to aid in fertilization. Thus, a similar mechanism is being proposed for the function of PCI in regulating APC and acrosin activities. PCI is present to keep these activities under control until they are needed. At this point a large excess of another protease is produced (thrombin or PSA), which effectively depletes the local PCI pool, allowing the APC or acrosin to function. These models account for the relative promiscuity of PCI in inhibiting multiple proteases and its low concentrations in plasma and other fluids.

5.3. Role in Prostate Cancer

The detection of PCI–urokinase, PCI–kallikrein and PCI–PSA complexes strengthens the proposal that PCI in some way mediates one or more of these proteases in vivo. The actual function of PCI in the male reproductive system is at this time unclear. The recently discovered serpin maspin is depleted in cells isolated from breast cancer patients. This suggests that maspin may be acting as an antitumor agent (Sheng et al., 1994; Zou et al., 1994). A similar role seems plausible for PCI. Perhaps PCI inhibits urokinase and PSA in the prostate, preventing activation of plasmin and thus initiation of metastasis. Once prostate cancer had begun to metastasize the existing pool of PCI would be overwhelmed and urokinase could activate plasmin. If this were the case, PCI levels would be expected to be depleted in patients with prostate cancer. Examination of prostate tissue from patients with prostate cancer did not reveal dramatic changes in PCI levels as measured by immunohistochemistry. Analysis of patients with benign hyperplasia and prostate cancer by Northern blot also shows no difference in PCI mRNA (Cooper and Church, unpublished results). While this does not rule out a potential role for PCI in mediating prostate cancer, the effect is not as dramatic as that seen for maspin.

6. WHAT DOES PCI STAND FOR?

Recent discoveries suggest that PCI may have many more functions than originally envisioned. Certainly the role of PCI in inhibiting APC is well established and makes this protein worthy of the name Protein C Inhibitor (PCI). Given its newly discovered function of inhibiting thrombin bound to thrombomodulin (a complex with protein C-ase activity) one could imagine an appropriate name of Protein C-ase Inhibitor (PCI). Its presence in prostate and ability to form complexes with two known markers of prostate cancer may also earn this protein the moniker Prostate Cancer Inhibitor (PCI) some day, although this name is premature at present. With the current pace of research in this field we should soon understand the function of PCI. The only remaining challenge at that time would be to adjust the name to fit the abbreviation.

ACKNOWLEDGMENTS

We thank Laura L. Neese, Stephanie L. Bialas, Charlotte W. Pratt, Jeanne E. Phillips, Rebecca A. Shirk, Tracy P. Jackson, Elizabeth E. Potter, Jennifer M. Boyd, Herbert C.

Whinna, Darla K. Liles, Michael N. DiCuccio, Alireza R. Rezaie, Charles T. Esmon, Joost C.M. Meijers, Marc G.L.M. Elisen, Dougald M. Monroe, and Maureane Hoffman for their significant contributions to many of the studies described in this review. This work was supported by a Grant-in-Aid from the American Heart Association-Sanofi Winthrop and Research Grants 5T32-HL07149 and HL-32656 from the National Institutes of Health.

REFERENCES

Björk, I., and Lindahl, U. (1982). Mechanism of the anticoagulant action of heparin. Mol. Cell. Biochem. *48*, 161–182.

Bourin, M., Boffa, M., Björk, I., and Lindahl, U. (1986). Functional domains of rabbit thrombomodulin. Proc. Natl. Acad. Sci. U.S.A. *83*, 5924–5928.

Cooper, S. T., and Church, F. C. (1995). Reactive site mutants of recombinant protein C inhibitor. Biochim. Biophys. Acta *1246*, 29–33.

Cooper, S. T., Neese, L. N., Dicuccio, M., Liles, D. K., Hoffman, M., and Church, F. C. (1996). Heparin-binding serpins are not released following intravenous heparin injection. Clinical and Applied Thrombosis/Hemostasis *2*, 185–191.

Cooper, S. T., Whinna, H. C., Jackson, T. P., Boyd, J. M., and Church, F. C. (1995). Intermolecular interactions between protein C inhibitor and coagulation proteases. Biochemistry *34*, 12991–12997.

Dahlback, B. (1994). Physiological anticoagulation: Resistance to activated protein C and venous thromboembolism. J. Clin. Invest. *94*, 923–927.

Dahlback, B., Carlsson, M., and Svensson, P. J. (1993). Familial thrombophilia due to a previously unrecognized mechanism characterized by poor anticoagulant response to activated protein C: Prediction of a cofactor to activated protein C. Proc. Natl. Acad. Sci. *90*, 1004–1008.

Dahlback, B., and Hildebrand, B. (1994). Inherited resistance to activated protein C is corrected by anticoagulant cofactor activity found to be a property of Factor V. Proc. Natl. Acad. Sci. *91*, 1396–1400.

Ecke, S., Geiger, M., Resch, I., Jerabek, I., Stingl, L., Maier, M., and Binder, B. R. (1992). Inhibition of tissue kallikrein by protein C inhibitor. Evidence for identity of protein C inhibitor with the kallikrein binding protein. J. Biol. Chem. *267*, 7048–7052.

Esmon, C. T. (1992). The protein C anticoagulant pathway. Arteriosclerosis and Thrombosis *12*, 135–145.

Esmon, C. T. (1989). The roles of protein C and thrombomodulin in the regulation of blood coagulation. J. Biol. Chem. *264*, 4743–4746.

Espana, F., Berrettini, M., and Griffin, J. H. (1989). Purification and characterization of plasma protein C inhibitor. Thromb. Res. *55*, 369–384.

Espana, F., Estelles, A., Fernandez, P. J., Gilabert, J., Sanchez-Cuenca, J., and Griffin, J. H. (1993). Evidence for the regulation of urokinase and tissue type plasminogen activators by the serpin protein C inhibitor, in semen and blood plasma. Thromb. Hemostasis *70*, 989–994.

Espana, F., Fink, E., and Witzgall, K. (1995). Complexes of tissue kallikrein with protein C inhibitor in human semen and urine. Eur. J. Biochem. *234*, 641.

Espana, F., Gilabert, J., Estelles, A., Romeu, A., Aznar, J., and Cabo, A. (1991). Functionally active protein C inhibitor/plaminogen activator inhibitor-3 (PCI/PAI-3) is secreted in seminal vesicles, occurs at high concentrations in human seminal plama and complexes with prostate-specific antigen. Thromb. Res. *64*, 309–320.

Espana, F., Vicente, V., Tabernero, D., Scharrer, I., and Griffin, J. H. (1990). Determination of plasma protein C inhibitor and of two activated protein C-inhibitor complexes in normals and in patients with intravascular coagulation and thrombotic disease. Thrombosis research *59*, 593–608.

Geiger, M., Huber, K., Wojta, J., Stingl, L., Espana, F., Griffin, J. H., and Binder, B. R. (1989). Complex formation between urokinase and plasma protein C inhibitor *in vitro* and *in vivo*. Blood *74*, 722–728.

Geiger, M., Priglinger, U., Griffin, J. H., and Binder, B. R. (1991). Urinary protein C inhibitor. Glycosaminoglycans synthesized by the epithelial kidney cell line TLC-598 enhance its interaction with urokinase. J. Biol. Chem. *266*, 11851–11857.

Heeb, M. J., Espana, F., and H., G. J. (1989). Inhibition and complexation of activated protein C by two major inhibitors in plasma. Blood *73*, 446–454.

Heeb, M. J., and Griffin, J. H. (1988). Physiologic inhibition of human activated protein C by α1-antitrypsin. J. Biol. Chem. *263*, 11613–11616.

Hofsteenge, J., Taguchi, H., and Stone, S. R. (1986). Effect of thrombomodulin on the kinetics of the interaction of thrombin with substrates and inhibitors. Biochem. J. *237*, 243–251.

Hoogendoorn, H., Nesheim, M. E., and Giles, A. R. (1990). A qualitative and quantitative analysis of the activation and inactivation of protein C *in vivo* in a primate model. Blood *75*, 2164–2171.

Hoogendoorn, H., Toh, C. H., Nesheim, M. E., and Giles, A. R. (1991). Alpha2-macroglobulin binds and inhibits activated protein C. Blood *78*, 2283–2290.

Kane, W. H., and Davie, E. W. (1988). Blood coagulation factors V and VIII: Structural and functional similarities and their relationship to hemorrhagic and thrombotic disorders. Blood *71*, 539–555.

Kisiel, W. (1979). Human plasma protein C: Isolation, characterization, and mechanism of activation by α–thrombin. J. Clin. Invest. *64*, 761–770.

Kuhn, L. A., Griffin, J. H., Fisher, C. L., Greengard, J. S., Bouma, B. N., Espana, F., and Tainer, J. A. (1990). Elucidating the structural chemistry of glycosaminoglycan recognition by protein C inhibitor. Proc. Natl. Acad. Sci. U.S.A. *87*, 8506–8510.

Laurell, M., Christensson, A., Abrahamsson, P.-A., Stenflo, J., and Lilja, H. (1992). Protein C inhibitor in human body fluids. Seminal plasma is rich in inhibitor antigen deriving from cells throughout the male reproductive system. J. Clin. Invest. *89*, 1094–101.

LeBonniec, B. F., Guinto, B. F., and Esmon, C. T. (1992). Interaction of thrombin desETW with antithrombin III, the kunitz inhibitors, thrombomodulin and protein C. J. Biol. Chem. *267*, 19341–19348.

LeBonniec, B. F., Guinto, E. R., MacGillivray, R. T. A., Stone, S. R., and Esmon, C. T. (1993). The role of thrombin's Tyr-Pro-Pro-Trp motif in the interaction with fibrinogen, thrombomodulin, protein C, antithrombin III and the kunitz inhibitors. J. Biol. Chem. *268*, 19055–19061.

Marlar, R. A., Endres-Brooks, J., and Miller, C. (1985). Serial studies of protein C and its plasma inhibitor in patients with disseminated intravascular coagulation. Blood *66*, 59–63.

Marlar, R. A., and Griffin, J. H. (1980). Deficiency of protein C inhibitor in combined factor V/VIII deficiency disease. J. Clin. Invest. *66*, 1186–1189.

Meijers, J. C. M., Kanters, D. H. A. J., Vlooswijk, R. A. A., van Erp, H. E., Hessing, M., and Bouma, B. N. (1988). Inactivation of human plasma kallikrein and factor XIa by protein C inhibitor. Biochemistry *27*, 4231–4237.

Minamikawa, K., Wada, H., Wakita, Y., Ohiwa, M., Tanigawa, M., Deguchi, K., Hiraoka, N., Huzioka, H., Nishioka, J., Hayashi, T., Shirakawa, S., Nakano, T., and Suziki, K. (1994). Increased activated protein C-protein C inhibitor complex levels in patients with pulmonary embolism. Thromb. Hemostasis *71*, 192–194.

Morito, F., Saito, H., Suzuki, K., and Hashimoto, S. (1985). Synthesis and secretion of protein C inhibitor by the human hepatoma-derived cell line, Hep G2. Biochimica et Biophysica Acta *844*, 209–215.

Olson, S. T., and Björk, I. (1992). Regulation of thrombin by antithrombin and heparin cofactor II. In Thrombin: Structure and Function, L. J. Berliner, ed. (New York, NY: Plenum Press), pp. 159–217.

Phillips, J. E., Cooper, S. T., Potter, E. E., and Church, F. C. (1994). Mutagenesis of recombinant protein C inhibitor reactive site residues alters target proteinase specificity. J. Biol. Chem. *269*, 16696–16700.

Pratt, C. W., and Church, F. C. (1992). Heparin binding to protein C inhibitor. J. Biol. Chem. *267*, 8789–8794.

Pratt, C. W., Macik, B. G., and Church, F. C. (1989). Protein C inhibitor: purification and proteinase reactivity. Thromb. Res. *53*, 595–602.

Pratt, C. W., Whinna, H. C., and Church, F. C. (1992). A comparison of three heparin-binding serine proteinase inhibitors. J. Biol. Chem. *267*, 8795–8801.

Preissner, K. T., Koyama, T., Muller, D., Tschopp, J., and Muller-Berghaus, G. (1990). Domain structure of the endothelial cell receptor thrombomodulin as deduced from modulation of its anticoagulant functions. Evidence for a glycosaminoglycan-dependent secondary binding site for thrombin. J. Biol. Chem. *265*, 4915–4922.

Radtke, K. P., Fernandez, J. A., Greengard, J. S., Tang, W. W., Wilson, C. B., Loskutoff, D. J., Scharrer, I., and Griffin, J. H. (1994). Protein C inhibitor is expressed in tubular cells of human kidney. J. Clin. Invest. *94*, 2117–2124.

Rezaie, A. R., Cooper, S. T., Church, F. C., and Esmon, C. T. (1995). Protein C inhibitor is a potent inhibitor of the thrombin-thrombomodulin complex. J. Biol. Chem. *270*, 25336–25339.

Rosenberg, R. D. (1977). Chemistry of the hemostatic mechanism and its relationship to the action of heparin. Fed. Proc. Fed. Am. Soc. Exp. Biol. *36*, 10–18.

Rosenberg, R. D., and Damus, P. S. (1973). The purification and mechanism of action of human antithrombin-heparin cofactor. J. Biol. Chem. *248*, 6490–6505.

Sheng, S., Pemberton, P. A., and Sager, R. (1994). Production, purification and characterization of recombinant maspin proteins. J. Biol. Chem. *269*, 30988–30993.

Shirk, R. A., Elisen, M. G. L. M., Meijers, J. C. M., and Church, F. C. (1994). Role of the H helix in heparin binding to protein C inhibitor. J. Biol. Chem. *269*, 28690–28695.

Shore, J. D., Olson, S. T., Craig, P. A., Choay, J., and Björk, I. (1989). Kinetics of heparin action. Ann. N. Y. Acad. Sci. *556*, 75–80.

Stenflo, J. (1976). A new vitamin K-dependent protein: Purification from bovine plasma and preliminary characterization. J. Biol. Chem. *251*, 355.

Suzuki, K., Deyashiki, Y., Nishioka, J., Kurachi, K., Akira, M., Yamamoto, S., and Hashimoto, S. (1987). Characterization of a cDNA for human protein C inhibitor. A new member of the plasma serine protease inhibitor superfamily. J. Biol. Chem. *262*, 611–616.

Suzuki, K., Deyashiki, Y., Nishioka, J., and Toma, K. (1989). Protein C inhibitor: structure and function. Thromb. Haemostas. *61*, 337–342.

Suzuki, K., Nishioka, J., and Hashimoto, S. (1983). Protein C inhibitor: purification from human plasma and characterization. J. Biol. Chem. *258*, 163–168.

Suzuki, K., Nishioka, J., and Hashimoto, S. (1983). Regulation of activated protein C by thrombin-modified protein S. J. Biochem. *94*, 699–705.

Tabernero, D., Espana, F., Vicente, V., Estelles, A., Gilabert, J., and Aznar, J. (1990). Protein C inhibitor and other components of the protein C pathway in patients with acute deep vein thrombosis during heparin treatment. Thromb. Haemost. *63*, 380–382.

Tanigawa, M., Wada, H., and Shiku, H. (1995). Decreased protein C inhibitor after percutaneous transluminal coronary angioplasty in patients with acute myocardial infarction. Am. J. Hematol. *49*, 1.

Travis, J., and Salvesen, G. S. (1983). Human plasma proteinase inhibitors. Annu. Rev. Biochem. *52*, 655–709.

van Deerlin, V. M. D., and Tollefsen, D. M. (1991). The N-terminal acidic domain of heparin cofactor II mediates the inhibition of α-thrombin in the presence of glycosaminoglycans. J. Biol. Chem. *266*, 20223–20231.

van der Meer, F. J. M., van Tilburg, N. H., van Winjgaarden, A., van der Linden, I. K., Briet, E., and Bertina, R. M. (1989). A second plasma inhibitor of activated protein C: alpha1-antitrypsin. Thromb. Hemostasis *62*, 756–762.

Zheng, X., Geiger, M., and Binder, B. R. (1994). Inhibition of acrosin by protein C inhibitor and localization of protein C inhibitor to spermatozoa. Am. J. Physiol. *267*, 1.

Zheng, X. L., Ecke, S., Geiger, M., Bielek, E., Schleuning, W.-D., and Binder, B. R. (1993). Physiological inhibition of acrosin by protein C inhibitor (PCI). Throm. Haemostas. *69 (abstract# 1275)*, 895.

Zou, Z., Anisowicz, A., Hendrix, M. J. C., Thor, A., Neveu, M., Sheng, S., Rafidi, K., Seftor, E., and Sager, R. (1994). Maspin, a serpin with tumor-suppressing activity in human mammary epithelial cells. Science *263*, 526–529.

6

MOLECULAR BASIS FOR THE RESISTANCE OF FIBRIN-BOUND THROMBIN TO INACTIVATION BY HEPARIN/SERPIN COMPLEXES

Debra L. Becker, Jim C. Fredenburgh, Alan R. Stafford
and Jeffrey I. Weitz[*]

Hamilton Civic Hospitals Research Centre
711 Concession St.
Hamilton, Ontario
Canada L8V 1C3

1. INTRODUCTION

Despite advances in the treatment of venous thrombosis with the use of better heparin dosing regimens (Cruickshank et al., 1991), the treatment of arterial thrombosis remains problematic. Not only does heparin have limitations in patients with acute coronary syndromes, but also hirudin, a potent and specific inhibitor of α-thrombin (thrombin), is of limited efficacy in this patient group (GUSTO Investigators, 1996). Thus, when used alone or as adjuncts to thrombolytic therapy, heparin and hirudin are unable to fully block thrombus growth and there is evidence of reactivation of coagulation when treatment is stopped. It is likely that both the resistance of acute arterial thrombi to these anticoagulants and rebound activation of coagulation after stopping treatment reflect the inability of heparin or hirudin to pacify the intense procoagulant activity of the thrombus. At least two factors contribute to the procoagulant activity of thrombi (figure 1); thrombin bound to fibrin (Hogg and Jackson, 1989; Weitz et al., 1990) and factor Xa bound to platelets within the thrombus (Eisenberg et al., 1993). In this paper, we (a) highlight the limitations of heparin and hirudin in the setting of acute arterial thrombosis, (b) discuss the factors responsible for the prothrombotic activity of thrombi, (c) review the current understanding of the mechanism of thrombin's interaction with fibrin, (d) discuss the consequences of the thrombin–fibrin interaction, and (e) outline the potential mechanisms by which thrombin bound to fibrin is protected from inactivation by heparin.

---

* Telephone: (905)-527-2299; Fax: (905)-575-2646; E-mail: weitzj@fhs.mcmaster.ca.

Chemistry and Biology of Serpins, edited by Church *et al.*
Plenum Press, New York, 1997

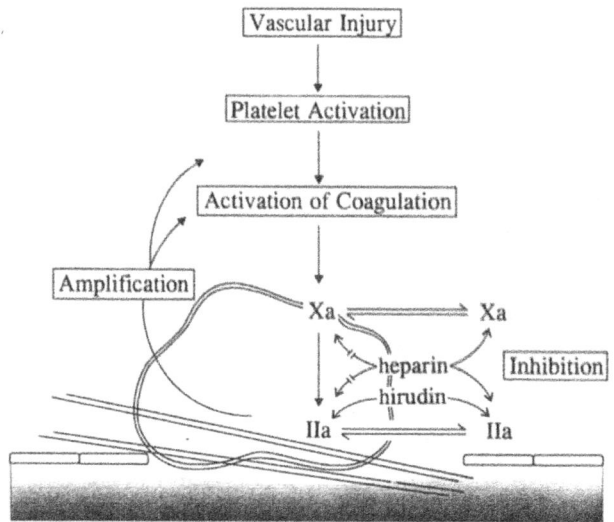

Figure 1. Prothrombotic activity of the arterial thrombus. Arterial injury leads to the formation of a thrombus that contains platelet-bound factor Xa and fibrin-bound thrombin, which unlike free factor Xa and thrombin, are resistant to inactivation by heparin. As a result, platelet-bound factor Xa triggers thrombin generation and fibrin-bound thrombin leads to thrombus growth by locally activating platelets and amplifying coagulation. Hirudin inactivates free thrombin and fibrin-bound thrombin equally well, but fails to block thrombin generation triggered by platelet-bound factor Xa. Consequently, there is reactivation of coagulation when hirudin treatment is stopped.

2. LIMITATIONS OF HEPARIN AND HIRUDIN IN ACUTE CORONARY SYNDROMES

Heparin and direct thrombin inhibitors have limitations in the treatment of acute coronary syndromes. In patients with unstable angina, there is a clustering of recurrent ischemic events after treatment with heparin is stopped (Theroux et al., 1992; Granger et al., 1996; Oldgren et al., 1996). A similar phenomenon occurs when treatment with hirudin is discontinued (GUSTO Investigators, 1996). Recurrent ischemic events are the result of reactivation of coagulation because there is an associated elevation in plasma levels of prothrombin fragment F1.2 (F1.2) and fibrinopeptide A (FPA), reflecting increased thrombin generation and thrombin activity, respectively (Granger et al., 1995).

In patients with acute myocardial infarction, thrombolytic therapy with tissue plasminogen activator (t-PA) or streptokinase induces a procoagulant state characterized by elevated levels of FPA (Eisenberg et al., 1987; Owen et al., 1988) which are only partially reduced by heparin (Galvani et al., 1994; Merlini et al., 1995). This may explain why adjunctive heparin does not reduce the incidence of recurrent ischemic events in patients receiving streptokinase and is of only questionable benefit in those given t-PA (Collins et al., 1996). Although hirudin is better than heparin both as an adjunct to thrombolytic therapy and in patients with non-Q wave infarction who do not receive thrombolytic agents, the early benefits of hirudin are lost within 30 days (GUSTO Investigators, 1996). These findings suggest that there is a persistent thrombogenic stimulus that is resistant to both heparin and hirudin.

Heparin and hirudin also have limitations in the setting of coronary angioplasty. Recurrent ischemic events occur in 6 to 8% of patients despite aspirin and high-dose heparin (Popma et al., 1995). Although hirudin is superior to heparin for the first 72 hours after successful coronary angioplasty, its benefits are lost by 30 days (Serruys et al., 1995). Similarly, at 7 days, Hirulog, a synthetic hirudin analogue (Maraganore et al., 1990), is better than heparin at preventing recurrent ischemic events in patients undergoing angioplasty for unstable angina after acute myocardial infarction; by 30 days, however, there is no difference between Hirulog and heparin (Bittl et al., 1995). It is likely that both the resistance of acute arterial thrombi to heparin and hirudin and the reactivation of coagulation that occurs when treatment is terminated reflect the inability of these antico-agulants to pacify the intense prothrombotic activity of the thrombus.

3. FACTORS RESPONSIBLE FOR THE PROTHROMBOTIC ACTIVITY OF THROMBI

Studies *in vitro* have attributed the procoagulant activity of arterial thrombi to (a) thrombin bound to fibrin (Hogg and Jackson, 1989; Weitz et al., 1990), or (b) factor Xa bound to platelets within the thrombi (Eisenberg et al., 1993). Fibrin-bound thrombin remains enzymatically active and can locally activate platelets (Kumar et al., 1995) and accelerate coagulation (Kumar et al., 1994) thereby inducing an intense procoagulant state. By triggering thrombin generation, factor Xa bound to platelets trapped within the thrombus augments the prothrombotic activity of the thrombus. The importance of plate-let-bound factor Xa is highlighted by the observation that the procoagulant activity of platelet-rich thrombi is attenuated by tick anticoagulant peptide (TAP) (Waxman et al., 1990), a potent inhibitor of factor Xa (Eisenberg et al., 1993; Prager et al., 1995). In con-trast, we have shown that the combination of hirudin and TAP abolishes the procoagulant activity of thrombi, indicating that both fibrin-bound thrombin and platelet-bound factor Xa contribute to this activity.

Fibrin-bound thrombin and platelet-bound factor Xa are resistant to inactivation by heparin (Hogg and Jackson, 1989; Weitz et al., 1990; Teitel and Rosenberg, 1983; Pieters et al., 1988) thereby explaining its limitations in the setting of acute arterial thrombosis (figure 1). Although unlike heparin, hirudin can inactivate fibrin-bound thrombin (Weitz et al., 1990), it does not block thrombin generation triggered by platelet-bound factor Xa. This may explain the reactivation of coagulation that occurs when treatment with hirudin is stopped. In this paper we focus on the phenomenon of thrombin binding to fibrin and the molecular basis for its resistance to inactivation by heparin.

4. MECHANISM OF THROMBIN BINDING TO FIBRIN

Fibrin-bound thrombin is a major contributor to the prothrombotic activity of thrombi. During the course of thrombin-mediated conversion of fibrinogen to fibrin, thrombin not only binds fibrinogen as a protein substrate, but also binds to fibrin during and after clot formation (Binnie and Lord, 1993). Binding of thrombin to fibrin was first demonstrated by Seegers et al. in 1945. Since then, numerous quantitative studies of thrombin–fibrin interactions have been reported. Thrombin binds to fibrin polymer with a dissociation constant (K_d) in the range of 0.3 to 1.7 µM (Liu et al., 1979; Hogg and Jack-son, 1990b; Banninger et al., 1994). Thrombin binds to fibrin II monomer (Fm), a solu-bilized form of fibrin, with slightly lower affinity, with K_d values ranging from 0.8 to 5 µM (Kaminski and McDonagh, 1983; Hogg and Jackson, 1989; Naski and Shafer, 1991; Hogg et al., 1996). The original binding experiments using ^{125}I-thrombin identified two classes of non-substrate binding sites on fibrin, one with high affinity ($K_d \sim 1.7$ µM) and the other with lower affinity ($K_d \sim 15$ µM) (Liu et al., 1979). Hogg and Jackson (1990a) also found two classes of binding sites on fibrin with K_d values of 0.3 µM and 33 µM. The low affinity site appears to be located in the central E domain (Kaminski and McDonagh, 1987; Fenton et al., 1988; Vali and Scheraga, 1988; Siebenlist et al., 1990). The amino-ter-minal b15–42 fibrin sequence may be an important constituent of this site because fibrin molecules lacking this sequence have reduced affinity for thrombin (Siebenlist et al., 1990). A more recent study indicates that the high affinity thrombin binding site is located on D domains of γ'-chains (Meh et al., 1996), which are anionic γ-chain variants found in up to 15% of the total fibrinogen population (Mosesson and Finlayson, 1963).

Studies of thrombin structure have provided valuable insight into the structural basis of thrombin interactions with fibrin. At least three important domains exist within the thrombin molecule. The catalytic site is a negatively charged region with defined subsites that contribute to substrate binding adjacent to the scissile bond. Two electropositive domains bordering the active site and in near opposition on the enzyme surface have been termed anion-binding exosites 1 and 2 (Fenton et al., 1988; Bode et al., 1992; Stubbs and Bode, 1993; Stone, 1995). These exosites mediate interactions of thrombin with specific protein substrates, inhibitors, and regulatory macromolecules. Specifically, exosite 1 binds fibrinogen, the fifth and sixth epidermal growth factor domains of thrombomodulin, the carboxy-terminal domain of hirudin, the amino-terminal domain of the thrombin receptor, the amino-terminal domain of heparin cofactor II, and clotting factors V and VIII (Stubbs and Bode, 1993; Tulinsky and Qiu, 1993; Whinna and Church, 1993; Stubbs and Bode, 1995; Esmon and Lollar, 1996). Exosite 2 binds heparin, the chondroitin sulphate moiety of thrombomodulin, fragment 2 of prothrombin, and may contribute to thrombin's interaction with factors V and VIII (Stubbs and Bode, 1993; Arni et al., 1993; Stubbs and Bode, 1995; Esmon and Lollar, 1996).

Thrombin binds to fibrin through a site distinct from its catalytic centre, thereby explaining why fibrin-bound thrombin remains enzymatically active (Seegers et al., 1945; Liu et al., 1979; Wilner et al., 1981; Berliner et al., 1985; Kaminski and McDonagh, 1987; Vali and Scheraga, 1988). In support of this concept, active-site-blocked thrombin binds to fibrin with an affinity similar to that of active thrombin, and D-Phe–Pro–Arg chloromethyl ketone (PPACK), which blocks the active centre of thrombin, does not displace ^{125}I-labeled thrombin from plasma clots (Weitz et al., 1990). In the absence of heparin, there is good evidence that thrombin's interaction with fibrin is mediated by exosite 1. Supporting this concept is the observation that hirudin, which binds to both exosite 1 and the active site of thrombin, displaces thrombin that is bound to Fm-Sepharose (Wilner et al., 1981; Kaminski and McDonagh, 1987), and that Hirugen, a synthetic analogue of the carboxy-terminal of hirudin which binds only to exosite 1 (Maraganore et al., 1989), displaces thrombin from polymeric fibrin I or fibrin II (Naski et al., 1990; Naski et al., 1991). Furthermore, we have shown that γ-thrombin (an autolytic thrombin derivative with cleavages in exosite 1) and Quick 1 dysthrombin, a naturally occurring thrombin mutant with an $Arg^{67} \rightarrow Cys$ point mutation in exosite 1 (Henriksen et al., 1980), do not bind to fibrin, whereas RA-thrombin, an exosite 2 mutant that has reduced affinity for heparin because of $Arg \rightarrow Ala$ substitutions at positions 93, 97, and 101 (Ye et al., 1994), binds to fibrin with an affinity similar to that of thrombin.

In the presence of heparin, the binding of thrombin to fibrin is more complex. Thus, when heparin is present, there are changes in both the amount of thrombin that binds to fibrin and the mode of thrombin's interaction with fibrin. We have shown that with heparin concentrations up to 250 nM, the amount of thrombin that binds to fibrin increases as does the apparent affinity of thrombin for fibrin; at higher heparin concentrations, however, thrombin binding and the affinity of thrombin for fibrin progressively decrease. These data extend the results of Hogg and Jackson (1990a) who demonstrated enhanced thrombin binding to fibrin with a fixed concentration of heparin.

The mode of thrombin binding also changes in the presence of heparin. Whereas thrombin binds to fibrin via exosite 1 in the absence of heparin, we have direct evidence

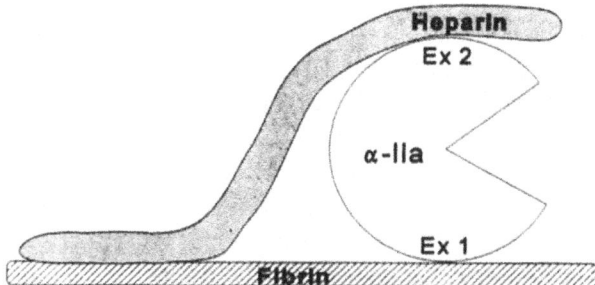

Figure 2. Ternary heparin–fibrin–thrombin complex. Thrombin within the complex is bound to fibrin via exosite 1 (Ex 1), whereas heparin binds to both fibrin and exosite 2 (Ex 2) on thrombin.

that the enhanced binding found in the presence of heparin is mediated by exosite 2. Thus, heparin augments the binding of γ-thrombin to fibrin to the same extent as thrombin, but has little effect on the binding of RA-thrombin, an exosite 2 mutant. Furthermore, excess thrombin bound in the presence of heparin is displaced with an antibody against exosite 2 or with prothrombin fragment 2, which, like heparin, also binds to exosite 2 (Arni et al., 1993). In contrast, Hirugen has no effect on heparin-dependent binding of thrombin.

We interpret these findings as indicating formation of a ternary heparin–fibrin–thrombin complex wherein thrombin binds to fibrin directly via exosite 1, and heparin binds to both fibrin and exosite 2 on thrombin (figure 2). Heparin binding domains on fibrin have been identified on the carboxy-termini of both the β and γ chains (Mohri et al., 1994), and more recently within the β15–42 fibrin sequence in the central E domain (Odr-ljin et al., 1996). The biphasic effect of heparin on thrombin binding supports the concept of ternary complex formation. Thus, heparin promotes thrombin binding to fibrin until the heparin binding sites on fibrin are saturated. With higher heparin concentrations, thrombin binding to fibrin decreases as nonproductive binary fibrin–heparin and thrombin–heparin complexes are formed. Quantitative studies of the binary binding interactions done in our laboratory indicate that thrombin binds Fm with the lowest affinity, having a K_d value of approximately 1 to 5 μM. In contrast, heparin binds to both thrombin and Fm with similar affinity; K_d values of 100 nM and 110 nM, respectively. These dissociation constants are qualitatively similar to those described by others for the interactions of thrombin and heparin with Fm (Hogg et al., 1994). In agreement with Hogg and Jackson (1990a), the affinities of thrombin and heparin for polymeric fibrin are approximately 10- to 20-fold higher than for Fm. Comparison of the dissociation constants for the three binary complexes indicates that heparin serves predominantly as a bridge between thrombin and fibrin, with thrombin–fibrin binary complexes making a secondary contribution to ternary complex assembly.

Hogg and Jackson (1990a) demonstrated that the ability of heparin to enhance binding of thrombin to fibrin polymer is independent of its affinity for antithrombin. In addition, they showed that only heparin chains with a molecular mass of greater than 11,200 promoted thrombin binding to fibrin suggesting that this process requires heparin chains that are long enough to bridge thrombin to fibrin.

5. CONSEQUENCES OF THROMBIN BINDING FIBRIN

Thrombin within the ternary heparin–fibrin–thrombin complex has been shown to be protected from inactivation by heparin/antithrombin complexes (Hogg and Jackson, 1989; Weitz et al., 1990). To extend these studies, we (a) investigated the effects of Fm on the rate of thrombin inactivation by antithrombin over a wide range of heparin concentrations, and (b) examined the effects of Fm on the uncatalyzed and heparin-catalyzed rates of thrombin inactivation by a second serpin, heparin cofactor II. In the absence of heparin, Fm has minimal effects on the second-order rate constant (k_2) for thrombin inactivation by antithrombin or heparin cofactor II (figure 3). However, in the presence of heparin, Fm reduces the rate at which thrombin is inactivated by heparin/antithrombin and heparin/HCII by as much as 58-fold and 247-fold, respectively, with the protective effect persisting at every heparin concentration tested. The inhibitory effects of Fm are concentration-dependent and saturable, with half-maximal inhibition occurring between 5 nM and 25 nM Fm.

For thrombin to be protected both of its exosites must be ligated; exosite 2 by heparin and exosite 1 by fibrin. Thus, when thrombin exosite variants were substituted for thrombin in inhibition assays, Fm had little effect on their heparin-catalyzed rates of inactivation by either antithrombin or heparin cofactor II (figure 4). The thrombin variants investigated included γ-thrombin, Quick 1 dysthrombin, and RA-thrombin. Thus, even though heparin enhances the binding of γ-thrombin and Quick 1 dysthrombin to fibrin by binding to their intact exosite 2 and bridging them to fibrin, neither is protected from inac-

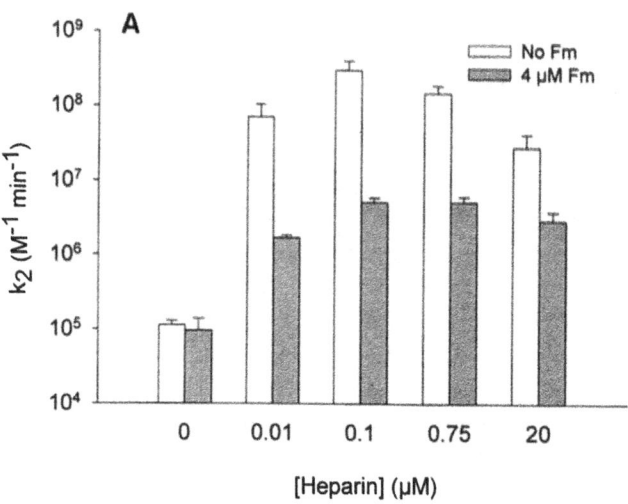

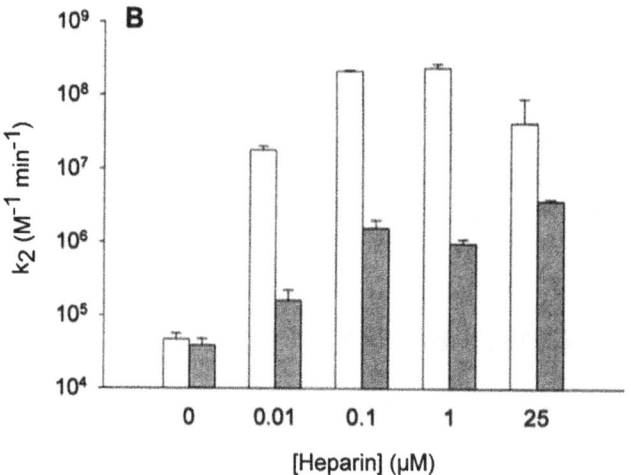

Figure 3. Effect of heparin on the rates of thrombin inactivation by antithrombin (A) or heparin cofactor II (B) in the absence or presence of 4 μM fibrin monomer (Fm). Each bar represents the mean of at least two experiments done in duplicate, while the lines represent the SD.

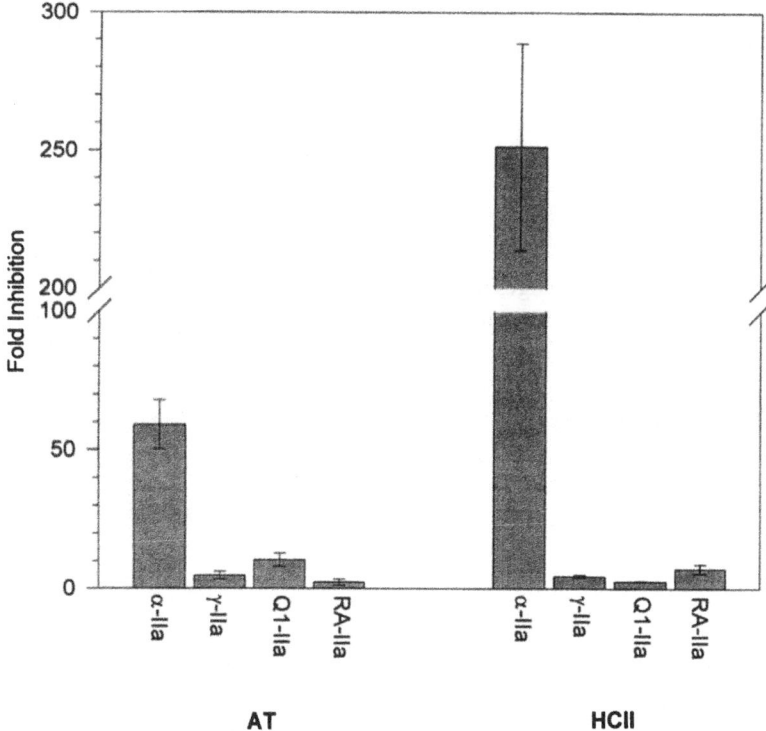

Figure 4. Inhibitory effect of 4 μM fibrin monomer on the heparin-catalyzed rates of thrombin or thrombin variant inactivation by antithrombin (AT) or heparin cofactor II (HCII). Fibrin monomer has only minimal inhibitory effects on the heparin-catalyzed rates of inactivation of the exosite 1 variants γ-thrombin (γ-IIa) or dysthrombin Quick 1 (Q1-IIa), or the exosite 2 mutant RA-thrombin (RA-IIa) by AT or HCII. Each bar represents the mean of at least two experiments done in duplicate, the lines represent the SD, and the fold inhibition was calculated as the ratio of the rate constant in the presence of fibrin monomer relative to that obtained in its absence.

tivation because their altered exosite 1 fails to interact with fibrin. Similarly, RA-thrombin also is susceptible to inactivation because even though it binds to fibrin with an affinity similar to that of thrombin, it has reduced affinity for heparin.

Collectively these results demonstrate that thrombin bound to fibrin is susceptible to inactivation by antithrombin and heparin cofactor II. However, when thrombin is bound within the ternary heparin–fibrin–thrombin complex, the ability of these inhibitors to inactivate it is greatly compromised. To be protected within the ternary complex, both exosites on thrombin must contribute to binding; exosite 1 to fibrin and exosite 2 to heparin. If either of these interactions is compromised, thrombin is susceptible to inactivation by its inhibitors.

6. MECHANISMS OF THROMBIN PROTECTION

It has been well established that thrombin bound to fibrin is protected from inactivation by heparin/serpin complexes through ternary complex formation, however, the mechanism responsible for this protection is unclear. There are two general mechanisms by which this could occur: allosteric modulation of thrombin's active site or spatial constraints that impair the formation of thrombin/inhibitor complexes (figure 5).

(A) ALLOSTERIC MODULATION

(i)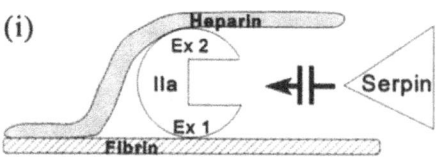

Allosteric changes at the active site of thrombin induced by
ternary complex formation limit thrombin's reactivity with serpins.

(B) SPATIAL LIMITATIONS

(i)   (ii)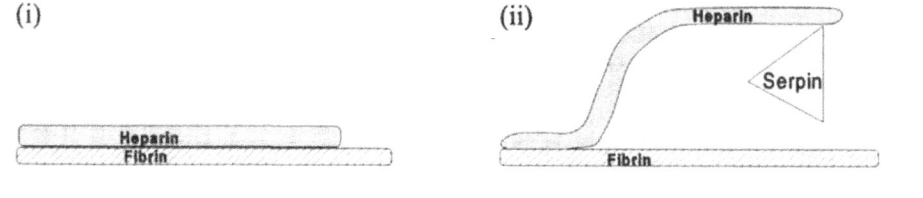

Sequestration of heparin by fibrin.   Sequestration of serpin by fibrin-bound heparin.

(iii)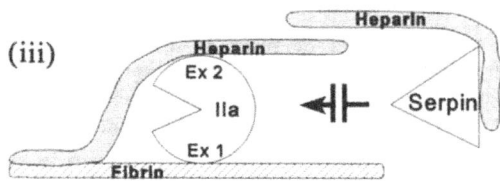

Steric hindrance: occupation of exosites 1 and 2 by fibrin and heparin,
respectively, limits thrombin's reactivity with heparin/serpin complexes.

Figure 5. Potential mechanisms by which fibrin-bound thrombin is protected from inactivation.

6.1. Allosteric Modulation of Thrombin's Active Site

Allosteric changes in the active site of thrombin induced by ternary complex formation may reduce its reactivity with serpins (figure 5Ai). Supporting this concept is the observation that thrombin within the ternary heparin–fibrin–thrombin complex has altered activity toward tripeptide substrates (Hogg and Jackson, 1989; Hogg et al., 1990b). Neither Fm nor heparin alone altered the ability of thrombin to hydrolyze the substrates, however, the combination of Fm and heparin increased the K_m values and decreased the k_{cat} values. We have obtained similar results with tosyl-Gly–Pro–Arg-pNA in which the K_m increases in the presence of Fm and heparin. In contrast, when either γ-thrombin or RA-thrombin is substituted for thrombin, no change in K_m is observed. These studies confirm that thrombin within the ternary heparin–fibrin–thrombin complex undergoes allosteric changes at its catalytic site, and demonstrate that ligation of both exosites is a prerequisite for these changes.

Allosteric modulation at the active site may also impair thrombin's reactivity with macromolecular substrates. Thus, Hogg and Jackson (1990b) demonstrated that Fm and heparin in combination decreased k_{cat}/K_m for thrombin cleavage of prothrombin by only two-fold. The relatively modest effects on the activity of thrombin with both chromogenic substrates and prothrombin suggest that ternary complex formation has only a minimal allosteric influence on thrombin activity. However, this does not exclude effects that might be unique to interactions with serpins.

Results of fluorescence studies provide evidence of allosteric changes at the catalytic site of thrombin occur when the ternary heparin–fibrin–thrombin complex is formed (Hogg et al., 1996). Using thrombin labeled at the catalytic site with a fluorescent chloromethyl ketone inhibitor, binding to Fm or to heparin alone causes changes in the environment of the thrombin catalytic site. The fluorescence enhancement accompanying ternary complex formation was greater, confirming conformational changes in the active site of thrombin are maintained within the ternary complex.

6.2. Spatial Constraints

Spatial constraints may impair the reactivity of thrombin within the ternary heparin–fibrin–thrombin complex with its inhibitors. This mechanism can be divided into two categories: (a) sequestration of the components necessary for thrombin/serpin complex formation, and (b) steric hindrance.

6.2.1. Fibrin-Mediated Sequestration of Components Necessary for Thrombin/Serpin Complex Formation. Heparin binding to fibrin may reduce the concentration of heparin available to interact with antithrombin or heparin cofactor II (figure 5Bi). Alternatively, antithrombin or heparin cofactor II may bind to fibrin-bound heparin, thereby decreasing the effective concentration of inhibitor (figure 5Bii). This latter mechanism would have a greater effect on antithrombin than heparin cofactor II because of the high affinity interaction between antithrombin and the pentasaccharide sequence on heparin ($K_d = 25$ nM). In contrast, heparin has no high affinity binding sequence for heparin cofactor II (Tollefsen, 1995). Our observation that Fm inhibits the heparin-catalyzed rate of factor Xa inactivation by antithrombin only 6-fold suggests that fibrin-mediated sequestration of heparin and/or antithrombin is unlikely to play a major role in the protection of thrombin.

6.2.2. Steric Hindrance. Occupation of exosites 1 and 2 on thrombin by fibrin and heparin, respectively, may physically limit access of serpins to thrombin within the ternary complex. This may occur because the molecular arrangement of heparin, fibrin, and thrombin within the ternary complex hinders the productive interaction of heparin/serpin complexes with thrombin (figure 5Biii). The relatively small magnitude of the rate decrease (two-fold) of k_{cat}/K_m for thrombin cleavage of prothrombin indicates that the accessibility of the active site to macromolecular substrates is not greatly restricted by heparin and fibrin binding (Hogg and Jackson, 1990b). Instead, it is more likely that the ability of heparin within the heparin/serpin complex to bridge the serpin to fibrin-bound thrombin is impaired because exosite 2 is occupied by the heparin molecule that bridges thrombin to fibrin. The occupation of exosite 1 by fibrin may cause the additional impairment of thrombin inhibition observed with heparin cofactor II compared to antithrombin, because the activity of heparin cofactor II is dependent on the amino-terminal domain which binds to exosite 1 on thrombin (Tollefsen, 1995). Experiments by Hogg et al. (1996) suggest that occupation of exosite 1 by fibrin also interferes with the interaction of thrombin with fibrinogen and hirudin, both of which use exosite 1 to bind to thrombin. Since both fibrinogen and hirudin interact with the active site as well as exosite 1 on thrombin, however, a role for allosteric changes at the catalytic site caused by heparin and fibrin binding cannot be excluded.

Finally, it is possible that fibrin compromises heparin's ability to serve as a template. If this is the case, antithrombin may bind to a pentasaccharide sequence on heparin that is not in close proximity to thrombin. Thrombin's mobility may be restricted when it is bridged to fibrin by heparin, preventing it from moving along the heparin chain to access antithrombin. Again, this mechanism may have limited application to heparin cofactor II since there is no high affinity binding sequence on heparin for heparin cofactor II (Tollefsen, 1995). The relative contribution of these mechanisms to the protection of thrombin from inactivation by heparin/serpin complexes requires further study.

7. FUTURE DIRECTIONS

Both heparin and hirudin have limitations in the setting of acute coronary syndromes. The limitations of heparin reflect, at least in part, its inability to inactivate fibrin-bound thrombin. Although considerable progress has been made in our understanding of the mechanisms by which thrombin bound to fibrin is protected from inactivation by heparin/ serpin complexes, further work is needed to delineate the relative importance of allosteric modulation of the active site of thrombin versus limited access. There is mounting evidence that thrombin generation triggered by platelet-bound factor Xa also contributes to the prothrombotic activity of acute arterial thrombi. This concept is supported by the limitations of direct thrombin inhibitors in recent clinical studies (Bittl et al., 1995; GUSTO Investigators, 1996) and by experimental animal data indicating that factor Xa inhibitors such as TAP attenuate acute arterial thrombosis (Schaffer et al., 1991; Sitko et al., 1992). Consequently, effective treatment of acute coronary syndromes will require antithrombotic strategies aimed at blocking thrombin generation as well as thrombin activity.

ACKNOWLEDGMENTS

This work was supported by a grant from the Medical Research Council of Canada. Dr. Weitz is a Career Investigator of the Heart and Stroke Foundation of Ontario.

REFERENCES

Arni, R.K., Padmanabhan, K., Padmanabhan, K.P., Wu, T.P., and Tulinsky, A. (1993) Structures of the non-covalent complexes of human and bovine prothrombin fragment 2 with human PPACK-thrombin. Biochemistry 32: 4727–4737.

Banninger, H., Lammle, B., and Furlan, M. (1994) Binding of α-thrombin to fibrin depends on the quality of the fibrin network. Biochem. J. 298: 157–163.

Berliner, L.J., Sugawara, Y., and Fenton, J.W. II. (1985) Human α-thrombin binding to non-polymerized fibrin-Sepharose: evidence for an anionic binding region. Biochemistry 24: 7005–7009.

Binnie, C., and Lord, S.T. (1993) The fibrinogen sequences that interact with thrombin. Blood 81: 3186–3192.

Bittl, J.A., Strony, J., Brinker, J.A., Ahmed, W.H., Meckel, C.R., Chaitman, B.R., Maraganore, J., Deutsch, E., and Adelman, B. (1995) Treatment with bivalirudin (Hirulog) as compared with heparin during coronary angioplasty for unstable or postinfarction angina. N. Engl. J. Med. 333: 764–769.

Bode, W., Turk, D., and Karshikov, A. (1992) The refined 1.9-Å x-ray crystal structure of D-Phe-Pro-Arg-chloromethyl ketone-inhibited human α-thrombin: structure analysis, overall structure, electrostatic properties, detailed active site geometry, and structure-function relationships. Protein Sci. 1: 426–471.

Collins, R., MacMahon, S., Flather, M., Baigent, C., Remvig, L., Mortensen, S., Appleby, P., Godwin, J., Yusuf, S., and Peto, R. (1996) Clinical effects of anticoagulant therapy in suspected acute myocardial infarction: systematic overview of randomized trial. BMJ 313: 652–659.

Cruickshank, M.K., Levine, M.N., Hirsh, J., Roberts, R., and Siguenza, M. (1991) A standard heparin nomogram for the management of heparin therapy. Arch. Intern. Med. 151: 333–337.

Eisenberg, P.R., Sherman, L.A., and Jaffe, A.S. (1987) Paradoxic elevation of fibrinopeptide A after streptokinase: Evidence for continued thrombosis despite intense fibrinolysis. J. Am. Coll. Cardiol. 10: 527–529.

Eisenberg, P.R., Siegel, J.E., Abendschein, D.R., and Miletich, J.P. (1993) Importance of factor Xa in determining the procoagulant activity of whole blood clots. J. Clin. Invest. 91: 1877–1883.

Esmon, C.T., and Lollar, P.T. (1996) Involvement of thrombin anion-binding exosites 1 and 2 in the activation of factor V and factor VIII. J. Biol. Chem. 271: 13882–13887.

Fenton, J.W. II, Olson, T.A., Zabinski, M.P., and Wilner, G.D. (1988) Anion-binding exosite of human alpha-thrombin and fibrin(ogen) recognition. Biochemistry 27: 7106–7112.

Galvani, M., Abendschein, D.R., Ferrini, D., Ottani, F., Rusticali, F., and Eisenberg, P.R. (1994) Failure of fixed dose intravenous heparin to suppress increases in thrombin activity after coronary thrombolysis with streptokinase. J. Am. Coll. Cardiol. 24: 1445–1452.

Granger, C.B., Hirsh, J., Califf, R.M., Woodlief, L.H., Bovill, E., Simes, R.J., and Topol, E.J. (1996) Activated partial thromboplastin time and outcome after thrombolytic therapy for acute myocardial infarction: results from the GUSTO Trial. Circulation 93: 870–888.

Granger, C.B., Miller, J.M., Bovill, E.G., Gruber, A., Tracy, R.P., Krucoff, M.W., Green, C., Berrios, E., Harrington, R.A., Ohman, E.M., et al. (1995) Rebound increase in thrombin generation and activity after cessation of intravenous heparin in patients with acute coronary syndromes. Circulation 91: 1929–1935.

GUSTO Investigators (1996) A comparison of recombinant hirudin with heparin for treatment of acute coronary syndromes. The Global Use of Strategies to Open Occluded Coronary Arteries (GUSTO) IIb Investigators. N. Engl. J. Med. 335: 775–782.

Henriksen, R.A., Owen, W.G., Nesheim, M.E., and Mann, K.G. (1980) Identification of a congenital dysthrombin, thrombin Quick. J. Clin. Invest. 66: 934–940.

Hogg, P.J., and Jackson, C.M. (1989) Fibrin monomer protects thrombin from inactivation by heparin-antithrombin III: implications for heparin efficacy. Proc. Natl. Acad. Sci. USA 86: 3619–3623.

Hogg, P.J., and Jackson, C.M. (1990a) Heparin promotes the binding of thrombin to fibrin polymer: quantitative characterization of a thrombin-fibrin polymer-heparin ternary complex. J. Biol. Chem. 265: 241–247.

Hogg, P.J., and Jackson, C.M. (1990b) Formation of a ternary complex between thrombin, fibrin monomer, and heparin influences the action of thrombin on its substrates. J. Biol. Chem. 265: 248–255.

Hogg, P.J., Jackson, C.M., Labanowski, J.K., and Bock, P.E. (1996) Binding of fibrin monomer and heparin to thrombin in a ternary complex alters the environment of the thrombin catalytic site, reduces affinity for hirudin, and inhibits cleavage of fibrinogen. J. Biol. Chem. 271: 26088–26095.

Kaminski, M., and McDonagh, J. (1983) Studies of the mechanism of thrombin interaction with fibrin. J. Biol. Chem. 258: 10530–10535.

Kaminski, M., and McDonagh, J. (1987) Inhibited thrombins: interactions with fibrinogen and fibrin. Biochem. J. 242: 881–887.

Kumar, R., Beguin, S., and Hemker, C. (1994) The influence of fibrinogen and fibrin on thrombin generation — Evidence for feedback activation of the clotting system by clot bound thrombin. Thromb. Haemost. 72: 713–721.

Kumar, R., Beguin, S., and Hemker, C. (1995) The effect of fibrin clots and clot-bound thrombin on the development of platelet procoagulant activity. Thromb. Haemost. 74(3): 962–968.

Liu, C.Y., Nossel, H.L., and Kaplan, K.L. (1979) The binding of thrombin by fibrin. J. Biol. Chem. 254: 10421–10425.

Maraganore, J.M., Bourdon, P., Jablonski, J., Ramachandran, K.L., and Fenton, J.W. II. (1990) Design and characterization of Hirulogs: a novel class of bivalent peptide inhibitors of thrombin. Biochemistry 29: 7095–7101.

Maraganore, J.M., Chao, B., Joseph, M.L., Jablonski, J., and Ramachandran, K.L. (1989) Anticoagulant activity of synthetic hirudin peptides. J. Biol. Chem. 264: 8692–8698.

Meh, D.A., Siebenlist, K.R., and Mosesson, M.W. (1996) Identification and characterization of the thrombin binding sites on fibrin. J. Biol. Chem. 271: 23121–23125.

Merlini, P.A., Bauer, K.A., Oltrona, L., Ardissino, D., Spinola, A., Cattaneo, M., Broccolino, M., Mannucci, P.M., and Rosenberg, R.D. (1995) Thrombin generation and activity during thrombolysis and concomitant heparin therapy in patients with acute myocardial infarction. J. Am. Coll. Cardiol. 25: 203–209.

Mohri, H., Iwamatsu, A., and Ohkubo, T. (1994) Heparin binding sites are located in a 40-kD γ-chain and a 36-kD β-chain fragment isolated from human fibrinogen. J. Thromb. Thrombolysis 1: 49–54.

Mosesson, M.W., and Finlayson, J.S. (1963) Subfractions of human fibrinogen. Preparation and analysis. J. Lab. Clin. Med. 62: 663–674.

Naski, M.C., Fenton, J.W., Maraganore, J.M., Olson, S.T., and Shafer, J.A. (1990) The COOH-terminal of hirudin. A exosite-directed competitive inhibitor of the action of α-thrombin on fibrinogen. J. Biol. Chem. 265: 13484–13489.

Naski, M.C., Lorand, and L., Shafer, J.A. (1991) Characterization of the kinetic pathway for fibrin promotion of α-thrombin-catalyzed activation of plasma factor XIII. Biochemistry 30: 934–941.

Naski, M.C., and Shafer, J.A. (1991) A kinetic model for the α-thrombin-catalyzed conversion of plasma levels of fibrinogen and fibrin in the presence of antithrombin III. J. Biol. Chem. 266: 13003–13010.

Odrljin, T.M., Shainoff, J.R., Lawrence, S.O., and Simpson-Haidaris, P.J. (1996) Thrombin cleavage enhances exposure of a heparin binding domain in the N-terminus of the fibrin β chain. Blood 88: 2050–2061.

Oldgren, J., Grip, L., and Wallentin, L. (1996) Reactivation after cessation of thrombin inhibition in unstable coronary artery disease occurs regardless of aspirin dose (abstract). Circulation 94 (suppl. 1) I-431.

Owen, J., Friedman, K.D., Grossmann, B.A., Wilkins, C., Berke, A.D., and Powers, E.R. (1988) Thrombolytic therapy with tissue-type plasminogen activator or streptokinase induces transient thrombin activity. Blood 72: 616–620.

Pieters, J., Willems, G., Hemker, C., and Lindhout, T. (1988) Inhibition of factor IXa, and factor Xa by antithrombin III/heparin during factor X activation. J. Biol. Chem. 263: 15313–15318.

Popma, J.J., Coller, B.S., Ohman, E.M., Bittl, J.A., Weitz, J., Kuntz, R.E., and Leon, M.B. (1995) Antithrombotic therapy in patients undergoing coronary angioplasty. Chest 108: 486–501.

Prager, N.A., Abendschein, D.R., McKenzie, C.R., and Eisenberg, P.R. (1995) Role of thrombin compared with factor Xa in the procoagulant activity of whole blood clots. Circulation 92: 962–967.

Schaffer, L.W., Davidson, J.T., Vlasuk, G.P., and Siegel, P.K.S. (1991) Antithrombotic efficacy of recombinant tick anticoagulant peptide. A potent inhibitor of coagulation factor Xa in a primate model of arterial thrombosis. Circulation 84: 1741–1748.

Seegers, W.H., Nieff, M., and Shafer, J.A. (1945) Note on the adsorption of thrombin on fibrin. Science (Wash. DC) 101: 521–521.

Serruys, P.W., Herrman, J.P., Simon, R., Rutsch, W., Bode, E., Laarman, G.J., vanDijk, R., vandenBos, A.A., Umans, V.A., Fox, K.A., et al. (1995) A comparison of hirudin with heparin in the prevention of restenosis after coronary angioplasty. N. Engl. J. Med. 333: 757–763.

Siebenlist, K.R., DiOrio, J.P., Budzynski, A.Z., and Mosesson, M.W. (1990) The polymerization and thrombin-binding properties of des-(Bβ1-42)-fibrin. J. Biol. Chem. 265: 18650–18655.

Sitko, G.R., Ramjit, D.R., Stabilito, I.I., Lehman, D., Lynch, J.J., and Vlasuk, G.P. (1992) Conjunctive enhancement of enzymatic thrombolysis and prevention of thrombotic reocclusion with the selective factor Xa inhibitor, tick anticoagulant peptide. Circulation 85: 805–815.

Stone, S.R. (1995) Thrombin inhibitors: a new generation of antithrombotics. Trends Cardio. Med. 5: 134–139.

Stubbs, M.T., and Bode, W. (1993) A player of many parts: the spotlight falls on thrombin's structure. Thromb. Res. 69: 1–58.

Stubbs, M.T., and Bode, W. (1995) The clot thickens: clues provided by thrombin structure. Trends Biochem. Sci. 20: 23–28.

Teitel, J.M., and Rosenberg, R.D. (1983) Protection of factor Xa from neutralization by the heparin-antithrombin III complex. J. Clin. Invest. 71: 1383–1391.

Theroux, P., Waters, D., Lam, J., Juneau, M., and McCans, J. (1992) Reactivation of unstable angina after the discontinuation of heparin in the acute phase of unstable angina. N. Engl. J. Med. 326: 141–145.

Tollefsen, D.M. (1995) Insight into the mechanism of action of heparin cofactor II. Thromb. Haemost. 74: 1209–1214.

Tulinsky, A., and Qiu, X. (1993) Active site and exosite binding of alpha-thrombin. Blood Coag. and Fibrinolysis 4: 305–312.

Vali, Z., and Scheraga, H.A. (1988) Localization of the binding site on fibrin for the secondary binding site of thrombin. Biochemistry 27: 1956–1963.

Waxman, L., Smith, D.E., Arcuri, K.E., and Vlasuk, G.P. (1990) Tick anticoagulant peptide: a highly selective inhibitor of blood coagulation and factor Xa. Science. 248: 593–596.

Weitz, J.I., Hudoba, M., Massel, D., Maraganore, J., and Hirsh, J. (1990) Clot-bound thrombin is protected from inhibition by heparin-antithrombin III but is susceptible to inactivation by antithrombin III-independent inhibitors. J. Clin. Invest. 86: 385–391.

Whinna, H.C., and Church, F.C. (1993) Interaction of thrombin with antithrombin, heparin cofactor II, and protein C inhibitor. J. Protein Chem. 12: 677–688.

Wilner, G.D., Danitz, M.P., Mudd, M.S., Hsieh, K.H., and Fenton, J.W. II. (1981) Selective immobilization of thrombin by surface bound fibrin. J. Lab. Clin. Med. 97: 403–411.

Ye, J., Rezaie, A.T., and Esmon, C.T. (1994) Glycosaminoglycan contributions to both protein C activation and thrombin inhibition involve a common arginine-rich site in thrombin that includes residues arginine 93,97, and 101. J. Biol. Chem. 269: 17965–17970.

REGULATION OF NEURONS AND ASTROCYTES BY THROMBIN AND PROTEASE NEXIN-1

Relationship to Brain Injury

Dennis D. Cunningham and Frances M. Donovan

Department of Microbiology and Molecular Genetics
College of Medicine, University of California
Irvine, California 92697-4025

1. REGULATION OF CELLULAR ACTIVITIES BY THROMBIN

Thrombin is best known for its pivotal role as the final protease in the blood coagulation cascade and its ability to cleave fibrinogen to fibrin and to cause platelet aggregation. Studies over the past twenty or so years have shown that in addition thrombin regulates a number of important activities of cells that may be important in repair processes following injury. The first demonstration of a cell regulatory role for thrombin came from landmark studies by Chen and Buchanan who showed that thrombin is a potent mitogen for cultured fibroblasts[1]. Subsequently, much effort was directed at understanding the mechanisms by which thrombin regulates certain cells. Early studies identified cell surface receptors for thrombin[2] and showed that the proteolytic activity of thrombin was required for cell activation[3]. A major advance in understanding the mechanism of cellular regulation by thrombin was the cloning of the cDNA for the thrombin receptor[4, 5]. These studies revealed that the thrombin receptor is a seven transmembrane G protein-coupled receptor that is activated by proteolytic cleavage. This novel activation mechanism produces a new extracellular amino terminus which serves as a tethered ligand which binds back to the receptor and activates it. This was demonstrated by the finding that a peptide representing the first six amino acids of the newly produced amino terminus of the receptor (SFLLRN) could fully activate the thrombin receptor[4]. The thrombin receptor has been shown subsequently to be present on a number of cell types and to mediate the cell regulatory effects of thrombin.

2. REGULATION OF NEURONS AND ASTROCYTES BY THROMBIN

The first evidence that thrombin is important in regulating cells from the central nervous system came from studies on the glial-derived nexin (GDN) which showed that it promotes neurite outgrowth from cultured neuroblastoma cells and that thrombin can prevent this neurite outgrowth activity[6,7]. GDN was subsequently shown to be identical to the serpin protease nexin-1 (PN-1), a potent inhibitor of thrombin[8,9]. Later studies showed that addition of thrombin to neuroblastoma cells[10] or to stellate astrocytes[11] leads to retraction of cellular processes and that inhibition of thrombin by PN-1 leads to outgrowth of these processes[12]. Thus, the primary regulation of cellular processes on neurons and astrocytes, at least in cell culture, is due to thrombin, and the regulation by PN-1 is a result of inhibiting thrombin. This thrombin effect is receptor-mediated since it can be produced by the thrombin receptor activating peptide SFLLRN[13,14]. The physiological or pathophysiological significance of the process retraction produced by thrombin is not clear, although it has been suggested that after injury to the brain low concentrations of thrombin could increase the plasticity of these cells and thus facilitate repair processes. On the other hand, higher concentrations of thrombin could produce excessive retraction of the processes and compromise repair processes[15].

In view of the known mitogenic effect of thrombin on fibroblasts, it is not surprising that thrombin is also mitogenic for cultured astrocytes[11]. This suggests that thrombin might be at least partly responsible for the astrocyte proliferation that sometimes occurs in gliosis following injury to the central nervous system.

3. REGULATION OF NEURON AND ASTROCYTE VIABILITY BY THROMBIN

In view of the above effects of thrombin on cultured neurons and astrocytes, and the fact that thrombin is produced almost immediately after injury, it seemed important to determine if thrombin might regulate other cellular activities that are related to restoration of normal function. Accordingly, we examined the hypothesis that thrombin might regulate the viability of neurons and astrocytes cultured under adverse conditions similar to those found at sites of brain trauma. The studies described below show that thrombin can protect cultured neurons and astrocytes from certain conditions that prevail upon injury to the central nervous system[16].

Trauma to the central nervous system leads to vascular damage and the establishment of localized areas of hypoglycemia and anoxia[17]. The data in Figure 1 show that when cultured primary rat astrocytes are placed in a hypoglycemic culture medium (0.5 mM glucose instead of the normal 25 mM glucose), only about 20% of the cells survive after 48 h. If, however, thrombin is added to the cultures at the same time they are switched to the low glucose medium, it can be seen that it produces a marked protective effect on the cells. Optimal concentrations of thrombin (about 0.1 nM) increase the cell viability to between 70 and 80%. Panel B of Figure 1 shows that most of the protective effect is due to activation of the thrombin receptor. The thrombin receptor activating peptide SFLLRN confers protection to the hypoglycemic astrocytes whereas a control peptide with the first two amino acids reversed does not confer protection[16]. The data in Figure 2 show that this protective effect of thrombin also is observed with cultured primary rat hippocampal neurons. When cultured neurons are placed in a hypoglycemic medium (0.2 mM glucose instead of the normal 25 mM glucose), virtually all of the cells die after 48 h. If, however, thrombin is added at the same time the cells are switched to low glucose, about 70 to 90% of the cells survive (Figure 2, panel A). Moreover, thrombin protects cultured neurons from the killing effects of growth factor removal (Figure 2, panel B). It should be pointed out that the thrombin concentrations required to protect the cultured neurons is much higher than the concentrations required to protect the cultured astrocytes. As with astrocytes, the protective effect of thrombin on cultured neurons is receptor mediated (Figure 2, panel C)[16].

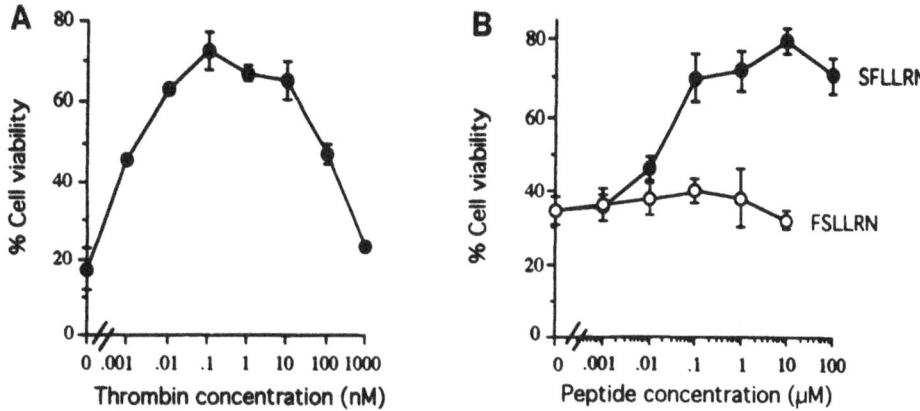

Figure 1. Thrombin and thrombin receptor activating peptide protect astrocytes from cell death produced by hypoglycemia. A, dose-response curve for the protective effect of thrombin. Rat astrocytes were incubated in medium containing 0.5 mM glucose in the absence or presence of the indicated concentrations of thrombin. After 48, cell viability was determined. B, Dose-response curve for the protective effect of thrombin receptor activating peptide. Rat astrocytes were treated as described for A except that the thrombin receptor activating peptide, SFLLRN or scrambled peptide FSLLRN were substituted for thrombin. Reprinted from Vaughan et al.[16] with permission of the publisher.

The above experiments were conducted with astrocytes and neurons that had been stressed by environmental insults similar to ones seen under conditons of brain injury. We also examined the effects of thrombin on cells that were cultured under normal conditions. Figure 3 shows that high concentrations of thrombin kill both neurons and astrocytes under these conditions. The dose-dependence of the cell killing is similar for both cell types[16]. In agreement with these studies Smith-Swintosky et al.[18] showed that PN-1 can protect cultured neurons under certain conditions from the killing action of thrombin.

4. POSSIBLE PHYSIOLOGICAL SIGNIFICANCE OF CELL CULTURE THROMBIN EFFECTS

Taken together, the above studies show that thrombin can either protect or kill cultured neurons and astrocytes, and that the thrombin effects depend on the physiological state of the cells as well as the concentration of thrombin added. In order to probe the significance of these results, it is important to consider several factors. First, is it likely that neurons and astrocytes would be exposed to thrombin concentrations employed in the above studies? Although it is not possible to answer this question with certainty, it should be noted that the concentration of prothrombin in plasma is about 1 to 5 micromolar. Thus, in areas of severe trauma in the central nervous system, cells could be exposed to very high thrombin concentrations. Lower thrombin concentrations would be found in peripheral areas and in areas of less severe trauma. A second point which suggests that the thrombin effects on neuron and astrocyte viability might be significant is the finding that the effects are receptor-mediated judged by the ability of the peptide SFLLRN to produce the same cellular effects as thrombin. As discussed below, thrombin receptor mRNA is broadly distributed throughout the brain although expression is very focal[19, 20].

In an effort to better understand the significance of the thrombin regulation of cultured neuron and astrocyte viability, we recently conducted cell culture experiments to further mimic conditions found in trauma to the brain. Under conditions of injury, the extracellular

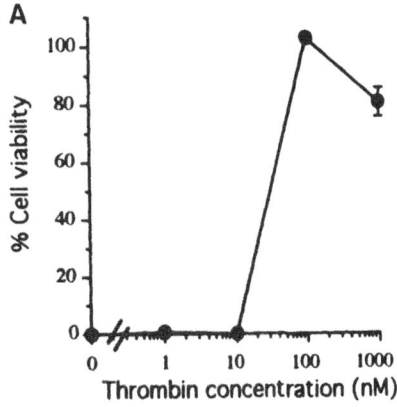

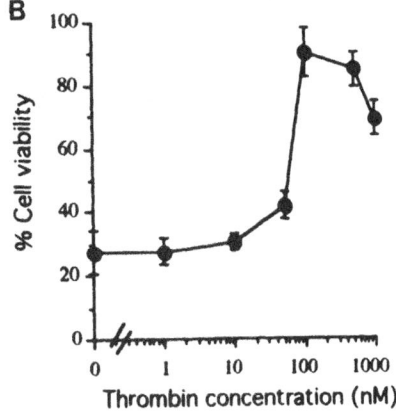

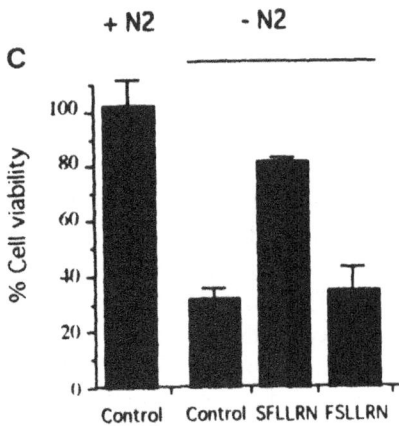

Figure 2. Thrombin and thrombin receptor activating peptide protect rat hippocampal neurons from cell death produced by hypoglycemia and growth supplement deprivation. A, Dose-response curve for the protective effect of thrombin against hypoglycemia. Neurons were incubated in medium containing 0.2 mM glucose in the absence or presence of the indicated concentrations of thrombin. After 48 h cell viability was determined. B, Dose-response curve for the protective effect of thrombin against growth supplement deprivation. One hour after plating, the indicated concentrations of thrombin were added and cell viability determined after 48 h. C. Protective effect of the thrombin receptor activating peptide. Freshly isolated neurons were plated in the absence of N2 supplements. One hour after plating the cells were either left untreated (control, −N2) or 1mM of the peptide SFFLRN or FSLLRN was added and cell viability determined after 48 h. Control cultures were maintained in medium with N2 supplements throughout the experiment (control, +N2). Reprinted from Vaughan et al.[16] with permission of the publisher.

matrix (ECM) is degraded and neurons and astrocytes can be partially or fully detached from their ECM. To determine if attachment to an ECM affected the regulation of cell viability by thrombin, we cultured primary rat astrocytes on either plastic or plastic coated with ECM and examined the effects of thrombin. These studies were conducted on cells under "normal" culture conditions as in Figure 3. We found that in the absence of an ECM, the killing effect was shifted to much lower concentrations of thrombin (Donovan and Cunningham, unpublished results). We speculate that partial or complete detachment of cells from their ECM might signal the cells that they are disrupted and that this, in the presence of thrombin, signals the cells

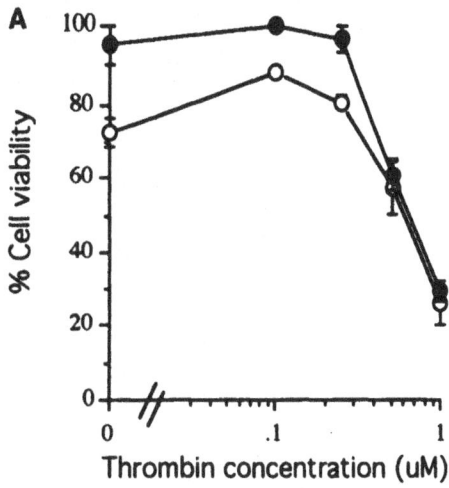

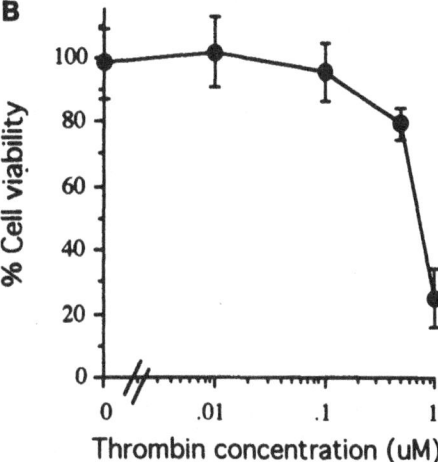

Figure 3. Thrombin kills rat neurons and astrocytes cultured under normal conditions. A, Differentiated (open circles) or undifferentiated (closed circles) rat astrocytes were incubated in medium containing 25 mM glucose in the absence or presence of the indicated concentrations of thrombin. Cell viability was determined 72 h later. B, Rat hippocampal neurons were plated in medium containing 25 mM glucose and N2 supplements. Twenty four hours after plating, the cells were treated with the indicated concentrations of thrombin. Cell viability was determined 48 h later. Reprinted from Vaughan et al.[16] with permission of the publisher.

to initiate programmed cell death. This could represent a mechanism to clear highly traumatized areas in the brain of cells that are so disrupted or damaged that it would be advantageous to kill them and clear them from the area. The diagram in Figure 4 illustrates some of these proposed events in the region of trauma to the brain.

To date, the ECM studies have only been conducted with cultured astrocytes but soon will be extended to cultured neurons. It will also be informative to examine whether the presence or absence of an ECM affects the protective effect of thrombin on cells that have been stressed by hypoglycemia or other injury-related conditions.

5. EVIDENCE FOR REGULATION OF NEURONS BY THROMBIN *IN VIVO*

Recent evidence from *in vivo* experiments has suggested that thrombin may play a role in regulating neuronal viability during development and also in injury. In an experimental chicken embryo model system which examines programmed motoneuron cell death during

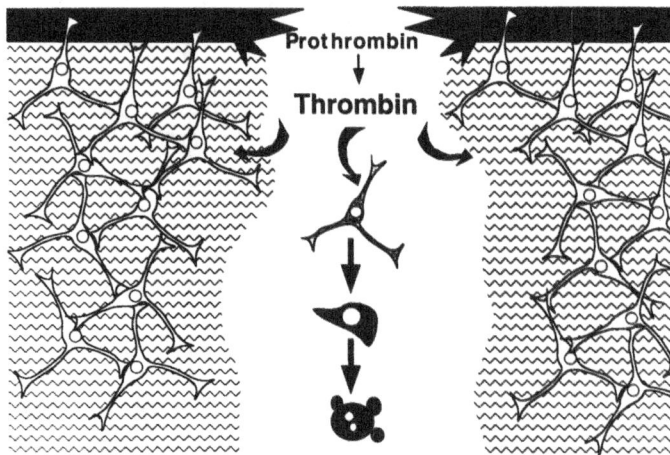

Figure 4. Diagrammatic representation of proposed thrombin-mediated cellular events in a region of brain trauma. Injury to the brain results in breakdown of the blood-brain-barrier, release of prothrombin and conversion of prothrombin to thrombin. Cells attached to an ECM are not killed by thrombin. In the region of severe trauma, the ECM is degraded. Cells not attached to an ECM undergo programmed cell death upon exposure to thrombin. The cell killing by thrombin could represent a mechanism to clear the badly traumatized area of cells that are severely disrupted or damaged.

development, Houenou et al. found that injection of PN-1 from E6 to E9 resulted in a substantial reduction of naturally occurring death of spinal motoneurons when examined on E10. The survival effect of PN-1 was comparable to that obtained with previously examined neurotrophic agents[21]. Moreover, in an injury model involving axotomy of the sciatic nerve in mice, these investigators found that implantation of a piece of gelfoam soaked in a solution containing PN-1 prevented the death of virtually all of the severed motoneurons. The authors suggested that these PN-1 effects were likely due to inhibition of thrombin, although other mechanisms were not ruled out[21]. This interpretation seems likely in view of the ability of thrombin, described above, to kill cultured neurons. Finally, it should be pointed out that studies on the distribution of thrombin receptor mRNA in rat brain have revealed that it is broadly distributed throughout the brain, although expression was very focal and often anatomically limited within specific neural structures[19,20]. In view of thrombin's ability to regulate the viability of cultured neurons, it is noteworthy that thrombin receptor mRNA was abundant in certain vulnerable populations of neurons. It was abundant in dopaminergic neurons of the substantia nigra which degenerate in Parkinson's Disease as well as pyramidal neurons of the hippocampus which are particularly vulnerable to ischemia[20].

The recent development of knockout mice which lack the SFLLRN-activated thrombin receptor provides new opportunities to explore the role of thrombin and its receptor in regulating cellular repair events after injury. Although about half of the embryos lacking the thrombin receptor die at E9 to E10, the other half survive to become grossly normal adult mice[22]. These mice should provide unique opportunities to compare specific groups of neurons and astrocytes in animals possessing or lacking the thrombin receptor using protocols that damage cells and disrupt the blood-brain-barrier.

6. STUDIES ON PN-1 REGULATION AFTER INJURY

Much of the PN-1 in brain tissue is localized around blood vessels where it is ideally situated to inhibit extravasated thrombin[23]. Thus, studies on PN-1 levels after injury to the central nervous system could provide important insights into thrombin regulation. Monard and colleagues addressed this question by examining the effect of transient forebrain ischemia on brain PN-1 levels in the gerbil[24,25]. Ischemia was induced by bilateral clamping of the common carotid arteries; this leads to alterations in the blood-brain-barrier and selective degeneration of hippocampal neurons after a delay of three to four days. The induced ischemia also led to increased levels of PN-1 in hippocampal astrocytes at about the same time that degeneration of hippocampal neurons occurred. Monard and colleagues also found that astrocytes in the substantia nigra also synthesize PN-1 after administration of toxins that preferentially kill dopaminergic neurons in this region of the brain[26]. Finally, it is noteworthy that cytokines known to be produced in the brain after injury (particularly IL-1, TGF-β, and TNF-α) markedly stimulate the synthesis and secretion of PN-1 by cultured astrocytes after about 10 h[27]. Together, these studies provide evidence of a regulatory mechanism to inhibit thrombin after injury.

7. CONCLUDING REMARKS

Virtually any injury to the brain involves some alteration of the blood-brain-barrier and release of plasma prothrombin into the brain tissue; the released prothrombin is rapidly converted to thrombin. Thus, thrombin is one of the first new molecules to appear at the site of trauma. In addition to its well known role in blood coagulation, it appears that thrombin may play an important role in initiating cellular activities involved in repair processes. Thrombin can retract processes on cultured neurons and astrocytes; it has been suggested that this may facilitate repair processes by increasing cellular plasticity. Thrombin is mitogenic for astrocytes, indicating that thrombin might be partly responsible for glial proliferation that sometimes accompanies gliosis following injury to the brain. Recent studies have shown that thrombin can protect cultured neurons and astrocytes from certain conditions, like hypoglycemia, which prevail after brain injury. This may be important in the critical moments after injury before more specific repair molecules, like neurotrophins, can be synthesized and secreted. On the other hand, high concentrations of thrombin can kill neurons and astrocytes. When astrocytes are not attached to an ECM, this killing occurs at lower thrombin concentrations. This suggests the hypothesis that a dual signalling mechanism (lack of an ECM and presence of thrombin) may signal cells in an area of relatively severe trauma that they are badly disrupted and that it would be advantageous if these cells were killed by apoptosis so they could be cleared from the area without initiating an inflammatory response. Although these suggestions are speculative, it is clear that thrombin can either protect or kill cultured neurons and astrocytes and that the thrombin action depends on the physiological state of the cell and the thrombin concentration. The increased synthesis of PN-1 following injury undoubtedly reflects a control mechanism to regulate thrombin levels and the cellular events described above. It will be informative to examine the role of thrombin and its receptor in cellular repair processes in the brain using *in vivo* models like recently developed knockout mice that lack the SFLLRN-activated thrombin receptor.

ACKNOWLEDGMENTS

This paper is dedicated to the memory of Dr. Patrick J. Vaughan who died May 6, 1995. Dr. Vaughan initiated the studies in this laboratory on regulation of cell viability by thrombin. He was a valuable and well liked colleague. These studies were supported by NIH grant AG 00538.

REFERENCES

1. Chen, L. B., & Buchanan, J. M. (1975) Mitogenic Activity of Blood Components. Thrombin and Prothrombin. Proceedings of the National Academy of Sciences USA 72: 131–135.
2. Carney, D. H., & Cunningham, D. D. (1978) Role of specific cell surface receptors in thrombin-stimulated cell division. Cell 15: 1341–9.
3. Glenn, K. C., Carney, D. H., Fenton, J. II., & Cunningham, D. D. (1980) Thrombin active site regions required for fibroblast receptor binding and initiation of cell division. Journal of Biological Chemistry 255: 6609–16.
4. Vu, T. K., Hung, D. T., Wheaton, V. I., & Coughlin, S. R. (1991) Molecular cloning of a functional thrombin receptor reveals a novel proteolytic mechanism of receptor activation. Cell 64: 1057–68.
5. Rasmussen, U. B., Vouret-Craviari, V., Jallat, S., Schlesinger, Y., Pages, G., Pavirani, A., Lecocq, J. P., Pouyssegur, J., & Van Obberghen-Schilling, E. (1991) cDNA cloning and expression of a hamster alpha-thrombin receptor coupled to Ca^{2+} mobilization. FEBS Letters 288: 123–8.
6. Monard, D., Niday, E., Limat, A., & Solomon, F. (1983) Inhibition of protease activity can lead to neurite extension in neuroblastoma cells. Progressive Brain Research 54:359–364.
7. Guenther, J., Nick, H., & Monard, D. (1985) A glia-derived neurite-promoting factor with protease inhibitory activity. EMBO Journal 4: 1963–6.
8. Gloor, S., Odink, K., Guenther, J., Nick, H., & Monard, D. (1986) A glia-derived neurite promoting factor with protease inhibitory activity belongs to the protease nexins. Cell 47: 687–93.
9. Sommer, J., Gloor, S. M., Rovelli, G. F., Hofsteenge, J., Nick, H., Meier, R., & Monard, D. (1987) cDNA sequence coding for a rat glia-derived nexin and its homology to members of the serpin superfamily. Biochemistry 26: 6407–10.
10. Gurwitz, D., & Cunningham, D. D. (1988) Thrombin modulates and reverses neuroblastoma neurite outgrowth. Proceedings of the National Academy of Sciences USA 85: 3440–4.
11. Cavanaugh, K., Gurwitz, D., Cunningham, D. D., & Bradshaw, R. (1990) Reciprocal modulation of astrocyte stellation by thrombin and Protease Nexin-1. Journal of Neurochemistry 54: 1735–1743.
12. Gurwitz, D., & Cunningham, D. D. (1990) Neurite outgrowth activity of Protease Nexin-1 on neuroblastoma cells requires thrombin inhibition. Journal of Cellular Physiology 142: 155–162.
13. Jalink, K., & Moolenaar, W. H. (1992) Thrombin receptor activation causes rapid neural cell rounding and neurite retraction independent of classic second messengers. Journal of Cell Biology 118: 411–9.
14. Suidan, H. S., Stone, S. R., Hemmings, B. A., & Monard, D. (1992) Thrombin causes neurite retraction in neuronal cells through activation of cell surface receptors. Neuron 8: 363–75.
15. Cunningham, D. D., Pulliam, L., & Vaughan, P. J. (1993) Protease nexin-1 and thrombin: injury-related processes in the brain. Thrombosis and Haemostasis 70: 168–171.
16. Vaughan, P. J., Pike, C. J., Cotman, C. W., & Cunningham, D. D. (1995) Thrombin receptor activation protects neurons and astrocytes from cell death produced by environmental insults. Journal of Neuroscience 15: 5389–5401.
17. Sochocka, E., Juurlink, B. H., Code, W. E., Hertz, V., Peng, L., & Hertz, L. (1994) Cell death in primary cultures of mouse neurons and astrocytes during exposure to and 'recovery' from hypoxia, substrate deprivation and simulated ischemia. Brain Research 638: 21–28.
18. Smith-Swintosky, V. L., Zimmer, S., Fenton, J. W. II., & Mattson, M. P. (1995) Protease Nexin-1 and Thrombin Modulate Neuronal Calcium Homeostasis and Sensitivity to Glucose Deprivation-Induced Injury. Journal of Neuroscience 15: 5840–5850.
19. Niclou, S., Suidan, H., Brown-Luedi, M., & Monard, D. (1994) Expression of the thrombin receptor mRNA in rat brain. Cellular and Molecular Biology 40: 421–428.
20. Weinstein, J. R., Gold, S. J., Cunningham, D. D., & Gall, C. M. (1995) Cellular localization of thrombin receptor mRNA in rat brain: expression by mesencephalic dopaminergic neurons and codistribution with prothrombin mRNA. Journal of Neuroscience 15: 2906–19.
21. Houenou, L. J., Turner, P. L., Linxi, L., Oppenheim, R. W., & Festoff, B. W. (1995) A Serine Protease Inhibitor, Protease Nexin 1, Rescues Motoneurons from Naturally Occuring and Axotomy-induced Cell Death. Proceedings of the National Academy of Sciences USA 92: 895–899.
22. Connolly, A. J., Ishihara, H., Kahn, M. L., Farese, R. V., & Coughlin, S. R. (1996) Role of the thrombin receptor in development and evidence for a second receptor. Nature 381: 516–519.

23. Choi, B. H., Suzuki, M., Kim, T., Wagner, S. L., & Cunningham, D. D. (1990) Protease nexin-1: localization in the human brain suggests a protective role against extravasated serine proteases. American Journal of Pathology 137: 741–747.

24. Hoffmann, M. C., Nitsch, C., Scotti, A. L., Reinhard, E., & Monard, D. (1992) The prolonged presence of Glia-derived Nexin, and endogenous protease inhibitor, in the hippocampus after ischemia-induced delayed neuronal death. Neuroscience 49: 397–408.

25. Nitsch, C., Scotti, A. L., Monard, D., Heim, C., & Sontag, K. H. (1993) The glia-derived protease nexin 1 persists for over 1 year in rat brain areas selectively lesioned by transient global ischaemia. European Journal of Neuroscience 5: 292–7.

26. Scotti, A. L., Monard, D., & Nitsch, C. (1994) Re-expression of glia-derived nexin/protease nexin 1 depends on mode of lesion-induction or terminal degeneration: observations after excitotoxin or 6-hydroxydopamine lesions of rat substantia nigra. Journal of Neuroscience Research 37: 155–68.

27. Vaughan, P. J., & Cunningham, D. D. (1993) Regulation of protease nexin-1 synthesis and secretion in cultured brain cells by injury related factors. Journal of Biological Chemistry 268: 3720–3727.

8

MASPIN

A Tumor Suppressing Serpin[*]

R. Sager,[1] S. Sheng,[1] P. Pemberton,[2] and M. J. C. Hendrix[3]

[1]Division of Cancer Genetics
Dana-Farber Cancer Institute
44 Binney Street, Boston, Massachusetts 02115
[2]LXR Biotechnology
1401 Marina Way South, Richmond, California 94804
[3]Department of Anatomy and Cell Biology
College of Medicine
University of Iowa
1115 Bowen Science Bldg., Iowa City, Iowa 52242-1107

1. INTRODUCTION

Maspin (mammary serpin) is a member of the serpin superfamily (Potempa et al. 1994). The gene was first identified by subtractive hybridization on the basis of its expression at the mRNA level in normal but not in tumor-derived mammary epithelial cells (Zou et al. 1994). The cloned and sequenced cDNA consists of 2584 nucleotides encoding a 42 kDa peptide with the overall structure of a serpin. Maspin has been localized to chromosome 18q21.3-q23 (Sager et al. 1994) closely linked to plasminogen activator inhibitor-2 (PAI-2), to BCL2, to the candidate tumor suppressor gene DCC (Sager et al. 1994), and to the candidate tumor suppressor gene SCCA (Schneider et al. 1995).

When normal cells become malignant, a crucial development is the acquisition of elevated protease activity, opening the door to invasion across the basement membrane and subsequently to metastasis. Several lines of evidence support the role of maspin as a strong tumor suppressor gene, acting at the level of invasion and metastasis. These include: (1) sequence similarity to other serpins with protease inhibitor activity; (2) functional studies in nude mice showing that tumor transfectants expressing maspin are inhibited in tumor growth and metastasis; (3) invasion assays showing that endogenous

---

[*] This article was originally published in "Attempts to Understand Metastasis Formation I: Metastasis-Related Molecules," edited by U. Günthert and W. Birchmeier. Copyright Springer-Verlag Berlin Heidelberg 1996. Reprinted by permission of the publisher, Springer-Verlag.

maspin produced by tumor transfectants, or exogenous recombinant maspin, inhibit invasion through basement membrane matrix in culture, and that this inhibition is reversible by anti-maspin antibodies; and (4) motility assays using membrane filters or direct video tracking, showing that motility can be blocked by endogenous or exogenous maspin, and that this inhibition is reversible with anti-maspin antibodies.

These results, discussed below, support the tumor suppressor activity of maspin protein. Consistent with the functional evidence is the loss of expression of maspin in invasive mammary carcinomas, shown by immunostaining of well-characterized tissue specimens (Zou et al. 1994). The protein is expressed in normal tissue and in some carcinoma in situ specimens, but infrequently in invasive or metastatic tumors. These results suggest the potential use of maspin expression as a positive indicator of a favorable prognosis in primary breast cancer.

On the biochemical side, however, the molecular basis of maspin's biological activity is unknown. Cleavage of the reactive center is sufficient to destroy all biological activity (Sheng et al. 1994), but maspin does not act as an inhibitor of proteinases such as trypsin, chymotrypsin, elastase, plasmin, thrombin, or of plasminogen activators in vitro (Pemberton et al. 1995). The molecule is not secreted in cell culture, but rather it is found in the cytoplasm and in the membrane fraction (S. Sheng, unpublished). Thus its action is probably distinct from that of the classical serine protease inhibitors. Its cellular target is as yet unidentified.

2. COMPARATIVE STRUCTURAL PROPERTIES OF MASPIN

Maspin cDNA was isolated from a normal human mammary epithelial cell library. The cDNA sequence contains 1125 nucleotides in the coding region with a polyadenylation signal located 16 nucleotides from the 3' terminal of the sequence shown in Fig. 1. The sequence as shown includes 75 nucleotides of the 5' untranslated region and 1381 nucleotides of 3' untranslated region. The initiation codon and surrounding nucleotides fit the Kozak consensus. The inferred protein consists of 375 amino acids with an NH_2-terminal methionine and COOH-terminal valine. Maspin contains eight cysteine residues and may utilize two or more disulfide bonds to stabilize its tertiary structure.

Multiple alignment studies using BLAST analyzed by the GCG Pileup Program demonstrate close homology of maspin to the serpin superfamily (Fig. 2). Serpins are a diverse family of proteins related by primary sequence homology spanning the entire molecule and varying from 15% to 50% at the amino acid identity level. The first serpin to be analyzed crystallographically, α1 protease inhibitor (α1PI), is the prototype (Huber and Carrell 1989). Overall amino acid similarities compared to maspin are about 40% for plasminogen activator inhibitor (PAI)-2, neutrophil elastase inhibitor and ovalbumin and about 30% for α1PI and PAI-1. Recently identified serpins include PTI, (placental thrombin inhibitor, now renamed PI6) which is related to neutrophil elastase and interacts with trypsin, thrombin, and urokinase type plasminogen activator (uPA) (Coughlin et al. 1993) and SCCA which had previously been identified at the protein level as squamous cell carcinoma antigen. Two tandem linked genes, SCCA-1 and -2, have been described (Schneider et al. 1995).

Most of the amino acid substitutions in maspin do not alter the overall serpin conformation. However, significant differences are located in the hinge region, a peptide stretch located nine to 15 residues NH_2-terminal to the P1-P1' peptide bond within the reactive site loop (RSL) (Carrell and Evans 1992). The reaction center is indicated in Fig. 2., with the putative reactive site marked by the arrow. In a typical reaction between a serine pro-

Figure 1 (cDNA and predicted amino acid sequence, rotated 90° on the page). The sequence is organized in rows; nucleotide positions are given at the left and right of each row and amino‑acid positions beneath.

```
1                                        GGCACGAGTGTGTCCCGCCGCTGCCTGTCCTTTCCACGCATTTCCAGGATAACTGACTCCAGGCCCGCA   75

76    ATG GAT GCC CTG CAA CTA GCA AAT TCG GCT TTT GCC GTT GAT CTG CTC TTT AAA CAA CTA TGT GAA AAG CCA CTG GGC AAT GTC CTC TTC   165
1     Met Asp Ala Leu Gln Leu Ala Asn Ser Ala Phe Ala Val Asp Leu Leu Phe Lys Gln Leu Cys Glu Lys Pro Leu Gly Asn Val Leu Phe   30

166   TCT CCA ATC TGT CTC ACC TCT CTG TCA CTT ... TCA TAC ... AAA ATC AAG   255
31    Ser Pro Ile Cys Leu Thr Ser Leu Ser Leu ... Ser Tyr ... Lys Ile Lys   60

256   GAA AAT GTC AAA GAT ATA CCC TTT GGA TTT ... TCA ... AAA CTA ATC AAG   345
61    Glu Asn Val Lys Asp Ile Pro Phe Gly Phe ... Ser ... Lys Leu Ile Lys   90

346   CGG CTC TAC GTA GAC GAC AAA TCT CTT CTG ... ACA TTG ... ACT GTT GAC   435
91    Arg Leu Tyr Val Asp Asp Lys Ser Leu ... Thr Leu ... Thr Val Asp   120

436   TTC AAA GAT AAA TTG GAA GAA ACG ... CAC TTT GAG GAA ... TTA GCT GAC   525
121   Phe Lys Asp Lys Leu Glu Glu Thr ... His Phe Glu Glu ... Leu Ala Asp   150

526   AAC AGT GTG AAC GAC ACC AAA ATC ... GTT GGC TAC ... TCA GAA ACA   615
151   Asn Ser Val Asn Asp Thr Lys Ile ... Val Gly Tyr ... Ser Glu Thr   180

616   AAA GAA TGT CCT TTT AGA CTC AAG ... ACA GTG ... AAC ATT GAC   705
181   Lys Glu Cys Pro Phe ... Thr Val ... Asn Ile Asp   210

706   AGT ATC TGT AAT CTT CCT TTT CAA AAT ... CAT AAG ... GAG GAG TCC   795
211   Ser Ile Cys Asn Leu Pro Phe Gln Asn ... His Lys ... Glu Glu Ser   240

796   ACA GGC TTG GAG AAG ATT TTG GAA GTG ... GAT GTG ... GCC AAG GTC AAA   885
241   Thr Gly Leu Glu Lys Ile Leu Glu Val ... Asp Val ... Ala Lys Val Lys   270

886   CTC TCC ATT CCA AAA TTT AAG GTG ATG ... CAT ATC ... AGT GAA GAC   975
271   Leu Ser Ile Pro Lys Phe Lys Val Met ... His Ile ... Ser Glu Asp   300

976   ACA TCT GAT TTC TCT GGA ATG TCA GAG ... AAT GTT ... ACT GAA GAT GGT   1065
301   Thr Ser Asp Phe Ser Gly Met Ser Glu ... Asn Val ... Thr Glu Asp Gly   330

1066  GGG GAT TCC ATA GAG GTG GTG CCA GCA CGG ... GAT GAA ... ATC ACC AGG CAC   1155
331   Gly Asp Ser Ile Glu Val Val Pro Ala Arg ... Asp Glu ... Ile Thr Arg His   360

1156  AAC AAA ACT CGA AAC ATC ATT TTC TTT GGC AAA TTC TGT TCT CCT TAA   1203
361   Asn Lys Thr Arg Asn Ile Ile Phe Phe Gly Lys Phe Cys Ser Pro End   376
```

```
1204  GTGGCATAGCCCATGTTAAGTCCTCCCGACTTTCTGTGATGCGATCCCGATTCTGTAAACTCGTAAACTCGCATCCAGAGATTCATTTCTGGATCAGGAAG   1323
1324  CCGCCAGTACTGTCATATGTAGCCTCACAGATAGCCCTTTTTTTCTCCAATCTACTCTTGTTCCTATATCTTGTTTCCCATAGTTATGAAAGGAATCA         1443
1444  CGTTAGAGAAAAATATTATGATCATGATTGTCAGACATATCGTCCCGGGAGTTCTCCAAAGAAATTCGAAGCATTGGAACGAAGATTTGGAGTCTTCCCAGCACT     1563
1564  ATGCTTTCCTTCTTTGGATAGAGAAATGTCCAGACACATCCCGCCTCGCCAGCCAGGTGTTTATTAAAATCTGAAACTGTAGTCCATGGGACCACAGTTGGGCACAT    1683
1684  GCTCAGCCTACTATAGGTCCAGAGAGTGCCTATGTTAAGCACGAACCAGAACCCAGGGCTCCAGTGAAACTGGGCAATAATTTCAAAGATAATATTTACACACTGTATGTATAG   1803
1804  AACTCATGGATCAGATCGGACGGAAGTGTGAGGAAATGAGCTTCAACACTCGTCGCAGAGCTTTCGATGTGATGAATGTGGACTAGTAGCCTAGTAGCAGGGGTCA          1923
1924  AAATTTGCTCCCAAATCGGTATGCCATGCCTTAATATAAACTCATTTTGATGCGTCCCAAACAAATGAGAAAAAATGGAATTAGAGCCTCTAGTAGCTGAAATGCAAGACCC      2043
2044  CAAGAGGAAGTTCAGATCTAATATAAAAACACTTGATACTCAGAACTTGTCTCTGAACTGTCCATCGGAACTTCTGATGGCAAGCGGGAAAAGACCGCTTCCAGCTAGACTCGCACAGGGATTCT  2163
2164  CACAATAGCCGATATCAGAATTTTGCACACATAGTTTTCAGTCTATGGGTTTAGTTACTTTGAGTAGCTGGCAGCAAGCATGAACTAACTATATTAATAATTGTAAAGGATTAAAGTGCTC    2283
2284  ACGTTACCTGACACATAGTTTTCAAATATATTCGCATAGGTTAATTTTGCAATTTGTGCACATCTCCTGCACAGCATGCATGTAAAGTGGATAAGAGTCATCCCTGTTGCCCGTT        2403
2404  CATCGATTACTTCTCTATAAAAAATATATTTGTGACAATCTCCATCTCTGCAAATTGTAATACCAAAAATAATTATTGACTACCAAAAAAAAAAAAAAA  2523
2524  AATTTCTCCCTATGCTATTGACAAAAATATTATTGACTACCAAAAAAAAAAAAAAAAA   2584
```

Figure 1. cDNA and predicted amino acid sequence of maspin. The polyadenylation signal is *underlined*.

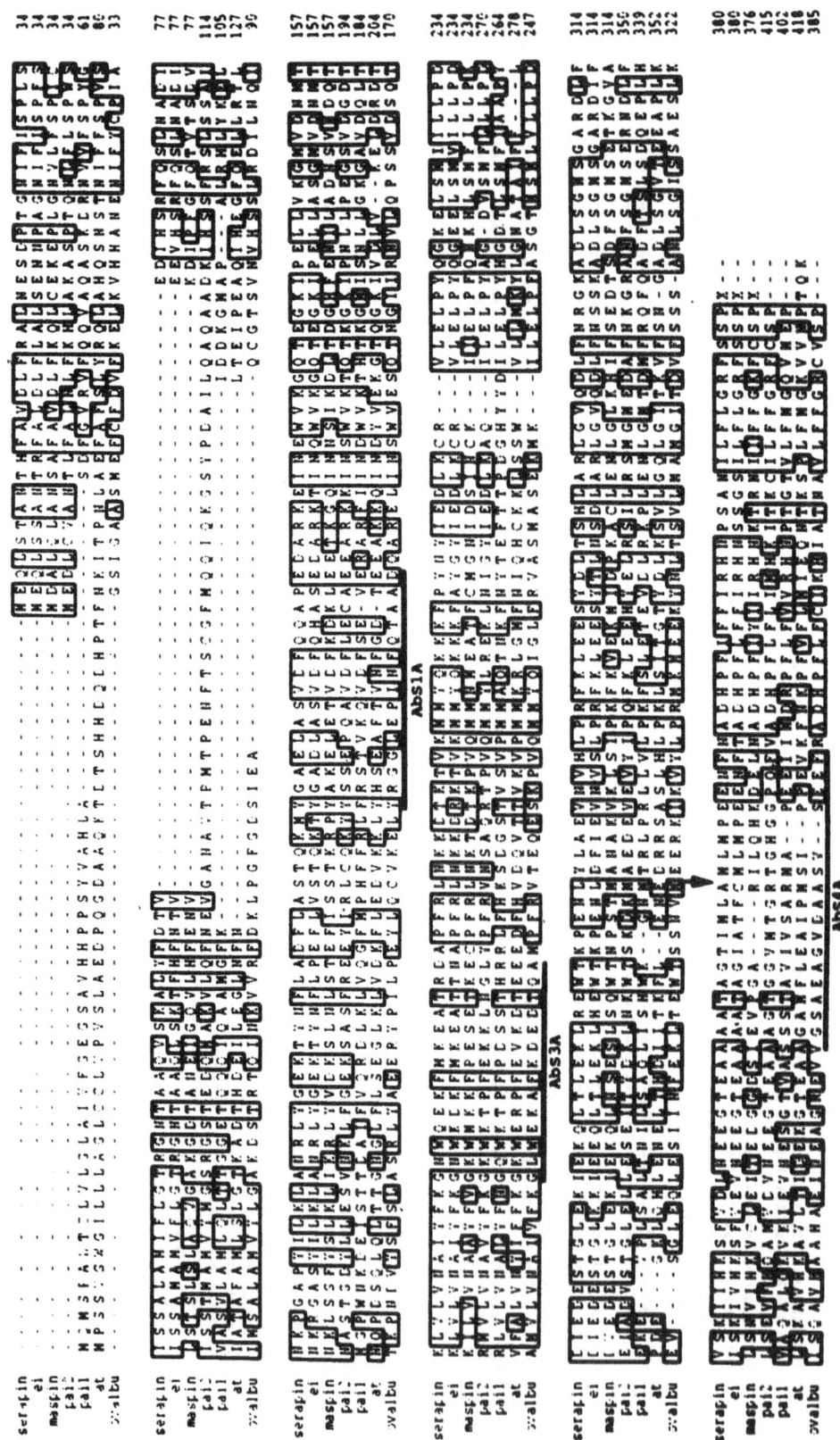

Figure 2. Sequence comparison of maspin with related serpins. Identical resides are *boxed*. Three regions used for antibody production are *underlined*. *Arrow* denotes the proposed reactive center of maspin. *Serapin*, horse serapin; *ei*, human monocyte/neutrophil elastase inhibitor; *pai2*, plasminogen activator inhibitor 2; *pai1*, plasminogen activator inhibitor 1; *at*, α1 protease inhibitor; *ovalb*, ovalbumin.

tease inhibitor and its protease target, the protease attempts to cleave the reactive site peptide bond but becomes trapped in a stable 1:1 stoichiometric complex in which the protease is inactive. Currently, all serpins of known inhibitory function (e.g. antitrypsin, antithrombin) undergo a transition from a stressed (S) to relaxed (R) stable molecule while serpins with no known inhibitory function (e.g., ovalbumin, angiotensinogen) do not undergo this transition. Thus the S-R transition is a useful predictor of serpin function.

Structurally, the stressed inhibitory (S) form was thought to result from partial insertion of the hinge region into β-pleated sheet A of the native molecule, following proteolysis with the RSL. The peptide sequence COOH-terminal to the hinge region then completely inserts into β-pleated sheet A to provide a more relaxed (R) stable structure. In maspin the amino acid substitutions in the hinge region are divergent from the consensus sequence usually observed in inhibitory serpins, and on this basis it has been suggested that maspin may not function as a protease inhibitor (Hopkins et al. 1994). However, the recent crystallographic structure of α1 anti-chymotrypsin has revealed that partial insertion is not a prerequisite for inhibitory function (Wei et al. 1994). Thus, knowledge of serpin hinge region sequences is not a sufficient criterion on which to base a prediction of inhibitory function.

It was recently shown that maspin does not behave as a classical inhibitory serpin (Pemberton et al. 1995). Maspin is not an inhibitor of serine proteinases including trypsin, chymotrypsin, elastase, plasmin, thrombin, and the plasminogen activators, tissue type plasminogen activator (tPA) and uPA, but it does act as a substrate for several of them. Importantly, maspin does not undergo the (S-R) transition predictive of inhibitory function. These findings support the view that the tumor suppressing activity of maspin is not based on a latent or intrinsic trypsin-like serine proteinase inhibitory activity. The question of its mode of action as a membrane-bound or cytoplasmic serpin remains to be determined.

In view of the close linkage between maspin, SCCA, and PAI-2 on chromosome 18q, as well as their sequence similarities, it is likely that these serpins evolved from a common precursor. The hinge region of the RSL represents one important site of evolutionary diversity, which has apparently led to functional diversity, since both SCCA and PAI-2 act as typical inhibitory serpins, whereas maspin apparently does not.

3. PRODUCTION AND CHARACTERIZATION OF RECOMBINANT MASPIN PROTEINS PRODUCED IN BACTERIA, INSECT, AND YEAST SYSTEMS

The rGST-maspin protein was produced in E. coli using the pGEX-2T/Mas vector (Sheng et al. 1994). The fusion protein contains two polypeptides covalently linked by a thrombin cleavage site plus a short sequence resulting from *Bam*H1/*Bcl*I ligation. rGST-maspin accounted for 30%–40% of the total cellular protein. The recombinant protein was purified to near 100% homogeneity in a single step on a glutathione affinity column (for details see Sheng et al. 1994). The glutathione S-transferase polypeptide could not be removed by thrombin cleavage without degrading maspin, and therefore the fusion protein was used in further studies.

rMaspin(i) was produced in insect cells (*Spodoptera frugiperda*) infected with the recombinant viral vector pVL 1393/mas. rMaspin(i) was purified by a combination of anion exchange and heparin affinity chromatography, and accounted for 30%–40% of the total extractable protein. The identity of the two recombinant proteins (molecular 42 mass kDa) was confirmed by western blot analysis using the polyclonal antibody preparation AbS4A (Zou et al. 1994).

The identification of the probable reaction center was determined by cleavage of rMaspin(i) with limiting concentrations of trypsin (Sheng et al. 1994), leading to loss of the 42 kDa band and appearance of a 38 kDa band. Concomitant appearance of a 4227 Da fragment was detected by mass spectometry. The NH_2-terminal sequence of the 4227 Da fragment was ILQHKDELNAD, in agreement with the published cDNA sequence of amino acids located 3' to arginine (Sheng et al. 1994). These results show that the reactive center of the recombinant protein is readily accessible to trypsin cleavage, and suggest that maspin is an Arg-serpin.

rMaspin(y) was produced in yeast cells transformed with the expression vector pYMV4 (Pemberton et al. 1995). Recombinant maspin comprised at least 40% of the total soluble yeast protein. After purification by anion exchange chromatography on Q-sepharose, size fractionation S100HR columns, and affinity chromatography on heparin superflow, the average yield was 13.5 mg per gram wet weight of yeast (Pemberton et al. 1995). Yeast maspin is fully active biologically in the invasion assay (Sager et al. 1994; Sheng et al. 1994).

4. ANTIBODIES TO MASPIN

Polyclonal antibodies were produced in rabbits directed to three poorly conserved sequences, using conjugation of the corresponding synthetic oligopeptides to keyhole limpet hemocyanin. The selected sequences are underlined in Fig. 1 as S1A, S3A, and S4A. AbS4A recognizes the RSL encompassing the reactive site. All three antibody preparations react with a 42 kDa band present in proteins in extracts of normal cells, tumor transfectants that express maspin mRNA, and recombinant maspin protein on western blots. No 42 kDa protein was detected in extracts of mammary carcinoma cell lines that do not express maspin mRNA. All three antibody preparations are effective in protocols for immunostaining of formalin-fixed, paraffin-embedded tissue sections. Because of its high specificity, AbS4A is being used in retrospective and prospective studies of maspin expression in tumors for application to diagnostic and prognostic clinical evaluation.

5. BIOLOGICAL PROPERTIES OF MASPIN

5.1. Inhibition of Invasion in the Matrigel Assay

Classically, invasion is measured in cell culture by use of the Boyden chamber assay, in which cells are introduced into upper wells of the double-welled device and allowed to invade through a defined membrane into lower wells, where they are counted at suitable time points. In the maspin studies, MICS (membrane invasion culture system) manifolds comprising sets of 12 Boyden-like chambers were used. The perforated polycarbonate support membrane separating the upper and lower chambers was coated with growth factor reduced Matrigel as described (Hendrix et al. 1987). Cells were removed from the bottom chambers at 24 or 48 h, fixed, stained, and counted microscopically.

The inhibitory effects of recombinant maspin added to tumor cells 1 h before transfer to MICS chambers are shown in the dose response graphs of Fig. 3. When the fusion protein rGST-maspin was used, a linear response curve was seen in the range from 0.04 to

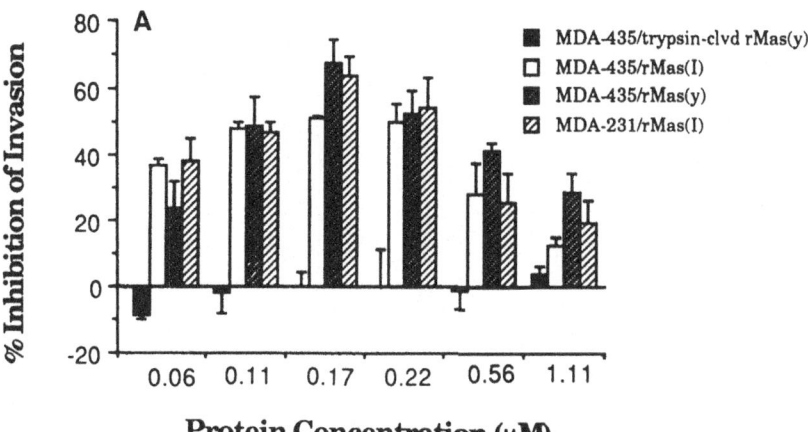

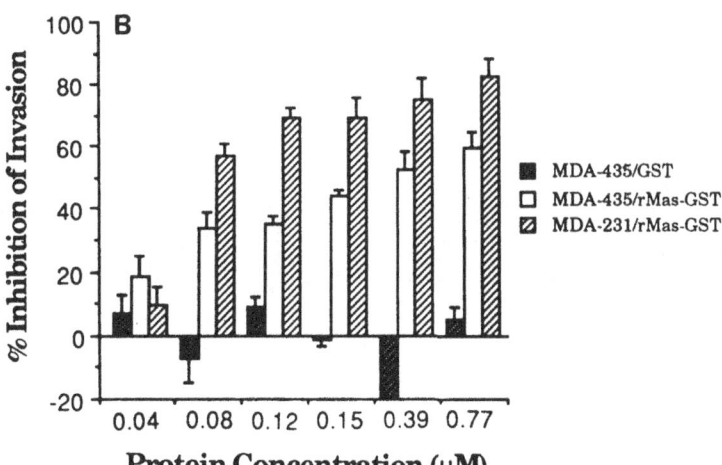

Figure 3. A, B. The effects of rMaspin(i) (A) and rGST-Maspin (B) on invasion by two mammary carcinoma cell lines, MDA-435 and MDA-231. The invasion data from untreated tumor cells were normalized as 100% and invasion data from treated tumor cells are expressed as a percentage of this control. The values of "% inhibition of invasion" were obtained by subtracting the invasion percentage of treated cells from 100%. Each value represents the average of the triplicate results and the error bars represents the S.E.M. A [solid box], invasion by MDA-435 in the presence of trypsin-cleaved rMaspin(y); [hatched box], invasion by MDA-435 in the presence of rMaspin(i); [transparent box], invasion by MDA-435 in the presence of rMaspin(y); and [hatched transparent box], invasion by MDA-231 in the presence of rMaspin(i). B [solid box], invasion by MDA-435 in the presence of recombinant glutathione S-transferase; [transparent box], invasion by MDA-435 in the presence of rGST-maspin; [hatched transparent box], invasion by MDA-231 in the presence of rGST-maspin.

0.77 μM maspin. The control, recombinant GST alone, showed no effect. However, using the recombinant maspins from insects (rMaspin(i)) or from yeast (rMaspin(y)), a bell-shaped curve was obtained with its peak at 0.17 μM. Maspin cleaved within the RSL by trypsin was totally inactive. The inhibitory effects of maspin were fully reversible by antibody AbS4A (Sheng et al. 1994). The basis for the bell-shaped curve is not known, but it is suggested that interaction between maspin and its target on the membrane may induce a conformational change leading to polymerization and concomitant decreased activity. Polymerization is favored by a decreased pH, as occurs in cell culture. The GST-maspin fusion protein may be restricted in its ability to undergo this conformational change.

A

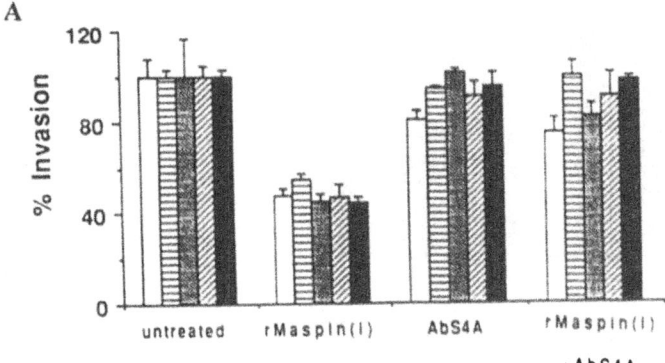

B

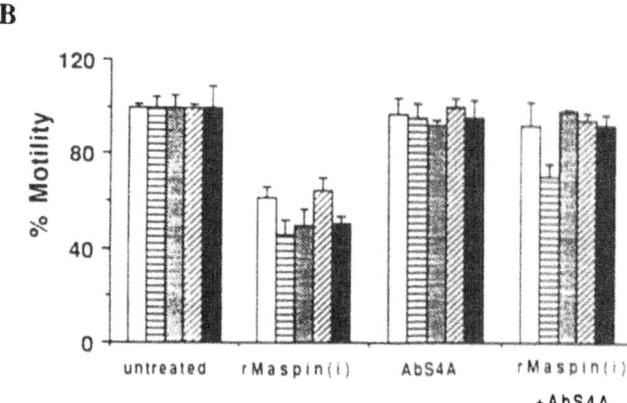

Figure 4. A, B. Effect of rMaspin(i) and AbS4A on invasion (A) and motility (B) of breast carcinoma cell lines 70N ([transparent box]), ZR75-1 ([horizontally hatched transparent box]), 21NT ([shaded box]), MDA-MB-231 ([diagonally hatched transparent box]), and MDA-MB-435 ([solid box]). Invasiveness and the motility rate of the untreated cells in each corresponding set of data were normalized as 100%, respectively, and the data from treated cells are expressed as a percentage of these controls. The data represent the average of three parallel experiments and the error bars represent the standard error.

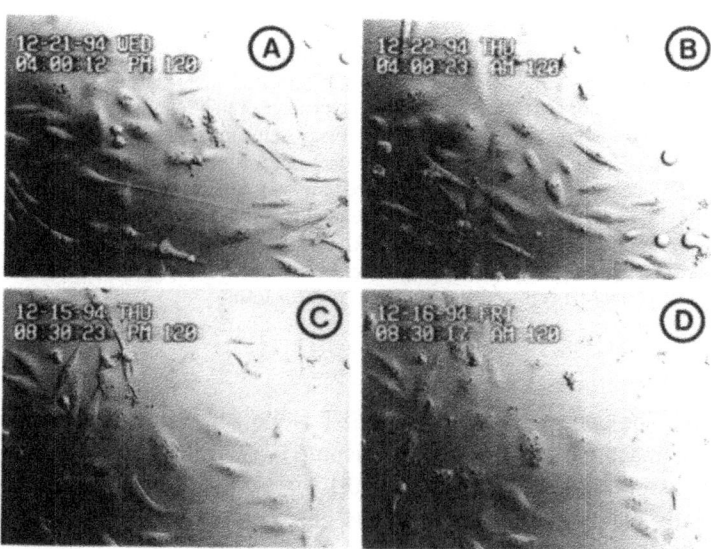

Figure 5. A–D. Photographs of video frames from time-lapse cinematography of MDA-MB-231 cells (motility data presented in Table 1) without rMaspin(i) (A and B) or with rMaspin(i) (C and D), recorded over a 12h interval in the Focht's environmental chamber. Note the relative inactivity of the cells treated with exogenous rMaspin(i) $(0.47 \mu M)$, × 580.

In further studies, the effect of rMaspin(i) on invasion was examined in a series of mammary carcinoma cell lines (ZR-75-1, 21NT, MDA-MB-231, and MDA-MB-435) and in normal cells (70N) (Sheng et al. 1995). Although the normal cells are not invasive in vivo, they show low level invasiveness in culture, which is inhibited by maspin. Fig. 4A shows inhibition of invasion by maspin and reversion by treatment with antibody.

5.2. Inhibition of Motility

Motility can be examined qualitatively by the Burk method in which a patch of cells in a confluent monolayer are removed, e.g., by scraping, and the movement of surrounding cells into the denuded area is observed. For quantitation, a modified Boyden assay or video time lapse recording can be used. All three methods have been used in maspin studies. The Burk assay readily revealed the inability of maspin treated cells to move. The modified Boyden assay gave quantitative results (Fig. 4B) showing the inhibition of motility measured as movement through a minimally coated Matrigel barrier in 6 h (Sheng et al. 1995).

For time-lapse video microscopy, cells were seeded onto coverslips coated with a combination of human laminin, collagen IV, and gelatin, and grown overnight, then treated with rMaspin(i) (or untreated) for 1 h prior to videotaping for 24 h. A Zeiss Axiovert 135 microscope equipped with a Focht environmental chamber and DIC optics was used for videotaping. Tapes were analyzed as described (Sheng et al. 1995).

As shown in Fig. 5 and summarized in Table 1, the migratory ability of tumor cell lines was reduced up to 75% by maspin compared with untreated controls over a 24 h period. The actual migration rate of cell line MDA-MB-231 was about three fold higher than that of MDA-MB-435, which is similar to their comparative metastatic potential in vivo (Thompson et al. 1992).

Two-dimensional migration across an extracellular matrix (ECM) involves attachment, detachment, and ultimately motility. Thus our data suggest that maspin, perhaps in combination with other proteins, inhibits this entire process.

5.3. Inhibition of Invasion and Motility in Prostate Cells

Because of the similarities of mammary and prostate tissues, both being secretory glands under steroid hormonal control, it seemed of interest to examine the effects of maspin on prostate cells in culture. Accordingly, invasion and motility assays were performed with three prostate cell lines: LNCaP, DU 145, and PC 3. The results (Fig. 6) show that invasion is strongly reduced by rMaspin(i) and that this inhibition is partially reversed by the AbS4A antibody. Motility is reduced about 40%–60%, similar to results with breast carcinoma cells, and reversible by antibody.

Table 1. Migration rate on extracellular matrix[a]

Cell line	−Maspin	+Maspin
MDA-MB-435	1.42 ± 0.14 (n=6)	0.23 ± 0.15 (n=9)
MDA-MB-231	4.48 ± 0.38 (n=7)	0.19 ± 0.12 (n=6)

[a]Data represent the average cell movement measured by video cinematography in μm/h with standard error(s).

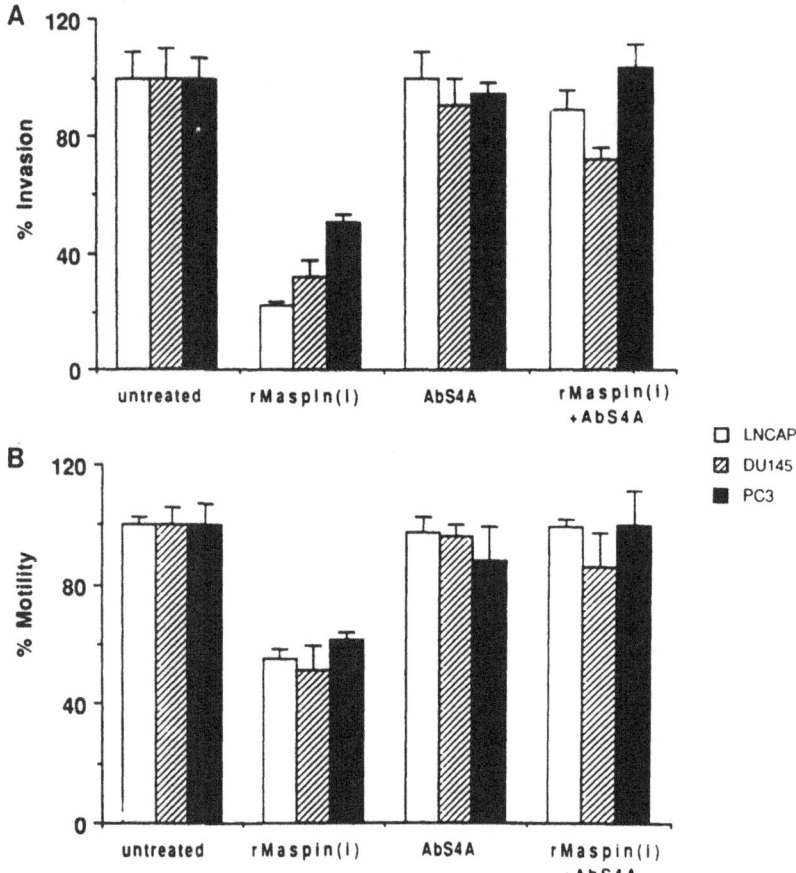

Figure 6. A, B. Effect of rMaspin(i) and AbS4A on the invasion (A) and motility (B) of prostate cancer cell lines LNCaP ([transparent box]), DU145 ([hatched box]) and PC 3 ([shaded box]). Invasiveness and the motility rate of the untreated cells in each corresponding set of data were normalized as 100%, respectively, and the data from treated cells are expressed as a percentage of these controls. The data represent the average of three parallel experiments and the error bars represent the standard error.

Table 2. Tumorigenicity of maspin-transfected MDA-MB-435 cells[a]

Cells	Median weight (g)[b]	*p* value[c]
Nn12	1.60	
Tn8	0.60	0.019
Tn13	0.45	0.024
Tn15	0.75	0.016
Tn16	0.36	0.0072
Tn45	0.60	0.190

[a]Transfected cells (5×10^6) were injected into the mammary fat pads of nude mice. Each mouse was injected at two sites and the tumor development was monitored weekly.
[b]Tumor weights were measured 12 weeks postinjection.
[c]*p*-values were obtained by Wilcoxon rank sum test using null hypothesis.

These results raise the novel possibility that maspin will have similar therapeutic potential in breast and prostate cancers. To follow up this lead, investigations have been initiated using recombinant maspin in animal testing. Preliminary studies in nude mice with breast tumor transfectants expressing maspin have been encouraging and will be presented below.

5.4. Inhibition of Tumor Growth and Metastasis by Maspin in Nude Mice

In initial studies (Zou et al. 1994) it was reported that tumor transfectants (MDA-MB-435 tumor cells expressing transfected maspin) were inhibited in tumor cell growth and metastasis compared with mock transfectants not expressing maspin in the nude mouse assay. In the initial study, three out of four transfectants expressed this inhibition. The fourth cell line showed increased tumor growth in this study and elevated invasion in MICS chamber studies, suggesting an elevated invasive potential in this clone. These initial studies have now been repeated with a new set of five transfectants of similar origin, and four out of five of these cell lines have shown inhibition of tumor growth compared with controls (Table 2). Thus it seems evident that expression of maspin in tumor transfectants inhibits the growth rate of the primary tumors as well as their metastatic potential.

5.5. Loss of Maspin Expression in Mammary Tumor Tissue

In initial studies of benign and malignant breast cancer tissues immunostained with maspin antibody, benign breast tissues ($n=6$) and benign epithelium adjacent to invasive carcinomas were maspin positive, with intense staining in myoepithelial cells and more heterogeneous staining in luminal cells. Inflammatory and stromal cells were negative. Twenty specimens of invasive primary carcinoma, lymph node and distant metastases were also evaluated. Most malignant cells in invasive carcinomas failed to express maspin but a minority of cells in well differentiated tumors expressed maspin focally. Maspin was undetectable or very weakly expressed in lymph node and distant metastases (Zou et al. 1994).

6. SUMMARY

Maspin, a serpin found in mammary epithelial cells, has been shown to have tumor suppressor activity. The gene is expressed in normal human mammary epithelial cells but down-regulated in invasive breast carcinomas. Similar patterns of expression at the RNA and protein levels are seen by Northern analysis with cells grown in culture and by immunostaining of tissues. Biological assays of invasion by tumor cells through Matrigel membranes and of motility have shown that recombinant maspin inhibits both processes, and that its inhibitory action is totally lost by a single cleavage at the reaction center. Tumor transfectants expressing maspin are inhibited in growth and metastasis in nude mice. Maspin is located in the cell membrane and extracellular matrix, and does not behave as a classical inhibitory serpin against any known target protease. Its mode of action is presently unknown.

ACKNOWLEDGMENTS

This work was supported by grant CA61253 to R.S. from the National Institutes of Health, and by LXR Biotechnology, Inc., Richmond, CA.

REFERENCES

1. Carrell R.W., Evans D.L.I. (1992) Serpins: mobile conformations in a family of proteinase inhibitors. Cur. Opin. Cell. Biol. 50: 438–446.
2. Coughlin P., Sun J., Cerruti L., Salem H.H., Bird P. (1993) Cloning and molecular characterization of a human intracellular serine protease inhibitor. Proc. Natl. Acad. Sci. USA 90: 9417–9421.
3. Hendrix M.J.C., Seftor E.A., Seftor R.E.B., Fidler I.J. (1987) A simple quantitative assay for studying the invasive potential of high low human metastatic variants. Cancer Lett. 38: 137–147.
4. Hopkins P.C.R., Whisstock J., Sager R. (1994) Function of maspin. Science 265: 1893–1894.
5. Huber R., Carrell R.W. (1989) Implications of the three-dimensional structure of α1-antitrypsin for structure and function of serpins. Biochemistry 28: 8951–8966.
6. Pemberton P.A., Wong D.T., Gibson H.L., Kiefer M.C., Fitzpatrick P.A., Sager R., Barr P.J. (1995) The tumor suppressor maspin does not undergo the stressed to relaxed transition or inhibit trypsin-like serine proteases. J. Biol. Chem. 270: 15832–15837.
7. Potempa J., Korzus E., Travis J. (1994) The serpin superfamily of proteinase inhibitors: structure, function and regulation. J. Biol. Chem. 269: 15957–15960.
8. Sager R., Sheng S., Anisowicz A., Sotiropoulou G., Zou Z., Stenman G., Swisshelm K., Chen Z., Hendrix M.J.C., Pemberton P., Rafidi K., Ryan K. (1994) RNA genetics of breast cancer: maspin as paradigm. Cold Spring Harb. Symp. Quant. Biol. 59: 537–546.
9. Schneider S.S., Schick C., Fish K.E., Miller E., Pena J.C., Treter S.D., Hui S.M., Silverman G.A. (1995) A serine proteinase inhibitor locus at 18q21.3 contains a tandem duplication of the human squamous cell carcinoma antigen gene. Proc. Natl. Acad. Sci. USA 92: 3147–3151.
10. Sheng S., Pemberton P.A., Sager R. (1994) Production purification and characterization of recombinant maspin proteins. J. Biol. Chem. 269: 30988–30993.
11. Sheng S., Carey J., Hendrix M.J.C., Seftor E.A., Dias L., Sager R. (1995) Maspin acts at the cell membrane to inhibit invasion and motility of mammary and prostatic cancer cells. Proc. Natl. Acad. Sci. USA 93: 11669–11674.
12. Thompson E.W., Paik S., Brunner N., Sommers C.L., Zugmaier G., Clarke R., Shima T.B., Torri J., Donahue S., Lippman M.E., Martin G.R., Dickson R.B. (1992) Association of increased basement membrane invasiveness with absence of estrogen receptor and expression of vimentin in human breast cancer cell lines. J. Cell. Physiol. 150: 534–544.
13. Wei A., Rubin H., Cooperman B.S., Christianson D.W. (1994) Crystal structure of an uncleaved serpin reveals the conformation of an inhibitory reactive loop. Struct. Biol. 1: 251–258.
14. Zou Z., Anisowicz A., Hendrix M.J.C., Thor A., Neveu M., Sheng S., Rafidi K., Seftor E., Sager R. (1994) Identification of a novel serpin with tumor suppressing activity in human mammary epithelial cells. Science 263: 526–529.

9

PROTEASES AND PROTEASE INHIBITORS IN TUMOR PROGRESSION

Yves A. DeClerck,[1,2*] Suzan Imren,[1] Anthony M. P. Montgomery,[3] Barbara M. Mueller,[3] Ralph A. Reisfeld,[3] and Walter E. Laug[1]

[1]Division of Hematology-Oncology, Department of Pediatrics
[2]Department of Biochemistry and Molecular Biology
Childrens Hospital Los Angeles
University of Southern California, Los Angeles, California 90027
[3]Department of Immunology
The Scripps Research Institute, La Jolla, California 92037

ABSTRACT

Our understanding of the role of matrix degrading proteases in cancer has dramatically expanded over the last two decades. From correlative observations linking proteases to cancer progression, we have accumulated evidence supporting a causal role for proteases in various steps of tumor progression and have become increasingly aware of the complex interactions that exist among proteases. Specific natural inhibitors of these proteases have also been identified and their role as potent cytostatic agents in cancer has been suggested. In this article some of the concepts on the role of proteases in cancer are discussed and examples of cooperation between matrix metalloproteinases and the plasmin/plasminogen activators system are presented. The role of protease inhibitors such as tissue inhibitor of metalloproteinases-2 (TIMP-2) and plasminogen activator inhibitor-2 (PAI-2) as inhibitors of tumor growth, invasion and metastasis is discussed.

1. OUR UNDERSTANDING OF THE ROLE OF PROTEASES IN CANCER HAS MARKEDLY EVOLVED OVER THE LAST TWO DECADES

The ability of cancer cells to locally invade tissues and to establish distant metastasis is an important aspects of tumor progression and a major cause of morbidity and mor-

---

* Correspondence should be addressed to: Yves A. DeClerck, MD, Division of Hematology-Oncology, MS #54, Childrens Hospital Los Angeles, 4650 Sunset Boulevard, Los Angeles, CA 90027. Telephone: 213-669-2150; Fax: 213-664-9455; E-mail: ydeclerck%smtpgate@chlais.usc.edu.

Chemistry and Biology of Serpins, edited by Church *et al.*
Plenum Press, New York, 1997

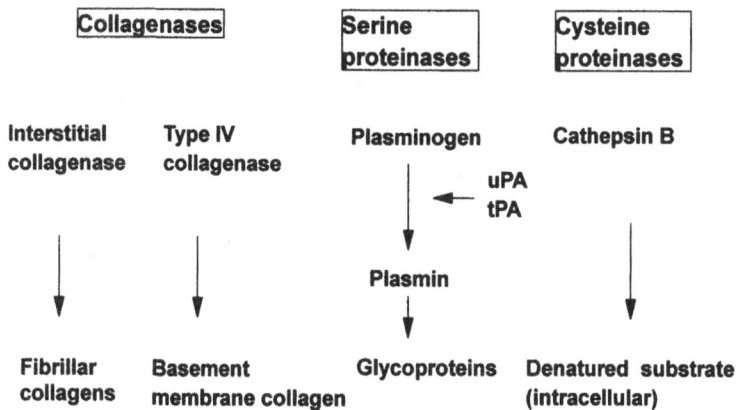

Figure 1. Concept on the role of proteases in tumor invasion that prevailed in the early eighties. Two major collagen degrading enzymes produced by tumor cells were described, interstitial collagenase and type IV collagenase. Plasmin generated from plasminogen by urokinase-type and tissue-type plasminogen activators produced by tumors cells was known to degrade glycoproteins. Cathepsin B was mainly recognized as a lysosomal cysteine proteinase involved in intracellular protein catabolism, and its role in the degradation of ECM proteins was unclear.

tality in cancer patients. Numerous studies published over the last two decades have pointed to the importance of tumor cell proteases that degrade the extracellular matrix (ECM) in allowing them to proteolytically modify their extracellular environment, to escape the control of the ECM, and to establish distant metastasis. In the early eighties, evidence linking proteases to tumor invasion and metastasis was obtained from experiments demonstrating a direct correlation between the amount of proteases produced by tumor cell lines and their ability to invade *in vitro* and to form lung metastatic colonies in mice when injected intravenously (experimental metastasis) or subcutaneously (spontaneous metastasis)[1,2]. Among the proteases involved in this process were 2 metalloproteases known as interstitial collagenase and type IV collagenase, a group of serine proteases including plasmin and urokinase type and tissue type plasminogen activators (uPA and tPA), and cathepsin B, a lysosomal cysteine proteinase[3,4] (Fig. 1). The prevailing concept was that collagenases were important for the degradation of native cross-linked collagens (fibrillar type I, II, III collagens and basement membrane type IV and V collagens) while serine proteases were important for the degradation of glycoproteins such as laminin and fibronectin. Cathepsin B, being a lysosomal enzyme with optimal activity at acid pH, was considered to play a role in the intracellular processing of proteolytically degraded components of the ECM[3]. Recently, our concept has dramatically expanded as many new proteases involved in cancer progression have been identified and as we became increasingly aware of the complex interactions that exist between the various families of proteases involved in degradation of the ECM. For example, collagenases belong to a family of 14 members known as matrix metalloproteinases (MMPs). These enzymes can be classified in several groups based on some specific structural features[5] (Fig. 2). Matrilysin represents the minimum prototype that is composed of a leader peptide, a prodomain containing a unique Cys residue that covers the active (Zn^{++} binding) site and a catalytic domain including a highly preserved HEXXHXXGXXH sequence that contains a Zn^{++}. A second group consists of 3 interstitial collagenases (MMP-1, MMP-8 and MMP-13), a metalloelastase (MMP-12), and 3 stromelysins (MMP-3, MMP-10 and MMP-11), which all have a hemopexin (or vitronectin)-like domain important for substrate recognition. A third group

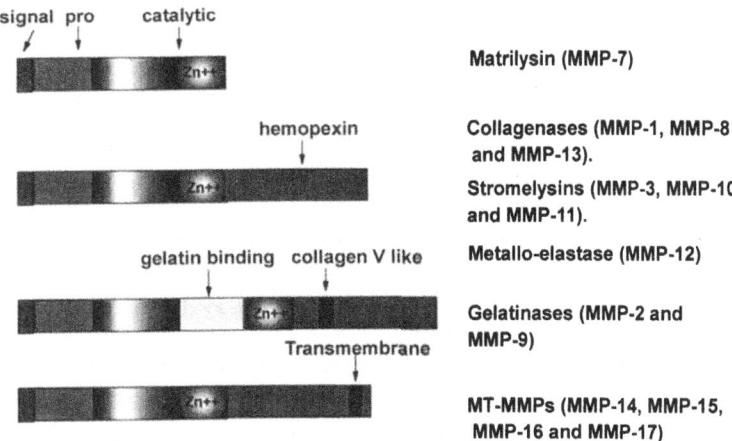

Figure 2. Classification of matrix metalloproteinases based on their structure. Matrilysin (MMP-7) represents the "minimum" enzyme and is composed of a leader peptide, a prodomain that is eliminated after activation and a catalytic domain that includes a Zn^{++} ion stabilized by 3 His residues positioned in a HEXXHXXGXXH motif highly preserved in all MMPs. Collagenases, stromelysins and metalloelastase contain in addition to these 3 domains, a C-terminal hemopexin (or vitronectin)-like domain that plays a role in substrate recognition. Gelatinases (MMP-2 and MMP-9) are characterized by the presence of a fibronectin-like gelatin binding domain located within the catalytic region (MMP-9 contains in addition a collagen V domain present in the hemopexin domain). The hemopexin-like domain of gelatinases is involved in binding TIMPs. MT-MMPs are characterized by the presence of a short transmembrane domain responsible for the localization of these MMPs at the cell surface.

consists of 2 type IV collagenases also known as gelatinases (MMP-2 and MMP-9), which typically contain a fibronectin-like gelatin binding domain. A fourth group is composed of 4 membrane-type MMPs (MT-MMP) that due to a short transmembrane domain are localized at the cell surface where they act as activators of proMMP-2. Many additional cathepsins, members of the serine (cathepsin G), cysteine (cathepsin B, H, L), or aspartyl (cathepsin D) classes of proteases were reported to be expressed in malignancies[6–9]. These lysosomal enzymes are constitutively expressed and secreted by several cells[10], can actively degrade the ECM in the extracellular milieu, and can control cell growth[11] and activate proMMPs and proPA[12,13]. As the cDNAs for these proteases and their inhibitors became available more direct evidence supporting a causal role of proteases in tumor progression were obtained, i.e. by using transfection experiments, it became possible to positively and negatively regulate protease activity in tumor cells and demonstrate a direct effect on malignant progression.

2. INTERACTIONS BETWEEN CLASSES OF PROTEASES

We have become increasingly aware of the complex nature of the interactions between various proteases involved in tumor invasion, and it is now clear that tumor cells rely on more than one protease and often on more than one class of proteases to degrade the ECM. This is not surprising if one considers the tremendous variety and complexity of the ECM that tumor cells will come in contact with during their metastatic journey. Although the details of such interactions are still poorly understood, several specific steps of cooperations between the plasmin/PA system and the MMPs have been identified and are briefly discussed here. Due to differences in substrate specificity, MMPs and plasmin can optimally collaborate to degrade various components of the extracellular matrix. Many MMPs such as

MMP-1 (interstitial collagenase) and MMP-2 (type IV collagenase) preferentially degrade native, cross-linked collagens whereas the substrates for plasmin are fibronectin and laminin. These glycoproteins that cover collagen fibers have been shown to protect collagen from being degraded by collagenase and therefore plasmin, because of its ability to eliminate glycoproteins, will facilitate the subsequent degradation of fibrillar collagen by MMPs. As an example illustrating this point, we have previously demonstrated that in human melanoma cells (M24 met) that secrete both uPA and MMPs[14], the degradation of ECM produced by smooth muscle cells is temporarily regulated. Seeded on these matrices, M24 met cells degraded first the glycoproteins by the plasmin-PA system and thereafter the collagens by the production of MMPs. Accordingly, inhibition of glycoprotein degradation by rPAI-2 also prevented the degradation of collagen.

Activation of proMMP represents a second level of interaction between MMPs and PAs. Plasmin activates proMMP-1 (interstitial collagenase) and proMMP-3 (stromelysin-1) *in vitro* by cleaving in the prodomain a peptide bond located upstream of the Cys residue that protects the active Zn^{++} binding domain of the MMP[15]. This cleavage results in destabilization of the proenzyme followed by autocatalytic cleavage and formation of a stably activated enzyme. Plasmin has been proposed as a physiological activator of proMMP-1 and proMMP-3, however, it appears that it is not a universal activator of proMMPs as initially anticipated[16]. In the case of gelatinase A (MMP-2), the proenzyme is activated by MT-MMPs[17] and plasmin has no activity on either proMMP-2 or the proform of MT-MMP-1[18] (Fig. 3). Consistently, there are no cleavage sequences for serine proteases in the prodomain of MMP-2 and MT-MMP such as those found in the prodomain of MMP-1 and MMP-3. While the prodomain of MMP-2 contains an Asn^{37}–Leu^{38} cleavage site for MT-MMP, the prodomain of proMT-MMP includes a RRKR motif recognized by furins, a family of trans-Golgi serine proteases[19].

A third line of interaction between MMPs and serine proteases exists at the level of their specific inhibitors. We have shown that plasmin and trypsin have proteolytic activity

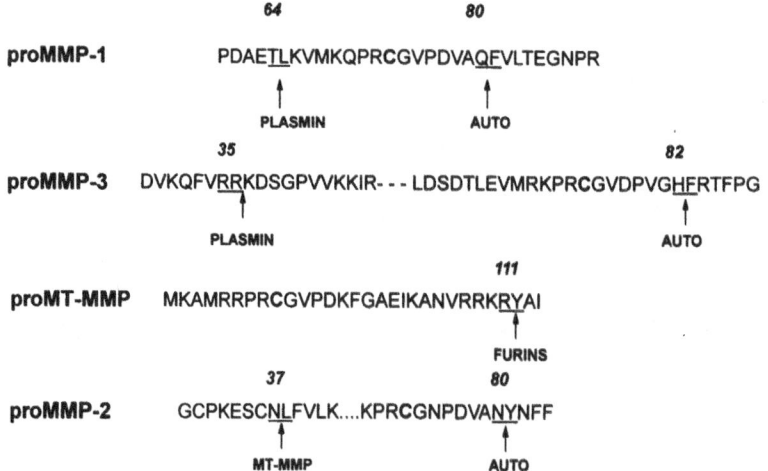

Figure 3. Comparison between the prodomain sequences of MMP-1, MMP-3, MMP-2 and MT-MMP. The sequences of the prodomains are shown with the Cys residue indicated in bold. Peptide bonds for specific cleavage by plasmin, furins, MT-MMP and for autocatalytic cleavage are underlined. Note the presence of specific cleavage sites for plasmin in the prodomains of MMP-1 and MMP-3 but their absence in the prodomain of the 2 other MMPs. Note the RRKR recognition sequence for furins on MT-MMP and the site of cleavage for MT-MMP in the prodomain of MMP-2.

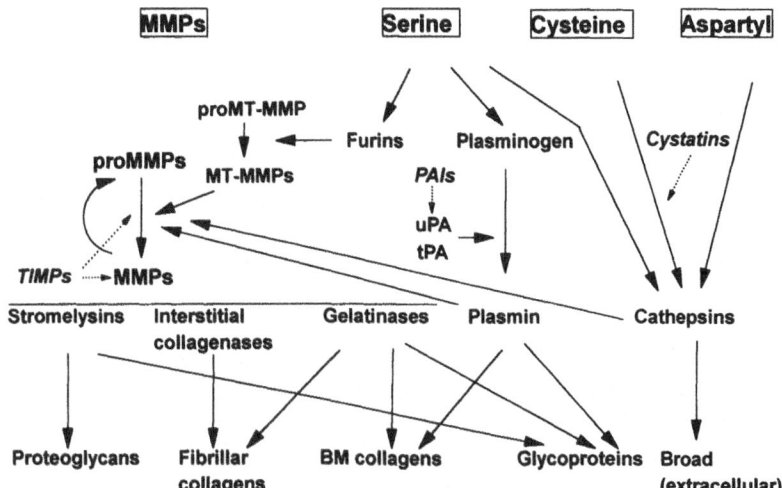

Figure 4. Current concept on the role of proteases in tumor progression. Collagenases have expanded to a family of 14 Matrix Metalloproteinases which degrade a broad spectrum of ECM substrate. Mechanisms of activation of MMPs include MT-MMPs, plasmin, cathepsin B and autocatalysis. Serine proteases entail the furins which activate MT-MMP and stromelysin 3 and the plasmin/plasminogen activator system. Plasmin has been shown to degrade not only glycoproteins but also type IV collagen. Many cathepsins, members of the serine, cysteine and aspartyl proteases have also been implicated. The spectrum of substrates for these latter enzymes is broad and not restricted to ECM. Specific natural inhibitors of these proteases (in italic) have been well characterized, including TIMPs for MMPs, PAI-1 and PAI-2 for PA and cystatins for cysteine proteinases.

for TIMP-2[20]. These proteases generate an initial cleavage in the C-terminal domain which leaves an intact and active N-terminal domain that over 16 hours, is further degraded to inactive proteolytic fragments. We have consistently observed that addition of plasminogen to the culture medium of tumor cells markedly accelerates the degradation of TIMP-2 and therefore contributes to the enhancement of MMP activity. Similarly, stromelysin-3 has been shown to degrade a breast carcinoma cell-derived serpin[21]. Thus there are many interactions between the various classes of proteases involved in the degradation of the ECM during tumor invasion that contribute to our current concept of the role played by proteases in tumor invasion (Fig. 4).

3. INHIBITORS OF PROTEASES IN TUMOR PROGRESSION

Our understanding of the role of proteases in cancer progression has expanded from initial correlative studies to more direct causal evidence for the involvement of proteases with studies that have examined the effect of specific protease inhibitors on tumor progression. Furthermore, as information on the structure of many of these proteases and their corresponding inhibitors becomes available via x-ray crystallographic and nuclear magnetic resonance studies, it will be possible to design synthetic inhibitors with increased specificity and higher affinity. Many of the lessons we have learned on the role of inhibitors of proteases in cancer have been obtained from studies that focused on natural inhibitors of proteases, in particular tissue inhibitors of metalloproteinases (TIMPs) and inhibitors of plasminogen activators (PAIs).

TIMPs are a family of 3 inhibitors (TIMP-1, 2 and 3)[22], the genes of which are located on different chromosomes (Table 1). Although these 3 inhibitors differ in the size

Table 1. Comparison between the 3 members of the TIMP family

	TIMP-1	TIMP-2	TIMP-3
Chromosome	X	17 q 25	22 q 13.1
mRNA (kB)	0.9	1.2 and 3.8	2.3 and 4.5
Mr (kDa)	28	21.5	21
Solubility	+	+	−
Binding	proMMP-9	proMMP-2	ECM
Growth promotion activity	yes	yes	unknown

of their mRNA, the molecular mass of their protein, and their solubility, they share in common the ability to inhibit all members of the MMP family. In addition, TIMP-1 and TIMP-2 possess a growth potentiating activity. TIMPs inhibit active MMPs by forming tight (K_i in the nM range) irreversible noncovalent complexes. These complexes involve interaction between the N-terminal domain of TIMPs and the catalytic (zinc binding domain) of MMPs. In addition, TIMP-1 and TIMP-2 have been shown to form complexes with proMMP-9 and proMMP-2, respectively[23,24]. These latter complexes involve interaction between the C-terminal domain of the inhibitor and the C-terminal domain of the MMPs and preserve the inhibitory activity of the TIMPs. Although the role of these TIMP-proenzyme complexes is not fully understood, they may play an important role in controlling activation and in the particular case of the TIMP-2/proMMP-2 complex, in promoting the interaction with MT-MMP[25]. TIMPs exert a two step control on MMPs by governing the activation process as well as inhibiting active MMPs[26]. PAIs are members of the serpins family and specifically interact with plasminogen activators. While PAI-1 is a major plasma inhibitor of tPA, both PAI-1 and PAI-2 inhibits uPA[27]. PAIs form SDS-stable 1:1 stoichiometric covalent complexes with PA which, after being formed, are rapidly cleared by internalization of uPA receptors[28].

We have examined the effect of overexpression of these inhibitors on tumor progression, in M24 met human melanoma cells since these cells are dependent on both uPA and MMPs to degrade ECM proteins[14]. M24 met cells were transfected with mammalian expression plasmids containing either a full length TIMP-2 or PAI-2 cDNA. Clones over expressing TIMP-2 or PAI-2 were selected and examined for tumorigenic, invasive and metastatic behavior after subcutaneous injection into CB17 scid/scid mice[29,30]. Overexpression of TIMP-2 had a significant inhibitory effect on the growth of the primary tumors but no influence on the incidence of spontaneous metastases to lymph nodes and lungs, nor on the number of lung foci. In contrast, overexpression of PAI-2 in these cells had no effect on primary tumor growth. However, clones that secreted the highest amount of PAI-2 showed a marked inhibition of the formation of spontaneous metastatic nodules. Thus in these cells, which depend on the PA/plasmin system to degrade glycoproteins and on MMPs to degrade collagen, the effect of alteration of the proteases/protease inhibitors balance on tumor progression was a function of the family of proteases affected. These observations also indicate that inhibitors of proteases such as TIMP-2 inhibit tumor growth. A similar growth inhibitory effect was in fact documented with TIMP-1 in murine B16 melanoma cells[31,32] and with a synthetic inhibitor of MMPs (BB94) in ovarian and colon cancer cells in mice[33,34].

Inhibitors of proteases have been proposed as cytostatic agents in the treatment of human cancer and clinical trials with BB94 were recently initiated[35]. Considering the important role played by proteases in many physiological processes, systemic administration of synthetic inhibitors is likely to be associated with side effects. Therefore, alteration of the expression of natural inhibitors of proteases by genetic manipulation (gene therapy) in tumor tissue, represents an attractive alternative that may be devoid of systemic effect. Furthermore, gene targeting of protease inhibitors to every single malignant cell may not be required assuming that these inhibitors exert their action in the ECM. We recently tested this hypothesis by using a murine Moloney based retroviral vector containing the TIMP-2 cDNA as a vehicle to target TIMP-2 to tumor cells *in vivo*[36]. In these experiments, tumor cells were co-injected *in vivo* with mouse fibroblasts that produce retroviral particles containing the TIMP-2 retroviral vector. Our results indicated that overexpression of TIMP-2 in a limited (less than 13%) number of tumor cells by retroviral mediated gene transfer, was sufficient to prevent local invasion and to limit tumor growth. Thus, if tumor specific and highly efficient vectors can be developed, this approach may be valuable in limiting growth and local invasion of primary tumors as well as established metastatic lesions.

It has also become increasingly apparent that inhibitors of proteases may have various functions. For example, TIMP-1 and TIMP-2 have been shown to promote the growth of a variety of normal and malignant cells including erythroid precursor cells[37–40]. Intracellular PAI-2 was reported to prevent apoptosis[41] and interaction between uPA and PAI at the cell surface regulates cell adhesion[42,43]. Whether these effects may counteract the inhibition of the degradation of ECM proteins during invasion and metastasis remains to be determined. It is likely that they will vary in function of several factors such as the presence of protease/protease inhibitor receptors, the secretory pattern of the inhibitor, and the nature of the pericellular matrix. To better understand these effects has become the focus of many investigations.

4. CONCLUSION

It has become clear that the interactions that exist between the various proteases that degrade the ECM and their inhibitors are much more complex than initially anticipated and that tumors cells rely on more than one class of proteases to invade and metastasize. Furthermore, proteases play an active role in tumor growth and, therefore, the concept of using inhibitors of extracellular proteases as cytostatic agents has emerged over the last few years[44]. In view of the data discussed in this manuscript, it is likely that the type of inhibitors used will vary depending on the cancer cells targeted and that inhibition of multiple classes of proteases will be required.

ACKNOWLEDGMENTS

This work was supported in part by grant CA42919 from the National Institutes of Health and grant BE84 from the American Cancer Society to YAD, by grant CA42508 from the National Institutes of Health to RAR, and by grant CA59692 from the National Institutes of Health to BMM. The authors wish to thank J. Rosenberg for typing the manuscript.

REFERENCES

1. Liotta LA, Tryggvason K, Garbisa S, Hart I, Foltz CM, Shafie S (1980) Metastatic potential correlates with enzymatic degradation of basement membrane collagen. Nature 284: 67–68.
2. Liotta LA, Stetler-Stevenson WG (1990) Metalloproteinases and cancer invasion. Semin Cancer Biol 1: 99–106.

3. Sloane BF (1990) Cathepsin B and cystatins: evidence for a role in cancer progression. Semin Cancer Biol 1: 137–152.

4. DeClerck YA, Laug WE (1993) The role of the extracellular matrix in tumor invasion, metastasis and angiogenesis. In: Teicher BA (ed) Drug Resistance in Oncology. Marcel Dekker, NY, pp 121–163.

5. Birkedal-Hansen H, Moore WG, Bodden MK, Windsor LJ, Birkedal-Hansen B, DeCarlo A, Engler JA (1993) Matrix metalloproteinases: a review. Crit Rev Oral Biol Med 4: 197–250.

6. Chambers AF, Colella R, Denhardt DT, Wilson SM (1992) Increased expression of cathepsins L and B and decreased activity of their inhibitors in metastatic, ras-transformed NIH 3T3 cells. Mol Carcinog 5: 238–245.

7. Gabrijelcic D, Svetic B, Spaic D, Skrk J, Budihna M, Dolenc I, Popovic T, Cotic V, Turk V (1992) Cathepsins B, H and L in human breast carcinoma. Eur J Clin Chem Clin Biochem 30: 69–74.

8. Hahnel R, Harvey J, Robbins P, Sterrett G (1993) Cathepsin-D in human breast cancer: correlation with vascular invasion and other clinical and histopathological characteristics. Anticancer Res 13: 2131–2135.

9. Brouillet JP, Spyratos F, Hacene K, Fauque J, Freiss G, Dupont F, Maudelonde T, Rochefort H (1993) Immunoradiometric assay of pro-cathepsin D in breast cancer cytosol: relative prognostic value versus total cathepsin D. Eur J Cancer 29A: 1248–1251.

10. Castiglioni T, Merino MJ, Elsner B, Lah TT, Sloane BF, Emmert Buck MR (1994) Immunohistochemical analysis of cathepsins D, B, and L in human breast cancer [see comments]. Hum Pathol 25: 857–862.

11. Briozzo P, Morisset M, Capony F, Rougeot C, Rochefort H (1988) In vitro degradation of extracellular matrix with Mr 52,000 cathepsin D secreted by breast cancer cells. Cancer Res 48: 3688–3692.

12. Murphy G, Ward R, Gavrilovic J, Atkinson S (1992) Physiological mechanisms for metalloproteinase activation. Matrix Suppl 1: 224–230.

13. Kobayashi H, Moniwa N, Sugimura M, Shinohara H, Ohi H, Terao T (1993) Effects of membrane-associated cathepsin B on the activation of receptor-bound prourokinase and subsequent invasion of reconstituted basement membranes. Biochim Biophys Acta 1178: 55–62.

14. Montgomery AM, DeClerck YA, Langley KE, Reisfeld RA, Mueller BM (1993) Melanoma-mediated dissolution of extracellular matrix: contribution of urokinase-dependent and metalloproteinase-dependent proteolytic pathways. Cancer Res 53: 693–700.

15. Suzuki K, Lees M, Newlands GFJ, Nagase H, Woolley DE (1995) Activation of precursors for matrix metalloproteinases 1 (interstitial collagenase) and 3 (stromelysin) by rat mast-cell proteinases I and II. Biochem J 305: 301–306.

16. Vassalli JD, Pepper MS (1994) Tumour biology. Membrane proteases in focus [news; comment]. Nature 370: 14–15.

17. Pei D, Weiss SJ (1996) Transmembrane-deletion mutants of the membrane-type matrix metalloproteinase-1 process progelatinase A and express intrinsic matrix-degrading activity. J Biol Chem 271: 9135–9140.

18. Lim YT, Sugiura Y, Laug WE, Sun B, Garcia A, DeClerck YA (1996) Independent regulation of matrix metalloproteinases and plasminogen activators in human fibrosarcoma cells. J Cell Physiol 167: 333–340.

19. Pei D, Weiss SJ (1995) Furin-dependent intracellular activation of the human stromelysin-3 zymogen. Nature 375: 244–247.

20. DeClerck YA, Yean TD, Lee Y, Tomich JM, Langley KE (1993) Characterization of the functional domain of tissue inhibitor of metalloproteinases-2 (TIMP-2). Biochem J 289: 65–69.

21. Pei D, Majmudar G, Weiss SJ (1994) Hydrolytic inactivation of a breast carcinoma cell-derived serpin by human stromelysin-3. J Biol Chem 269: 25849–25855.

22. Denhardt DT, Feng B, Edwards DR, Cocuzzi ET, Malyankar UM (1993) Tissue inhibitor of metalloproteinases (TIMP, aka EPA): structure, control of expression and biological functions. Pharmacol Ther 59: 329–341.

23. Goldberg GI, Marmer BL, Grant GA, Eisen AZ, Wilhelm S, He CS (1989) Human 72-kilodalton type IV collagenase forms a complex with a tissue inhibitor of metalloproteases designated TIMP-2. Proc Natl Acad Sci U S A 86: 8207–8211.

24. Goldberg GI, Strongin A, Collier IE, Genrich LT, Marmer BL (1992) Interaction of 92-kDa type IV collagenase with the tissue inhibitor of metalloproteinases prevents dimerization, complex formation with interstitial collagenase, and activation of the proenzyme with stromelysin. J Biol Chem 267: 4583–4591.

25. Strongin AY, Collier I, Bannikov G, Marmer BL, Grant GA, Goldberg GI (1995) Mechanism of cell surface activation of 72-kDa type IV collagenase.Isolation of the activated form of the membrane metalloprotease. J Biol Chem 270: 5331–5338.

26. DeClerck YA, Yean TD, Lu HS, Ting J, Langley KE (1991) Inhibition of autoproteolytic activation of interstitial procollagenase by recombinant metalloproteinase inhibitor MI/TIMP-2. J Biol Chem 266: 3893–3899.

27. Kruithof EK, Baker MS, Bunn CL (1995) Biological and clinical aspects of plasminogen activator inhibitor type 2. Blood 86: 4007–4024.

28. Conese M, Blasi F (1995) Urokinase/urokinase receptor system: internalization/degradation of urokinase-serpin complexes: mechanism and regulation. Biol Chem Hoppe Seyler 376: 143–155.

29. Montgomery AM, Mueller BM, Reisfeld RA, Taylor SM, DeClerck YA (1994) Effect of tissue inhibitor of the matrix metalloproteinases-2 expression on the growth and spontaneous metastasis of a human melanoma cell line. Cancer Res 54: 5467–5473.

30. Mueller BM, Yu YB, Laug WE (1994) Overexpression of plasminogen activator inhibitor 2 in human melanoma cells inhibits spontaneous metastasis in scid mice. Proc Natl Acad Sci USA: 205–209.

31. Koop S, Khokha R, Schmidt EE, MacDonald IC, Morris VL, Chambers AF, Groom AC (1994) Overexpression of metalloproteinase inhibitor in B16F10 cells does not affect extravasation but reduces tumor growth. Cancer Res 54: 4791–4797.

32. Khokha R (1994) Suppression of the tumorigenic and metastatic abilities of murine B16-F10 melanoma cells in vivo by the overexpression of the tissue inhibitor of the metalloproteinases-1. J Natl Cancer Inst 86: 299–304.

33. Davies B, Brown PD, East N, Crimmin MJ, Balkwill FK (1993) A synthetic matrix metalloproteinase inhibitor decreases tumor burden and prolongs survival of mice bearing human ovarian carcinoma xenografts. Cancer Res 53: 2087–2091.

34. Wang X, Fu X, Brown PD, Crimmin MJ, Hoffman RM (1994) Matrix metalloproteinase inhibitor BB-94 (Batimastat) inhibits human colon tumor growth and spread in a patient-like orthotopic model in nude mice. Cancer Res 54: 4726–4728.

35. Brown PD (1994) Preclinical and clinical studies on the matrix metalloproteinase inhibitor, batimastat (BB-94). Clin Experimental Metastasis 12: 23(Abstract).

36. Imren S, Kohn DB, Shimada H, Blavier L, DeClerck YA (1996) Overexpression of tissue inhibitor of metalloproteinases-2 in vivo by retroviral mediated gene transfer inhibits tumor growth and invasion. Cancer Res 56: 2891–2895.

37. Hayakawa T, Yamashita K, Tanzawa K, Uchijima E, Iwata K (1992) Growth-promoting activity of tissue inhibitor of metalloproteinases-1 (TIMP-1) for a wide range of cells. A possible new growth factor in serum. FEBS Lett 298: 29–32.

38. Hayakawa T, Yamashita K, Ohuchi E, Shinagawa A (1994) Cell growth-promoting activity of tissue inhibitor of metalloproteinases-2 (TIMP-2). J Cell Sci 107: 2373–2379.

39. Stetler-Stevenson WG, Bersch N, Golde DW (1992) Tissue inhibitor of metalloproteinase-2 (TIMP-2) has erythroid-potentiating activity. FEBS Lett 296: 231–234.

40. Chesler L, Golde DW, Bersch N, Johnson MD (1995) Metalloproteinase inhibition and erythroid potentiation are independent activities of tissue inhibitor of metalloproteinases-1. Blood 86: 4506–4515.

41. Gan H, Newman GW, Remold HG (1995) Plasminogen activator inhibitor type 2 prevents programmed cell death of human macrophages infected with Mycobacterium avium, serovar 4. J Immunol 155: 1304–1315.

42. Wei Y, Waltz DA, Rao N, Drummond RJ, Rosenberg S, Chapman HA (1994) Identification of the urokinase receptor as an adhesion receptor for vitronectin. J Biol Chem 269: 32380–32388.

43. Waltz DA, Sailor LZ, Chapman HA (1993) Cytokines induce urokinase-dependent adhesion of human myeloid cells. A regulatory role for plasminogen activator inhibitors. J Clin Invest 91: 1541–1552.

44. DeClerck YA, Imren S (1994) Protease inhibitors: role and potential therapeutic use in human cancer. Eur J Cancer 30A: 2170–2180.

10

THE ROLE OF REACTIVE-CENTER LOOP MOBILITY IN THE SERPIN INHIBITORY MECHANISM

Daniel A. Lawrence[*]

American Red Cross
Jerome H. Holland Laboratory, Department of Biochemistry
15601 Crabbs Branch Way, Rockville, Maryland 20855

1. INTRODUCTION

The serpins are a gene family that encompasses a wide variety of protein products, including many of the proteinase inhibitors in plasma[1]. However, in spite of their name, not all serpins are proteinase inhibitors. They include steroid binding globulins, the prohormone angiotensinogen, the egg white protein ovalbumin, and barley protein Z, a major constituent of beer. The serpins are thought to share a common tertiary structure[2] and to have evolved from a common ancestor[3]. Proteins with recognizable sequence homology have been identified in vertebrates, plants, insects and viruses but not, thus far, in prokaryotes[1,4,5]. Current models of serpin structure are based largely on seminal X-ray crystallographic studies of one member of the family, α-1-antitrypsin (α1AT), also called α-1-proteinase inhibitor[1]. The structure of a modified form of α1AT, cleaved in its reactive center, was solved by Loebermann and coworkers in 1984[6]. An interesting feature of this structure was that the two residues normally comprising the reactive center (Met-Ser), were found on opposite ends of the molecule, separated by almost 70 Å (Figure 1). Loebermann and coworkers proposed that a relaxation of a strained configuration takes place upon cleavage of the reactive center peptide bond, rather than a major rearrangement of the inhibitor structure. In this model, the native reactive center is part of an exposed loop, also called the strained loop[6-8]. Upon cleavage, this loop moves or "snaps back", becoming one of the central strands in a major β–sheet structure (β-sheet A). This transformation is accompanied by a large increase in thermal stability[9-12].

Recent crystallographic structures of several native serpins, with intact reactive center loops, have confirmed Loebermann's hypothesis that the overall native serpin structure is very similar to cleaved α1AT, but that the reactive center loop is exposed above the plane of the molecule[13-16]. Additional evidence for this model has come from studies

[*] Tel. (301) 517–0356; Fax (301) 738–0794; E-mail: lawrencd@usa.redcross.org.

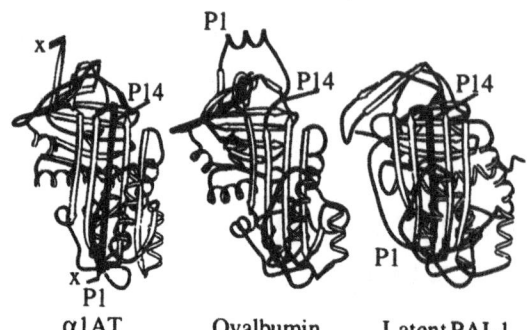

Figure 1. Backbone conformations of cleaved α1AT, native ovalbumin and latent PAI-1 are shown. Residues of the reactive-center loop from P14 to P1 are shown as black ribbons. The approximate positions of the P1 and P1′ residues in α1AT are marked by an x, and the P1 and P14 residues in each structure are indicated (Adapted from S. R. Sprang with permission[8]).

where synthetic peptides, homologous to the reactive center loops of α1AT, antithrombin III (ATIII), or PAI-1 when added in *trans*, incorporate into their respective molecules, presumably as a central strand of β-sheet A[17–21]. This leads to an increase in thermal stability similar to that observed following cleavage of a serpin at its reactive center, and converts the serpin from an inhibitor to a substrate for its target proteinase. A third serpin structural form has also been identified, the so-called latent conformation (Figure 1). In this structure the reactive center loop is intact, but instead of being exposed, the entire amino-terminal side of the reactive center loop is inserted as the central strand into β–sheet A[22]. This accounts for the increased stability of latent PAI-1[23] as well as its lack of inhibitory activity[24]. The ability to adopt this conformation is not unique to PAI-1, but has also now been shown for ATIII and α1AT[14,25]. Together, these data have led to the hypothesis that active serpins have mobile reactive center loops, and that this mobility is essential for inhibitor function[12,20,26–31]. The large increase in thermal stability observed with loop insertion, is presumably due to reorganization of the five stranded β-sheet A from a mixed parallel-antiparallel arrangement to a six stranded, predominantly antiparallel β-sheet[9–11,23]. This dramatic stabilization has led to the suggestion that native inhibitory serpins may be metastable structures, kinetically trapped in a state of higher free energy than their most stable thermodynamic state[29,32]. Such an energetically unfavorable structure would almost certainly be subject to negative selection, and thus its retention in all inhibitory serpins implies that it has been conserved for functional reasons.

The serpins act as "suicide inhibitors" that react only once with a target proteinase forming an SDS-stable complex. They interact by presenting a "bait" amino acid residue, in their reactive center, to the enzyme. This bait residue is thought to mimic the normal substrate of the enzyme and to associate with the specificity crevice, or S1 site, of the enzyme[1,7,33]. The bait amino acid is called the P1 residue, with the amino acids toward the N-terminal side of the scissile reactive center bond labeled in order P1 P2 P3 etc. and the amino acids on the carboxyl side labeled P1′ P2′ etc.[7] The reactive center P1–P1′ residues, appear to play a major role in determining target specificity. This point was dramatically illustrated by the identification of a unique human mutation, α1AT "Pittsburgh", in which a single amino acid substitution of Arg for Met at the P1 residue converted α1AT from an inhibitor of elastase to an efficient inhibitor of thrombin, resulting in a unique and ultimately fatal bleeding disorder[34]. Numerous mutant serpins have been constructed, demonstrating a wide range of changes in target specificity, particularly with substitutions at P1[26,35–38].

The exact structure of the complex between serpins and their target proteinases has been controversial. Originally it was thought that the complex was covalently linked via an ester bond between the active site serine residue of the proteinase and the new carboxyl-terminal end of the P1 residue, forming an acyl-enzyme complex[39–42]. However, in the late 1980s and early 1990s it was suggested that this interpretation was incorrect, and that the serpin-proteinase complex is instead trapped in a tight non-covalent association similar to the so called standard mechanism inhibitors of the Kazal and Kunitz family[43–45]. Alternatively, one study suggested a hybrid of these two models where the complex was frozen in a covalent but un-cleaved tetrahedral transition state configuration[46]. Recently however, new data by several groups have suggested that the debate has come full circle, with various studies using independent methods indicating that the inhibitor is indeed cleaved in its reactive-center and that the complex is most likely trapped as a covalent acyl-enzyme complex[12,28–31,47,48].

2. FLUORESCENCE ANALYSIS OF SERPIN-PROTEINASE COMPLEXES

Many studies have utilized plasminogen activator inhibitor-1 (PAI-1) as a model for serpin function. PAI-1 can be expressed to high level in *E. coli*[49], and this has facilitated the development of rapid and efficient systems to create and analyze site-directed mutants of PAI-1[23]. Recently two independent groups, Dr. Shore's[28], and Dr. Ny's[30], have utilized this capability to develop systems to monitor the local state of the reactive-center loop (RCL) in complex with a proteinase. Both laboratories have taken advantage of the fact that PAI-1 has no Cys residues. By placing Cys mutations at any desired location in the PAI-1 molecule it becomes possible to tag this position with environmentally responsive probes. In one study, PAI-1 mutants containing a substitution of Cys at either the P9 or the P1′ position of the RCL were constructed, and the free sulfhydryl of each Cys side chain labeled with the environmentally sensitive probe N-((2-(iodoacetoxy)-ethyl)-N-methyl)-amino-7-nitrobenz-2-oxal, 3-diazole (NBD). This probe shows a large change in fluorescence depending upon the hydrophobicity of its environment. The P9 position was chosen because in a model of active PAI-1[50], the P9 side-chain was predicted to be fully exposed to the solvent, while, in cleaved[51] or latent PAI-1[22], where the reactive-center loop is fully inserted into β-sheet A, the P9 side-chain is shielded from the solvent by α-helix H and the adjoining extended loop. The purified, NBD labeled mutants were fully functional with inhibitory activities against tPA and uPA nearly identical to wtPAI-1. For the P9-NBD variant, conversion to the latent structure, cleavage of the RCL by a non-target proteinase, or complex formation with either uPA or tPA resulted in an approximate 6-fold increase in fluorescence in each case[28]. These data indicate that the NBD-labeled P9-PAI-1 provides an excellent reporter for the local environment of the P9 residue. The nearly identical change in fluorescence upon complex formation with uPA and tPA to that observed upon conversion to the latent state, or cleavage by a non-target proteinase suggests that in the stable complex the RCL is also inserted into β-sheet A. The observed rate of loop insertion with tPA was also relatively fast with a limiting rate for this reaction of ~ 4 s^{-1}. In contrast, reaction of the P1′-NBD PAI-1 with tPA resulted in a 30% reduction in relative fluorescence occurring at approximately the same rate[29]. This quench indicates that unlike the P9 position of the RCL, the P1′ side-chain is exposed to a more hydrophilic environment upon reaction with tPA. Although such a shift in position could result from a minor conformational change in the RCL, this explanation seems unlikely given the close association of the serpin and proteinase via the directly adjacent P1 Arg residue of PAI-1 and the S1 sub-site of tPA. Alternatively, cleavage of the PAI-1 RCL by tPA between the P1 and P1′ residues could permit the NBD reporter group to move away from the enzyme into a more aqueous environment. Consistent with this hypothesis reaction of the P9-NBD-PAI-1 with catalytically inactive anhydrotrypisn resulted in only a minor increase in fluorescence, indicating that an active catalytic mechanism is necessary to induce RCL insertion[31]. A second independent study utilizing Cys mutations at the P3 or P1' positions, and with a different environmentally sensitive probe (BDYIA)[30] yielded similar conclusions. This latter study indicated that when in complex with a proteinase the rotational freedom of the P3 probe decreased while that of the P1' probe increased[30]. The increased freedom of the P1' probe was similar to that seen with RCL cleaved PAI-1 suggesting, as above, that in the stable complex the P1–P1' bond is cleaved[30]. Furthermore, complex formation with inactive anhydrourokinase resulted in a decrease in the rotational freedom of the P1' probe[30] indicating that as with the P9-NBD PAI-1 in complex with anhydrotrypsin[31], an active catalytic mechanism is necessary to induce the observed conformational change associated with a stable complex.

3. CLEAVAGE OF THE SERPIN RCL IN THE COMPLEX WITH A PROTEINASE

To distinguish between cleavage of the P1–P1' peptide bond versus a simple conformational change in the intact RCL, the PAI-1–PA complex was reacted with limiting amounts of the amino specific ^{125}I Bolton-Hunter reagent (^{125}I-BH)[52]. Two different plasminogen activators were used, tPA and uPA. These experiments were performed with trace amounts of ^{125}I-BH under nondenaturing conditions, followed by treatment of the unreacted label with glycine, and removal prior to SDS-PAGE analysis. This procedure should report the presence of a cleaved RCL in the complex by the appearance of a novel labeled peptide fragment of the correct size. Furthermore, since the unreacted Bolton-Hunter reagent was blocked with glycine and removed before the samples were denatured by exposure to SDS, any labeled peptide must have been formed while the complexes were in their native state. Although all accessible amines could potentially be labeled, including ε-NH_3^+ groups of internal Lys residues, the only position that can incorporate label in the PAI-1 RCL C-terminal peptide would be its amino-terminus, since PAI-1 contains no Lys residues in the 33 residue peptide produced by cleavage of the P1–P1' bond. SDS-PAGE analysis of the labeled complexes shown in figure 2, demonstrated a unique ~3.0 kDa band with both PAI-1–PA complexes which was not present with PAI-1, or either PA alone (panel B). The observed mobility of this novel peptide is in agreement with the predicted molecular mass (3.8 kDa) of the PAI-1 C-terminal peptide. As a positive control for cleavage of the RCL, and the labeling efficiency of the C-terminal peptide, a mutant PAI-1 that was previously shown to be a pure substrate for plasminogen activators, and is completely cleaved at its RCL P1–P1' peptide bond was also tested[12]. Consistent with previous observations, the mutant PAI-1 fails to form stable complexes with either PA and is instead completely cleaved (compare figure 2, panel A, lanes 2 & 3 with lanes 7 & 8). Furthermore, a labeled peptide identical to that observed with wtPAI-1–PA complexes is also seen (compare figure 2, panel B, lanes 2 & 3 with lanes 7 & 8). With both the wtPAI-1 and the mutant the extent of labeling observed in the ~3 kDa peptide with uPA is less than that seen with tPA. This is consistent with the limiting amounts of ^{125}I-BH employed. In each sample the concentration of ^{125}I-BH utilized was 100 nM, compared to 1 μM each for the PAI-1–enzyme complex. This 1:10 ratio means that that in each complex there will be a competition for binding the ^{125}I-BH. If the ε-NH_3^+ groups of internal Lys residues are also considered then this results in an approximate 500-fold excess of potential labeling sites relative to the ^{125}I-BH. Thus, it is not surprising that the newly generated amino-termini in

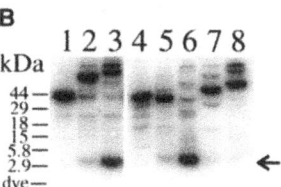

Figure 2. SDS-PAGE analysis of [125]I Bolton-Hunter reagent treated PAI-1 ± uPA or tPA. Panel A: 20% homogeneous gel stained with Coomassie blue; lane 1 wtPAI-1 only; lane 2 wtPAI-1 + uPA; lane 3, wtPAI-1 + tPA; lane 4, uPA only; lane 5, tPA only; lane 6, substrate PAI-1 only; lane 7, substrate PAI-1 + uPA; lane 8, substrate PAI-1 + tPA. Panel B: autoradiography of the gel in panel A. The numbers at the left indicate the position of molecular mass standard proteins, and the arrow marks the position of the labeled C-terminal peptide. The experiments were performed as described[29]. Reprinted from [29] with permission.

each complex label with different efficiency. Indeed, the nearly identical efficiency of peptide labeling for the mutant PAI-1 cleaved by either uPA or tPA compared to wtPAI-1 in association with the different enzymes indicates that the RCL within the stable serpin-proteinase complex is cleaved, and suggests that this cleavage is as complete in the stable complex as it is in the substrate mutant.

4. QUANTITATION OF FREE AMINO-TERMINAL RESIDUES

To establish that cleavage of the RCL in the PAI-1–PA complex is complete, and exclude substrate behavior by a subset of the inhibitor molecules, the extent of RCL cleavage was directly quantitated by microsequencing of the PAI-1–uPA complex. Since complex denaturation during the sequencing reaction could potentially induce cleavage, the extent of cleavage in native complexes was determined by a subtractive method. PAI-1–uPA complexes were first reacted with the amino-specific reagent N-hydroxysuccinimide acetic acid (NHS) under nondenaturing, physiological conditions. This compound is similar to Bolton-Hunter reagent in its reactivity and covalently binds to free amines. Treatment of PAI-1–uPA complexes with NHS should therefore block available amino-termini in a dose dependent manner. The excess NHS was then reacted with Tris and removed by ultrafiltration prior to direct amino-terminal sequence analysis of remaining unreacted amino-termini. This analysis is quantitative and the relative reactivity of natural amino-termini from both the inhibitor and proteinase serve as internal controls for NHS reactivity and sequencing efficiency. The results of this analysis are shown in figure 3A. The yield of RCL peptide amino-terminus at each dose of NHS is very similar to those of the natural PAI-1 and uPA amino-termini. Thus the RCL peptide amino-terminus generated upon complex formation is as reactive as the natural amino-termini and therefore is very likely to be fully cleaved and exposed, consistent with the ^{125}I labeling results. Interestingly, the least NHS reactive amino-terminus tested is that of the uPA proteinase domain (uPA PD, Figure 3A), a relatively hydrophobic sequence Ile-Ile-Gly-Gly which is likely to be oriented toward the interior of the molecule[53]. The latter observation suggests that this approach is quite sensitive to amino-termini solvent accessibility, further supporting the conclusion that the PAI-1 RCL must be completely cleaved when in complex with uPA. Similar results were also obtained with PAI-1–tPA complexes, and these data are shown in figure 3B.

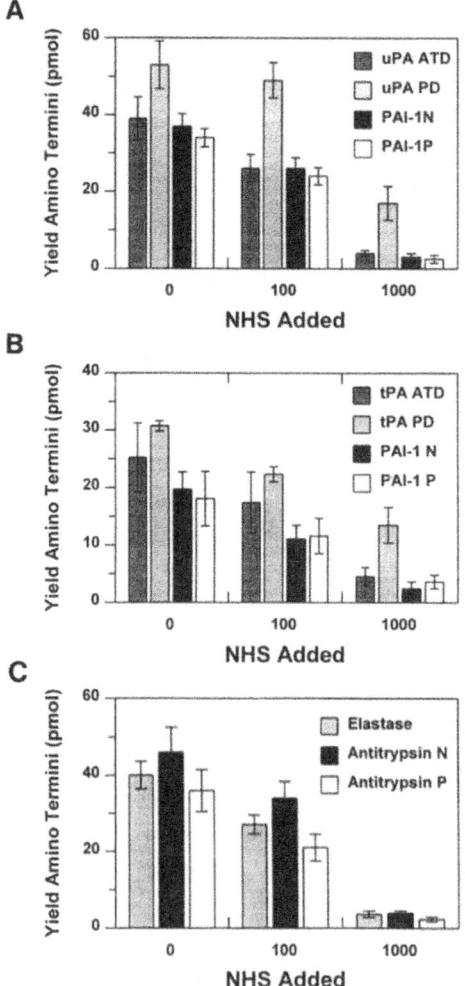

Figure 3. Amino-terminal sequence analysis of PAI-1–uPA complexes ± pretreatment with NHS. Amino-termini yields are given in pmol and represent the average yield of the first 5 residues in each sequence. The NHS concentration in the pretreatment step is shown on the x axis. ATD (dark gray bars), enzyme amino-terminal domain; PD (light gray bars), enzyme proteinase domain; N (black bars), the natural amino-terminus of the serpin; P (white bars), the newly generated serpin carboxyl-terminal peptide beginning with the P1' residue. Panel A, complex between PAI-1 and uPA; panel B, complex between PAI-1 and tPA; panel C complex between elastase and α1AT. The experiments were performed as described[29]. The data in Panels A and C are taken from[29] with permission.

To test the general relevance of these observations for other serpin-proteinase complexes, the α1AT-elastase complex was also treated with NHS and subjected to microsequencing. The results shown in figure 3C are similar to the data obtained with the PAI-1–PA complexes, demonstrating that α1AT is also cleaved in its RCL when in complex with elastase. Taken together with the known stability of these complexes to SDS-PAGE, these observations strongly suggest that the serpin-proteinase complex is trapped in the form of a covalent acyl-enzyme intermediate. Similar results were also recently described by Wilczynska et al.[47] using a different amino-terminal blocking agent. In addition, these authors also examined the complex between plasmin and α2-antiplasmin and demonstrated that this complex too is cleaved at its P1–P1' bond. Previous studies had suggested that the serpin-proteinase complex was reversible[43–46]. However, it is unlikely that a serpin cleaved in its reactive-center could be reversible. Since serpin inhibition appears to be a two step process with an initial reversible encounter complex followed by formation of an apparently irreversible stable complex[28,31,54], it is likely that the previous studies suggesting complex reversibility were observing this first step. Alternatively, it is known that even the stable serpin complex slowly turns over yielding free active enzyme and cleaved inhibitor.

It is also possible that in other studies it was this process that was observed. It has also been suggested that the complex is a stable tetrahedral-intermediate[46]. Peptide bond cleavage by serine proteinases is known to proceed through two tetrahedral-intermediates[55]. Although these data rule out a complex frozen at the point of the first tetrahedral intermediate, prior to RCL cleavage, it is possible that the serpin-proteinase complex is frozen at the point of the second tetrahedral, following RCL cleavage, or that it is trapped in a distorted conformation intermediate between these two forms.

5. A MODEL OF SERPIN FUNCTION

Recently, three groups have almost simultaneously proposed similar mechanisms for serpin inhibition[29,47,56], and a schematic representation by Wright and Scarsdale of this so-called branched pathway mechanism is shown in figure 4. This model suggests that upon encountering a target proteinase, a serpin binds to the enzyme forming a reversible complex that is similar to a Michaelis complex between an enzyme and substrate. Next, the proteinase cleaves the P1–P1' peptide bond resulting in formation of a covalent acyl-enzyme intermediate. This cleavage is coupled to a rapid insertion of the RCL into β-sheet A at least up to the P9 position. Since the RCL is covalently linked to the enzyme via the active-site Ser, this transition should also affect the proteinase, significantly changing its position relative to the inhibitor. If, during this transition, the RCL is prevented from attaining full inser-

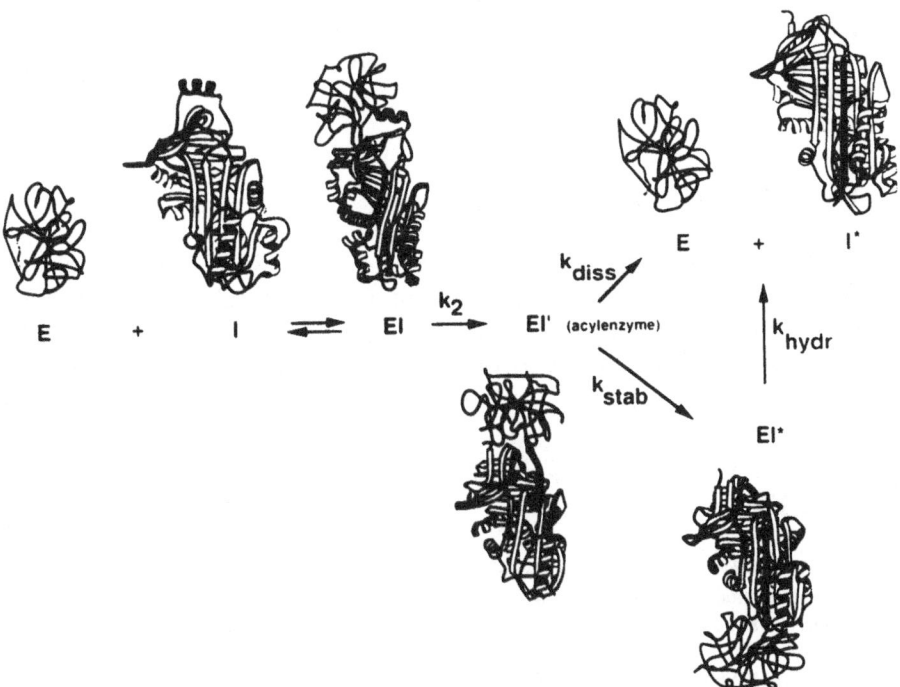

Figure 4. Schematic representation of the serpin mechanism by Wright and Scarsdale. E, free enzyme; I, active serpin inhibitor; EI, Michaelis-like encounter complex; EI', first covalent acyl-intermediate prior to loop insertion; EI*, RCL inserted stable inhibited complex; I*, cleaved inactive serpin following turnover and release from the enzyme. It should be noted that the exact position of the enzyme relative to the inhibitor can not be determined. Reprinted from[56] with permission.

tion because of its association with the enzyme, and the complex becomes locked, with the RCL only partially inserted, then the resulting stress might be sufficient to distort the active site of the enzyme. This distortion would then prevent efficient deacylation of the acyl-enzyme intermediate, thus trapping the complex. However, if RCL insertion is prevented, or if deacylation occurs before RCL insertion then the cleaved serpin is turned over as a substrate and the active enzyme released. This means that what determines whether a serpin is an inhibitor or a substrate is the ratio of k_{diss} to k_{stab}. If deacylation (k_{diss}) is faster than RCL insertion (k_{stab}) then the substrate reaction predominates. However, if RCL insertion and distortion of the active site can occur before deacylation then the complex is frozen as a covalent acyl-enzyme. A similar model was first proposed in 1990[26] and is consistent with studies demonstrating that RCL insertion is not required for proteinase binding but is necessary for stable inhibition[12] as well as the observation that only an active enzyme can induce RCL insertion[31]. Very recently, direct evidence for this model was provided by Plotnick *et al.*, who by NMR observed an apparent distortion of an enzyme's catalytic site in a serpin-enzyme complex[48]. In conclusion, these data suggest that serpins act as molecular springs where the native structure is kinetically trapped in a high energy state. Upon association with an enzyme some of the energy liberated by RCL insertion is used to distort the active site of the enzyme, preventing deacylation and trapping the complex.

ABBREVIATIONS

Reactive center loop, RCL; α1-antitrypsin, α1AT; plasminogen activator inhibitor 1, PAI-1; N-((2-(iodoacetoxy)-ethyl)-N-methyl)-amino-7-nitrobenz-2-oxal, 3-diazole, NBD; urokinase, uPA; tissue-type plasminogen activator, tPA; ^{125}I Bolton-Hunter reagent, ^{125}I-BH; N-hyroxysuccinimide acetic acid, NHS.

ACKNOWLEDGMENTS

I want to thank Dr. Church for organizing an outstanding symposium, S. Muhammad for excellent technical assistance, and M. Sandkvist, D. Ginsburg, J. D. Shore, K. Ingham and S. T. Olson for helpful discussions. Portions of this manuscript have been adapted from[29,57]. This work was supported in part by the National Institutes of Health grant HL 55374.

REFERENCES

1. Huber R, Carrell RW (1989): Implications of the three-dimensional structure of alpha 1-antitrypsin for structure and function of serpins. Biochem 28:8951–8966.
2. Doolittle RF (1983): Angiotensinogen is related to the antitrypsin-antithrombin-ovalbumin family. Science 222:417–419.
3. Hunt LT, Dayhoff MO (1980): A surprising new protein superfamily containing ovalbumin, antithrombin III, and alpha1-proteinase inhibitor. Biochem Biophys Res Commun 95:864–871.
4. Sasaki T (1991): Patchwork-structure serpins from silkworm (Bombyx mori) larval hemolymph. Eur J Biochem 202:255–261.
5. Komiyama T, Ray CA, Pickup DJ, Howard AD, Thornberry NA, Peterson EP, Salvesen G (1994): Inhibition of interleukin-1β converting enzyme by the cowpox virus serpin CrmA. An example of cross-class inhibition. J Biol Chem 269:19331–19337.

6. Loebermann H, Tokuoka R, Deisenhofer J, Huber R (1984): Human α_1-proteinase inhibitor. Crystal structure analysis of two crystal modifications, molecular model and preliminary analysis of the implications for function. J Mol Biol 177:531–557.

7. Carrell RW, Boswell DR (1986): Serpins: the superfamily of plasma serine proteinase inhibitors. In: Proteinase Inhibitors, AJ Barrett, G Salvesen, eds. Elsevier Science Publishers (Biomedical Division), Amsterdam, pp. 403–420.

8. Sprang SR (1992): The latent tendencies of PAI-1. Trends Biochem Sci 17:49–50.

9. Carrell RW, Owen MC (1985): Plakalbumin, alpha-1-antitrypsin, antithrombin and the mechanism of inflammatory thrombosis. Nature 317:730–732.

10. Gettins P, Harten B (1988): Properties of thrombin- and elastase-modified human antithrombin III. Biochem 27:3634–3639.

11. Bruch M, Weiss V, Engel J (1988): Plasma serine proteinase inhibitors (serpins) exhibit major conformational changes and a large increase in conformational stability upon cleavage at their reactive sites. J Biol Chem 263:16626–16630.

12. Lawrence DA, Olson ST, Palaniappan S, Ginsburg D (1994): Serpin reactive-center loop mobility is required for inhibitor function but not for enzyme recognition. J Biol Chem 269:27657–27662.

13. Schreuder HA, de Boer B, Dijkema R, Mulders J, Theunissen HJM, Grootenhuis PDJ, Hol WGJ (1994): The intact and cleaved human antithrombin III complex as a model for serpin-proteinase interactions. Nature Structural Biology 1:48–54.

14. Carrell RW, Stein PE, Fermi G, Wardell MR (1994): Biological implications of a 3 Å structure of dimeric antithrombin. Structure 2:257–270.

15. Stein PE, Leslie AGW, Finch JT, Turnell WG, McLaughlin PJ, Carrell RW (1990): Crystal structure of ovalbumin as a model for the reactive centre of serpins. Nature 347:99–102.

16. Wei A, Rubin H, Cooperman BS, Christianson DW (1994): Crystal structure of an uncleaved serpin reveals the conformation of an inhibitory reactive loop. Nature Structural Biology 1:251–258.

17. Björk I, Ylinenjärvi K, Olson ST, Bock PE (1992): Conversion of antithrombin from an inhibitor of thrombin to a substrate with reduced heparin affinity and enhanced conformational stability by binding of a tetradecapeptide corresponding to the P_1 to P_{14} region of the putative reactive bond loop of the inhibitor. J Biol Chem 267:1976–1982.

18. Björk I, Nordling K, Larsson I, Olson ST (1992): Kinetic characterization of the substrate reaction between a complex of antithrombin with a synthetic reactive-bond loop tetradecapeptide and four target proteinases of the inhibitor. J Biol Chem 267:19047–19050.

19. Schulze AJ, Baumann U, Knof S, Jaeger E, Huber R, Laurell C (1990): Structural transition of α_1-antitrypsin by a peptide sequentially similar to β-strand s4A. Eur J Biochem 194:51–56.

20. Carrell RW, Evans DL, Stein PE (1991): Mobile reactive centre of serpins and the control of thrombosis. Nature 353:576–578.

21. Kvassman J, Lawrence D, Shore J (1995): The acid stabilization of plasminogen activator inhibitor-1 depends on protonation of a single group that affects loop insertion into β-sheet A. J Biol Chem 270:27942–27947.

22. Mottonen J, Strand A, Symersky J, Sweet RM, Danley DE, Geoghegan KF, Gerard RD, Goldsmith EJ (1992): Structural basis of latency in plasminogen activator inhibitor-1. Nature 355:270–273.

23. Lawrence DA, Olson ST, Palaniappan S, Ginsburg D (1994): Engineering plasminogen activator inhibitor-1 (PAI-1) mutants with increased functional stability. Biochem 33:3643–3648.

24. Hekman CM, Loskutoff DJ (1985): Endothelial cells produce a latent inhibitor of plasminogen activators that can be activated by denaturants. J Biol Chem 260:11581–11587.

25. Lomas DA, Elliot PR, Chang W-SW, Wardell MR, Carrell RW (1995): Preparation and characterization of latent α1-antitrypsin. J Biol Chem 270:5282–5288.

26. Lawrence DA, Strandberg L, Ericson J, Ny T (1990): Structure-function studies of the SERPIN plasminogen activator inhibitor type 1: analysis of chimeric strained loop mutants. J Biol Chem 265:20293–20301.

27. Carrell RW, Evans DLI (1992): Serpins: mobile conformations in a family of proteinase inhibitors. Curr Opin Struct Biol 2:438–446.

28. Shore JD, Day DE, Francis-Chmura AM, Verhamme I, Kvassman J, Lawrence DA, Ginsburg D (1994): A fluorescent probe study of plasminogen activator inhibitor-1: Evidence for reactive center loop insertion and its role in the inhibitory mechanism. J Biol Chem 270:5395–5398.

29. Lawrence DA, Ginsburg D, Day DE, Berkenpas MB, Verhamme IM, Kvassman J-O, Shore JD (1995): Serpin-Protease Complexes are Trapped as Stable Acyl-Enzyme Intermediates. J Biol Chem 270:25309–25312.

30. Fa M, Karolin J, Aleshkov S, Strandberg L, Johansson LB-Å, Ny T (1995): Time-Resolved Polarized Fluorescence Spectroscapy Studies of Plasminogen Activator Inhibitor Type 1: Conformational Changes of the Reactive Center upon Interations with Target proteases, Vitronectin and Heparin. Biochem 34:13833–13840.

31. Olson ST, Bock PE, Kvassman J, Shore JD, Lawrence DA, Ginsburg D, Björl I (1995): Role of the catalytic serine in the interactions of serine proteinases with protein inhibitors of the serpin family. J Biol Chem 270:30007–30017.

32. Lee KN, Park SD, Yu M-H (1996): Probing the native strain in α_1-antitrypsin. Nature Structural Biology 3:497–500.

33. Bode W, Huber R (1994): Proteinase - Protein Inhibitor Interactions. Fibrinolysis 8:161–171.

34. Owen MC, Brennan SO, Lewis JH, Carrell RW (1983): Mutation of antitrypsin to antithrombin: alpha1-antitrypsin Pittsburgh (358 Met-Arg), a fatal bleeding disorder. N Engl J Med 309:694–698.

35. York JD, Li P, Gardell SJ (1991): Combinatorial mutagenesis of the reactive site region in plasminogen activator inhibitor I. J Biol Chem 266:8495–8500.

36. Strandberg L, Lawrence DA, Johansson LB, Ny T (1991): The oxidative inactivation of plasminogen activator inhibitor type 1 results from a conformational change in the molecule and does not require the involvement of the P1' methionine. J Biol Chem 266:13852–13858.

37. Shubeita HE, Cottey TL, Franke AE, Gerard RD (1990): Mutational and immunochemical analysis of plasminogen activator inhibitor 1. J Biol Chem 265:18379–18385.

38. Sherman PM, Lawrence DA, Yang AY, Vandenberg ET, Paielli D, Olson ST, Shore JD, Ginsburg D (1992): Saturation mutagenesis of the plasminogen activator inhibitor-1 reactive center. J Biol Chem 267:7588–7595.

39. Moroi M, Yamasaki M (1974): Mechanism of the interaction of bovine trypsin with human α1-antitrypsin. Biochim Biophys Acta 359:130–141.

40. Owen WG (1975): Evidence for the formation of an ester between thrombin and heparin cofactor. Biochim Biophys Acta 405:380–387.

41. Cohen AB, Gruenke LD, Craig JC, Geczy D (1977): Specific lysine labeling by ^{18}OH- during alkaline cleavage of the α-1-antitrypsin-trypsin complex. Proc Natl Acad Sci USA 74:4311–4314.

42. Nilsson T, Wiman B (1982): On the structure of the stable complex between plasmin and α_2-antiplasmin. FEBS Lett 142:111–114.

43. Longstaff C, Gaffney P, J. (1991): Serpin-serine protease binding kinetics: alpha-2-antiplasmin as a model inhibitor. Biochem 30:979–986.

44. Shieh BH, Potempa J, Travis J (1989): The use of alpha 2-antiplasmin as a model for the demonstration of complex reversibility in serpins. J Biol Chem 264:13420–13423.

45. Potempa J, Korzus E, Travis J (1994): The serpin superfamily of proteinase inhibitors: structure, function, and regulation. J Biol Chem 269:15957–15960.

46. Matheson NR, van Halbeek H, Travis J (1991): Evidence for a tetrahedral intermediate complex during serpin-proteinase interactions. J Biol Chem 266:13489–13491.

47. Wilczynska M, Fa M, Ohlsson P-I, Ny T (1995): The Inhibition Mechanism of Serpins: Evidence that the mobile reactive center loop is cleaved in the native protease-inhibitor complex. J Biol Chem 270:29652–29655.

48. Plotnick MI, Mayne L, Schechter NM, Rubin H (1996): Distortion of the active site of chymotrypsin complexed with a serpin. Biochem 35:7586–7590.

49. Lawrence D, Strandberg L, Grundström T, Ny T (1989): Purification of active human plasminogen activator inhibitor 1 from *Escherichia coli*. Comparison with natural and recombinant forms purified from eucaryotic cells. Eur J Biochem 186:523–533.

50. Berkenpas M, B., Lawrence DA, Ginsburg D (1995): Molecular evolution of plasminogen activator inhibitor-1: functional stability. EMBO J 14:2969–2977.

51. Aertgeets K, De Bondt HL, De Ranter CJ, Declerck PJ (1995): Mechanisms Contributing to the Conformational and Functional Flexibility of Plasminogen Activator Inhibitor-1. Nature Structural Biology 2:891–897.

52. Bolton AE, Hunter WM (1973): The labeling of proteins to high specific radioactivities by conjugation to a ^{125}I-containing acylating agent. Biochem J 133:529–539.

53. Freer ST, Kraut J, Robertus JD, Wright HT, Xuong HH (1970): Chymotrypsinogen: 2.5-Å Crystal Structure, Comparison with α-Chymortrypsin, and Implications for Zymogen Activation. Biochem 9:1997–2009.

54. Cooperman BS, Stavridi E, Nickbarg E, Rescorla E, Schechter NM, Rubin H (1993): Antichymotrypsin interaction with chymotrypsin. Partitioning of the complex. J Biol Chem 268:23616–23625.

55. Kraut J (1977): Serine Proteases: Structure and Mecahnism of Catalysis. Annu Rev Biochem 46:331–358.

56. Wright HT, Scarsdale JN (1995): Structural basis for serpin inhibitor activity. Proteins 22:210–225.

57. Lawrence DA, Ginsburg D (1995): Plasminogen Activator Inhibitors. In: Molecular Basis of Thrombosis and Hemostasis, 256th Ed., KA High, HR Roberts, eds. Marcel Dekker, Inc. New York, pp. 517–543.

11

SUBSTRATE SPECIFICITY OF TISSUE TYPE PLASMINOGEN ACTIVATOR

Edwin L. Madison

The Scripps Research Institute
Department of Vascular Biology
La Jolla, California 92037

In this chapter I will describe our recent studies designed to characterize and increase understanding of molecular determinants of the remarkably stringent substrate specificity of tissue-type plasminogen activator (t-PA). Ancillary to that discussion, however, is a very brief review of the standard terminology used to refer to specific subsites of a protease substrate[1]. The scissile bond is indicated by the arrow in Figure 1. Residues that are C-terminal of this bond are referred to as "primed" subsites, while those N-terminal of this bond are referred to as "unprimed" subsites. Each residue is also assigned a number based on its distance from the scissile bond. Residues forming the scissile bond are assigned the number "1", adjacent residues are assigned "2", and so forth, counting in both directions from the scissile bond. Each specific subsite, therefore, is uniquely identified by the assignment of a number and the designation as primed or unprimed.

The results of a kinetic analysis of plasminogen activation by t-PA, the isolated protease domain of t-PA, or trypsin are presented in Table 1[2]. In the presence of the co-factor fibrin, t-PA is a very active plasminogen activator with a k_{cat}/K_m value of approximately 1.2×10^7 $M^{-1}sec^{-1}$ at 37° C. Even in the absence of a co-factor, t-PA maintains reasonably high enzymatic activity towards plasminogen with a k_{cat}/K_m of approximately $2.9 \times 10^4 M^{-1} sec^{-1}$. This level of catalysis, in the absence of a co-factor, is an inherent property of the protease domain of t-PA. The isolated protease domain of t-PA exhibits a very similar catalytic efficiency for plasminogen activation, 5.9×10^4 $M^{-1}sec^{-1}$. Thus, even in the absence of a co-factor, t-PA is a 2 to 5-fold more efficient plasminogen activator than the related enzyme trypsin whose catalytic efficiency for this reaction is $1.2 \times 10^4 M^{-1}sec^{-1}$.

When these three enzymes are assayed towards the same target primary sequence, this time, however, in the context of synthetic linear or cyclic peptides rather than native plasminogen, a very different picture emerges, as summarized in Table 2[2]. Peptide 1 is the exact P6 to P8' sequence of plasminogen. We manipulated this peptide so that the intramolecular disulfide bond forms as it does in native plasminogen. Formation of the intramolecular disulfide bond was verified by HPLC and mass spectroscopy analysis. Peptide 2 is identical to peptide 1 except that the two cysteine residues were replaced by serine, so that the peptide remained linear. Trypsin displayed a slight preference for the plasminogen

Chemistry and Biology of Serpins, edited by Church *et al.*
Plenum Press. New York. 1997

109

P4-P3-P2-P1↓P1'-P2'-P3'-P4' $\begin{bmatrix} \text{Epitope} \\ \text{Tag} \end{bmatrix}$ - SSGGSG - [X$_6$] - LVPG-gene III

Figure 1.   Standard nomenclature used to refer   **Figure 2.**   Primary structure of gene III fusion protein in sub-
to specific subsites of a protease substrate (1).   strate phage.

target sequence in this context with k_{cat}/K_m values for hydrolysis of the peptides that were 4 to 7 times greater than its activity towards native plasminogen. This behavior is in very stark contrast to that of t-PA, which is a truly abysmal enzyme towards both of these peptides. The k_{cat}/K_m values of t-PA for these peptides are on the order of 1 M^{-1} sec^{-1}. To place this surprisingly low catalytic efficiency in context, it is similar to that observed with a variant of trypsin lacking Asp-102, one member of the catalytic triad[3]. Taken together, data in Figure 1 and Table 1 indicate the evolution of remarkable substrate specificity by t-PA. In striking contrast to trypsin, a closely related enzyme, t-PA showed a 4 to 7 order of magnitude difference in activity towards the same primary sequence in two distinct structural contexts.

These observations raise a number of interesting questions. For example, as the free enzyme exists in solution, are the active site and substrate binding determinants of t-PA properly formed and aligned before interaction with plasminogen? Or, alternatively, is it possible, even though plasminogen is the only known efficient substrate for t-PA *in vivo*, that the target sequence in plasminogen actually represents a poor match to optimal subsite occupancy for t-PA? To address these questions, we asked directly a third, related question. Can we find peptide sequences that are cleaved significantly more efficiency by t-PA than the plasminogen-derived peptides?

To address this question experimentally, we adapted substrate phage protocols that have been described by Jim Wells, Marc Navre, and their co-workers[4,5]. This effort began with the construction of an fd library whose members contained gene III fusion proteins with the primary sequence shown in Figure 2[5]. The epitope for a high affinity anti-peptide monoclonal antibody was placed at the N-terminus of the fusion proteins. The epitope tag was followed by a 6 amino acid flexible linker sequence, a random hexamer sequence, and the full length gene III coding sequence. The library used in these studies contained greater than 200 million independent recombinants so we expect that this library contains a very high percentage of the 64 million theoretically possible random hexamer sequences. We then screened this library using the protocol outlined in Figure 3[6]. The

Table 1.   Kinetic analysis of plasminogen activation by t-PA,
the isolated protease domain of t-PA (PD), or trypsin (Tn)

Substrate	Enz	k_{cat} (s^{-1})	K_m(μM)	k_{cat}/K_m (M^{-1} s^{-1})
Plg w. fibrin	t-PA	0.2±0.03	0.018±0.006	1.2 (10)^7
Plg w/o fibrin	t-PA	0.2±0.04	7.6±0.9	2.9 (10)^4
Plg	PD	0.44±0.11	7.8±0.7	5.9 (10)^4
Plg	Tn	0.023±0.0032	2.0±0.5	1.2 (10)^4

Table 2. Catalysis of synthetic peptides by t-PA, the isolated protease domain of t-PA (PD), and trypsin (Tn). Entries in the final column are the ration of the k_{cat}/K_m for each reaction and the k_{cat}/K_m of t-PA towards peptide (2). The underlined and boldfaced cysteine indicate a disulfide linkage

Substrate	Enz	k_{cat} s^{-1}	K_m(μM)	k_{cat}/K_m (M^{-1} s^{-1})	Ratio
(1) KK**C**PGRVVGG**C**VAH	t-PA	0.009±.0002	5900±70	1.52	4.6
(2) KKSPGRVVGGSVAH	t-PA	0.0012±.00008	3600±700	0.33	1
(1) KK**C**PGRVVGG**C**VAH	Tn	44±1.4	1100±60	4(10)^4	120,000
(2) KKSPGRVVGGSVAH	Tn	61±0.1	870±80	7(10)^4	210,000
(1) KK**C**PGRVVGG**C**VAH	PD	0.0053±.0008	3600±500	1.47	4.5
(2) KKSPGRVVGGSVAH	PD	0.0008±.00005	6500±300	0.12	0.36

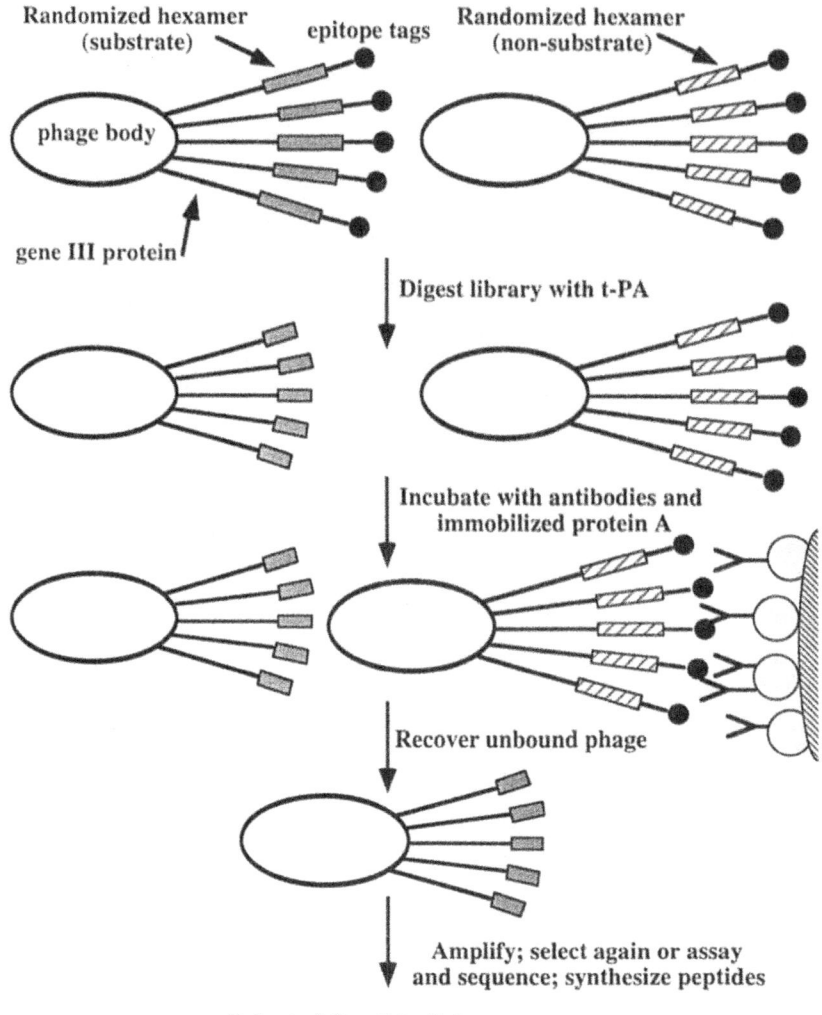

Figure 3. Outline of protocol used to isolate substrate phage.

phage library was first incubated with highly purified t-PA. Phage containing a random hexamer sequence that could serve as a substrate for t-PA were cleaved and separated from the epitope tag. Phage that were not cleaved were removed by adding to the solution the high affinity monoclonal antibody and immobilized protein-A and centrifuging down the resulting ternary complexes. This procedure leaves in solution a population of phage greatly enriched for those containing substrate sequences for t-PA. We repeated this screening procedure four times, and at each step we increased the stringency of the digestion by decreasing the concentration of t-PA, or the time of digestion, or both.

Using this protocol we isolated 44 independent substrate phage for t-PA. The primary sequences of the randomized hexamers in these phage are shown in Table 3[6]. Each of the 44 sequences was distinct. Given the trypsin-like primary specificity of t-PA, it was reassuring that all of the phage contained an arginine residue. In fact, half of the phage

Table 3. Amino acid sequences of the randomized hexamers in 44 isolated t-PA substrate phage clones. Peptide sequences have been shifted to the right or left to align corresponding subsites of each individual hexamer sequence. Residues in parentheses are flanking residues from the gene III fusion protein

clone #	amino	acid	sequences							
1-11			R	G	R	G	A	G		
1-17			L	G	R	R	R	S		
1-21		I	R	G	R	W	A			
1-22			A	G	R	S	A	F		
1-33			(G)	G	R	T	D	F	W	
1-35			I	K	R	N	I	T		
1-38		P	R	G	R	G	G			
1-42	N	K	Y	A	R	Q	(L)			
1-56	Y	H	I	K	R	S	(L)			
2-1			V	L	R	S	A	R		
2-3		Q	Y	L	R	Y	G			
2-5		K	D	A	R	R	A			
2-8		W	L	P	R	R	A			
2-9			I	L	R	A	A	Y		
2-11		L	R	G	R	T	A			
3-1		G	F	A	R	R	A			
3-7			F	G	R	R	R	T		
3-11		T	R	A	R	R	A			
4-1		Q	R	G	R	K	A			
4-6			I	V	R	R	A	E		
4-9			V	A	R	R	A	A		
5-2			I	G	R	R	A	Q		
5-4		S	Y	L	R	R	A			
5-6		L	K	G	R	R	A			
5-7			R	G	R	R	A	R		
5-9		S	R	A	R	K	A			
5-24		F	T	G	R	D	I			
5-26		P	Y	S	R	M	A			
5-28		T	F	A	R	R	A			
5-31	K	A	I	G	R	M	(L)			
5-34		W	L	G	R	R	G			
5-38		G	F	A	R	A	A			
5-40			V	A	R	R	A	A		
5-43			(S	G)	R	Y	A	R	S	A
5-44			R	G	R	S	A	G		
5-52			(S	G)	R	Y	A	R	R	L
5-57		M	R	G	R	R	G			
5-61		T	V	M	R	R	A			
5-63		P	F	G	R	S	A			
5-64		M	R	L	R	R	A			
5-66		R	R	G	R	R	A			
5-67		Y	I	G	R	R	G			
5-68			L	G	R	K	A	T		
5-72		Q	L	G	R	K	A			

Table 4. Comparison of k_{cat}, K_m, and K_{cat}/K_m values for the hydolysis by t-PA of peptides derived from a substrate phage screen (IV–VIII) or modeled after the native cleavage sequence in plasminogen (I) or the reactive loops of PAI-I and PAI-III (II–III). The identity of the hydrolyzed peptide fragments, and therefore also the cleavage site in the initial peptide, was determined by Mass Spectral analysis. Residues present in the randomized hexamer region of selected, substrate phage are in boldface. Arrows denote the locations of the cleavage sites between P1 and P1′

Substrate (Pn,.P3, P2, P1,↓P1′, P2′, P3′..Pn′)	k_{cat} s^{-1}	K_m (μM)	k_{cat}/K_m ($M^{-1} s^{-1}$)	Ratio* rel to (I)
(I) YKKSPGR↓VVGGSKY	0.0043	15000	0.29	1
(II) TAVIVSAR↓MAPEEII	0.25	1900	130	448
(III) TGGVMTGR↓TGHGGPQ	0.036	3900	9.2	32
(IV) GGSG**SRAR**↓**KA**LVPE	0.061	22000	2.8	9.6
(V) GGSG**WLGR**↓**RG**LVPE	2.0	10000	200	700
(VI) GGSG**YIGR**↓**RG**LVPE	1.6	7300	220	750
(VII) GGSG**PFGR**↓**SA**LVPE	3.3	2200	1500	5300
(VIII) TAVI**PFGR**↓**SA**PEEI	1.4	2300	610	2100

contained more than one arginine residue. Significantly more reassuring, however, was the emergence of a clear consensus among the selected substrate phage; specifically, glycine > alanine at P2, arginine at P1, P1′ could be several different residues but was most often arginine, and alanine > glycine at P2′. We noted less stringent conservation at P3, which was usually either a large hydrophobic residue or arginine.

Although these experiments were quite informative, we were not completely satisfied at this point because the data were not quantitative. Consequently, we synthesized peptides containing sequences found in several of the selected phage substrates and then analyzed the kinetics of cleavage of these peptides by t-PA. These data are summarized in Table 4[6]. As described above, linear plasminogen-derived peptides are poor substrates of t-PA. The least labile of the selected substrates was cleaved an order of magnitude more efficiently than the plasminogen derived peptide. Two of the selected peptides were cleaved at approximately 700-fold higher catalytic efficiency than the control peptide. The most labile selected peptide was cleaved at greater than 5,000 times higher catalytic efficiency than the plasminogen derived peptide.

These data have several interesting implications. First, the answer to the question of whether we can find peptides that are cleaved by t-PA more efficiently than the plasminogen derived peptides is clearly yes. Second, the answer to the question of whether the cleavage site in plasminogen actually represents a poor match to optimal subsite occupancy is also yes. And, third, these data strongly suggest that the active site of t-PA is properly formed before interaction with plasminogen.

There are also more general implications. For example, these results clearly caution that even for a highly specific enzyme such as t-PA, the target sequence in a natural substrate is not necessarily a reasonable lead compound for the design of small molecule inhibitors or substrates for the enzyme. Finally, these results have intriguing implications regarding the evolution of efficient protease cascades. Comparison of the plasminogen target sequence with the consensus sequence derived from the selected substrates indicates that plasminogen matches the consensus very well on the unprimed side of the scissile bond but diverges dramatically on the primed side. We believe that the reason for this divergence lies in the mechanism of zymogen activation of chymotrypsin-family serine proteases. After activation cleavage, the P1' and P2' residues fit into the activation pocket of the enzyme where they form a number of key hydrophobic interactions and an important new salt bridge with Asp 194, interactions that are essential for the development of the active conformation of the mature enzyme[7–10]. This critical role after activation cleavage places severe functional constraints on the P1' and P2' residues and prevents them from evolving simply to interact optimally with the activating enzyme. We believe that the necessity to overcome this obstacle, and therefore accelerate catalysis of substrates containing non-optimal subsite occupancy on the primed side of the scissile bond, may well have provided the initial impetus for the widespread evolution of co-factors to participate in protease cascades. The much discussed role of these co-factors in localizing proteolytic activity to specific micro-environments may actually have evolved later.

The availability of the kinetic assay of peptide cleavage set the stage for a detailed, quantitative analysis of the effect of optimal or sub-optimal occupancy of specific subsites on catalysis by t-PA. We began these studies by synthesizing and analyzing the kinetics of cleavage of the sixteen peptides listed in Table 5.[6] The most labile, selected hexamer sequence spanned the P4–P2' subsites and differed from the plasminogen sequence at four of these six positions, specifically, P4, P3, P1' and P2'. Data in Table 5 (peptides IX–XII) indicate that all four of those changes contributed to the enhanced catalysis observed for the selected peptide. Individual replacement of any of these four residues in the plasminogen peptide with the amino acid found in the selected peptide increased the rate of catalysis by t-PA by factors varying from 6.5 to 27. These data also indicate that the functional constraints on the P1' and P2' residues mentioned above are, in fact, a significant barrier to catalysis. Replacement of the P1' and the P2' residue of the selected peptide with the valine-valine sequence found in the plasminogen target sequence (peptide XXI) reduced catalysis by t-PA by a factor of approximately 50. In a protease cascade involving multiple activation steps, therefore, this would very quickly become a formidable barrier.

Data in Table 5 also indicate the selectivity of our screening protocol as indicated graphically in Figure 4. The 16 peptides from the previous figure are listed on the ordinate. The k_{cat}/K_m value for cleavage of each of these peptides, relative to that for the plasminogen derived peptide, is plotted on the abscissa in closed bars for t-PA and open bars for trypsin. Some of the peptides are cleaved more efficiently by trypsin than the plasminogen derived peptide. However, the enhancement of catalysis is orders of magnitude greater for t-PA than for trypsin.

These data also indicate that the contributions of optimal or sub-optimal occupancy of specific subsites to catalysis are independent of one another and, therefore, additive. In Figure 5 the change in free energy for stabilization of the transition state during the hydrolysis of multiply substituted peptides by t-PA is plotted versus the sum of free energy changes for singly substituted peptides containing the same set of mutations[11]. Least squared regression analysis of these data results in a linear plot with a slope of 0.98 and an R^2 value of 0.98.

Data in Table 6 indicate that the conservation of specific residues at a particular subsite during the selection of substrate phage is, in fact, predictive of optimal subsite occupancy[11]. Moreover, these results indicate that the screening method is actually remarkably sensitive. The P2 and P2' subsites were the most conserved non-P1 residues in our substrate phage panning. We therefore synthesized, and analyzed the kinetics of cleavage by t-PA, peptides containing 6 different residues at each of these two locations. In both cases, the residue found most frequently in the substrate phage, glycine at P2 and alanine at P2', was, in fact, the optimal residue. In addition, it is interesting to note that cleavage of a

Table 5. Comparison of kcat, Km, and kcat/Km for the hydrolysis by t-PA or trypsin of peptides of varying similarity to the most labile selected substrate (VII) and the native plasminogen sequence (I)

Substrate (Pn..P3, P2, P1,↓P1', P2', P3'..Pn')	t-PA k_{cat} s^{-1}	K_m (μM)	k_{cat}/K_m ($M^{-1}s^{-1}$)	k_{cat}/K_m rel. to (I)	TRYPSIN k_{cat} s^{-1}	K_m (μM)	k_{cat}/K_m ($M^{-1}s^{-1}$)	k_{cat}/K_m rel. to (I)
(I) YKKSPGR↓VVGGSKY	0.0043	15,000	0.29	1.0	25	790	$3.2(10)^4$	1.0
(IX) YKKSPGR↓VAGGSKY	0.058	22,000	2.6	9.0	28	420	$6.6(10)^4$	1.0
(X) YKKPPGR↓VVGGSKY	0.033	28,000	1.9	6.5	26	600	$4.3(10)^4$	1.3
(XI) YKKSFGR↓VVGGSKY	0.066	21,000	3.1	11	33	81	$4.1(10)^5$	12.8
(XII) YKKSPGR↓SVGGSKY	0.048	6,200	7.8	27	130	520	$2.5(10)^5$	7.8
(XIII) YKKPFGR↓VVGGSKY	0.41	8,700	48	170	120	640	$1.9(10)^5$	5.9
(XIV) YKKPFGR↓SAGGSKY	5.2	9,200	570	2,000	130	200	$6.5(10)^5$	15
(VIIa) GGSGPFGR↓SALVPEE	4.2	3,100	1350	4,700	220	103	$2.1(10)^6$	66
(XVI) GGSGSFGR↓SALVPEE	1.9	2,700	720	2,500	370	230	$1.6(10)^6$	50
(XVII) GGSGPFGR↓SVIVPEE	0.52	1,200	430	1,500	340	160	$2.1(10)^6$	66
(XVIII) GGSGPFGR↓VALVPEE	0.12	290	400	1,400	50	80	$6.0(10)^5$	21
(XIX) GGSGPPGR↓SALVPEE	0.26	3,200	82	290	220	320	$7.0(10)^5$	22
(XX) GGSGSPGR↓SALVPEE	0.22	3,700	60	210	60	330	$1.8(10)^5$	5.6
(XXI) GGSGPFGR↓VVIVPEE	0.04	1,500	27	82	61	97	$6.3(10)^5$	20
(XXII) GGSGPFGK↓SALVPEE	0.07	4,400	16	55	111	370	$3.0(10)^5$	9.3
(XXIII) GGSSPGR↓VVGLVPEE	0.0033	1,700	2	6.7	64	1,500	$4.2(10)^4$	1.3

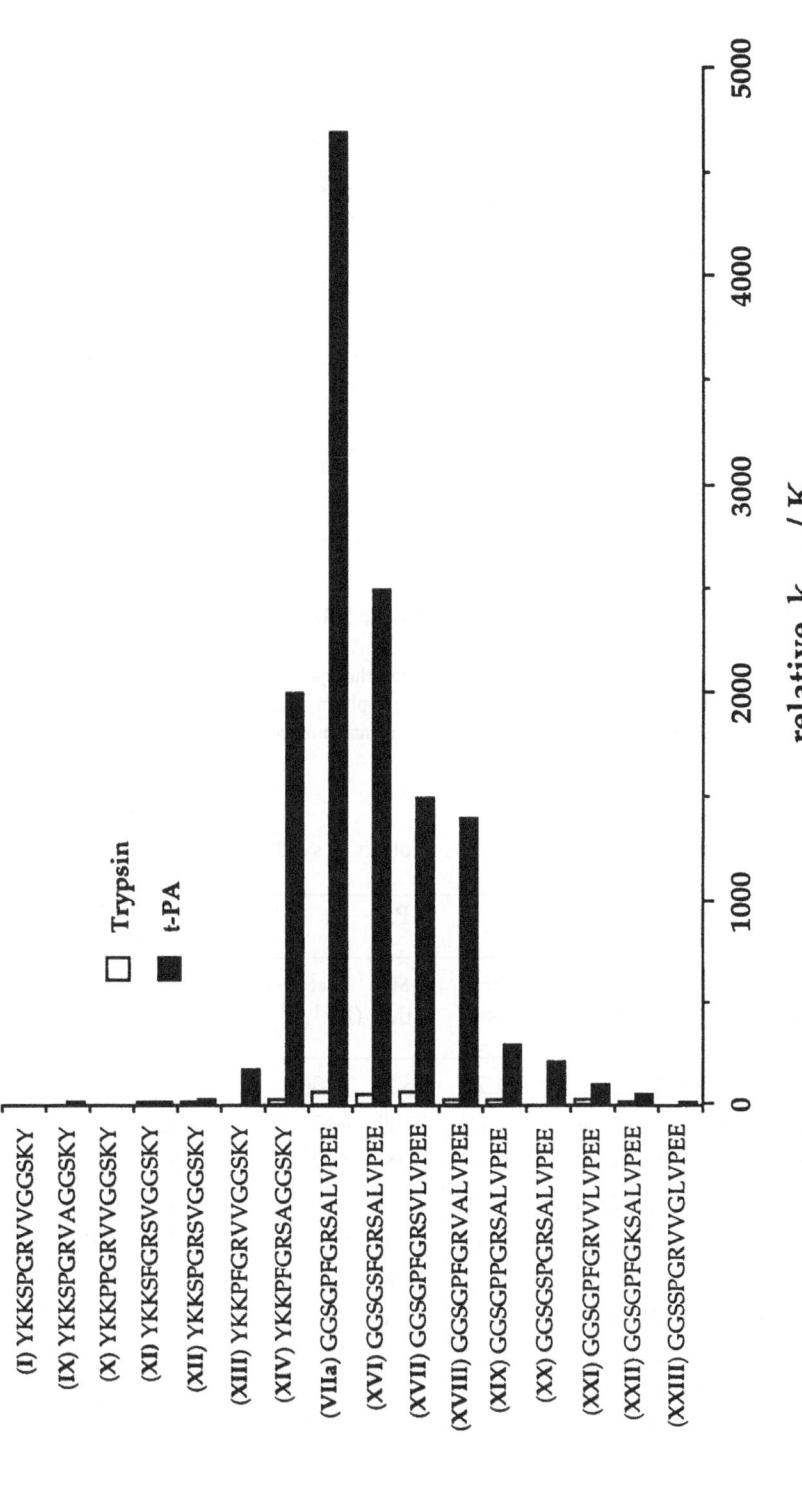

Figure 4. Catalytic efficiencies for cleavage of peptides, relative to that of peptide (I), by t-PA and trypsin, relative to that of peptide (I), for trypsin (open bars) and t-PA (closed bars). x-axis: kcat/Km relative to peptide (I), by t-PA and trypsin. x-axis: kcat/Km relative to peptide (I) for trypsin (open bars) and t-PA (closed bars). y-axis: peptides listed in Table 5.

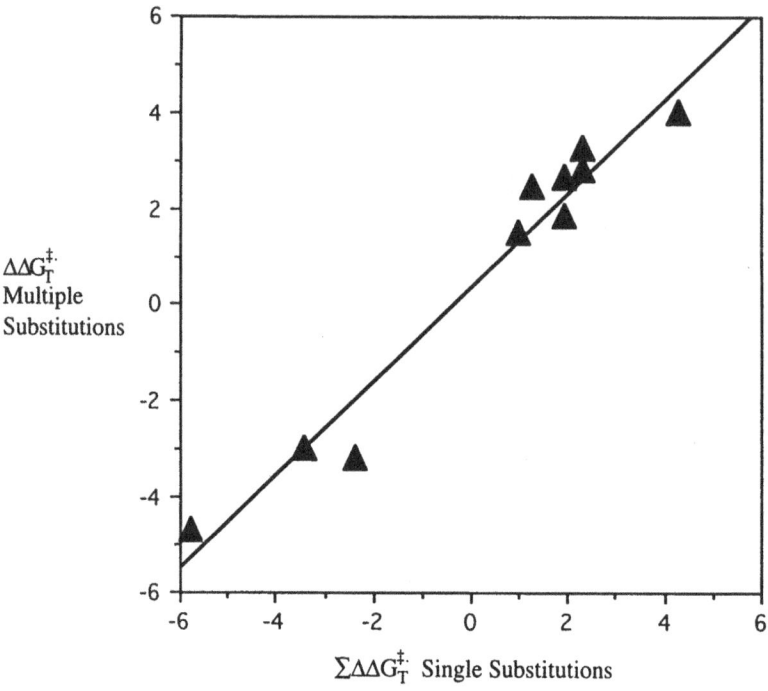

Figure 5. $\Delta\Delta$ G for stabilization of the transition state during the hydrolysis of peptides containing 1, 2, or 4 mutations compared to peptides containing the target sequence in plasminogen or the most labile selected hexamer sequence plotted against the sum of $\Delta\Delta$ G's for hydrolysis of the matching combination of peptides containing single mutations.

Table 6. Effect of variation at P2 and P2′ on the hydrolysis of synthetic peptides by t-PA and trypsin

	Substrate (Pn,P3,P2,P1↓P1′P2′P3′Pn′)	t-PA			TRYPSIN		
		k_{cat} (s^{-1})	K_M (μM)	k_{cat}/K_M (M^{-1},s^{-1})	k_{cat} (s^{-1})	K_M (μM)	k_{cat}/K_M (M^{-1},s^{-1})
(I)	GGSGPF**GR**↓SALVPEE	4.20	3500	1200	220	110	2.0(10)^6
(II)	GGSGPF**A**R↓SALVPEE	2.60	3100	840	200	58	3.4(10)^6
(III)	GGSGPF**I**R↓SALVPEE	0.73	1800	410	86	76	1.1(10)^6
(IV)	GGSGPF**F**R↓SALVPEE	0.17	1100	150	130	88	1.5(10)^6
(V)	GGSGPF**K**R↓SALVPEE	2.10	4500	470	98	54	1.8(10)^6
(VI)	GGSGPF**D**R↓SALVPEE	0.02	2900	5.2	39	310	1.3(10)^5
(VII)	GGSGPFGR↓S**G**LVPEE	2.20	3200	690	180	240	7.5(10)^5
(VIII)	GGSGPFGR↓S**I**LVPEE	0.51	2200	230	240	93	2.6(10)^6
(IX)	GGSGPFGR↓S**F**LVPEE	0.67	2300	290	96	58	1.7(10)^6
(X)	GGSGPFGR↓S**K**LVPEE	0.10	2000	50	81	62	1.3(10)^6
(XI)	GGSGPFGR↓S**D**LVPEE	0.31	2500	124	2.6	1700	1.5(10)^3

Table 7. Effect of subsite occupancy on the kinetics of peptide hydrolysis by t-PA and trypsin. "ac" indicates that the amino acid terminus of a peptide has been acylated while "am" indicates that the carboxyl terminus has been amidated

Substrate (Pn,P3,P2,P1↓P1'P2'P3'Pn')	t-PA			TRYPSIN		
	k_{cat} (s^{-1})	K_M (μM)	k_{cat}/K_M (M^{-1},s^{-1})	k_{cat} (s^{-1})	K_M (μM)	k_{cat}/K_M (M^{-1},s^{-1})
(XII) PFGR↓SAL	1.8	9500	190	62	110	$5.6(10)^5$
(XIII) PFGR↓SA	0.00032	4400	0.073	0.025	1700	15
(XIV) PFGR↓S	0.00053	8100	0.065	0.021	1000	21
(XV) R↓SALV	0.00086	27000	0.032	0.19	4800	40
(XVI) GR↓SALV	0.0025	17000	0.15	70	3700	$1.9(10)^4$
(XVII) FGR↓SALV	0.41	4200	98	84	300	$2.8(10)^5$
(XVIII) PFGR↓SALV	1.0	1800	560	130	80	$1.6(10)^6$
(XIX) FGR↓SAL	0.77	4600	170	300	390	$7.7(10)^5$
(XX) acFGR↓SALam	1.8	4300	420	87	150	$5.8(10)^5$
(XXI) acFGR↓SAam	2.8	12000	230	110	190	$5.8(10)^5$
(XXII) acFGR↓Sam	13	10000	1300	69	280	$2.0(10)^5$
(XXIII) acR↓SALam	0.021	9300	2.2	17	3000	$5.7(10)^3$
(XXIV) acGR↓SALam	0.38	44000	8.6	45	820	$5.5(10)^4$

peptide containing a glycine at P2 (peptide I) occurs at less than 2-fold greater catalytic efficiency than a related peptide containing alanine at this position (peptide II). Nevertheless, this small difference in catalytic efficiency was sufficient to yield 3 times more glycines than alanines at P2 in the substrate phage. This very high sensitivity is almost certainly related to the requirement for multiple cleavage events during each round of selection of polyvalent phage.

We have also analyzed hydrolysis by trypsin of the 11 peptides shown in Table 6[11]. Although optimal subsite occupancy of the P2 and P2' positions was not identical for the two enzymes, their preferences were qualitatively similar. Both enzymes prefer either alanine or glycine at these positions, and both cleave peptides containing acidic residues in either of these positions inefficiently. With only one exception (Asp at P2'), however, t-PA was substantially more discriminating against suboptimal subsite occupancy than was trypsin. Consequently, differences in the relative rates of catalysis among the peptides listed in Table 6 are much greater for t-PA than for trypsin.

Another approach to examining the contribution to catalysis of occupancy of specific subsites is illustrated by the experiment summarized in Table 7[11]. To begin this experiment we synthesized an 8 residue peptide that spanned the P4 to P4' positions of the most labile selected peptide. We then synthesized related peptides in which single residues were deleted sequentially from either termini, and we examined the effect of these deletions on catalysis by both t-PA and trypsin. For the most part, the relative reduction in catalysis due to loss of occupancy of a specific subsite was similar for the two enzymes. However, there was one exception to this observation and that involved occupancy of the P3 position. The loss of P3 is significantly more detrimental to catalysis by t-PA than by trypsin.

In conclusion, we would like to suggest that the concomitant use of several distinct mechanisms contributes to the exquisitely stringent substrate specificity of t-PA. We have shown that t-PA displays a greater dependence than trypsin on occupancy of the P3 subsite. We have also shown that t-PA exerts substantially enhanced discrimination against sub-optimal occupancy at each subsite from P4 to P2'. Both of these effects are mediated predominantly by alterations in k_{cat}. In experiments that I did not describe, we have also obtained evidence for specific, and also relatively nonspecific, productive protein-protein interactions between t-PA and protein substrates that are more distant from the active site[11]. The nonspecific interactions appear to reduce K_m while the specific interactions enhance k_{cat}. Finally, unique to the reaction between t-PA and plasminogen, fibrin greatly enhances catalysis, primarily by reducing K_m.

ACKNOWLEDGMENTS

I am very fortunate to have outstanding scientists in the laboratory and also extremely talented collaborators. Dr. Song Ke, Dr. Leif Strandberg and Kathy Tachias (Scripps Research Institute); Dr. Marc Navre and his co-workers Dr. Matt Smith and Dr. Li Ding (Affymax Research Institute); and Dr. David Corey and his colleague Gary Coombs all made essential contributions to the work described in this chapter.

REFERENCES

1. Berger, A. and Schecter, I. (1970) Mapping the active site of papain with the aid of peptide substrates and inhibitors. Phil. Trans. Roy. Soc. Lond. 257:249–264.

2. Madison, E. L., Coombs, G. S., and Corey, D. R. (1995) Substrate specificity of tissue type plasminogen activator. Characterization of the fibrin independent specificity of t-PA for plasminogen. J. Biol. Chem. 270(13):7558–7562.
3. Corey, D. R. and Craik, C. S. (1992) An investigation into the minimum requirements for peptide hydrolysis by mutation of the catalytic triad of trypsin. J. Am. Chem. Soc. 114(5):1784–1790.
4. Matthews, D. J. and Wells, J. A. (1993) Substrate phage: selection of protease substrates by monovalent phage display. Science 260:1113–1117.
5. Smith, M. M., Shi, L., and Navre, M. (1995) Rapid identification of highly active and selective substrates for stromelysin and matrilysin using bacteriophage peptide display libraries. J. Biol. Chem. 270(12):6440–6449.
6. Ding, L., Coombs, G. S., Strandberg, L., Navre, M., Corey, D. R., and Madison, E. L. (1995) Origins of the specificity of tissue-type plasminogen activator. Proc. Natl. Acad. Sci. U.S.A. 92:7627–7631.
7. Freer, S. T., Kraut, J., Robertus, J. D., Wright, H. T., and Xuong, N. H. (1970) Chymotrypsinogen: 2.5 Å Structure, Comparison with Chymotrypsin, and Implications for Zymogen Activation. *Biochemistry* **9**, 1997–2009.
8. Kerr, M. A., Walsh, K. A., and Neurath, H. (1976) A Proposal for the Mechanism of Chymotrypsinogen Activation. *Biochemistry* **15**, 5566–5570.
9. Fehlhammer, H., Bode, W., and Huber, R. (1977) Crystal Structure of Bovine Trypsinogen at 1.8 Å Resolution II. Crystallographic Refinement, Refined Crystal Structure, and Comparison with Bovine Trypsin *J. Mol. Biol.* **111**, 415–438.
10. Huber, R., and Bode, W. (1978) Structural Basis of the Activation and Action of Trypsin. *Acc. Chem. Res.* **11**, 114–122.
11. Coombs, G. S., Dang, A. T., Madison, E. L., and Corey, D. R. Distinct mechanisms contribute to stringent substrate specificity of tissue-type plasminogen activator. J. Biol. Chem. 271(8):4461–4467, 1996.

12

PLASMINOGEN ACTIVATOR INHIBITOR TYPE-2

A Spontaneously Polymerizing Serpin that Exists in Two Topological Forms[*]

Tor Ny[†] and Peter Mikus

Department of Medical Biochemistry and Biophysics
Umeå University
S-90187 Umeå, Sweden

1. INTRODUCTION

Plasminogen activator inhibitor type-2 (PAI-2) is a quite unusual Serpin. Unlike most other Serpins, which are secreted proteins with well characterized target proteinases and function, PAI-2 exists in both cytosolic and secreted forms; its physiological role is not well understood.

PAI-2 belongs to a subgroup of Serpins denoted the ovalbumin related Serpin (OV-Serpin) family, which reveal a particularly high degree of amino-acid sequence homology and seem to have a similar gene organization. OV-Serpins lack cleavable NH_2-terminal signal sequences and some of them utilize unconventional internal non-cleavable secretion signals for their translocation to the extracellular compartment. For several members of the OV-Serpin family, a functional role remains to be established.

The internal non-cleaved translocation signal of PAI-2 seems to be inefficient by design thereby allowing the synthesis of both intracellular and extracellular forms of the protein. In the extracellular environment, PAI-2 is thought to play an important role in the control of pericellular proteolysis by inhibiting urokinase-type plasminogen activator (uPA) and the two-chain type of tissue-type PA (two-chain tPA).

The functional role of intracellular PAI-2 is a subject for speculation as no intracellular target protein has been identified. However, recent studies suggest that PAI-2 may play a role in protecting cells against programmed cell death. In this review we will summarize recent findings on the properties of PAI-2 and discuss functional and biochemical aspects of PAI-2, especially relating to its secretion pattern and its ability to spontaneously

---

[*] This work was supported by the Swedish Natural Science Foundation Research Grants NFR BU 8473-308.
[†] To whom correspondence should be addressed at the Department of Medical Biochemistry and Biophysics. Umeå University. S-90187 Umeå Sweden. Telephone: 46-90-166565; Fax: 46-90-136465.

form polymers. In addition, extensive reviews covering biological and clinical aspects of PAI-2, as well as its regulation, have also been published recently (Dear & Medcalf, 1995; Kruithof *et al.*, 1995).

2. BIOCHEMISTRY AND STRUCTURE OF PAI-2

The presence of uPA-inhibitor activity in partially purified extracts from placenta was first reported by Kawano and coworkers (Kawano *et al.*, 1968; Kawano *et al.*, 1970). This activity was initially named placental-type plasminogen activator inhibitor after the source of its discovery but was later renamed PAI-2.

PAI-2 exists in two distinct molecular mass forms (47 kDa and 60 kDa) that differ in their topology and in their degree of glycosylation (Genton *et al.*, 1987; Wohlwend *et al.*, 1987).

In most cellular sources used for purification the majority of PAI-2 remains intracellular and appears to be unglycosylated. This low molecular mass form of PAI-2, which has an isoelectric point of 5.0, (Kruithof *et al.*, 1986) has been purified and characterized from human placenta (Åstedt *et al.*, 1985; Wun & Reich, 1987), phorbol ester stimulated human U937 cells (Kruithof *et al.*, 1986), and from recombinant bacterial (Mikus *et al.*, 1993) and yeast expression systems (Steven *et al.*, 1991).

For the production and purification of the secreted high molecular mass form of PAI-2, an artificial signal sequence was in-frame fused to the coding sequence of PAI-2 in order to direct a larger portion of PAI-2 protein to the secretory pathway (Mikus *et al.*, 1993). By using this expression system the secreted glycosylated form of PAI-2 could be produced and biochemically characterized. Apart from differences in molecular weight, isoelectric point and glycosylation, both forms of PAI-2 appear to have similar characteristics (Kruithof *et al.*, 1986; Åstedt *et al.*, 1985; Mikus *et al.*, 1993; Thorsen *et al.*, 1988; Wun & Reich, 1987; Steven *et al.*, 1991). Both secreted and intracellular forms of PAI-2 inhibit PAs by forming SDS-resistant complexes. They are efficient inhibitors of uPA (second order rate constants of 10^6 M^{-1} s^{-1}), and two-chain tPA (second order rate constants of 2×10^5 M^{-1} s^{-1}) but react very poorly with single-chain tPA (second order rate constants of 10^4 M^{-1} s^{-1}). As a comparison, PAI-1 inhibits both tPA and uPA much faster than PAI-2 (second order rate constants of 10^7 M^{-1} s^{-1}). The difference in rate is especially large with single-chain tPA, which makes it unlikely that PAI-2 contributes to the inhibition of this activator *in vivo*.

PAI-2 forms 1:1 SDS-resistant complexes with target proteases and treatment with nucleophilic agents yields reactive center cleaved PAI-2 (Kruithof *et al.*, 1986 and Kiso *et al.*, 1988). Although the structure of the native inhibitory complex has not been studied in detail, the P1-P1′ peptide bond of the reactive center of PAI-2 is most likely cleaved, leaving the inhibitor in an acyl-complex with cognate protease (Wilczynska *et al.*, 1995; Lawrence *et al.*, 1995). In accordance with that proposed for PAI-1 and other Serpins, the mechanism of PAI-2 action therefore most likely involves complex formation and reactive center cleavage. The subsequent rapid insertion of the reactive center loop then induces the conformational changes required to lock the PAI-2–protease complex (Lawrence *et al.*, 1990; Fa *et al.*, 1995; Wilczynska *et al.*, 1995; Lawrence *et al.*, 1995).

Fibrin bound tPA appears to be protected from inhibition by PAI-2 (Leung *et al.*, 1987). It is therefore unlikely that PAI-2 plays any major role in tPA-mediated vascular fibrinolysis. Although it is generally accepted that PAI-2 functions primarily as a uPA inhibitor and plays only a minor role in controlling tPA mediated proteolysis, the finding of complexes between PAI-2 and tPA in gingival crevicular fluid indicates that PAI-2 can modulate tPA mediated proteolysis at local sites *in vivo* (Kinnby *et al.*, 1991). PAI-2 also contains transglutamination sites and can be covalently cross-linked to cellular and extracellular structures by tissue transglutaminase (Jensen *et al.*, 1994).

The 1.9 kb PAI-2 message encodes a 415 amino acid protein with a molecular mass of 46.6 kDa and the corresponding cDNA has been isolated from several different cellular sources including the histiocytic lymphoma cell line U937, monocytes and placenta. (Ye *et al.*, 1987; Webb *et al.*, 1987; Schleuning *et al.*, 1987; Antalis *et al.*, 1988; Ny *et al.*, 1989).

In agreement with the Arg-specificity of its target proteases, tPA and uPA, Arg-380 has been identified as the reactive center P1 residue of PAI-2 (Kiso *et al.*, 1988).

PAI-2 has three potential N-linked glycosylation sites (Asn-75, Asn-115, and Asn-339) that all appear to be glycosylated in PAI-2 secreted from U937 cells (Ye *et al.*, 1988). Two common variants of PAI-2 (type A and B) have been identified. Type B has an extra unpaired cysteine, which may mediate dimerization or the formation of cysteine bridges to other proteins.

3. SECRETION PROPERTIES OF PAI-2

The first methionine initiator codon in PAI-2 mRNA is located at the position that corresponds to the start of helix A of α_1-antitrypsin and sequencing of labeled PAI-2 has revealed that the secreted form of PAI-2 starts with this methionine. Together this demonstrates that PAI-2 lacks a cleavable NH_2-terminal signal sequence (Ye *et al.*, 1988). The internal secretion signal of human PAI-2 has been mapped to two mildly hydrophobic regions (H1 and H2) near the NH_2-terminus (von Heijne *et al.*, 1991). Mutational analysis of H1 and H2 revealed that the translocation efficiency of PAI-2 into microsomes *in vitro*, as well as the secretion efficiency from cells *in vivo*, can be enhanced by point mutations that increase their hydrophobicity (von Heijne *et al.*, 1991; Mikus, 1995). Together these studies reveal that the PAI-2 secretion signal provides a unique example of a translocation signal that, by virtue of its poor efficiency, allows the synthesis of both an extracellular and an intracellular form of the protein.

Interestingly, the secretion efficiency of PAI-2 also seems to vary among different cell lines. In fourteen different cell lines the secretion efficiency was found to vary up to 17-fold. Only 4% is secreted in the human embryonic kidney cell line 293, while 70% is secreted in the human promyelocytic cell line HL 60 (Mikus, 1995). This suggests that the internal translocation signal of PAI-2 may function in a cell-specific manner.

4. SPONTANEOUS POLYMERIZATION OF PAI-2

The reactive loops of Serpins can adopt varying conformations and a mobile reactive center loop seems to be required for inhibitory activity (reviewed by Carrell & Evans, 1992; Gettins *et al.*, 1993). For the inhibitory Serpin α_1-antitrypsin, it has been demonstrated that exposure to mild denaturing conditions at elevated temperatures, particularly at higher protein concentrations, leads to the formation of non-covalent polymers (Schulze *et al.*, 1990; Evans, 1991; Mast *et al.*, 1992; Mast *et al.*, 1992). These polymers are formed by a mechanism denoted loop-sheet polymerization, whereby the reactive center loop of one molecule is inserted into the gap in the β-sheet of another molecule. The naturally occurring Z variant form of α_1-antitrypsin, for which 4% of Northern Europeans are heterozygotes, contains a mutation at the base of the reactive center loop of the molecule

(Carrell, 1986). This polymorphic variant is even more prone to polymerization than wild-type and forms polymers even under physiological conditions (Lomas *et al.*, 1992). In patients that are homozygous for the Z allele only about 15% of the mutated α_1-antitrypsin is secreted into plasma while the remaining 85% accumulates in the endoplasmic reticulum of hepatocytes which causes two major clinical sequels: the formation of intracellular inclusions associated with hepatocellular damage (Sharp *et al.*, 1969) and the deficiency of circulating α_1-antitrypsin which predisposes to emphysema (Laurell *et al.*, 1963). (for review see (Carrell, 1986). The accumulation of the Z form of α_1-antitrypsin in the secretory pathway is caused by spontaneous loop-sheet polymerization of mutated antitrypsin (Lomas *et al.*, 1992). Siiyama antitrypsin is another abnormal variant of α_1-antitrypsin that accumulates in the endoplasmic reticulum of hepatocytes and for which an even more extensive degree of polymerization has been reported (Lomas *et al.*, 1993).

During purification and biochemical characterization of recombinant PAI-2, we noted that PAI-2 readily polymerized even in the absence of denaturing agents (Mikus *et al.*, 1993). Polymerization of the intracellular form of PAI-2 was also observed in the cytosol of cells and in tissues that normally produce PAI-2, e.g. human placenta and phorbol ester stimulated U 937 cells. This indicates that intracellular polymerization of PAI-2 may occur also in cells that express PAI-2 at physiological levels.

To our knowledge, PAI-2 is the only "normal" Serpin so far described that spontaneously forms polymers *in vitro* as well as intracellularly under physiological conditions. It is interesting to note that the wild-type form of PAI-2 seems to be much more prone to form polymers than the pathologic Z-form of antitrypsin and also that all attempts to induce a locked latent like conformation have resulted in polymerization (Mikus *et al.*, 1993).

Since PAI-2 readily forms polymers it is possible that cells that store PAI-2 intracellularly have mechanisms that can dissociate the polymers to regain active inhibitor molecules. Future studies will reveal whether the tendency to polymerize is significant for the physiological function of PAI-2 in haemostasis and inflammation or is just a fortuitous consequence of its structural organization.

5. GENE STRUCTURE AND CHROMOSOMAL LOCALIZATION

The PAI-2 gene (PLANH2) contains eight exons and seven introns and is located on the long arm of chromosome 18 at position 18q21.3 (Ye *et al.*, 1989; Webb *et al.*, 1987; Webb *et al.*, 1994). This is an interesting and well characterized chromosomal region, which contains four OV-Serpin genes. In this region, the maspin gene and the two squamous cell carcinoma antigen genes are located within 300 kbp from the PAI-2 gene (Schneider *et al.*, 1995). In addition, the bcl-2 gene, an important regulator of apoptosis, also maps in this region only 600 kbp from the PAI-2 gene (Schneider *et al.*, 1995; Silverman *et al.*, 1991). Deletions near this region are also frequently found in colorectal tumors, associated with the DCC (deleted in colon cancer) gene (Cho *et al.*, 1994) and in B-cell lymphoma and familial expansile osteolysis (van Kessel *et al.*, 1994). However, it is not known if any of these pathologies are due to deletions in any of the OV-Serpin genes that map to this region.

6. RELATION TO OTHER MEMBERS OF THE SERPIN FAMILY

Although PAI-1 and PAI-2 have a close target enzyme specificity, they are phylogenetically distant from each other. As previously mentioned, PAI-2 belongs to a subgroup of Serpins denoted OV-Serpins (Remold-O'Donnell, 1993) that are characterized by a high degree of amino acid sequence homology, similarities in gene structure and the lack of a cleavable NH_2-terminal translocation signal sequence. In addition to PAI-2, the OV-Serpins include the following members: ovalbumin, the chicken gene Y product, squamous cell carcinoma antigen (Suminami et al., 1991), leukocyte elastase inhibitor (Remold O'Donnell et al., 1992); (Dubin et al., 1992), (Teschauer et al., 1993), placental thrombin inhihitor (Coughlin et al., 1993), maspin, a serpin with tumor-supressing activity (Zou et al., 1994), two limulus intracellular coagulation inhibitors, (Miura et al., 1995), CrmA, a cowpox-virus-encoded Serpin (Pickup et al., 1986) that inhibits interleukin-1-β converting enzyme and which may have anti-apoptotic activity (Ray et al., 1992). Many of these proteins are mainly intracellular and the physiological function of most of them remains to be established.

7. REGULATION OF PAI-2 BIOSYNTHESIS

Regulation of PAI-2 biosynthesis has been studied in primary cultures and in different tumor cell lines in culture. Similar to the other components of the PA system, PAI-2 reveals a complex regulatory pattern which may be a reflection of its involvement in many physiological processes. In most of these cell types, the basal PAI-2 expression is low or undetectable. However, following stimulation by a variety of agents, PAI-2 often becomes a major cellular protein. Modulation of PAI-2 biosynthesis has been observed by use of numerous agents including growth factors, cytokines, vasoactive agents, toxins and steroids.

Several lines of evidence suggest that PAI-2 plays a role in inflammation. Consequently, PAI-2 is induced in many cell types including monocytes/macrophages by inflammatory mediators such as lipopolysaccharide (LPS), interleukin-1 (IL-1) or tumor necrosis factor (TNF). PAI-2 expression is also suppressed by anti-inflammatory agents such as dexamethasone. Regulation of PAI-2 often takes place at the transcriptional level (Medcalf et al., 1988a); (Medcalf et al., 1988b) but in some cases the mRNA stability also seems to be modulated (Schwartz & Bradshaw, 1992). The PAI-2 promoter has been extensively studied. Several putative regulatory elements have been identified and functional studies have identified elements required for basal and induced transcription (Kruithof & Cousin, 1988; Cousin et al., 1991; Antalis et al., 1993). For extensive review on PAI-2 regulation and references, see (Kruithof et al., 1995; Dear & Medcalf, 1995).

8. DISTRIBUTION AND FUNCTION OF PAI-2 *IN VIVO*

On the basis of initial studies it was believed that PAI-2 expression was limited to the placenta and to monocytes/macrophages. However, recent studies indicate that the distribution of PAI-2 is more widespread. Under normal conditions, PAI-2 levels in plasma are below detection, but the PAI-2 level increases during pregnancy to about 100 to 300 ng/ml during final stages, and then quickly decreases post partum (Lecander & Åstedt, 1986). PAI-2 is also detectable in amniotic fluid and umbilical cord plasma (Lecander & Åstedt, 1987). In placenta, PAI-2 is synthesized by the syncytiotrophoblasts (Åstedt et al., 1986). The functional role of PAI-2 during pregnancy is not well understood. PAI-2 may contribute to the regulation of the invasive potential of trophoblasts. The increased concentration of PAI-2 in the placenta may also prevent abruptio placenta, and secure haemostasis at separation of the placenta in the third stage of labor (Åstedt et al,. 1987).

The increase in PAI-2 biosynthesis in monocytes in response to inflammatory mediators suggests that PAI-2 plays an important role in the inflammatory response. PAI-2 could be involved in protection from inflammatory stress by inhibition of PA-dependent pericellular proteolysis either through the secretion of PAI-2 or through the release of stored intracellular unglycosylated PAI-2 through cell suffering (Belin *et al.*, 1989).

Gingivitis is an inflammatory reaction that develops in the gingival tissue when bacteria accumulate on the tooth adjacent to the gingiva. During this process, a fluid denoted gingival crevicular fluid accumulates in the gingival crevice. This fluid contains 2–8 mg/L of PAI-2, which is an order of magnitude higher than the level found in pregnancy plasma (Kinnby *et al.*, 1991).

PAI-2 expression has also been identified in vascular endothelium (Schleef *et al.*, 1988), vascular smooth muscle cells (Laug *et al.*, 1989), granulosa cells (Piquette *et al.*, 1993), and in fibroblasts from a variety of sources (Kruithof *et al.*, 1995).

In the skin, PAI-2 is expressed in most epidermal layers and seems to be the predominant epidermal PA inhibitor. The expression of PAI-2 by epidermal keratinocytes also seems to be regulated in a differentiation-dependent manner (Jensen *et al.*, 1995); (Lyons Giordano & Lazarus, 1994). In addition, synthesized PAI-2 can be cross-linked to the cornified envelope (Jensen *et al.*, 1995). Together these findings indicate that PAI-2 can be incorporated into the cornified envelope during terminal differentiation of keratinocytes and may act to modulate protease activity during formation of cornified cells (Jensen *et al.*, 1995).

Many *in vitro* experiments and clinical studies suggest that uPA may play a role in tumor metastasis. Correlations exist between elevated levels of uPA and PAI-1 in tumor extracts and degree of invasion, higher incidences of cancer relapse, and shorter overall survival. Despite the abundant literature implicating a role of uPA in metastasis, few studies have addressed the role of endogenous PAI-2 in malignancy. The results obtained are also sometimes contradictory with high PAI-2 correlating with a poor prognosis in ovarian cancer, whereas in human pancreatic carcinomas elevated PAI-2 is associated with a higher survival than negative or low PAI-2 levels (Takeuchi *et al.*, 1993). However, in animal studies, significantly less metastases in lung and lymph nodes were found when transfected cells that over express PAI-2 were injected into immunodeficient SCID mice, as compared to after the injection of mock transfected control melanoma cells (Mueller *et al.*, 1995). Although this study, as well as other similar studies, suggests that PAI-2 might be an agent that decreases tumor growth, angiogenesis and metastasis, the effect of PAI-2 in tumor related disease needs to be studied in more detail.

Recent studies by Kumar and colleges suggest that PAI-2 may have an alternative function unrelated to inhibition of PAs (Kumar & Baglioni, 1991). Over expression of PAI-2 in HT 1080 cells was found to protect these cells from TNF-induced cytolysis. PAI-2 gradually disappeared during the treatment suggesting that PAI-2 may interact with an as of yet unidentified intracellular protein. In a similar fashion, PAI-2 can also prevent Mycobacterium avium-induced apoptosis in human macrophages (Gan *et al.*, 1995) and TNF-induced apoptosis in Hela cells (Dickinson *et al.*, 1995). In Hela cells, the apoptosis pathway blocked by PAI-2 is independent of bcl-2. Although the mechanism of protection is independent of uPA, the reactive center P1-Arg residue of PAI-2 is required (Dickinson *et al.*, 1995). Emerging evidence suggests that proteases play an important role in regulating apoptosis (Kumar, 1995). Future studies on the role of intracellular PAI-2 and its possible interaction with an intracellular target protease will reveal the possible role of PAI-2 in apoptosis.

ACKNOWLEDGMENTS

We thank Malgorzata Wilczynska for critical reading and Ingrid Råberg for help with the manuscript.

REFERENCES

Antalis, T. M., Clark, M. A., Barnes, T., Lehrbach, P. R., Devine, P. L., Schevzov, G., Goss, N. H., Stephens, R. W. & Tolstoshev, P. (1988) *Proc. Natl. Acad. Sci. U.S.A. 85*, 985–989.

Antalis, T. M., Godbolt, D., Donnan, K. D. & Stringer, B. W. (1993) *Gene 134*, 201–208.

Åstedt, B., Hagerstrand, I. & Lecander, I. (1986) *Thromb. Haemost. 56*, 63–65.

Åstedt, B., Lecander, I., Brodin, T., Lundblad, A. & Low, K. (1985) *Thromb. Haemost. 53*, 122–125.

Åstedt, B., Lecander, I. & Ny, T. (1987) *Fibrinolysis 1*, 203–208.

Belin, D., Wohlwend, A., Schleuning, W. D., Kruithof, E. K. & Vassalli, J. D. (1989) *EMBO J. 8*, 3287–3294.

Carrell, R. W. (1986) *J. Clin. Invest. 78*, 1427–1431.

Carrell, R. W. & Evans, D. L. I. (1992) *Current Opinion in Structural Biology 2*, 438–446.

Cho, K. R., Oliner, J. D., Simons, J. W., Hedrick, L., Fearon, E. R., Preisinger, A. C., Hedge, P., Silverman, G. A. & Vogelstein, B. (1994) *Genomics 19*, 525–531.

Coughlin, P., Sun, J., Cerruti, L., Salem, H. H. & Bird, P. (1993) *Proc. Natl. Acad. Sci. U.S.A. 90*, 9417–9421.

Cousin, E., Medcalf, R. L., Bergonzelli, G. E. & Kruithof, E. K. (1991) *Nucleic Acids Res. 19*, 3881–3886.

Dear, A. E. & Medcalf, R. L. (1995) *Fibrinolysis 9*, 321–330.

Dickinson, J. L., Bates, E. J., Ferrante, A. & Antalis, T. M. (1995) *J. Biol. Chem. 17*, 27894–27904.

Dubin, A., Travis, J., Enghild, J. J. & Potempa, J. (1992) *J. Biol. Chem. 267*, 6576–6583.

Evans, D. L. (1991) *Heparin Activatable Serpins,* Ph.D. thesis, Cambridge University, UK.

Fa, M., Karolin, J., Aleshkov, S., Strandberg, L., Johansson, L. B.-Å. & Ny, T. (1995) *Biochemistry 34*, 13833–13840.

Gan, H., Newman, G. W. & Remold, H. G. (1995) *J. Immunol. 155*, 1304–1315.

Genton, C., Kruithof, E. K. & Schleuning, W. D. (1987) *J. Cell Biol. 104*, 705–712.

Gettins, P., Patston, P. A. & Schapira, M. (1993) *Bioessays 15*, 461–467.

Jensen, P. H., Schueler, E., Woodrow, G., Richardson, M., Goss, N., Hojrup, P., Petersen, T. E. & Rasmussen, L. K. (1994) *J. Biol. Chem. 269*, 15394–15398.

Jensen, P. J., Wu, Q., Janowitz, P., Ando, Y. & Schechter, N. M. (1995) *Exp. Cell Res. 217*, 65–71.

Kawano, T., Morimoto, K. & Uemura, Y. (1968) *Nature 217*, 253–254.

Kawano, T., Morimoto, K. & Uemura, Y. (1970) *J. Biochem. (Tokyo) 67*, 333–342.

Kinnby, B., Lecander, I., Martinsson, G. & Åstedt, B. (1991) *Fibrinolysis 5*, 239–242.

Kiso, U., Kaudewitz, H., Henschen, A., Åstedt, B., Kruithof, E. K. & Bachmann, F. (1988) *FEBS Lett. 230*, 51–56.

Kruithof, E. K., Baker, M. S. & Bunn, C. L. (1995) *Blood 1*, 4007–4024.

Kruithof, E. K. & Cousin, E. (1988) *Biochem. Biophys. Res. Commun. 156*, 383–388.

Kruithof, E. K., Vassalli, J. D., Schleuning, W. D., Mattaliano, R. J. & Bachmann, F. (1986) *J. Biol. Chem. 261*, 11207–11213.

Kumar, S. (1995) *Trends Biochem. Sci. 20*, 198–202.

Kumar, S. & Baglioni, C. (1991) *J. Biol. Chem. 266*, 20960–20964.

Laug, W. E., Aebersold, R., Jong, A., Rideout, W., Bergman, B. L. & Baker, J. (1989) *Thromb. Haemost. 61*, 517–521.

Laurell, C.-B. & Eriksson, S. (1963) *Scand. J. Clin. Lab. Invest. 15*, 132–140.

Lawrence, D. A., Ginsburg, D., Day, D. E., Berkenpas, M. B., Verhamme, I. M., Kvassman, J. O. & Shore, J. D. (1995) *J. Biol. Chem. 270*, 25309–25312.

Lawrence, D. A., Strandberg, L., Ericson, J. & Ny, T. (1990) *J. Biol. Chem. 265*, 20293–20301.

Lecander, I. & Åstedt, B. (1986) *Br. J. Haematol. 62*, 221–228.

Lecander, I. & Åstedt, B. (1987) *J. Lab. Clin. Med. 110*, 602–605.

Leung, K., Byatt, J. A. & Stephens, R. W. (1987) *Thromb. Res. 46*, 767–777.

Lomas, D. A., Evans, D. L., Finch, J. T. & Carrell, R. W. (1992) *Nature 357*, 605–607.

Lomas, D. A., Finch, J. T., Seyama, K., Nukiwa, T. & Carrell, R. W. (1993) *J. Biol. Chem. 268*, 15333–15335.

Lyons Giordano, B. & Lazarus, G. S. (1994) *Experimental Dermatology 3*, 85–88.

Mast, A. E., Enghild, J. J. & Salvesen, G. (1992) *Biochemistry 31*, 2720–2728.

Medcalf, R. L., Kruithof, E. K. & Schleuning, W. D. (1988a) . *Exp. Med. 168*, 751–759.

Medcalf, R. L., Van den Berg, E. & Schleuning, W. (1988b) *J. Cell Biol. 106*, 971–978.

Mikus, P. (1995) Ph.D. thesis, Umeå University, Sweden.

Mikus, P., Urano, T., Liljestrom, P. & Ny, T. (1993) *Eur. J. Biochem. 218*, 1071–1082.

Miura, Y., Kawabata, S., Wakamiya, Y., Nakamura, T. & Iwanaga, S. (1995) *J. Biol. Chem. 270*, 558–565.

Mueller, B. M., Yu, Y. B. & Laug, W. E. (1995) *Proc. Natl. Acad. Sci.U.S.A. 92*, 205–209.

Ny, T., Hansson, L., Lawrence, D., Leonardsson, G. & Åstedt, B. (1989) *Fibrinolysis 3*, 189–196.

Pickup, D. J., Ink, B. S., Hu, W., Ray, C. A. & Joklik, W. K. (1986) *Proc. Natl. Acad. Sci. U.S.A. 83*, 7698–7702.

Piquette, G. N., Crabtree, M. E., el-Danasouri, I., Milki, A. & Polan, M. L. (1993) *J. Clin. Endocrinol. Metab. 76*, 518–523.

Ray, C. A., Black, R. A., Kronheim, S. R., Greenstreet, T. A., Sleath, P. R., Salvesen, G. S. & Pickup, D. J. (1992) *Cell 69*, 597–604.

Remold O'Donnell, E., Chin, J. & Alberts, M. (1992) *Proc. Natl. Acad. Sci. USA 89*, 5635–5639.

Remold-O'Donnell, E. (1993) *FEBS Lett. 315*, 105–108.

Schleef, R. R., Wagner, N. V. & Loskutoff, D. J. (1988) *J. Cell Physiol. 134*, 269–274.

Schleuning, W. D., Medcalf, R. L., Hession, C., Rothenbuhler, R., Shaw, A. & Kruithof, E. K. (1987) *Mol. Cell Biol. 7*, 4564–4567.

Schneider, S. S., Schick, C., Fish, K. E., Miller, E., Pena, J. C., Treter, S. D., Hui, S. M. & Silverman, G. A. (1995) *Proc. Natl. Acad. Sci. USA. 11*, 3147–3151.

Schulze, A. J., Baumann, U., Knof, S., Jaeger, E., Huber, R. & Laurell, C. B. (1990) *Eur. J. Biochem. 194*, 51–56.

Schwartz, B. S. & Bradshaw, J. D. (1992) *J. Biol. Chem. 267*, 7089–7094.

Sharp, H. L., Bridges, R. A., Krivit, W. & Freier, E. F. (1969) *J. Lab. Clin. Med. 73*, 934–939.

Silverman, G. A., Jockel, J. I., Domer, P. H., Mohr, R. M. & Taillon-Miller, P. (1991) *Genomics 9*, 219–228.

Steven, J., Cottingham, I. R., Berry, S. J., Chinery, S. A., Goodey, A. R., Courtney, M. & Ballance, D. J. (1991) *Eur. J. Biochem. 196*, 431–438.

Suminami, Y., Kishi, F., Sekiguchi, K. & Kato, H. (1991) *Biochem. Biophys. Res. Commun. 181*, 51–58.

Takeuchi, Y., Nakao, A., Harada, A., Nonami, T., Fukatsu, T. & Takagi, H. (1993) *Am. J. Gastroenterol. 88*, 1928–1933.

Teschauer, W. F., Mentele, R. & Sommerhoff, C. P. (1993) *Eur. J. Biochem. 217*, 519–526.

Thorsen, S., Philips, M., Selmer, J., Lecander, I. & Åstedt, B. (1988) *Eur. J. Biochem. 175*, 33–39.

van Kessel, A. G., Straub, R. E., Silverman, G. A., Gerken, S. & Overhauser, J. (1994) *Cytogenet. Cell. Genet. 65*, 142–165.

von Heijne, G., Liljestrom, P., Mikus, P., Andersson, H. & Ny, T. (1991) *J. Biol. Chem. 266*, 15240–15243.

Webb, A. C., Collins, K. L., Snyder, S. E., Alexander, S. J., Rosenwasser, L. J., Eddy, R. L., Shows, T. B. & Auron, P. E. (1987) *J. Exp. Med. 166*, 77–94.

Webb, G., Baker, M. S., Nicholl, J., Wang, Y., Woodrow, G., Kruithof, E. & Doe, W. F. (1994) *J. Gastroenterol. Hepatol. 9*, 340–343.

Wilczynska, M., Fa, M., Ohlsson, P. I. & Ny, T. (1995) *J. Biol. Chem. 270*, 29652–29655.

Wohlwend, A., Belin, D. & Vassalli, J. D. (1987) *J. Exp. Med. 165*, 320–339.

Wun, T. C. & Reich, E. (1987) *J. Biol. Chem. 262*, 3646–3653.

Ye, R. D., Ahern, S. M., Le Beau, M. M., Lebo, R. V. & Sadler, J. E. (1989) *J. Biol. Chem. 264*, 5495–5502.

Ye, R. D., Wun, T. C. & Sadler, J. E. (1987) *J. Biol. Chem. 262*, 3718–3725.

Ye, R. D., Wun, T. C. & Sadler, J. E. (1988) *J. Biol. Chem. 263*, 4869–4875.

Zou, Z., Anisowicz, A., Hendrix, M. J., Thor, A., Neveu, M., Sheng, S., Rafidi, K., Seftor, E. & Sager, R. (1994) *Science 263*, 526–529.

13

OF MICE AND MEN

The Function of Plasminogen Activator Inhibitors (PAIs) *in Vivo*

Daniel T. Eitzman[*] and David Ginsburg

Howard Hughes Medical Institute
University of Michigan Medical Center
MSRB 1, Room 4520
1150 W. Medical Center Drive
Ann Arbor, Michigan 48109-0650

As illustrated in figure 1, hemostasis is a delicate balance between the coagulation system on one side, giving rise to the insoluble fibrin blood clot, and the fibrinolytic system on the other, which leads to dissolution of the clot. The center of both systems is a cascade of proteolytic enzymes, each activating the next, with the final proteases in each cascade, thrombin and plasmin, being the final effector molecules. Thrombin proteolytically cleaves soluble circulating fibrinogen to form the insoluble fibrin clot and plasmin cleaves fibrin into soluble degradation products[1]. One of the primary regulatory points in the fibrinolytic system is thought to occur at the step of the plasminogen activators and their interaction with a specific group of inhibitors, the plasminogen activator inhibitors (PAI's)[2]. The two known, natural plasminogen activators in mammals are tissue plasminogen activator (tPA) and urokinase type plasminogen activator (uPA). tPA is found primarily in the vascular space while uPA is generally thought to be of more importance in the extravascular milieu. The primary PAI regulating uPA and tPA function is PAI-1. PAI-1 interacts with high affinity with both tPA and uPA, forming the typical stable serpin/protease complex[1,3]. PAI-1 is unique among serpins in its lability, spontaneously transforming to an inactive or latent conformation with a t1/2 of approximately 90 minutes[3]. PAI-1 in plasma is predominantly bound to an abundant plasma protein, vitronectin which stabilizes PAI-1 in the active form[4]. Vitronectin is also present in the extracellular matrix of many tissues and it may serve to localize PAI-1 function to specific sites.

A second PAI, PAI-2 is found predominantly in monocytes and macrophages, in the placenta, and in the plasma of pregnant women (for review see[5,6]). It is an efficient inhibitor of uPA and only a poor inhibitor of tPA. The exact function of PAI-2 is unknown,

* Fax: 313-936-2888.

Chemistry and Biology of Serpins, edited by Church *et al.*
Plenum Press. New York. 1997

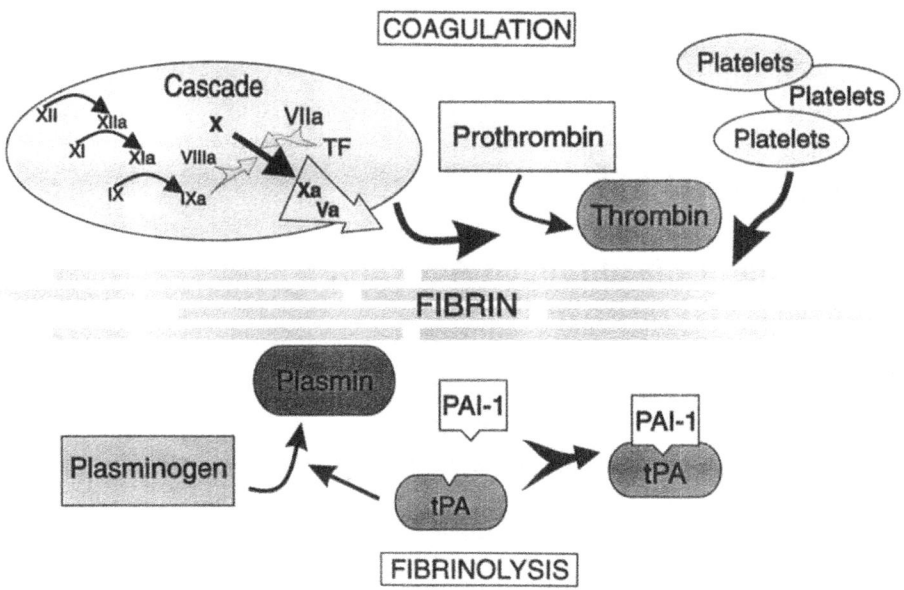

Figure 1. The coagulation cascade is initiated by the "intrinsic" pathway (via factor XII) or "extrinsic" pathway (via tissue factor (TF)) leading to the conversion of prothrombin to thrombin with subsequent cleavage of fibrinogen to form the fibrin clot. Plasmin-mediated fibrinolysis occurs following the conversion of plasminogen to plasmin by plasminogen activators, urokinase-type plasminogen activator (uPA) or tissue-type plasminogen activator (tPA). Both plasminogen activators are rapidly inhibited by plasminogen activator inhibitor-1 (PAI-1). (From Principles and Practice of Medical Genetics — Rimoin, Connor and Pyeritz, 3rd Edition In Press, Churchill Livingstone.)

though it is thought to play a major role in placental maintenance, tumor invasion and metastasis, tissue remodeling, and a variety of inflammatory processes involving monocytes and macrophages. PAI-2 exists in both a glycosylated secreted form as well as a nonglycosylated intracellular form. Approximately 80% of the protein is in the latter intracellular pool. The function of intracellular PAI-2 is unknown, though it has been proposed that it may play a role in regulating apoptosis[7,8].

This chapter will review recent results from our laboratory that address the function of the plasminogen activator axis *in vivo*.

PAI-1 DEFICIENCY IN MICE AND HUMANS

Considerable data collected over the past 10 years has demonstrated a complex pattern for PAI-1 expression and gene regulation in numerous cell types, influenced by a diverse set of modulators including a variety of cytokines and inflammatory mediators[9,10]. The proposed role of PAI-1 in the regulation of plasminogen activation in the processes of embryo implantation and development, tumor invasion and metastasis, wound healing, and ovulation, as well as the regulation of fibrinolysis, suggested that deficiency of PAI-1 might result in a severe or early embryonic lethal phenotype with pleomorphic manifestations. Rare patients were identified with partial deficiency of PAI-1 associated with significant clinical bleeding[11,12]. Only a single patient with complete deficiency of PAI-1 has been reported[13]. This patient, a 9 year old girl with several episodes of life-threatening hemorrhage following minor trauma or surgery, was found to have 2-base pair insertion at the end of PAI-1 exon 4, resulting in a frameshift which removes the carboxy terminal half of the protein, including the reactive center. The resulting mRNA was shown to be very unstable as was the recombinant mutant protein expressed in *E. coli*. The patient was homozygous for this defect, having inherited the same allele from both parents, and had undetectable levels of PAI-1 in platelets and plasma. The heterozygous parents and four heterozygous siblings all had intermediate PAI-1 levels, though within the wide range of normal. The lack of any apparent abnormality in development, wound healing, or the response to infection and inflammation was surprising, given the wide-range of presumed actions of the PA system, and suggested that the primary role of PAI-1 *in vivo* is restricted to the regulation of fibrinolysis.

Generation of "knockout" mice in which the PAI-1 gene had been disrupted by gene targeting, confirmed these observations[14]. PAI-1 deficient mice demonstrated normal development, fertility, and survival, though an enhanced basal state of fibrinolytic activity could be demonstrated. The mice were also relatively resistant to the prothrombotic effect of endotoxin injected into the foot pad[15].

An important point should be made in comparing the results of these mouse model experiments to human disease, an issue which is relevant not just for the PAI-1 knockouts but also for the other transgenic models to be discussed below. A number of investigators have described the PAI-1 deficient mice as having a "milder" phenotype than the human. However, what appears to be a milder phenotype may simply be a bias introduced by our much more limited ability to detect functional abnormalities in the mouse. Though the human patient did indeed have life-threatening hemorrhage, her first episode was at age 3, associated with minor head trauma. "Lack of significant bleeding" in knockout mice is generally based on observation for one year or less in a small cage in a pathogen free environment in the laboratory. Thus, the phenotypes may really not be very different at all and any description of knockout mice having "no phenotype" should be viewed with caution.

Analysis of recent "knockout" mice generated to be completely deficient in tPA and uPA also yielded surprising results. Again, these mice were fertile and developed normally. Even doubly deficient mice survived and were fertile, though the deficiency of both proteins resulted in serious illness and decreased survival associated with increased thrombosis[16]. The unmasking of this thrombotic phenotype with the double deficiency of both proteins illustrates the increasingly recognized phenomenon of partial overlap in function among gene products and the ability of one gene to compensate for loss of another. Such complex interrelated functions frequently makes prediction of phenotype in the deficient mouse (or human) very difficult, based only on *in vitro* studies and indirect *in vivo* observations.

VITRONECTIN DEFICIENCY

Recent analysis of knockout mice generated to be completely deficient in vitronectin provide another case in point[17]. Vitronectin is a major component of plasma and the extracellular matrix of many tissues[18,19]. Vitronectin is thought to play a major role in cell attachment and spreading and thereby contribute to organogenesis, development, and wound healing. In addition, vitronectin is an important cofactor for PAI-1. Nearly all PAI-1 in plasma is found in complex to vitronectin, resulting in increased stability with a

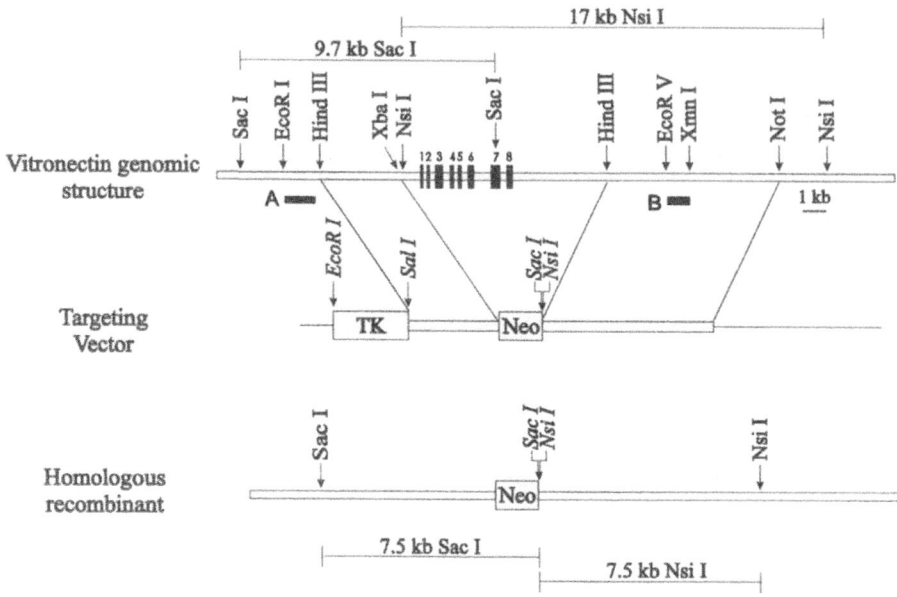

Figure 2. Targeted disruption of the vitronectin gene. The murine vitronectin genomic structure is depicted at the top with the exons indicated by vertical black bars. Probe A (*EcoR*I/*Hind*III genomic fragment) and probe B (*EcoR*V/*Xmn*1 fragment) are indicated with the black bar below the vitronectin genomic structure. The targeting vector, shown in the middle, contains two homologous fragments derived from vitronectin genomic sequences and a phosphoglycerate kinase (PGK) neomycin (Neo) expression cassette that replaces the entire vitronectin locus along with approximately 3 kb of downstream sequences. A PGK thymidine kinase (TK) cassette was inserted at the 5' end of the construct. Restriction sites denoted in italics are present in the vector. Recombination events within the two homologous segments should produce targeted deletion of the vitronectin gene, resulting in the structure depicted in the lower panel. (Adapted from Zheng et al[17], with permission.)

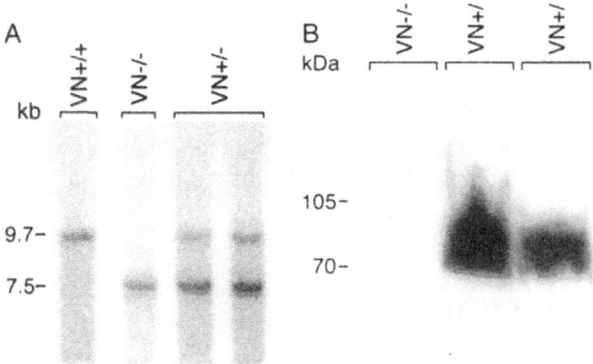

Figure 3. Deletion of the vitronectin (VN) gene. (*A*) Southern blot analysis of *Sac*1-digested tail DNA from VN $^{+/+}$, VN $^{+/-}$, and VN $^{-/-}$ mice using probe A (Fig. 2); the wild-type *Sac*1 fragment is 9.7 kb in length, with the targeted allele generating a novel 7.5-kb band. (*B*) Western blot analysis of VN in plasma samples prepared from VN $^{+/+}$, VN $^{+/-}$, and VN $^{-/-}$ mice; plasma proteins were separated on a 12.5% SDS/PAGE gel, electro-blotted, and detected with polyclonal rabbit antimouse VN antisera, using an ECL chemiluminescence kit. (Re-printed from Zheng et al [17], with permission.)

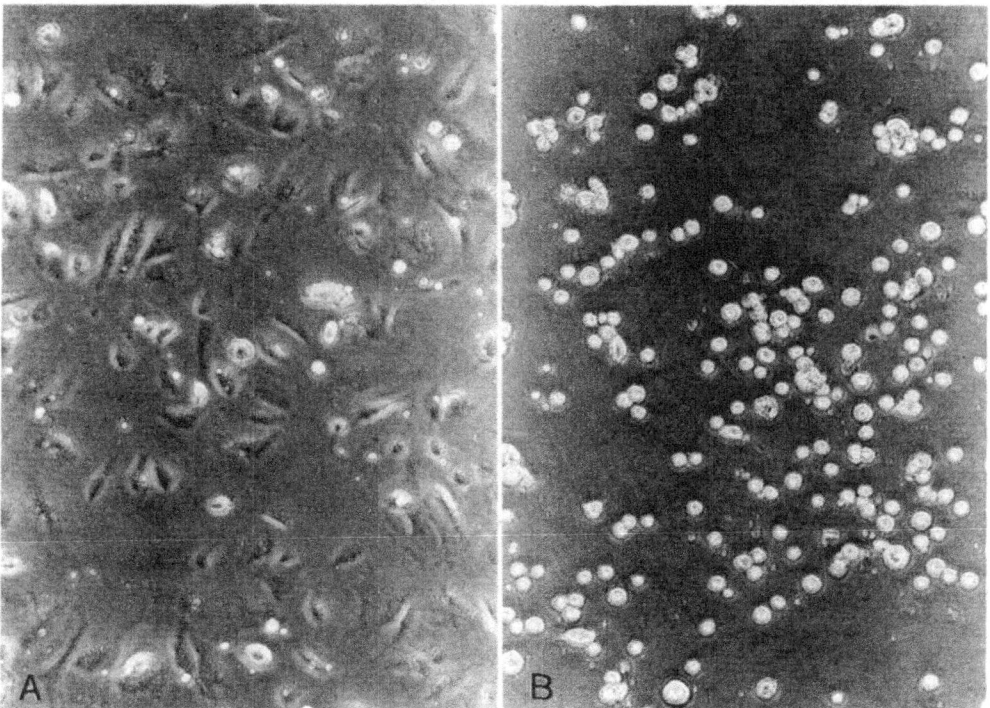

Figure 4. Effect of serum on COS cell spreading in tissure culture. Cells were plated in medium containg 5% serum obtained from either VN ⁺/+ (*A*) or VN ⁻/₋ (*B*) mice. Photographs were taken 2 hr after plateing. (×110). (Reprinted from Zheng et al[17], with permission.)

near doubling of half life[20]. Vitronectin may also serve to localize PAI-1 to specific sites in the extracellular matrix.

The strategy used to generate vitronectin deficienct mice by gene targeting is illustrated in figure 2[17]. Successful homologous recombination between the targeting vector and the endogenous vitronectin gene should result in replacement of an approximately 8 kb genomic segment containing the entire vitronectin coding sequence (exons 1–8) by a neomycin selectable gene expression unit. The detection of successful targeting is demonstrated in figure 3a. A unique 7.5 kb *Sac*I fragment is generated by the successful targeting event and a single copy of the wild-type 9.7 kb and 7.5 kb targeted allele is evident in heterozygous VN ⁺/₋ mice. VN ⁻/₋ mice demonstrate only the targeted allele. As predicted, a Western blot of plasma from targeted mice, shown in figure 3b, demonstrates complete absence of vitronectin from the homozygote deficient mice, with reduced levels in the heterozygotes. Sera from vitronectin deficient mice was unable to support the cell attachment and spreading of tissue culture cells observed in normal mouse sera (figure 4). This critical requirement of vitronectin for cell growth *in vitro* stands in marked contrast to the surprising *in vivo* observations of normal survival and fertility in VN ⁻/₋ mice. This striking contrast between the *in vivo* and *in vitro* setting, again emphasizes the complex nature of the intact organism. VN ⁻/₋ sera also exhibits complete absence of PAI-1 binding activity, demonstrating that vitronectin is indeed the major PAI-1 binding protein in blood. This deficiency is not associated with any obvious increase in bleeding tendency. However, it is important to emphasize that our observations are still limited and subtle abnormalities cannot be excluded until more detailed studies have been conducted.

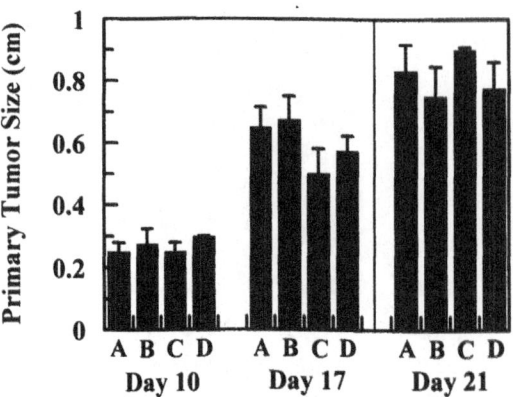

Figure 5. Effect of plasminogen activator inhibitor-1 (PAI-1) on primary tumor size. Mice underwent right hindlimb footpad injections with 1×10^6 melanoma cells in 0.25 ml of Hanks balanced salt solution (HBSS). The maximal foot pad dimension was measured on days 10, 17 and 21. No significant differences were observed between the genetically altered mice and their respective control groups at each time point. A = control mice carrying 2 normal PAI-1 alleles (PAI-1 +/+); B = mice deficient in PAI-1 (PAI-1 −/−); C = PAI-1 overexpressing mice carrying PAI-1 transgene (PAI-1 TG +); D = control mice without transgene (PAI-1 TG −). (Reprinted from Eitzman et al[32], with permission.)

THE ROLE OF PAI-1 IN TUMOR INVASION AND METASTASIS

Extensive evidence has been reported over the years for the central role of proteolysis in tumor cell invasion and metastasis[21]. The plasminogen activator uPA is widely expressed by a variety of neoplastic cells and is thought to be critical to the tumor invasive phenotype[10,22–30]. We studied the effects of genetic variation of host PAI-1 levels in a mouse model of metastatic melanoma. The B16 melanoma cells used in these experiments are known to produce high levels of uPA and tumor metastasis can be attenuated by pretreatment with antibodies to uPA[23]. However, despite this dramatic effect following pretreatment of melanoma cells *in vitro*, and the reported inhibitory effects on metastasis of PAI-2 transfected into melanoma cells[31], we could demonstrate no effect of host PAI-1 on B16 melanoma tumor cell behavior[32]. B16 melanoma cells were analyzed for local tumor growth and pulmonary metastasis after injection into transgenic mice engineered to overexpress murine PAI-1 in multiple tissues, including lung. Tumor cells were also administered to knockout mice completely deficient in PAI-1. The tumor size at days 10, 17, and 21 following injection of 1×10^6 melanoma cells into the footpad is shown in figure 5. No significant differences were observed in tumor size among the various genetic backgrounds despite the marked differences in endogenous PAI-1 levels. Similarly, the mean number of tumor nodules per lung observed after the tail vein injection did not differ significantly between control animals and transgenic PAI-1 overexpressing mice (figure 6). Finally, no difference in survival could be demonstrated among the various groups (figure 7). Although these results suggest that prior

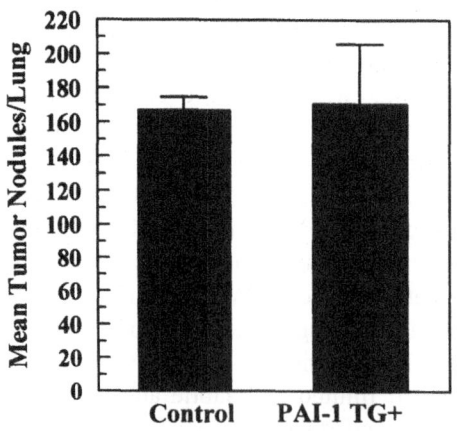

Figure 6. Effect of PAI-1 overexpression on pulmonary metastases. Littermate control mice without transgene (Control) and mice carrying the PAI-1 transgene (PAI-1 TG +) underwent tail vein injection with 3×10^5 melanoma cells in 0.5 ml buffer solution. 14 days later, mice were sacrificed and lungs retrieved for enumeration of pulmonary nodules. No significant differences were observed in the number of nodules (p = NS). (Reprinted from Eitzman et al[32], with permission.)

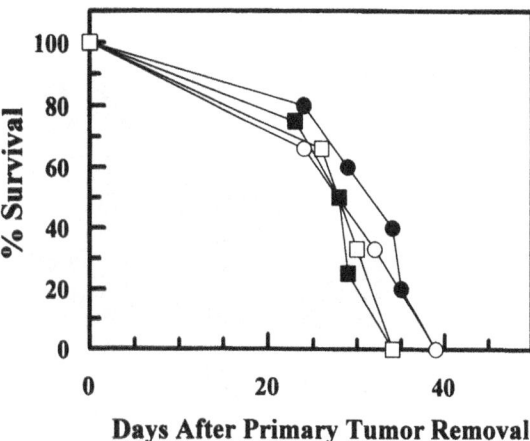

Figure 7. Effect of PAI-1 status on survival. At day 21 following right foot pad injection, a right hip disarticulation was performed to remove the primary tumor mass. Mice surviving this procedure were then followed for survival. Mean surviving days following removal of the primary tumor mass was similar between the 4 groups, as were survival curves. ● = PAI-1 deficient mice, ■ = PAI-1 control mice for PAI-1 deficiency, ○ = PAI-1 overexpressing mice, □ = PAI-1 control mice for overexpressors. (Reprinted from Eitzman et al[32], with permission.)

reports may have overestimated the critical role of plasminogen activation in tumor invasion and metastasis, these results should also be viewed with some caution. The number of animals was small in these studies and the analysis of larger numbers might reveal more subtle effects. In addition, considerable variation may be observed among different isolates of the B16 melanoma cell line and behavior of other tumors could vary considerably. Finally, these studies only examine the effect of host PAI-1 levels and cannot exclude a critical regulatory role for PAI-1 produced by the tumor itself, or high local concentrations of PA produced by the tumor cells which were protected from host PAI-1 levels.

THE REGULATION OF PLASMINOGEN ACTIVATION IN PULMONARY FIBROSIS

A number of diverse pulmonary disorders share a common pathway of acute and chronic inflammation resulting in pulmonary fibrosis. In a number of these conditions, including ARDS, idiopathic pulmonary fibrosis, sarcoidosis, and broncho-pulmonary dysplasia, fibrinolytic activity has been noted to be depressed in bronchoalveolar lavage (BAL) fluid[33-37]. Similar patterns have been observed in animal models of lung injury[38]. Elevated levels of PAI-1 have been observed in BAL specimens from patients with ARDS[34] and may explain the reduced fibrinolytic activity observed in a number of pulmonary disease states. Though increased PAI-1 levels have been associated with pulmonary fibrosis in earlier studies, we sought to test the hypothesis that fibrinolytic balance was critical to the recovery from lung injury in a transgenic mouse model[39]. For these experiments, pulmonary injury was induced by intratracheal injection of bleomycin, a well established model for lung injury in the mouse and rat[38,40,41]. Again, the studied animals included PAI-1 knockout mice with complete deficiency of PAI-1, wild-type mice, and transgenic mice expressing high levels of murine PAI-1 in multiple tissues. Hydroxyproline content of lung, as an indicator of total collagen, was determined two weeks after bleomycin exposure. The results are shown in figure 8. A highly significant increase in hydroxyproline content was evident in the PAI-1 transgenic overexpressing mice following bleomycin injury. Consistent with this data, PAI-1 deficient mice showed protection from bleomycin-induced collagen deposition, even at the higher bleomycin concentration (figure 9), as the hydroxyproline content of homozygous deficient mice did not significantly differ from control mice that received only intratracheal saline. Consistent with this close correlation between endogenous genetically determined PAI-1 level

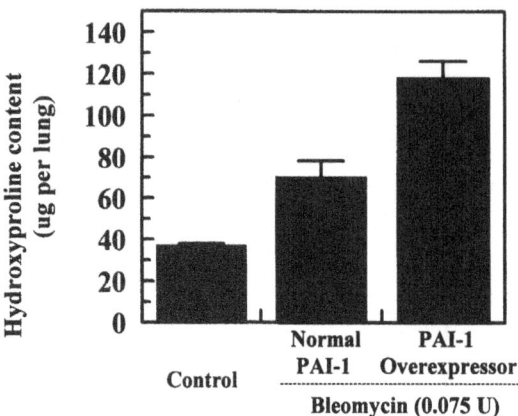

Figure 8. Effect of intratracheal bleomycin on lung hydroxyproline content in PAI-1 overexpressing mice and littermates: Lungs from all mice that received intratracheal bleomycin (0.075 units) contained a significantly greater hydroxyproline content than mice that received only PBS (p < 0.002). Mice overexpressing PAI-1 contained more lung hydroxyproline than littermate normal controls (p < 0.005). (Reprinted from Eitzman et al[39], with permission.)

and pulmonary fibrosis, heterozygous PAI-1 deficient mice showed an intermediate level of hydroxyproline content. Figure 10 shows illustrative lung sections with severe fibrosis in transgenic overexpressing mice at lower bleomycin concentrations and marked protection from lung injury in homozygous deficient mice at the higher bleomycin concentration.

These results are consistent with a model of a common pathway in which many forms of lung injury would result in intra-alveolar hemorrhage and thrombosis. The fibrin meshwork would then serve as a provisional matrix for the invasion of fibroblasts with subsequent collagen deposition. Prompt clearing of the fibrin meshwork by the fibrinolytic system would minimize the amount of connective tissue deposition and thus, reduce fibrosis, whereas failure of fibrinolysis would tip the balance toward more fibrosis.

This model for tissue injury and repair could also apply to the pathogenesis of other disease processes. For example, chronic injury and low level fibrin clot formation along the vessel wall at sites of arterial injury could form a provisional matrix for cellular migration and the development of atherosclerosis. In this setting as well, enhanced fibrinolysis could conceivably slow disease progression. This hypothesis is also being tested using available transgenic mouse reagents.

SUMMARY AND FUTURE DIRECTIONS

Since the discovery of the PAIs in the early 1980s, considerable progress has been made toward understanding the molecular mechanism for action of these serpins, as dis-

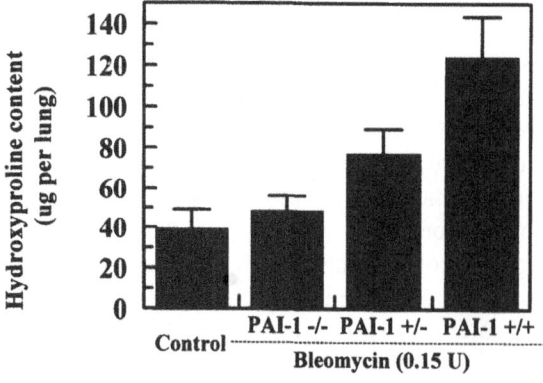

Figure 9. Effect of intratracheal bleomycin on lung hydroxyproline content in PAI-1 deficient mice and littermates: Lungs from all mice receiving bleomycin (0.15 units) contained more hydroxyproline than did mice that received only PBS (p < 0.002). Lungs from normal mice (PAI-1 +/+) contained more hydroxyproline than both PAI-1 heterozygotes (PAI-1 +/−) (p < 0.02) and PAI-1 deficient (PAI-1 −/−) (p < 0.0006) mice. (Reprinted from Eitzman et al[39], with permission.)

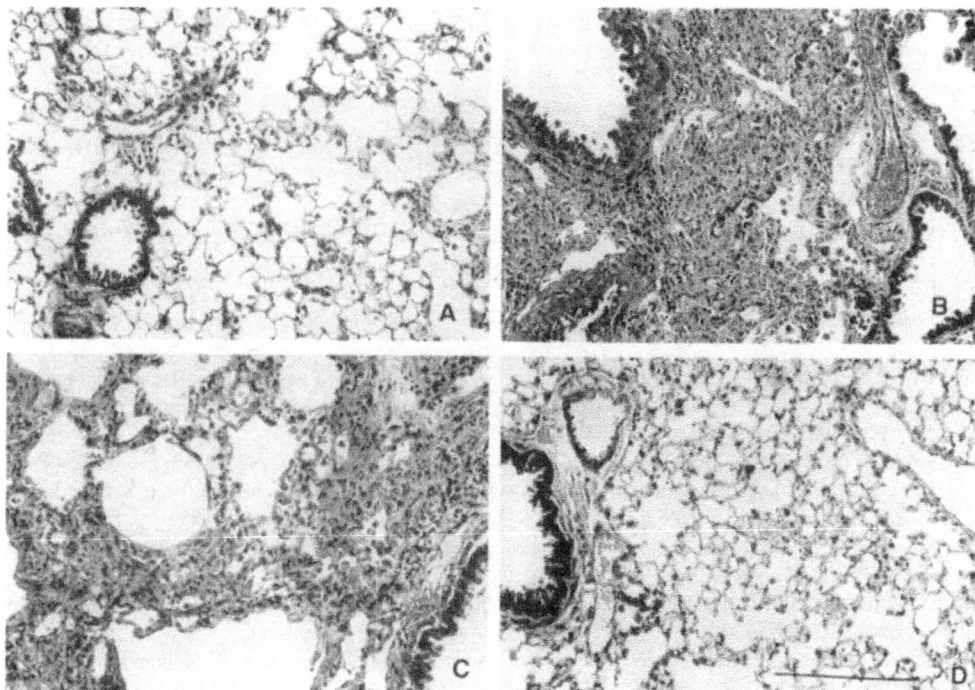

Figure 10. Histology: Photomicrographs of Masson trichrome-stained sections of lung tissue from bleomycin-treated mice. A) Normal mice 2 weeks after receiving 0.075 U bleomycin. B) PAI-1 overexpressing mice 2 weeks after receiving 0.075 U bleomycin. C) Normal mice 3 weeks after receiving 0.15 U bleomycin. D) PAI-1 homozygous deficient mice 3 weeks after receiving 0.15 U bleomycin. (Reprinted from Eitzman et al[39], with permission.)

cussed in other chapters of this monograph. Extensive *in vitro* studies have elucidated the transcriptional and post-transcriptional regulation of PAI gene expression and suggested potential interactions which might contribute to a number of *in vivo* processes and disease states. Over the past two years, remarkable progress in molecular genetics has made available unique animal reagents with which to dissect the true function of the PAIs *in vivo*. Early results have served to put previous *in vitro* studies into perspective. Current evidence suggest that the primary role of PAI and the PA's is in the regulation of fibrinolysis and that the system may not be as essential for other biologic processes such as reproduction, tissue remodeling, and embryonic development, as previously thought. However, more detailed studies are beginning to suggest an important role for PAI-1 as a regulator of net fibrinolytic balance in a variety of inflammatory disease states. These results also suggest a number of potential future avenues for the development of novel therapies and animal systems in which to test these hypotheses.

Modification of fibrinolytic balance may be a valuable approach for the development of new therapies. PAI-1 is a particularly attractive target, given its predicted wide therapeutic index; i.e. — complete deficiency of PAI-1 in humans and mice appears to be associated with only a mild increased bleeding tendency and marked overexpression of PAI-1 in a mouse model does not produce serious complications in unchallenged animals. Despite this tolerance for wide variations in PAI-1 levels in the normal host, alterations in endogenous PAI-1 may have dramatic effects on the development of chronic tissue damage following injury. It is conceivable that novel therapies targeted at PAI-1[42], based on the sophisticated understanding of PAI-1 function and serpin action in general, could find application in the treatment of a variety of pulmonary fibrotic disorders as well as some types of vascular disease.

REFERENCES

1. Francis CW, Marder VJ: Physiologic regulation and pathologic disorders of fibrinolysis, in Colman RW, Hirsh J, Marder VJ, Salzman EW (eds): Hemostasis and Thrombosis. Basic Principles and Clinical Practice, Philadelphia, J. B. Lippincott Company, 1987, p 358
2. Loskutoff DJ, Sawdey M, Mimuro J: Type 1 plasminogen activator inhibitor. Prog Hemost Thromb 9:87, 1989
3. Lawrence D, Strandberg L, Grundström T, Ny T: Purification of active human plasminogen activator inhibitor 1 from *Escherichia coli*. Comparison with natural and recombinant forms purified from eucaryotic cells. Eur J Biochem 186:523, 1989
4. Declerck PJ, De Mol M, Alessi MC, Baudner S, Pâques E-P, Preissner KT, Müller-Berghaus G, Collen D: Purification and characterization of a plasminogen activator inhibitor 1 binding protein from human plasma. Identification as a multimeric form of S protein. J Biol Chem 263:15454, 1988
5. Kruithof EKO, Baker MS, Bunn CL: Biological and clinical aspects of plasminogen activator inhibitor type 2 (PAI-2). Blood 86:4007, 1995
6. Dear AE, Medcalf RL: The cellular and molecular biology of plasminogen activator inhibitor type-2. Fibrinolysis 9:321, 1995
7. Gan H, Newman GW, Remold HG: Plasminogen activator inhibitor type 2 prevents programmed cell death of human macrophages infected with *Mycobacterium avium*, serovar 4[1]. J Immunol 155:1304, 1995
8. Dickinson JL, Bates EJ, Ferrante A, Antalis TM: Plasminogen activator inhibitor type 2 inhibits tumor necrosis factor α-induced apoptosis. J Biol Chem 270:27894, 1995
9. Saksela O, Rifkin DB: Cell-associated plasminogen activation: regulation and physiological functions. Annu Rev Cell Biol 4:93, 1988
10. Dano K, Andreasen PA, Grondahl-Hansen J, Kristensen P, Nielsen LS, Skriver L: Plasminogen activators,tissue degradation,and cancer. Adv Cancer Res 44:139, 1985
11. Schleef RR, Higgins DL, Pillemer E, Levitt LJ: Bleeding diathesis due to decreased functional activity of Type 1 plasminogen activator inhibitor. J Clin Invest 83:1747, 1989
12. Diéval J, Nguyen G, Gross S, Delobel J, Kruithof EKO: A lifelong bleeding disorder associated with a deficiency of plasminogen activator inhibitor type I. Blood 77:528, 1991
13. Fay WP, Shapiro AD, Shih JL, Schleef RR, Ginsburg D: Complete deficiency of plasminogen-activator inhibitor type 1 due to a frame-shift mutation. N Engl J Med 327:1729, 1992
14. Carmeliet P, Kieckens L, Schoonjans L, Ream B, Van Nuffelen A, Prendergast GC, Cole MD, Bronson R, Collen D, Mulligan RC: Plasminogen activator inhibitor-1 gene-deficient mice. I. Generation by homologous recombination and characterization. J Clin Invest 92:2746, 1993
15. Carmeliet P, Stassen JM, Schoonjans L, Ream B, van den Oord JJ, De Mol M, Mulligan RC, Collen D: Plasminogen activator inhibitor-1 gene-deficient mice. II. Effects on hemostasis, thrombosis, and thrombolysis. J Clin Invest 92:2756, 1993
16. Carmeliet P, Schoonjans L, Kieckens L, Ream B, Degen JL, Bronson R, De Vos R, van den Oord JJ, Collen D, Mulligan RC: Physiological consequences of loss of plasminogen activator gene function in mice. Nature 368:419, 1994
17. Zheng X, Saunders TL, Camper SA, Samuelson LC, Ginsburg D: Vitronectin is not essential for normal mammalian development and fertility. Proc Natl Acad Sci USA 92:12426, 1995
18. Preissner KT: Structure and biological role of vitronectin. Annu Rev Cell Biol 7:275, 1991
19. Tomasini BR, Mosher DF: Vitronectin. Prog Hemost Thromb 10:269, 1991
20. Lawrence DA, Ginsburg D: Plasminogen Activator Inhibitors, in High KA, Roberts HR (eds): Molecular Basis of Thrombosis and Hemostasis, New York, Marcel Dekker, Inc. 1995, p 517
21. Vassalli J-D, Pepper MS: Membrane proteases in focus. Nature 370:14, 1994
22. Cajot JF, Kruithof EK, Schleuning WD, Sordat B, Bachmann F: Plasminogen activators, plasminogen activator inhibitors and procoagulant analyzed in twenty human tumor cell lines. Int J Cancer 38:719, 1986
23. Hearing VJ, Law LW, Corti A, Appella E, Blasi F: Modulation of metastatic potential by cell surface urokinase of murine melanoma cells. Cancer Res 48:1270, 1988

24. Pyke C, Kristensen P, Ralfkiaer E, Grondahl-Hansen J, Eriksen J, Blasi F, Dano K, Grndahl-Hansen J, Dan K: Urokinase-type plasminogen activator is expressed in stromal cells and its receptor in cancer cells at invasive foci in human colon adenocarcinomas. Am J Pathol 138:1059, 1991
25. Mignatti P, Robbins E, Rifkin DB: Tumor invasion through the human amniotic membrane: requirement for a proteinase cascade. Cell 47:487, 1986
26. Ossowski L: *In vivo* invasion of modified chorioallantoic membrane by tumor cells: the role of cell surface-bound urokinase. J Cell Biol 107:2437, 1988
27. Crowley CW, Cohen RL, Lucas BK, Liu G, Shuman MA, Levinson AD: Prevention of metastasis by inhibition of the urokinase receptor. Proc Natl Acad Sci USA 90:5021, 1993
28. Sordat B, Reiter L, Cajot J-F: Modulation of the malignant phenotype with the urokinase-type plasminogen activator and the type 1 plasminogen activator inhibitor. Cell Differentiation and Development 32:277, 1990
29. Ossowski L, Reich E: Antibodies to plasminogen activator inhibit human tumor metastasis. Cell 35:611, 1983
30. Ossowski L: Plasminogen activator dependent pathways in the dissemination of human tumor cells in the chick embryo. Cell 52:321, 1988
31. Mueller BM, Yu YB, Laug WE: Overexpression of plasminogen activator inhibitor 2 in human melanoma cells inhibits spontaneous metastasis in *scid/scid* mice. Proc Natl Acad Sci USA 92:205, 1995
32. Eitzman DT, Krauss JC, Shen T, Cui J, Ginsburg D: Lack of plasminogen activator inhibitor-1 effect in a transgenic mouse model of metastatic melanoma. Blood 87:4718, 1996
33. Bertozzi P, Astedt B, Zenzius L, Lynch K, LeMaire F, Zapol W, Chapman HJ: Depressed bronchoalveolar urokinase activity in patients with adult respiratory distress syndrome. N Engl J Med 322:890, 1990
34. Idell S, James KK, Levin EG, Schwartz BS, Manchanda N, Maunder RJ, Martin TR, McLarty J, Fair DS: Local abnormalities in coagulation and fibrinolytic pathways predispose to alveolar fibrin deposition in the adult respiratory distress syndrome. J Clin Invest 84:695, 1989
35. Chapman HA, Allen CL, Stone OL: Abnormalities in pathways of alveolar fibrin turnover among patients with interstitial lung disease. Am Rev Respir Dis 133:437, 1986
36. Hasday JL, Bachwich PR, Lynch JP, Sitrin RG: Procoagulant and plasminogen activator activities of bronchoalveolar fluid in patients with pulmonary sarcoidosis. Exp Lung Res 14:261, 1988
37. Viscardi RM, Broderick K, Sun CC, Yale LA, Hessamfar A, Taciak V, Burke KC, Koenig KB, Idell S: Disordered pathways of fibrin turnover in lung lavage of premature infants with respiratory distress syndrome. Am Rev Respir Dis 146:492, 1992
38. Idell S, James KK, Gillies C, Fair DS, Thrall RS: Abnormalities of pathways of fibrin turnover in lung lavage of rats with oleic acid and bleomycin-induced lung injury support alveolar fibrin deposition. Am J Pathol 135:387, 1989
39. Eitzman DT, McCoy RD, Zheng X, Fay WP, Shen T, Ginsburg D: Bleomycin-induced pulmonary fibrosis in transgenic mice that either lack or overexpress the murine plasminogen activator inhibitor-1 gene. J Clin Invest 97:232, 1996
40. Schrier DJ, Phan SH, McGarry BM: The effects of the nude (nu/nu) mutation of bleomycin-induced pulmonary fibrosis. A biochemical evaluation. Am Rev Respir Dis 127:614, 1983
41. Idell S, Gonzales KK, MacArthur CK, Gillies C, Walsh PN, McLarty J, Thrall RS: Bronchoalveolar lavage procoagulant activity in bleomycin-induced lung injury in marmosets. Characterization and relationship to fibrin deposition and fibrosis. Am Rev Respir Dis 136:124, 1987
42. Eitzman DT, Fay WP, Lawrence DA, Francis-Chmura AM, Shore JD, Olson ST, Ginsburg D: Peptide-mediated inactivation of recombinant and platelet plasminogen activator inhibitor-1 *in vitro*. J Clin Invest 95:2416, 1995

14

BIOLOGY OF PROGESTERONE-INDUCED UTERINE SERPINS

Peter J. Hansen and Wen-Jun Liu

Department of Dairy and Poultry Sciences
University of Florida
Gainesville, Florida 32611-0920

1. OCCURRENCE OF SERPINS IN THE PROGESTERONE-DOMINATED UTERUS

Progesterone is a pleiotropic regulator of uterine function required for the mainte-nance of pregnancy in mammals. A major action of progesterone is to induce expression of genes encoding for secretory proteins of the uterine endometrium. These proteins partici-pate in several roles deemed essential for survival of the conceptus and include enzymes, transport proteins, and regulatory proteins. Several serpins have been identified as being part of the milieu of secretory proteins in the progesterone-dominated uterus. Among these are plasminogen activator inhibitor-1, induced by progesterone in cultured human en-dometrial cells (Miyauchi et al., 1995; Lockwood et al., 1995), and α_1-antitrypsin, which is synthesized by human endometrium during pregnancy (Fay et al., 1990) and also can enter uterine fluid as a serum transudate (Bany and McRae, 1992). These proteins likely inhibit serine proteinases secreted by the uterus and placenta and may limit remodeling of the endometrium, participate in local hemostasis and regulate placental invasiveness. In addition to these well-known serpins, the uterus of certain species of ungulates each pro-duce one or more members of a subfamily of serpins that are characterized by endometrial site of synthesis, sequence homology to each other, and weak or no known antiproteinase activity. This chapter will concentrate on the most well-characterized of these uterine-spe-cific serpins — a glycoprotein of the sheep uterus called uterine milk protein (UTMP) — and will compare the properties of this protein to other uterine serpins. Uterine milk pro-tein is the most abundant steroid-induced protein yet described and gram quantities can be obtained from the uterus of a single unilaterally-pregnant ewe (Moffatt et al., 1987a). There is evidence to implicate UTMP in inhibition of lymphocyte funciton and in binding interactions with placental secretory proteins. It is also possible that UTMP acts as a car-rier protein to transport maternal products across the placenta.

Chemistry and Biology of Serpins, edited by Church *et al.*
Plenum Press, New York, 1997

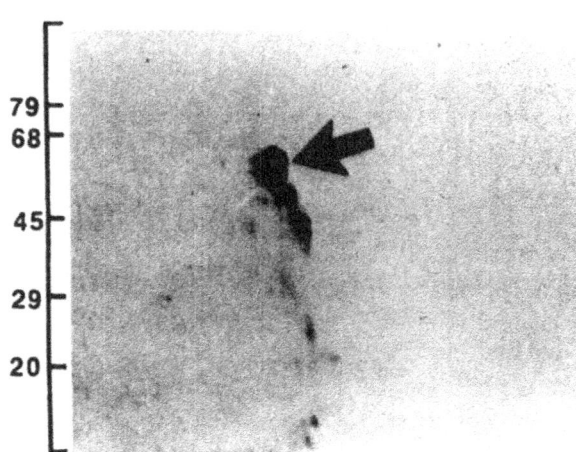

Figure 1. Analysis of proteins in uterine fluid of a unilaterally-pregnant ewe at day 120 of gestation. Proteins were separated in the first dimension by non-equilibrium pH gradient electrophoresis (origin at left) and in the second dimension by SDS-polyacrylamide gel electrophoresis (migration of molecular weight standards are depicted on the vertical axis). Proteins were visualized by staining with Coomassie Blue R-250. The major protein in the gel is uterine milk protein (UTMP), which exists in two major forms of $M_r = 55,000$ and $57,000$ (identified by the arrow). The other lower molecular weight proteins on the gel are believed to be breakdown products of UTM-proteins. Data are reproduced from Moffatt et al. (1987b) with permission from *Biology of Reproduction*.

2. SECRETION OF UTMP BY THE UTERUS

2.1. Patterns of Secretion during Pregnancy

Pregnancy proteins of the sheep uterus were originally characterized by placing a ligature over one uterine horn to create a unilaterally-pregnant ewe and allow accumulation of uterine fluid in the ligated horn (Bazer et al., 1979). This fluid, which can reach a liter in volume, has a milky-white appearance due to the presence of precipitated protein. Uterine milk protein was identified as a pair of proteins of similar molecular weight (55,000 and 57,000) and identical, basic pI that together make up the major proteins identifiable in both the soluble and particulate fractions of uterine fluid (Bazer et al., 1979; Moffatt et al., 1987b) (see Figure 1). The presence of UTMP is not an artifact of the unilaterally-pregnant model because UTMP and UTMP mRNA can be identified as products of endometrium from normal pregnant animals (Ing et al.,1989). Immunohistochemical analysis has revealed that UTMP is synthesized by glandular and lumenal epithelial cells (Moffatt et al., 1987b; Ing et al., 1989; Leslie et al., 1990; Leslie and Hansen, 1991).

Concentrations of UTMP are very low in the uterine lumen during the estrous cycle and do not accumulate in large amounts unless the animal is pregnant. Endometrial mRNA for UTMP has been detected as early as day 14 of pregnancy (Ing et al., 1989) and the protein itself can first be identified in uterine secretions at day 16 of pregnancy (Ing et al., 1989). By day 30 of pregnancy, and continuing until term, UTMP is the predominant protein present on electrophoretograms of uterine proteins (Moffatt et al., 1987b; Stephenson et al., 1989).

2.2. Role of Progesterone in Regulation of UTMP Secretion

The induction of endometrial UTMP secretion during pregnancy is a result of activation of UTMP gene expression by progesterone. Treatment of ovariectomized ewes with progesterone induces endometrial secretion of UTMP and accumulation of UTMP mRNA

(Hansen et al., 1986; Moffatt et al., 1987b; Ing et al., 1989; Stephenson and Hansen, 1990; Leslie and Hansen, 1991). Uterine milk protein is first synthesized in deeper areas of uterine glands, with the site of synthesis subsequently spreading first to other areas of glandular epithelium and then to the lumenal epithelium (Moffatt et al., 1987b; Ing et al., 1989; Leslie and Hansen, 1991). Probably as a result, prolonged exposure to progesterone increases secretion of UTMP: amounts in uterine flushings after 10 days of progesterone treatment are small compared to those from ovariectomized ewes treated for 30 days (Leslie and Hansen, 1991).

3. MOLECULAR BIOLOGY OF UTMP AND RELATED PROTEINS

Cloning of the UTMP gene and genes for related serpins in the pig and cow was conducted in the laboratory of R.M. Roberts at the University of Missouri. The inferred amino acid sequence for UTMP is shown in Figure 2. There are two putative start signals for translation that lead to unprocessed transcripts of 429 and 424 amino acids. Mature UTMP is 404 amino acids in length. The amino acid sequence of UTMP has homology with many serpins; greatest homology is with baboon α1-antitrypsin (31% sequence identity). Moreover, most of the protein domains present in UTMP are serpin domains (Figure 3). The exception are for two separate regions (the N-terminal portion of the protein and amino acids 286–310) which represent domains unique to the UTMP subfamily. The putative reactive center for UTMP is Ala^{390} (Ala^{365} in the mature protein).

```
         10         20         30         40         50         60
MSHRRMQLALSLVFILCGLFNSIFCEKQQHSQQHANLVLLKKISAFSQKMEAHPKAFAQE

         70         80         90        100        110        120
LFKALIAENPKKNIIFSPAAMTITLATLSLGIKSTMSTNHPEDLELELKLLDAHKCLHHL

        130        140        150        160        170        180
VHLGRELVKQKQLRHQDILFLNSKMMANQMLLHQIRKLQKMDIQMIDFSDTEKAKKAISH

        190        200        210        220        230        240
HVAEKTHTKIRDLITDLNPETILCLVNHIFFKGILKRAFQPNLTQKEDFFLNDKTKVQVD

        250        260        270        280        290        300
MMRKTEQMLYSRSEELFATMVKMPFKGNVSLILMLPDAGHFDNALKKLTAKRAKLQKISN

        310        320        330        340        350        360
FRLVHLTLPKFKITFDINFKHLLPKINLKHLLPKIDPKHTLTTTASSQHVTLKAPLPNLE

        370        380        390        400        410        420
ALHQVEIELSEHALTTDTAIHTDNLLKVPANTKEVPVVVKFNRPFLLFVEDEITQTDLFV

        429
GQVLNPQVE
```

Figure 2. Inferred amino acid sequence of sheep uterine milk protein. The signal sequence is underlined and start sites are marked by the boxed arrowheads. Putative glycosylation sites are marked with an asterisk and the putative reactive center is identified with an arrow. The sequence is from Ing and Roberts (1989) and was accessed from SWISS-PROT (Accession No. P21814).

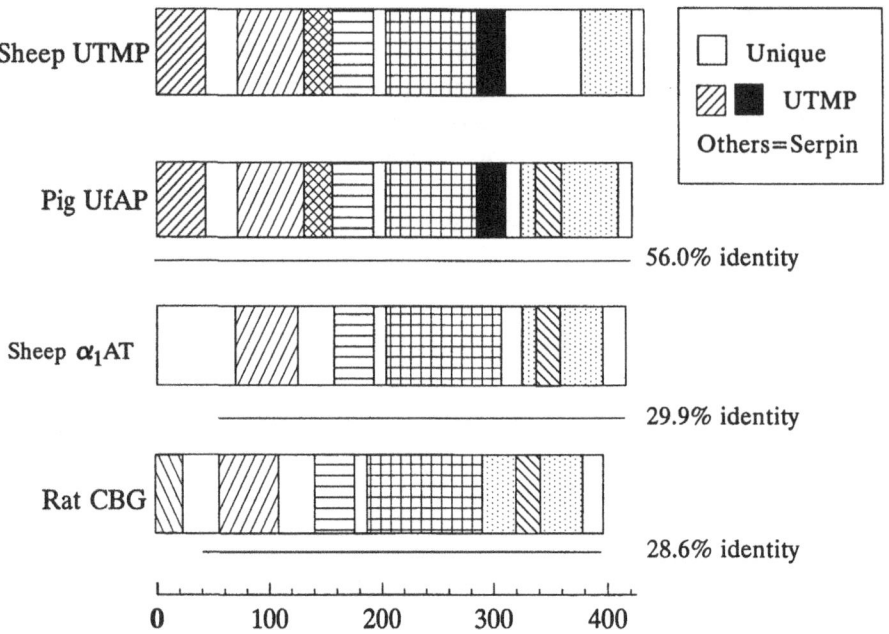

Figure 3. Comparison of the domain structure of sheep uterine milk protein (UTMP) and pig uteroferrin-associated protein (UfAP) with sheep α_1-antitrypsin (α_1AT) and rat corticosteroid binding globulin (CBG). All sequences are from the SWISS-PROT data bank (Bairoch and Boeckmann, 1991) and domains were obtained from the PRODOM database (Sonnhammer and Kahn, 1994). Sequence identities were calculated using the SIM local similarity program (Huang and Miller, 1991). Not shown here are two additional uterine serpins: a second pig UfAP which has 88.8% sequence identity with the UfAP shown here and a bovine UTMP gene that is 81.4% identical to the sheep UTMP.

Uterine proteins related to UTMP are found in other species. The pregnant bovine uterus secretes bovine UTMP (Leslie et al., 1990; Leslie and Hansen, 1991) while the uterus of the pregnant pig secretes several forms of a protein related to UTMP that is called uteroferrin-associated protein (UfAP) (Baumbach et al., 1986). The bovine gene has an inferred amino acid sequence (unpublished sequence of N. Mathialagan, T. Hansen and R.M. Roberts, SWISS-PROT Accession Number P46201) that is 81.4% identical to sheep UTMP. Two genes for UfAP have been cloned [Malathy et al., 1990 (SWISS-PROT Accession No. P16708) and an unpublished sequence of N. Mathialagan, P.V. Malathy and R.M. Roberts, (SWISS-PROT Accession No. P46202)] that are 55.6–56.0% identical to UTMP and 88.8% identical to each other. The gene for UfAP also contains two start codons (Malathy et al., 1990).

4. BIOCHEMICAL PROPERTIES OF UTMP

As illustrated in Figure 1, UTMP exists in uterine fluid and endometrium-conditioned culture medium as a pair of highly basic proteins that have a $M_r = 55,000$ and 57,000 as determined by two-dimensional SDS polyacrylamide gel electrophoresis. Based on the inferred amino acid sequence (Ing and Roberts, 1989), the protein has a pI of 9.47. N-terminal sequencing indicates only one N-terminal amino acid sequence (Hansen et al., 1987a; Ing and Roberts, 1989). Moreover, results of pulse-chase labeling experiments indicate that UTMP-57 and UTMP-55 are formed from a common tissue form of M_r = 53,000 (Hansen et al., 1987a). Thus, the two forms are likely to represent products of post-translational processing. Both forms of UTMP possess N-linked carbohydrates; the protein is approximately 5.6–5.7% carbohydrate by weight with one or two oligosaccharide chains per molecule that have been processed at least partially to complex-type forms (Hansen et al., 1987a,b). However, there is no evidence to indicate that the two forms of UTMP differ in the type of oligosaccharide conjugated to the protein core (Hansen et al., 1987a). Since there are two potential sites for N-glycosylation (Figure 2), one possibility is that UTMP-57 has two oligosaccharide chains while UTMP-55 has only one chain.

Uterine milk protein contains mannose 6-phosphate on its oligosaccharide chains (Hansen et al., 1987b). This sugar, which is also found on pig UfAP (Murray et al., 1989), is involved in targeting lysosomal enzymes to lysosomes and is also expressed on certain secretory proteins such as renin (Faust et al., 1987) and thyroglobulin (Herzog et al., 1987). No enzymatic activity characteristic of lysosomal enzymes has been found for UTMP (Hansen et al., 1987a). Also, UTMP becomes dephosphorylated after secretion because, while UTMP produced during tissue culture is phosphorylated, phosphate is largely absent from UTMP in uterine secretions (Hansen et al., 1987b).

In addition to the major forms of UTMP, uterine secretions and endometrium-conditioned medium contains other proteins that react with antibodies to UTMP (Moffatt et al., 1987b; Stephenson et al., 1989; Leslie et al., 1990; Leslie and Hansen, 1991). These lower molecular weight forms, the most prominent of which are at M_r ~ 47,000–49,000, are probably proteolytic breakdown products. It is not known whether cleavage is associated with a change in biological activity as occurs for some serpins (Pemberton et al., 1988). Uterine milk protein can also form large aggregates. Crude uterine fluid from unilaterally-pregnant ewes contains large amounts of colloidal or precipitated protein that is composed of UTMP aggregates that can be dissolved by reducing agents (Moffatt et al., 1987b). Storage of solutions of UTMP also results in their aggregation (Hansen and Liu, 1994).

5. PROTEINASE ACTIVITY

The P1–P14 region of the reactive center loop is important for antiproteinase activity of serpins (Carrell et al., 1987; Carrell et al., 1991). In particular, P1 can confer inhibitory specificity and the P10–P14 region at the base of the reactive center loop is important for whether a serpin can undergo partial loop insertion and possess inhibitory activity. The putative P1 for UTMP is alanine, an amino acid which is also at the P1 site of ovalbumin but is not found for any serpins with functional proteinase activity (Table 1). The P10–P14 region of UTMP is lacking large numbers of amino acids such as alanine, threonine and serine that are important for loop insertion. Thus, UTMP may not be an active proteinase inhibitor. In fact, UTMP did not display antiproteinase activity against a variety of serine proteinases or peptidases including elastase, trypsin, chymotrypsin, plasmin, thrombin, plasminogen activator or dipeptidyl peptidase IV (Ing and Roberts, 1989; Liu and Hansen, 1995a). Since UTMP binds heparin (Zhang et al., 1989), and heparin activates antithrombin III, heparin cofactor II, protease nexin I and protein C inhibitor, it remains possible that activation of UTMP requires heparin or other activator.

Recently, it has been reported that UTMP is a weak inhibitor of pepsin A and C (see Mathialagan et al., 1995 and the abstract by Mathialagan and Roberts in this volume). Thus, UTMP may be similar to CrmA, a serpin that inhibits a cysteine proteinase (Komiyama et al., 1994).

Table 1. Comparison of the amino acids at P_{14}-P_1' for uterine milk protein and uteroferrin-associated protein with other serpins[a]

	P_{14}	P_{13}	P_{12}	P_{11}	P_{10}	P_9	P_8	P_7	P_6	P_5	P_4	P_3	P_2	P_1	P_1'
Serpin consensus	T	E	A	A	A	T	X	X	X	X	X	X	X		
Sheep α_1AT	T	E	A	A	G	A	T	F	L	E	A	I	P	M	S
Mouse contrapsin	E	A	A	A	A	T	G	V	I	G	G	I	R	K	A
Bovine ATIII	S	E	A	A	A	S	T	V	I	S	I	A	G	R	S
Human HCII	T	Q	A	T	T	V	T	T	V	G	F	M	P	L	S
Human PCI	T	R	A	A	A	A	T	G	T	I	F	T	F	R	S
Bovine PAI-1	T	L	A	S	S	S	T	A	L	V	V	S	A	R	M
Human PAI-2	T	E	A	A	A	G	T	G	G	V	M	T	G	R	T
Human PN-I	T	K	A	S	A	A	T	T	A	I	L	I	A	R	S
Chicken ovalbumin	R	E	V	V	G	S	A	E	A	G	V	D	A	A	S
Human maspin	T	E	D	G	G	D	S	I	E	V	P	G	A	R	I
Mouse HSP47	N	P	F	D	Q	D	I	Y	G	R	E	E	L	R	S
Sheep UTMP	D	T	A	I	H	T	D	N	L	L	K	V	P	A	N
Pig UfAP	N	A	A	K	D	K	D	-	F	W	K	V	P	V	D

6. REGULATION OF LYMPHOCYTE FUNCTION BY UTMP

6.1. Characteristics

Like another serpin, α_1-antitrypsin (Bata and Revillard, 1981; Breit et al., 1983), UTMP can inhibit lymphocyte function. In particular, UTMP depresses lymphocyte proliferation induced by phytohemagglutinin, concanavalin A, mixed lymphocyte reactions and *Candida albicans* antigen (Skopets and Hansen, 1993; Skopets et al., 1995; Liu and Hansen, 1995a,b), inhibits natural killer activity of cultured sheep lymphocytes and mouse splenocytes (Liu and Hansen, 1993), reduces antibody response to ovalbumin in ewes (Skopets et al., 1995) and blocks NK-cell mediated abortion in mice (Liu and Hansen, 1993). The concentrations of UTMP required to inhibit lymphocyte function are high (~100–300 µg/ml) but well within the range of concentrations of UTMP in uterine fluid. Interestingly, the dose-response curve for inhibition of proliferation by UTMP is very similar to that of α_1-antitrypsin (Figure 4).

6.2. Mechanism

It is not clear whether UTMP and α_1-antitrypsin inhibit lymphocyte proliferation by similar mechanisms. The inhibitory effects of α_1-antitrypsin on lymphocyte function (Bata

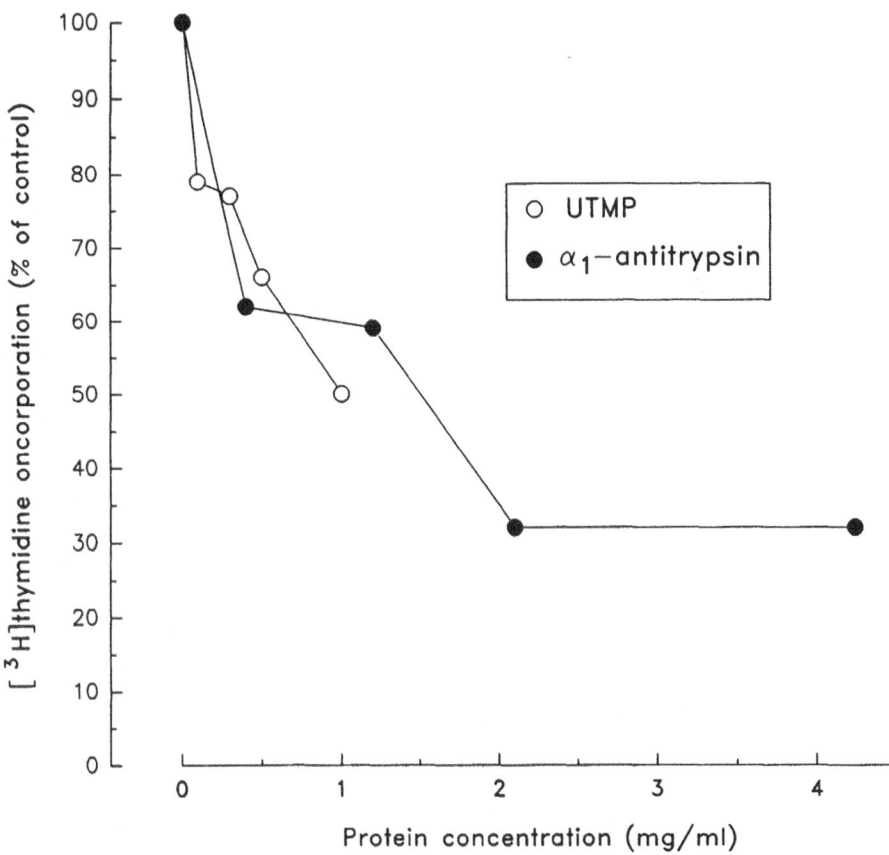

Figure 4. Comparison of lymphocyte inhibitory activity of uterine milk protein (UTMP) and α_1-antitrypsin. Peripheral blood lymphocytes stimulated with PHA were cultured with various concentrations of UTMP (sheep lymphocytes; Skopets et al., 1993) or human α_1-antitrypsin (human lymphocytes; Breit et al., 1983). Stimulation of proliferation was monitored by determining incorporation of [³H]thymidine.

and Revillard, 1981; Breit et al., 1983) is probably a reflection of the importance of proteinases or peptidases for lymphocyte function (Tchórzewski et al., 1995). Specific binding of UTMP to lymphocytes has been identified (Liu and Hansen, 1996) but it is not known whether this binding represents the presence of a specific UTMP receptor or interactions of UTMP with a proteinase or other surface protein required for lymphocyte activation (see model in Figure 5). One peptidase that plays an important role in lymphocyte activation is dipeptidyl peptidase IV (CD26) but UTMP does not inhibit enzymatic or lymphocyte co-stimulatory activity of this enzyme (Liu and Hansen, 1995a). Binding of UTMP to lymphocytes does not appear to represent interaction with a serpin-enzyme complex receptor characteristic of hepatocytes, monocytes and neutrophils (Perlmutter et al., 1990) because binding was not inhibited by α_1-antitrypsin-trypsin complexes.

Lectins or antigens stimulate lymphocytes through tyrosine kinase dependent activation of phospholipase C followed by protein kinase C activation and Ca^{2+} mobilization (Weiss and Littleman, 1994). Among the genes activated as a result are interleukin-2 (IL-2) and IL-2 receptor; further proliferation occurs as a result of IL-2 regulated events. Uterine milk protein blocked phorbol myristate acetate-induced lymphocyte proliferation but did not affect proliferation induced by interleukin-2 and did not block IL-2 receptor formation (Liu and Hansen, 1995b). Accordingly, UTMP appears to act at a point in lym-

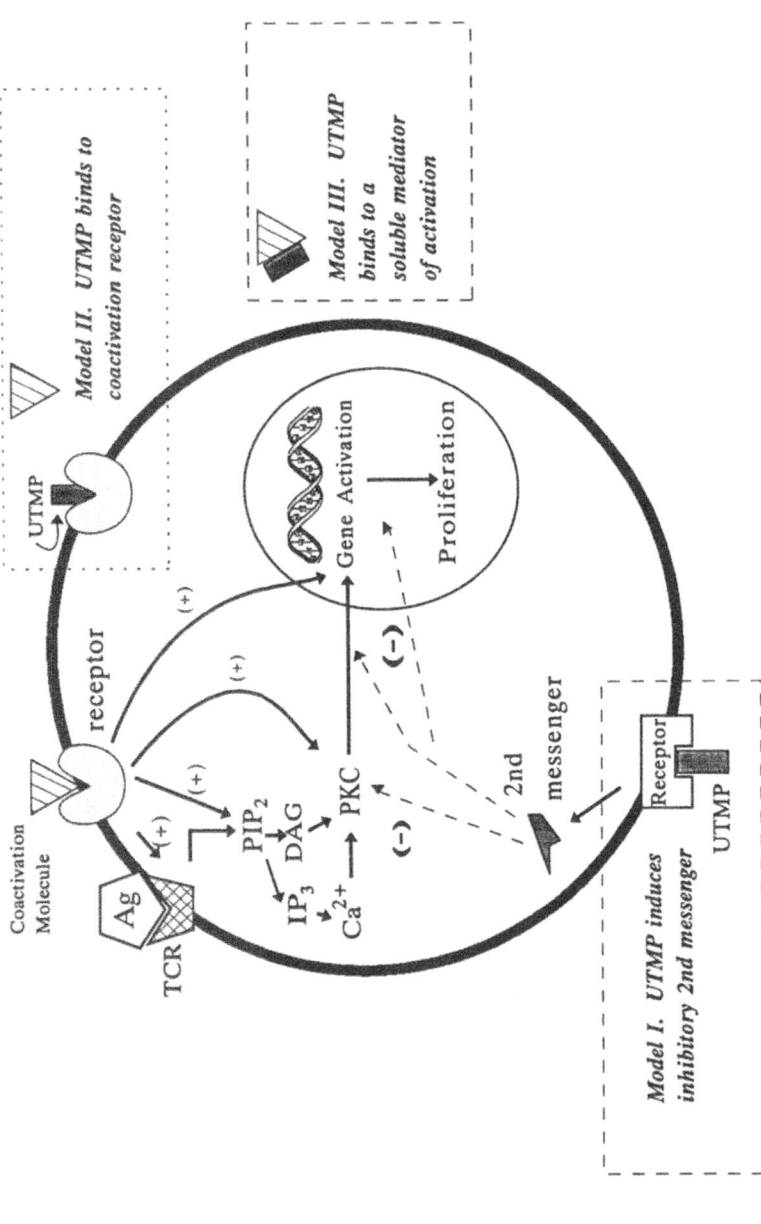

Figure 5. Alternative models for mechanisms for inhibition of lymphocyte proliferation by uterine milk protein (UTMP). It is known that UTMP blocks activation at a point at or distal to protein kinase C (PKC) activation by diacylglycerol (DAG) and calcium and that UTMP binds specifically to lymphocytes. One possibility (Model I) is that UTMP binds to a specific receptor and that the signal transduction signal generated as a result of binding blocks activation. A second possibility (Model II) is that binding of UTMP to lymphocytes represents binding and inactivation of a receptor for a coactivation molecule required for lymphocyte activation. It is also possible that UTMP binds a proteinase or other soluble molecule involved in lymphocyte activation (Model III). Other abbreviations are TCR (T-cell receptor), Ag (antigen), PIP_2 (phospatidylinositol-4,5-bisphosphate), and IP_3 (inositol-1,4,5-trisphosphate).

phocyte activation equal to or distal from diacylglycerol activation of protein kinase C and through mechanisms that do not involve decreased responsiveness to IL-2 (Figure 5).

6.3. Significance

Among its many effects on the uterus during pregnancy, progesterone inhibits uterine immune responses. In sheep, for example, progesterone prolongs skin graft survival *in utero* (Hansen et al., 1986) and causes a reduction in uterine lymphocyte numbers (Gottshall and Hansen, 1992). Inhibition of immune function at the fetal-maternal interface is probably an important aspect of progesterone's role as a pregnancy hormone because it prevents immunological rejection of the antigenically-foreign conceptus during pregnancy (see Hansen, 1995). While progesterone can directly inhibit lymphocyte proliferation (Low and Hansen, 1988), concentrations required are lower than achieved in studies in which uterine immune function was altered by progesterone (Hansen et al., 1986; Gottshall and Hansen, 1992), Thus, many effects of progesterone on lymphocytes resident in the uterine endometrium are probably mediated indirectly through secretion of UTMP and possibly other molecules.

7. OTHER POSSIBLE FUNCTIONS

Uterine milk protein is a likely candidate for serving as a transport protein to transfer poorly-soluble molecules across the placenta to the fetus. Large amounts of the protein are produced and potentially available for transport. Also, UTMP is present in colostrum (Hansen and Foti, 1986). which is another fluid involved in transport of nutrients to the offspring. Moreover, UTMP crosses the placenta and can be found in allantoic and amniotic fluid (Moffatt et al., 1987a; Newton et al., 1989). Despite this circumstantial evidence that UTMP serves some transport function, no appropriate ligand for UTMP has been identified. Uterine milk protein does not appear to bind fatty acids, calcium or prostaglandins (Moffatt et al., 1987a; P.J. Hansen, unpublished observations) and, while UTMP binds IgA (Hansen and Newton, 1988), immunoglobulins are not transported across the ruminant placenta.

Recently, it has been reported that UTMP binds to a group of placental proteins called pregnancy associated glycoproteins (PAG) (Mathialagan et al., 1995). The role of PAG, which are a family of inactive aspartyl proteinases, is undefined but UTMP may modify clearance or biological activity of PAG.

8. FUNCTION OF OTHER MEMBERS OF THE UTMP FAMILY

Uteroferrin-associated protein was originally described because of its association with the iron containing acid phosphatase of the pig uterus called uteroferrin. This protein exists at several molecular weights of approximately 50,000, 46,000 and 39,000 (Baumbach et al., 1986; Murray et al., 1989). The multiple forms of UfAP are a result of multiple genes (at least two), differential glycosylation and proteolytic cleavage (Murray et al., 1989; Malathy et al., 1990). One possible function of UfAP is to stabilize uteroferrin: association with UfAP maintains uteroferrin's acid phosphatase activity and pink color when exposed to air (Baumbach et al., 1986). Uteroferrin itself probably serves as a transport protein to carry iron to the fetus (Buhi et al., 1982).

The function of bovine UTMP is not known. Like sheep UTMP, it is a progesterone-induced product of the uterine endometrium (Leslie and Hansen, 1991). Progesterone induces immunosuppressive activity in uterine flushes of ovariectomized cows (Lander Chacin et al., 1990) but it has not been tested whether this activity is caused by UTMP.

9. CONCLUSIONS

The serpin superfamily is a broadly-distributed group of proteins that are involved in many biological processes such as fibrinolysis, hemostasis, inflammation, tumor suppression, chemotaxis, heat shock responses, and hormone transport. Clearly, the uterus should be considered as one of the arenas in which serpins perform their roles. This organ secretes serpins characteristic of other tissues (e.g., plasminogen activator inhibitor-1 and α_1-antitrypsin) as well as novel serpins of limited tissue distribution. Available evidence suggests that the progesterone-induced uterine serpins of the ungulates perform specialized functions required for maintenance of pregnancy. Further efforts to describe the species distribution of the UTMP family and the role these proteins play during pregnancy will result in enhanced recognition of the importance of serpins in uterine biology.

ACKNOWLEDGMENTS

Research was supported by Grants from NIH (HD 20671 and HD 26421). This is Journal Series No. R-05202 of the Florida Agricultural Experiment Station.

REFERENCES

Bairoch, A. and Boeckmann, B. (1991) The SWISS-PROT protein sequence data bank. Nucl. Acids Res. 19: 2247–2249.

Bany, B.M. and McRae, A.C. (1992) Uterine uptake of α_2-macroglobulin and α_1-proteinase inhibitor from the blood during early implantation in the mouse. Biol. Reprod. 47: 514–519.

Bata, J. and Revillard, J.P. (1981) Interaction between α_1 antitrypsin and lymphocyte surface proteases: immunoregulatory effects. Agents Actions 11: 614–617.

Baumbach, G.A., Ketcham, C.M., Richardson, D.E., Bazer, F.W. and Roberts, R.M. (1986) Isolation and characterization of a high molecular weight form of uteroferrin from uterine secretions and allantoic fluid of pigs. J. Biol. Chem. 261: 12869–12878.

Bazer, F.W., Roberts, R.M., Basha, S.M.M., Zavy, M.T., Caton, D. and Barron, D.H. (1979) Method for obtaining uterine secretions from unilaterally pregnant ewes. J Anim Sci 49: 1522–1527.

Breit, S.N., Luckhurst, E. and Penny, R. (1983) The effect of α_1 antitrypsin on the proliferative response of human peripheral blood lymphocytes. J. Immunol. 130: 681–686.

Buhi, W.C., Ducsay, C.A., Bazer, F.W. and Roberts, R.M. (1982) Iron transfer between the purple phosphatase uteroferrin and transferrin and its possible role in iron metabolism of the fetal pig. J. Biol. Chem. 257: 1712–1723.

Carrell, R.W., Evans, D.L. and Stein, P.E. (1991) Mobile reactive centre of serpins and the control of thrombosis. Nature 353: 576–578.

Carrell, R.W., Pemberton, P.A. and Boswell, D.R. (1987) The serpins: evolution and adaptation in a family of protease inhibitors. Cold Spring Harbor Symp. Quant. Biol. 52: 527–535.

Fay, T.N., Lindenberg, S., Teisner, B., Westergaard, L.G., Westergaard, J.G. and Grudzinskas, J.G. (1990) Identification of specific serum proteins synthesized de novo by monolayer cultures of glandular cells of gestational endometrium. Hum. Reprod. 5: 14–18.

Faust, P.L., Chirgwin, J.M. and Kornfeld, S. (1987) Renin, a secretory glycoprotein, acquires phosphomannosyl residues. J. Cell Biol. 105: 1947–1955.

Gottshall, S.L. and Hansen, P.J. (1992) Regulation of leucocyte subpopulations in the sheep endometrium by progesterone. Immunology 76: 636–641.

Hansen, P.J. (1995) Interactions between the immune system and the ruminant conceptus. J. Reprod. Fertil. Suppl. 49: 69–82.

Hansen, P. J., Bazer, F. W. and Segerson, E. C. (1986) Skin graft survival in the uterine lumen of ewes treated with progesterone. Am. J. Reprod. Immunol. Microbiol. 12: 48–54.

Hansen, P.J. and Foti, S.A. (1986) Proteins similar to the major proteins in uterine secretions of pregnant ewes are also found in colostrum. J. Anim. Sci. 63 (Suppl. 1): 53–54 (abstr).

Hansen, P.J., Ing, N.H., Moffatt, R.J., Baumbach, G.A., Saunders, P.T.K., Bazer, F.W. and Roberts, R.M. (1987a) Biochemical characterization and biosynthesis of the uterine milk proteins of the pregnant sheep uterus. Biol Reprod 36: 405–418.

Hansen, P.J. and Liu, W.-J. (1994) Biochemical/physiological properties of endometrial serpin-like proteins. Adv. Contracept. Delivery Sys. 10: 339–353.

Hansen, P.J. and Newton, G.R. (1988) Binding of immunoglobulins to the major progesterone-induced proteins of the sheep uterus. Arch. Biochem. Biophys. 260: 208–217.

Hansen, P.J., Segerson, E.C. and Bazer, F.W. (1987b) Characterization of immunosuppressive substances in the basic protein fraction of uterine secretions from pregnant ewes. Biol Reprod 36: 393–404.

Herzog V., Neumuller, W. and Holzmann, B. (1987) Thyroglobulin, the major and obligatory exportable protein of thyroid follicle cells, carries the lysosomal recognition marker mannose-6-phosphate. EMBO J. 6:555–560.

Huang, X. and Miller, W. (1991) A time-efficient, linear-space local similarity algorith. Adv. Appl. Math. 12: 337–357.

Ing, N.H. and Roberts, R.M. (1989) The major progesterone-modulated proteins secreted into the sheep uterus are members of the serpin superfamily of serine protease inhibitors. J. Biol. Chem. 264: 3372–3379.

Ing N.H., Francis H., Mc Donnell J.J., Amann J.F. and Roberts, R.M. (1989) Progesterone induction of the uterine milk proteins: major secretory proteins of sheep endometrium. Biol. Reprod. 41: 643–654.

Komiyama, T., Ray, C.A., Pickup, D.J., Howard, A.D., Thornberry, N.A., Peterson, E.P. and Salvesen, G. (1994) Inhibition of interleukin-1β converting enzyme by the cowpox virus serpin CrmA. An example of cross-class inhibition. J. Biol. Chem. 269: 19331–19337.

Lander Chacin, M.F., Hansen, P.J. and Drost, M. (1990) Effects of stage of the estrous cycle and steroid treatment on uterine immunoglobulin content and polymorphonuclear leukocytes in cattle. Theriogenology 34: 1169–1184.

Leslie, M.V., Hansen, P.J. and Newton, G.R. (1990) Uterine secretions of the cow contain proteins that are immunochemically related to the major progesterone-induced proteins of the sheep uterus. Domest. Anim. Endocrinol. 7: 517–526

Leslie M.V. and Hansen, P.J. (1991) Progesterone-regulated secretion of the serpin-like proteins of the ovine and bovine uterus. Steroids 56: 589–597

Low, B. G. and Hansen, P.J. (1988) Immunosuppressive actions of steroids and prostaglandins secreted by the placenta and uterus of the cow and sheep. Am. J. Reprod. Immunol. Microbiol. 18: 71–75.

Liu, W.-J. and Hansen, P.J. (1993) Effect of the progesterone-induced serpin-like proteins of the sheep endometrium on natural-killer cell activity in sheep and mice. Biol. Reprod. 49: 1008–1014.

Liu, W.-J. and Hansen, P.J. (1995a) Progesterone-induced secretion of dipeptidyl peptidase-IV (cell differentiation antigen-26) by the uterine endometrium of the ewe and cow that costimulates lymphocyte proliferation. Endocrinology 136: 779–787.

Liu, W.-J. and Hansen, P.J. (1995b) Pathways for inhibition of lymphocyte activation by uterine milk proteins. Am. J. Reprod. Immunol. 33: 445 (abstr.).

Liu, W.J. and Hansen, P.J. (1996) Binding of endometrial serpin-like proteins to peripheral blood lymphocytes in sheep. Am. J. Reprod. Immunol., 453 (abstr.).

Lockwood, C.J., Krikun, G., Papp, C., Aigner, S. and Schatz, F. (1995) Biological mechanisms underlying the clinical effects of RU 486: modulation of cultured endometrial stromal cell plasminogen activator and plasminogen activator inhibitor expression. J. Clin. Endocrinol. Metab. 80: 1100–1105.

Malathy, M.-V., Imakawa, K., Simmen, R.C.M. and Roberts, R.M. (1990) Molecular cloning of the uteroferrin-associated protein, a major progesterone-induced serpin secreted by the porcine uterus, and the expression of its mRNA during pregnancy. Mol. Endocrinol. 4: 428–440.

Mathialagan, N., Paul, L. and Roberts, R.M. (1995) Sheep uterine serpins are inhibitors of aspartic proteinases rather than serine proteinases. Biol. Reprod. 52 (Suppl. 1): 186 (abstr.).

Mathialagan, N., Hansen, T.R. and Roberts, R.M. (1997) Pepsin inhibitory activity of the uterine serpins. In: Chemistry and Biology of Serpins (F.C. Church, D.D. Cunningham, D. Ginsberg, and M. Hoffman, eds.) Plenum, New York. (abstr.)

Miyauchi, A., Osuga, Y. and Taketani, Y. (1995) Effects of steroid hormones on fibrinolytic system in cultured human endometrial cells. Endocrine J. 42: 57–62.

Moffatt R.J., Bazer, F.W., Hansen P.J., Chun, P.W. and Roberts, R.M. (1987b) Purification, secretion and immunocytochemical localization of the uterine milk proteins, major progesterone-induced proteins in uterine
secretions of the sheep. Biol Reprod 36: 419–430.

Moffatt, R.J., Bazer, F.W., Roberts, R.M. and Thatcher, W.W. (1987a) Secretory function of the ovine uterus:
effects of gestation and steroid replacement therapy. J. Anim. Sci. 65: 1400–1410

Murray, M.K., Malathy, P.V., Bazer, F.W. and Roberts, R.M. (1989) Structural relationship, biosynthesis, and immunocytochemical localization of uteroferrin-associated basic glycoprotein. J Biol Chem 264: 4143–4150

Newton, G. R., Hansen, P. J., Bazer, F. W., Leslie, M. V., Stephenson, D. C. and Low, B. G. (1989) Presence of the
major progesterone-induced proteins of the sheep endometrium in fetal fluids. Biol. Reprod. 40: 417–424.

Pemberton, P.A., Stein, P.E., Pepys, M.B., Potter, J.M. and Carrell, R.W. (1988) Hormone binding globulins
undergo serpin conformational change in inflammation. Nature (Lond) 336: 257–258.

Perlmutter, D.H., Glover, G.I., Rivetna, M., Schasteen, C.S. and Fallon, R.J. (1990) Identification of a serpin-
enzyme complex receptor on human hepatoma cells and human monocytes. Proc. Natl. Acad. Sci. USA.
87: 3753–3757.

Skopets, B. and Hansen, P.J. (1993) Identification of the predominant proteins in uterine fluids of unilaterally
pregnant ewes that inhibit lymphocyte proliferation. Biol. Reprod. 49: 997–1007.

Skopets, B., Liu, W.-J. and Hansen, P.J. (1995) Effects of endometrial serpin-like proteins on immune responses in
sheep. Am. J. Reprod. Immunol. 33: 86–93.

Sonnhammer, E.L.L. and Kahn, D. (1994) Modular arrangement of proteins as inferred from analysis of homology. Protein Sci. 3: 482–492.

Stephenson D.C. and Hansen, P.J. (1990) Induction by progesterone of immunosuppressive activity in uterine
secretions of ovariectomized ewes. Endocrinology 126: 3168–3178

Stephenson D.C., Low, B.G., Newton, G.R., Leslie, M.V., Hansen, P.J. and Bazer, F. W. (1989) Secretion of the
major progesterone-induced proteins of the sheep uterus by caruncular and intercaruncular endometrium of
the pregnant ewe from days 20–140 of gestation. Domest. Anim. Endocrinol. 6: 349–362

Tchórzewski, H., Fornalczyk, E. and Paśnik, J. (1995) Protease inhibitors diminish lymphocyte stimulation in
vitro. FEBS Lett. 46: 237–240.

Weiss, A. and Littleman, D. (1994) Signal transduction by lymphocyte antigen receptors. Cell 76: 631-634.

Zhang, X., Miller, B.G. and Stone, G.M. (1989) Protein secretion by the endometrium during pregnancy in the
ewe. Reprod. Fertil. Dev. 1: 15–30.

15

SERPINS FROM AN INSECT, *Manduca sexta*

Michael R. Kanost and Haobo Jiang

Department of Biochemistry
Kansas State University
Manhattan, Kansas 66506

1. SERPINS IN INVERTEBRATES

Members of the serpin superfamily have been identified in several invertebrate species, primarily as proteins in the hemolymph of animals from the phylum Arthropoda[9]. Serpin cDNAs have been cloned from horseshoe crabs[14,15], a crayfish[13], two species of lepidopteran insects[6,10,17,20,24], the fruit fly *Drosophila melanogaster*[3], trematodes from the genus *Schistosoma*[2], and from a parasitic nematode[25]. The presence of serpins with differing proteinase inhibitory activities in the hemolymph of horseshoe crabs and insects suggests that serpins may function in several physiological processes in invertebrates, as they do in mammals. The number of invertebrate species vastly exceeds the number of vertebrates, with far greater biological diversity. Thus, it may be predicted that as biochemical studies on invertebrates progress, a large number of serpins with diverse activities and functions will be discovered. In our research on serpins from the tobacco hornworm, *Manduca sexta*, we have identified a family of serpins with different inhibitory activities, and a unique serpin gene structure for generating reactive site diversity.

2. SERPINS FROM HEMOLYMPH OF *Manduca sexta* AND *Bombyx mori*

The tobacco hornworm (*M. sexta*) and the silkworm (*B. mori*) are large caterpillars that are convenient subjects for biochemical research on insects, because they are easily reared in the laboratory and, in comparison to smaller insects, they provide reasonable amounts of experimental material. Serpins from insects were first characterized from *B. mori* by the group of Sasaki[21], with the isolation of a trypsin inhibitor and a chymotrypsin inhibitor. These 43 kDa proteins were found to behave similarly to serpins, with a reactive site near their carboxyl terminus[19] and formation of a stable complex in SDS[22]. Serpins with inhibitory activity against elastase, chymotrypsin, and elastase were also isolated from hemolymph of *M. sexta*[8,10].

Chemistry and Biology of Serpins, edited by Church *et al.*
Plenum Press, New York, 1997

Figure 1. Variable and constant regions in *M. sexta* serpins and corresponding cDNAs. Solid boxes indicate cDNA or amino acid sequences that are identical in all of the serpin-1 sequences. Open boxes indicate the variable region present near the 3' end of the cDNA and at the carboxyl terminus of the serpin-1 proteins. The variable region is 39–46 amino acid residues long. This sequence includes the reactive site, beginning at residue P7 to P5 in different variants.

The amino acid sequences of these insect proteins confirmed that they are members of the serpin superfamily[10,17,20,24]. In amino acid sequence alignments, the hornworm and silkworm serpins are 12–35% identical with various mammalian serpins. The cDNA sequences include a signal sequence that is cleaved upon secretion of the proteins into the hemolymph. The primary tissue source for the serpins in *M. sexta* is the fat body[11], a tissue functionally analogous to vertebrate liver. However, we have recently identified a second serpin gene in *M. sexta* (serpin gene-2) that is an intracellular protein expressed in granular hemocytes, which are neutrophil-like blood cells (H. Gan and M.R. Kanost, unpublished results). The serpin-1 proteins are present in hemolymph at 200–600 µg/ml, with concentration changing during development, partly under control of the steroid hormone 20-hydroxyecdysone[11].

3. ALTERNATE EXON SPLICING TO PRODUCE REACTIVE SITE VARIANTS

When we sequenced additional serpin clones from our *M. sexta* cDNA libraries, we were surprised to find that, although their sequences at the 5' and 3' ends were all identical, there was a region of approximately 130 nucleotides that differed between individual clones[6]. This highly variable region is near the 3' end of the serpin mRNA and includes the stop codon for the open reading frame (Fig. 1). The amino acid sequence encoded by the variable region begins at residue 337 of the mature protein and includes the serpin reactive site region. We have now isolated cDNA clones with twelve different versions of the variable sequence.

The mechanism for producing the *M. sexta* serpin variants was apparent when we determined the sequence of the serpin-1 gene (Fig. 2). The gene is composed of 10 exons, which extend over 20.5 kb of genomic DNA. The constant regions of the serpin-1 mRNA

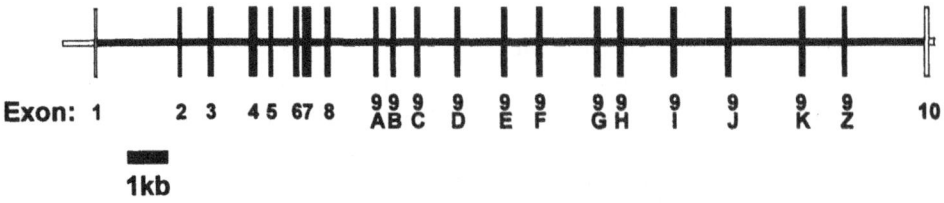

Figure 2. Structure of *M. sexta* serpin gene-1. Exons are indicated by vertical bars and introns by a horizontal line. Exons are numbered 1 to 10. The untranslated exon 10 is shown as an open bar. Alternate versions of exon 9 are numbered 9A, 9B, 9C, etc. Mature serpin-1 mRNAs contain exons 1–8, one of the exon 9 alternates, and exon 10.

are encoded by exons 1–8 and exon 10. The variable regions correspond with twelve alternate versions of exon 9, which are present in an 11.6 kb stretch of DNA between exons 8 and 10. Mutually exclusive alternate exon splicing results in selection of one version of exon 9 for each mature serpin mRNA. This requires a mechanism for accurate recognition and splicing of only a single exon 9 to exons 8 and 10. In over 40 cDNAs sequenced we have identified only one instance of inaccurate splicing, in which exon 9B is followed by 9C and then exon 10[6]. However, since each exon 9 contains its own translation stop codon, such mis-spliced mRNAs still encode serpin proteins of normal size, and the result is that the mRNA has a longer untranslated 3′ sequence that includes the extra exon 9.

A combination of cDNA library screening and reverse transcription-PCR experiments have shown that all twelve of the exon 9 variants are used and are represented in accurately spliced serpin mRNAs from fat body and hemocytes. However, there is apparently some tissue-specificity in exon 9 selection by the fat body. The majority of serpin-1 cDNA clones isolated from our larval fat body library contain exon 9F[6].

We envision that this region of *M. sexta* serpin-1 gene evolved through duplication of an ancestral exon 9, followed by several events of unequal crossing over, each giving rise to more copies of exon 9. Sequence divergence of these repeated exons has resulted in the variation now observed in the twelve copies of exon 9. A similar gene structure is very likely for a serpin gene from the silkworm. Two sequences obtained for serpins from *B. mori* differ in the same way as the *M. sexta* serpin variants, with the variable region located at precisely the same position in the sequence[20]. Thus, this molecular mechanism for producing serpins with different reactive sites on the same protein scaffold may have occurred long ago in the evolution of at least lepidopteran insects. It will be interesting to learn how widespread in the insects this phenomenon might be.

4. INHIBITORY ACTIVITY OF THE *M. sexta* SERPIN-1 VARIANTS

To investigate the proteinase inhibitory activities of the serpin-1 variants, we have expressed them as recombinant proteins in *E. coli*, beginning with serpin-1B[7]. We found that serpin-1B inhibits both pancreatic and neutrophil elastase. Its P1 residue is Ala 343, as predicted from sequence alignments with vertebrate serpins. Site-directed mutation of Ala 343 to Lys, changed the protein to an inhibitor of trypsin, plasmin and thrombin. Mutation of residue 343 to Phe produced a chymotrypsin inhibitor with weak activity against trypsin.

We have now expressed all twelve of the variants in *E. coli* and have begun to characterize their inhibitory activities (Fig. 3). We have located the scissile bond of three of the variants that are chymotrypsin inhibitors (H, K, and L), all of which have a Tyr residue in the P1 position. Seven of the variants are active against at least one of the mammalian enzymes we have assayed: porcine pancreatic and human neutrophil elastase, bovine trypsin, a-chymotrypsin, plasmin, and thrombin, and human neutrophil cathepsin G. It may be that the other five variants have activity as inhibitors of enzymes we have not yet tested or that some variants lack a function as proteinase inhibitor.

5. POTENTIAL FUNCTIONS OF SERPINS IN INSECTS

To understand the physiological functions of serpins in insects, we must identify proteinases that are their natural targets for inhibition. There are several systems in insect development and antimicrobial defense in which serine proteinases are important and are likely to be regulated by inhibitors.

SEQUENCE ENCODED BY EXON 9　　　　　　　　　　ENZYMES INHIBITED

```
A  AFFITRQARL ....DIRYFV ANKPFIFLLR FNGLALFNGV FKA       trypsin, plasmin
B  AFGIVPASLI LY....PEVH IDRPFYFELK IDGIPMFNGK VIEP      elastase, cathepsin G
C  AFFIIESYS. SYEPVVPVFD IDKPFYFNIR ANGQSLFNGL CFQP
D  .VVRGIRPRP SVRPPTPKFE ADRPFLFYLK TNDQTLFNGI CMQP
E  .VIRVVKKKF RVIPPVLKFH VDRPFFFNLK ANDQSLFNGI CLQP
F  AFIAVVDSID IFERTI.EFH ADRPFFFNLK ANGQSLFNGI CVMPML    elastase
G  AFFIVGITSI QFEPPVIEFH VNRPFFFNLK ASGQSLFNGI CVQP      chymotrypsin, cathepsin G
H  AFITYVESID NFVPTI.EFD VNRPFYFNLK ANDLYLFNGI CVQPKLQ   elastase, chymotrypsin, cathepsin G
I  EFGIV.ALSL EFSLNEIKFV VNKPFYFNIR SNGQHLFNGI CFQP
J  AFILTDRCCS DYDDNI.EFD VNRPFYFNLR TNEHLLFSGI CIQPEI
K  AFKITTYSFH FV....PKVE INKPFFFSLK YNRNSMFSGV CVQP      chymotrypsin, cathepsin G
Z  AFGIAYLSAV I...RSPVFN ADHPFVFFLR QDKTTLFSGV FQS       chymotrypsin, cathepsin G
```

Figure 3. Reactive site sequences and inhibitory activities of *M. sexta* serpin-1 variants. On the left is an alignment of the amino acid sequences encoded by the twelve alternate versions of exon 9. These sequences begin at approximately the P7 residue of the reactive site. For variants in which the scissile bond has been identified, the P1 residue is underlined. On the right is a list of the proteinases known to be inhibited by each serpin-1 variant.

In embryonic development of *Drosophila*, the pathway that establishes dorsal-ventral polarity includes at least three genes that encode serine proteinases[16]. These proteins, named nudel, snake, and easter, participate in a cascade that results in cleavage of a protein called Spätzle. The processed form of Spätzle binds to a membrane-bound receptor (Toll) related to the interleukin-1 receptor. The proteinases in this pathway may need to regulated by inhibitors, to achieve the necessary concentration and distribution of Spätzle for correct dorsal-ventral pattern signaling.

A second point in insect development involving proteinase activity is metamorphosis. At metamorphosis there is a spectacular tissue remodeling and reorganization, with participation of proteinases that have not been well characterized. Programmed cell death of larval tissues takes place, concomitant with their replacement with newly differentiating adult tissues from imaginal discs. The eversion and growth of imaginal discs, which give rise to adult structures such as legs, wings, eyes, and antennae, is associated with proteolysis that can be inhibited by serine proteinase inhibitors[18]. We have found that serpin-1 is present during development of antennal imaginal discs in *M. sexta* [4].

Responses of arthropods to bacterial and fungal infections include serine proteinase cascades analogous in some ways to blood coagulation and the complement system in vertebrates. The hemolymph coagulation system in horseshoe crabs has been very well characterized[5]. It includes a pathway that is activated by bacterial lipopolysaccharide and a cascade of serine proteinases. Two of these proteinases, Factor B and the proclotting enzyme, have an amino-terminal regulatory domain similar to those found in snake and easter from the *Drosophila* dorsal pathway. Two serpins from hemocytes of horseshoe crabs are inhibitors of Factor C and the clotting enzyme, and are likely to be regulators of this pathway[14,15].

A second arthropod defensive pathway that contains a serine proteinase cascade is the activation of a zymogen for phenoloxidase in response to wounding or infection[23]. This pathway is not as well understood as the horseshoe crab clotting system, but appears to involve at least two serine proteinases that are activated by bacterial peptidoglycan or fungal glucans. A specific proteolytic cleavage then activates prophenoloxidase. Phenoloxidase catalyzes the production of quinones, which polymerize to form melanin. The quinones may be toxic to invading organisms, while the melanin forms capsules around parasites and can also assist in sealing wounds. This pathway is probably regulated by proteinase inhibitors, including serpins, present in hemolymph. We have found in preliminary experiments that *M. sexta* serpin-1J can inhibit activation of prophenoloxidase.

The study of proteinases from insect hemolymph is in its infancy. Only one such serine proteinase has been purified so far, from hemolymph of *B. mori*[12]. This enzyme is activated by peptidoglycan or β-1,3 glucan and has trypsin-like specificity, but apparently is not part of the prophenoloxidase activating pathway. It is inhibited by one of the serpins (antitrypsin) from silkworm hemolymph[1], but its physiological function is not yet known. In our laboratory we have isolated cDNA clones for four serine proteinases from a *M. sexta* hemocyte cDNA library. Two of these enzymes have sequences related to horseshoe crab clotting Factor B, proclotting enzyme, and *Drosophila* snake and easter. The other two hemocyte proteinase sequences are similar to vertebrate granzymes. However, we do not yet know the functions of any of these enzymes.

6. CONCLUSION

Studies on serpins from only two insect species have uncovered some very interesting molecular strategies for proteinase inhibition. The structural and biochemical characteristics of insect serpins are very similar to those of vertebrate serpins. However, lepidopteran insects have evolved a mechanism for generating serpins with diverse reactive sites by alternate exon splicing of a cassette including the reactive site loop onto a constant protein framework. The physiological roles serpins in insects may be anticipated to be as interesting and varied as those identified in mammals.

ACKNOWLEDGMENTS

This work was supported by National Institutes of Health grant GM41247 and by a Kansas Health Foundation Fellowship (to H.J.). This is contribution 97-13-B from the Kansas Agricultural Experiment Station.

REFERENCES

1. Ashida, M. and Sasaki, T. (1994) A target protease activity of serpins in insect hemolymph. Insect Biochem. Mol. Biol. 24: 1037–1041.
2. Blanton, R. E., Licate, L. S. and Aman, R. A. (1994) Characterization of a native and recombinant *Schistosoma haematobium* serine protease inhibitor gene product. Mol. Biochem. Parasitol. 63: 1–11.
3. Coleman, S., Drähn, B., Petersen, G., Stolorov, J., and Kraus, K. (1995) A *Drosophila* male accessory gland protein that is a member of the serpin superfamily of proteinase inhibitors is transferred to females during mating. Insect Biochem. Mol. Biol. 25: 203–207.
4. Hanneman, E. and Kanost, M.R. (1992) Differential alaserpin expression during development of the antennae in the tobacco hawkmoth, *Manduca sexta*. Arch. Insect Biochem. Physiol. 19: 39–52.
5. Iwanaga, S. (1993) Primitive coagulation systems and their message to modern biology. Thromb. Haemostas. 70: 48–55.
6. Jiang, H., Wang, Y. and Kanost, M. (1994) Mutually exclusive exon use and reactive center diversity in insect serpins. J. Biol. Chem. 269: 55–58.
7. Jiang, H.B., Mulnix, A.B. and Kanost, M.R. (1995) Expression and characterization of recombinant *Manduca sexta* serpin-1B and site-directed mutants that change its inhibitory selectivity. Insect Biochem. Molec. Biol. 25: 1093–1100.
8. Kanost, M.R. (1990) Isolation and characterization of four serine proteinase inhibitors (serpins) from hemolymph of *Manduca sexta*. Insect Biochem. 20: 141–147.
9. Kanost, M.R. and Jiang, H. (1996) Proteinase inhibitors in invertebrate immunity. pp. 155–174, In "New Directions in Invertebrate Immunology" (Eds., Sēderhäll, K., Iwanaga, S., and Vasta, G.R.) SOS Publications, Fair Haven, NJ.
10. Kanost, M., Prasad, S. and Wells, M. (1989) Primary structure of a member of the serpin superfamily of proteinase inhibitors from an insect, *Manduca sexta*. J. Biol. Chem. 264: 965–972.
11. Kanost, M., Prasad, S., Huang, Y. and Willott, E. (1995) Regulation of serpin gene-1 in *Manduca sexta*. Insect Biochem. Mol. Biol. 25: 285–291.
12. Katsumi, Y., Kihara, H., Ochiai, M. and Ashida, M. (1995) A serine protease zymogen in insect plasma. Purification and activation by microbial cell wall components. Eur. J. Biochem. 228: 870–877.
13. Liang, Z. and Sēderhäll, K. (1995) Isolation of cDNA encoding a novel serpin of crayfish hemocytes. Comparative Biochemistry and Physiology B, 112: 385–391.
14. Miura, Y., Kawabata, S. and Iwanaga, S. (1994) A *Limulus* intracellular coagulation inhibitor with characteristics of the serpin superfamily. J. Biol. Chem. 269: 542–547.
15. Miura, Y., Kawabata, S., Wakamiya, Y., Nakamura, T. and Iwanaga, S. (1995) A Limulus intracellular coagulation inhibitor type 2. Purification, characterization, cDNA cloning, and tissue localization. J. Biol. Chem. 270: 558–565.
16. Morisato, D. and Anderson, K.V. (1995) Signaling pathways that establish the dorsal-ventral pattern of the *Drosophila* embryo. Annu. Rev. Genetics 29: 371–399.
17. Narumi, H., Hishida, T., Sasaki, T., Feng, D. and Doolittle, R. (1993) Molecular cloning of silkworm (*Bombyx mori*) antichymotrypsin. A new member of the serpin superfamily of proteins from insects. Eur. J. Biochem. 214: 181–187.

18. Pino-Heiss, S. and Schubiger, G. (1989) Extracellular protease production by *Drosophila* imaginal discs. Dev. Biol. 132: 282–291.

19. Sasaki, T. (1985) The reactive site of silkworm hemolymph antichymotrypsin is located at the COOH-terminal region of the molecule. Biochem. Biophys. Res. Commun. 132: 320–326.

20. Sasaki, T. (1991) Patchwork-structure serpins from silkworm (*Bombyx mori*) larval hemolymph. Eur. J. Biochem. 202: 255–261.

21. Sasaki, T. and Kobayashi, K. (1984) Isolation of two novel proteinase inhibitors from hemolymph of silkworm larva, *Bombyx mori*. Comparison with human serum proteinase inhibitors. J. Biochem. 95: 1009–1017.

22. Sasaki, T., Kohara, A., Shimidzu, T. and Kobayashi, K. (1990b) Single site proteolysis in silkworm antitrypsin causes structural changes in behavior against denaturing reagents. Agric. Biol. Chem. 54: 139–145.

23. Sёderäll, K., Cerenius, L., and Johansson, M.W. (1994) The prophenoloxidase activating system and its role in invertebrate defense. Annal. New York. Acad. Sci. 712: 155–161.

24. Takagi, H., Narumi, H., Nakamura, K. and Sasaki, T. (1990) Amino acid sequence of silkworm (*Bombyx mori*) hemolymph antitrypsin deduced from its cDNA nucleotide sequence: Confirmation of its homology with serpins. J. Biochem. 108: 372–378.

25. Yenbutr, P. and Scott, A.L. (1995) Molecular cloning of a serine proteinase inhibitor from *Brugia malayi*. Infect. Immun. 63: 1745–1753.

16

THE SIGNIFICANCE OF SERPINS IN THE REGULATION OF PROTEASES IN THE MALE GENITAL TRACT

Anders Christensson,[1*] Anders Bjartell,[2] and Hans Lilja[3]

[1]Department of Vascular and Renal Diseases
[2]Department of Urology
[3]Department of Clinical Chemistry
Lund University, Malmö University Hospital
S-205 02 Malmö, Sweden

INTRODUCTION

The male genital tract is rich in proteases, delivered by the male accessory sex glands, that are delicately balanced in their action by serpins, non-serpin class protease inhibitors and other regulatory mechanisms. Still, the biological function of the serpins and their target enzymes in the male genital tract and possible involvement in the regulation of normal reproductive function mainly remains to be elucidated. However, it is important with careful control of the catalytic activity of serine proteases, in particular in the different extracellular compartments, where they may produce significant potential hazards for biological structures. Immunochemical measurements of the serine protease prostate-specific antigen (PSA) in serum have gained widespread use in the monitoring and detection of prostate cancer. Moreover, the rapidly growing body of data on the disease-related variations in the proportion of different forms of PSA in serum which relate to the covalent complex formation between the serpin α_1-antichymotrypsin and PSA has significantly improved the diagnostic specificity in blood testing for early detection of prostate cancer.

PROTEASES

Prostate-Specific Antigen (PSA)

The long arm on chromosome 19 contains the three structurally similar human glandular kallikrein genes; pancreatic and renal tissue kallikrein (prKK or hK1), prostate-

---

*Correspondence to: Anders Christensson MD, PhD, Department of Vascular and Renal Diseases, Lund University. Malmö University Hospital. S-205 02 Malmö. Sweden. Tel +46-40-331000; Fax +46-40-337043.

specific antigen (PSA or hK3), and human glandular kallikrein-1 (hGK-1 or hK2) (1). The PSA encoding gene of 6 kb contains four introns and five exons. Both PSA and hK2 appear to almost uniquely be produced by benign and malignant prostate epithelium (2–5), and both are under androgen regulation (6,7). However, localization of PSA-like immunoreactivity in malignant breast tumors and endometrium has been demonstrated (8,9).

PSA is one of the three most abundant prostate derived proteins and by far the most abundant protease in human seminal fluid, occuring at concentrations of 10–150 μmol/L (i.e. 0.5–2.0 mg/mL) (10,11). In the prostate, PSA mRNA is found exclusively in the secretory epithelial cells, as shown with the *in situ* hybridization technique (4), as opposed to neuroendocrine, basal cells and stromal cells that are devoid of PSA expression.

PSA is a single chain glycoprotein with a molecular mass of 26 kDa (10,12,13). A putative site for glycosylation has been found at Asn-45 (*N*-linked) (13). The PSA mRNA codes for a signal peptide (17 amino acids), a short propiece (7 amino acids) and the mature protein of 237 amino acid residues (13,14).

The mature form of PSA has been shown to manifest enzyme activity and to be a kallikrein-like protease belonging to the large group of extracellular serine proteases (15,16). The protease activity of PSA is mainly responsible for the fragmentation of semenogelin I and II, the major gel-forming proteins in human semen derived from the seminal vesicles (16–19). The generation of multiple soluble fragments of the major gel-forming proteins results in liquefaction of the freshly ejaculated semen and release of progressively motile spermatozoa (20,21).

The substrate specificity of PSA is similar to that of chymotrypsin as PSA has been shown to hydrolyze peptide bonds carboxy terminal of certain tyrosine and leucine residues. However, the enzyme activity of PSA is distinct from that of chymotrypsin as shown with chromogenic substrates (16,22). The presently reported data strongly suggest that PSA, isolated from the seminal fluid, does not exhibit any trypsin-like activity (22). Earlier preparations of PSA also manifested detectable trypsin-like activity (15,16). This may have been due to trace amounts of hK2 (22,23). In view of the high degree of sequence identity and suggested similarities in biochemical and immunogenic properties, it is possible that hK2 may have contaminated earlier PSA preparations and been responsible for the trypsin-like activity detected (22,23).

hK2

In 1987, a new human glandular kallikrein gene (hGK-1 or hK2) was described (24). The tissue expression of hK2 appears to be similar to that of PSA (25). In normal prostate tissue the content of hK2 mRNA is estimated to be 10–20% of that of PSA mRNA (26). This is to be compared with the results of *in situ* hybridization studies showing the content of hK2 mRNA in human benign prostatic hyperplasia tissues to be approximately half that of PSA mRNA (25).

The coding sequence of the hK2 gene predicted a 261 amino acid preprotein with a mature protein of 237 amino acid residues. In 1995, fragments of hK2 were sequence analysed, demonstrating the existence of hK2 in seminal plasma (27). These authors suggest that at protein level, the difference between hK2 and PSA concentrations in seminal plasma are probably 100- to 1000-fold in favour of PSA.

The amino acid sequence derived from the gene suggests hK2 to be a kallikrein-like serine protease with trypsin-like activity (24).

The primary structure of hK2 is very similar to that of PSA, showing 80% sequence identity (13,24). In view of the high degree of similarity of PSA and hK2 the immunogenic properties of the two proteins may be related. However, it has been shown that several monoclonal anti-PSA IgGs can be used to distinguish the two proteins, also when PSA is complexed to α_1-antichymotrypsin (28). This has also made it possible to design assays specific for hK2 (29).

The physiological function of hK2 remains to be elucidated.

Acrosin

As deduced from the cDNA sequence acrosin is synthesized as a preproenzyme, preproacrosin (30). The gene has been mapped to chromosome 22 (31). According to the exon-intron structure, preproacrosin is suggested to be closely related to the serine proteinase subfamily containing trypsin and kallikrein. However, the light chain of proacrosin seems to be similar to that of chymotrypsin (32).

Proacrosin is a single-chain polypeptide stored in the acrosomal vesicle on the sperm head (33). The human proacrosin, is composed of 402 amino acids, 23 in the light chain and 379 in the heavy chain and has a molecular mass around 43.9 kDa. The mature acrosin is a disulfide-bonded two-chain glycoprotein with trypsin-like enzymatic activity (34,35). It is proposed that proacrosin is activated to the mature enzyme by limited autoproteolysis at the time of the acrosome reaction (36). The acrosome reaction, release of the acrosomal contents, is a prerequisite for the spermatozoa to penetrate the zona pellucida of the ovum. This reaction occurs concomitantly with the conversion of proacrosin to acrosin which involves proteolysis at both the N- and C-terminal ends of proacrosin (37). A specific feature for acrosin, compared to other serine proteases, is a 125-residue C-terminal extension containing a stretch of 23 consecutive proline residues that plays a role in the activation and maturation process of proacrosin (38).

PROTEASE INHIBITORS

Alpha$_1$-Antichymotrypsin

The gene for α_1-antichymotrypsin is located on chromosome 14, is 12 kb in length and contains five exons and four introns (39,40). It codes for a molecule of 408 amino acid residues and a 25 amino acid signal peptide (40–42). The native molecule is a glycoprotein of between 55 and 68 kDa, the variation in size being attributed to microheterogeneity in glycosylation (43). Amino-terminal sequence heterogeneity is reflected in two isoforms, one of which lacks two amino acid residues at the amino terminus (44).

Alpha$_1$-antichymotrypsin and α_1-antitrypsin manifest very marked similarity in gene structure, amino acid sequence and crystal structure of the cleaved inhibitors, indicating that many of the known properties of α_1-antitrypsin may also be valid for α_1-antichymotrypsin (41,42,45). The overall identity of the primary structures of the two proteins is 45% (41). The active site residues leucine-serine in α_1-antichymotrypsin are located at positions 358 and 359 (41). Alpha$_1$-antichymotrypsin inhibits chymotrypsin-like proteases such as neutrophile cathepsin G, mast cell chymase and chymotrypsin (46,47).

As being an acute phase reactant, α_1-antichymotrypsin has been suggested to be synthesized predominantly in the liver (48). However, the major source of α_1-antichymotrypsin has been debated (49). A widespread localization of α_1-antichymotrypsin has been described in many other organs and tumor cells (50–53). Extrahepatic localization of α_1-antichymotrypsin has been demonstrated in breast epithelial cells (54), and in prostatic tissue and tumors (55,56). Human seminal plasma has been reported to contain α_1-antichymotrypsin (57). Based on immunocytochemical methods with monoclonal antibodies and *in situ* hybridization it has been shown specific production of α_1-antichymotrypsin in the epithelial cells in the prostate (55,56).

After α_1-antitrypsin, α_1-antichymotrypsin is the second most predominant serine antiprotease in human plasma, where it occurs at a concentration of 7 µmol/L. It is an acute phase reactant, reaching 4–5 fold higher concentrations in response to trauma or surgery (43,58). So far, no patient with homozygous deficiency has been identified.

PROTEIN C INHIBITOR

The gene coding for human protein C inhibitor (PCI) (11.5 kb) has been localized to chromosome 14 and consists of five exons separated by four introns (59). The gene for PCI (60) shows a high degree of homology with the genes coding for α_1-antitrypsin (61) and α_1-antichymotrypsin (40).

PCI, first purified from human plasma, is a single-chain glycoprotein with a molecular mass of 57 kDa (62). The mature PCI protein consists of 387 amino acid residues and belongs to the superfamily of serpins (63). PCI contains only one Cys residue and has five potential glycosylation sites. The reactive site peptide bond in PCI is located between arginine-354 and serine-355 (63). However, two additional reactive site peptide bonds have been proposed, one at arginine-357–leucine-358 and one at arginine-362–leucine-363 (64).

PCI was first suggested to be a regulator of activated protein C, a vitamin K-dependent anticoagulant plasma protein (62,65) exerting its anticoagulant activity by degrading two important cofactors in the regulation of haemostasis (66,67). However, the inhibition of activated protein C by PCI is a rather slow process and its physiological significance is unclear (68,69). PCI also inhibits such plasma proteases as trypsin, thrombin, factor Xa, chymotrypsin (68), plasma kallikrein, factor XIa (59), glandular kallikrein (70), urokinase (urinary plasminogen activator) (71), acrosin (72,73), PSA (74), and tissue plasminogen activator but not plasmin (69).

PCI has affinity for heparin and the secondary structure of PCI is altered after binding of heparin to the peptide sequence 264–283 (75). Heparin increases the rate of protease inhibition by PCI, in some cases up to several hundred-fold (59,68,72,74).

PCI is synthesized in several tissues. Production was first observed in the liver (63), and the normal plasma PCI concentration is about 5 mg/L (80 nmol/L) (62,69). Later, PCI was detected in 40-fold higher levels in human seminal plasma (76,77). Specific PCI immunoreactivity and Northern blot analysis revealed that the inhibitor is mainly synthesized in the seminal vesicles, but also in the testes, epididymis glands, and the prostate (76). Furthermore, PCI immunoreactivity has been shown on the acrosomal head of human spermatozoa (73).

Alpha$_2$-Macroglobulin

The gene for human alpha$_2$-macroglobulin (α_2–M) is located on chromosome 12 (78). Alpha$_2$-macroglobulin is a high molecular weight plasma protease inhibitor (720 kDa) of 1451 amino acid residues, that contains four identical subunits (180 kDa) and has the ability to inhibit proteases of all subclasses (79). The subunits are held together pairwise by two interchain disulphide bonds per dimer. The dimers, the functional units of α_2–M, are non-covalently linked together to form a tetramer (Fig. 1).

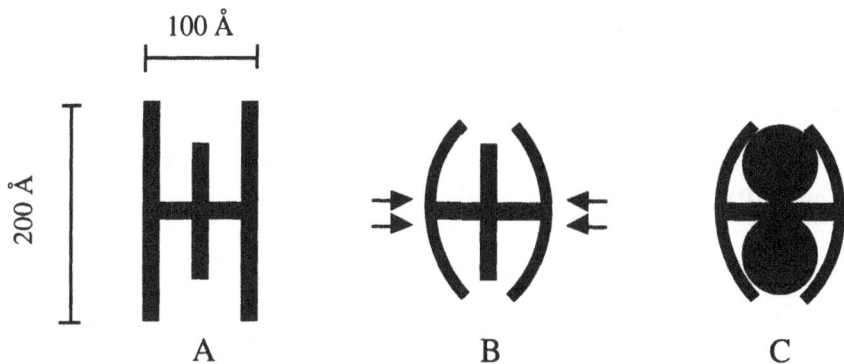

Figure 1. Schematic representation of α_2–M derived from electron micrographs. The α_2–M molecule consists of four subunits. The tetramer comprises two identical dimers, each containing two long arms and two short arms (one is hidden behind the one in the front). (A) Represents native α_2–M, (B) the compact form of methylamine-inactivated α_2–M, and (C) the protease-α_2–M complex in which each dimer accomodates one protease molecule. The arrows denote putative receptor-recognition sites (80).

Alpha$_2$-macroglobulin is a protease inhibitor that inhibits all four classes of proteases. Therefore, it may act as a 'backup' inhibitor under conditions where primary inhibitors have been consumed. Alpha$_2$-macroglobulin enables the activity of proteases to be controlled by steric shielding and rapid clearance. After binding, the active site of the enzyme remains intact and preserves its ability to interact with chromogenic substrates (79). Each α_2–M molecule generally binds two protease molecules (81). Moreover, the bound enzyme is sterically hindered from interacting with antibodies, which renders the enzyme impossible to detect by immunochemical methods in non-denaturing systems (79).

These features have resulted in the 'trap' hypothesis for α_2–M-protease complex formation. Alpha$_2$-macroglobulin recognizes proteases by presenting a peptide stretch, the 'bait' region, which is cleaved by the protease (79). Cleavage anywhere within the 'bait' region initiates a pronounced conformational change in α_2–M (82). This is followed by cleavage of an internal thioester bond, resulting in covalent binding of the active protease to α_2–M (81). Cleavage of the 'bait' region and thioester bond also results in exposure of previously concealed areas that are recognized by high affinity receptors on cells in the reticuloendothelial system, resulting in rapid clearance of the α_2–M-protease complex (83,84). The α_2–M receptor, found on hepatocytes and placenta tissue (85), is identical with the low-density-lipoprotein receptor-related protein (85). The half-life of the circulating complexes is short (2–9 min), while the clearance of native α_2–M is much slower (86).

Alpha$_2$-macroglobulin is synthesized in the liver (87), but can also be produced by macrophages and monocytes in the lungs (88). Alpha$_2$-macroglobulin is not an acute phase protein and no dramatic changes of its concentrations have been observed in the presence of disease (87).

Pregnancy-Zone Protein

The primary structure of pregnancy zone protein (PZP) is very similar to that of α_2–M (89). PZP is also similar to α_2–M in its mechanism of enzyme inhibition (90). Despite their manifest structural similarity, PZP and α_2–M differ in substrate specificity (90) as PZP readily forms complexes with chymotrypsin, but not with trypsin and plasmin (91).

PZP is synthesized in hepatocytes (92) and is a trace protein in normal plasma but in women it may reach concentrations of 1000 mg/L during pregnancy (93). Its function is presently unknown.

Regulation of PSA Activity in Extracellular Space and Disease Related Changes of PSA Fractions in Serum

Despite the extremely high concentration of active PSA in the seminal fluid (0.5–2.0 g/L) PSA occurs in the blood at only 10^{-6} times these values (<1mg/L) (10,11). The measurement of PSA in serum has served as a marker for prostate cancer since the beginning of the 80s and is now the most useful test for early prostate cancer detection. Elevated serum levels are found in several different prostatic diseases but the ability of the test to differentiate between benign prostatic hyperplasia and prostatic cancer is not sufficient. Approximately 25 to 30% of men with benign prostatic hyperplasia (BPH) and 80% with proven prostate cancer have serum PSA levels above 4.0 ng/mL (94). Thus, 20% of men with diagnosed prostate cancer have levels lower than 4.0 ng/mL. Using a cutoff level of 4.0 ng/mL gives 65% false positive and 20% false negative results (94). The appreciable false positive and negative rates are drawbacks to PSA testing that have led to a search for other means of increasing the accuracy of PSA testing.

Active PSA released from the prostate gland is inactivated by different mechanisms but the exact location for inactivation of PSA is not clarified. It becomes evident that release of enzymatically active PSA may require inactivation of its protease activity in order not to cause tissue damage. *In vitro* studies have shown that purified PSA slowly forms stable complexes with several protease inhibitors in serum, α_1-antichymotrypsin, α_2–M, and PZP (22). Complex formation with PCI has also been described (74). There have been no data suggesting that PSA in serum is complexed in appreciable amount to other serine protease inhibitors, except for small amounts of complexes between PSA and α_1-antitrypsin (95).

The reaction between PSA and α_1-antichymotrypsin results in a 1:1 molar ratio complex with a molecular mass of about 90 kDa (22). It also induces an internal peptide-bond cleavage in α_1-antichymotrypsin between lysine-358 and serine-359 (22). This cleavage position has previously been demonstrated for the complex formation between chymotrypsin and α_1-antichymotrypsin (96).

Complex formation with α_1-antichymotrypsin results in a loss of PSA-activity due to blockage of the active site pocket. However, PSA still exposes a number of antigenic determinants detectable with immunoassays (97). The rate of complex formation is rather slow compared with other protease/protease inhibitor reactions. Approximately 55% of PSA-mediated hydrolysis is inhibited 1 min after incubation with a tenfold molar excess of α_1-antichymotrypsin (22).

PSA forms a 2:1 molar ratio complex with α_2–M. In contrast to the reaction with α_1-antichymotrypsin, PSA does not expose any antigenic determinants after complex formation with α_2–M (Fig. 2) (22). This is explained by an ability of α_2–M to enclose the target enzyme in a cage-like structure that renders all PSA epitopes inaccessible in denaturing conditions (79). Thereby, immunodetection of PSA with antibodies is impossible and PSA will be unable to interact with high molecular mass protein substrates (22). However, the complex can be immunodetected by PSA antibodies after denaturation, as shown by SDS-PAGE followed by Western blot procedures. This is in accord with previously reported results of protease reactions with α_2–M (81).

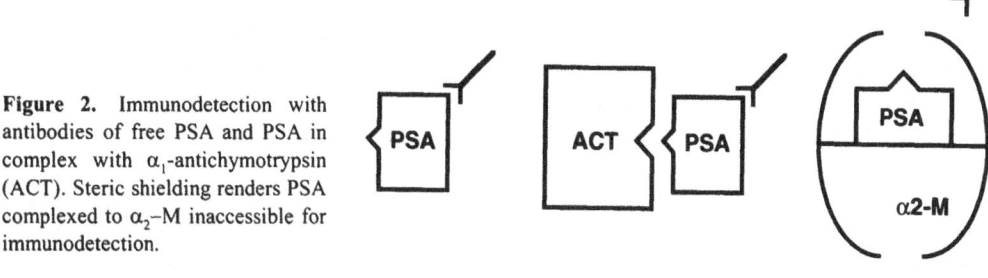

Figure 2. Immunodetection with antibodies of free PSA and PSA in complex with α_1-antichymotrypsin (ACT). Steric shielding renders PSA complexed to α_2–M inaccessible for immunodetection.

Purified PSA and PCI form a 1:1 molar ratio complex with a molecular mass of about 90 kDa. This complex formation is slow compared to that reported for PCI and activated protein C, but is enhanced up to sixfold in the presence of heparin (62,74).

Initial tests did not distinguish between different molecular forms of PSA in serum. To ascertain whether the estimated immunoreactivity in serum represents the PSA zymogen, the active protease, or PSA inactivated by complex formation with extracellular protease inhibitors (α_1-antichymotrypsin and α_2–M), different immunoassays for PSA were constructed. Any complex formation between PSA and α_2–M is not detectable by these methods, as these reactions result in encapsulation of PSA and complete loss of PSA epitopes (22,98). Furthermore, complexes between PCI and PSA have not been found in detectable amounts in serum. Therefore, different monoclonal antibodies were combined to detect free PSA, PSA complexed to α_1-antichymotrypsin, and both free and complexed PSA (Fig. 3) (98). It has been shown that approximately 70–90% of PSA immunoreactivity in serum is constituted of PSA complexed to α_1-antichymotrypsin (95,98,99). Among men with total PSA levels in the 4.0 to 10 ng/mL range, the percentage of free PSA varies from 4 to 50% (95,98,100,101). These findings were supported by the results of gel filtration chromatography of sera from patients (95,98).

Still, there is no information as to whether the non-complexed PSA immunoreactivity constitutes the PSA zymogen, active PSA or PSA inactivated as a result of internal peptide bond cleavage. The minor non-complexed molecular form of PSA seems unlikely to be enzymatically active (in which case it would react with protease inhibitors). Preferrably it represents the PSA zymogen or an internally cleaved inactive form that may be similar to that identified in the seminal fluid (22).

The existence of complexes between PSA and α_1-antichymotrypsin in serum shows that PSA is released in an active form from the prostate gland. To get access to the blood

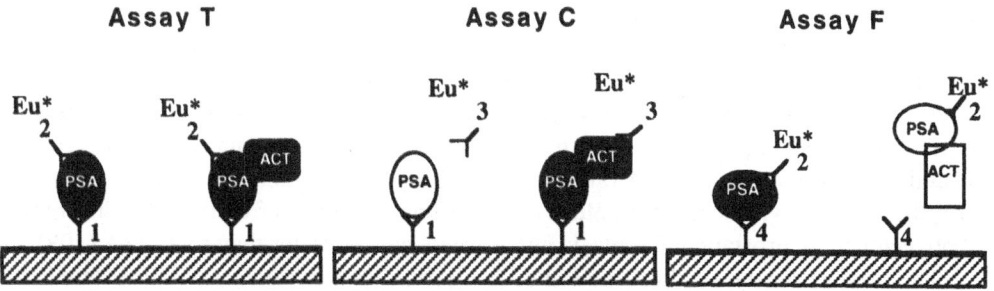

Figure 3. Principles of PSA-assays using Europium-labelled detection antibodies. Monoclonal antibodies are given numbers 1–4. Assay T measures both free and complexed PSA. Assays C detects only complexed PSA, and assay F measures free PSA.

plasma, PSA must pass the following barriers in the prostate: the basal cell layer, the epithelial basement membrane, intervening stroma with interstitial fluid, the capillary basement membrane and the endothelial cell (102). Apart from reaction in blood plasma, PSA may react with α_1-antichymotrypsin along its way from the cellular level through the above mentioned extracellular compartments. Complex formation within the prostatic tissue is supported by recent findings of local production of α_1-antichymotrypsin in the prostate (55).

PSA probably also occurs in the circulation complexed to α_2-M, but escapes immunodetection due to the encapsulation of the PSA molecule and loss of the PSA epitopes. This is supported by a deficit in PSA-concentration after addition of PSA to serum. Therefore, the significance of complex formation with α_2-M remains unclear. The molar ratio between α_1-antichymotrypsin and α_2-M is higher in interstitial fluids than in blood plasma. Therefore, complex formation between PSA and α_1-antichymotrypsin may be the predominant reaction in interstitial fluids.

Complex formation between the target enzyme and both α_1-antichymotrypsin and α_2-M results in conformational changes of the protease inhibitor and subsequently increased affinity for receptors and rapid clearance from plasma (84,103). Half-lives for several complexes of both α_1-antichymotrypsin and α_2-M are a few minutes, as compared to the much longer half-life of the native inhibitors (84,103). Receptors for complexes between serine proteases and serpins have been reported, SEC-receptors (104), as well as receptors for complexes between proteases and α_2-M, LDL-receptor-related proteins (85). Complexes between PSA and α_1-antichymotrypsin and between PSA and α_2-M, might be cleared via SEC-receptors, LDL-receptor-related proteins or similar receptors.

There has been described a biphasic disappearance pattern of total PSA from serum with biological half-lifes of 1.6 hours and 4.6 days, respectively (105). This is in accordance with specific measurements of different molecular forms of PSA (106). It has been speculated that free PSA may be excreted by glomerular filtration, while renal clearance for complexed PSA with a molecular mass of 80 to 90 kDa is unlikely and rather may be removed from the circulation by the liver (107,108).

Measurement of free PSA is of clinical value, as it has been shown, in several independent reports, that the free-to-total PSA ratio (PSA-F/PSA-T) is significantly lower in patients with prostatic carcinoma compared to those with benign prostatic hyperplasia (95,100). Several other research groups have confirmed these results (101,109). Accordingly, the proportion of complexed PSA (PSA-C/PSA-T) is higher in patients with prostatic carcinoma compared to those with benign prostatic hyperplasia. However, the PSA-F/PSA-T ratio gives a higher level of significance in prostate cancer detection (100). The differences in degree of complex formation is independent both of the PSA and the α_1-antichymotrypsin concentration.

The combination of measurement of total PSA and the PSA-F/PSA-T ratio has increased the specificity and sensitivity of PSA in the diagnosis of prostate cancer (95,100). In one study in men with serum PSA concentrations from 4 to 20 ng/mL the specificity of screening was improved from 55% to 73% while maintaining sensitivity at 90% and using a free to total PSA ratio cut off at 0.18 (100). In another study it has been shown that, in men with total PSA values of 4 to 10 ng/mL, the % free PSA did differentiate between benign prostatic hyperplasia and prostate cancer better than total PSA (109).

The reasons for the smaller fraction of free serum PSA in patients with prostate cancer is not clarified. However, the much higher expression of α_1-antichymotrypsin in prostate tumors than in nodules of benign hyperplasia may reflect the findings in blood of a higher proportion of PSA complexed to α_1-antichymotrypsin in patients with prostate cancer than in those with benign prostatic hyperplasia (56). Another explanation may be different tissue expression of the free form of PSA in prostate tumors compared to benign tissue.

Regulation of PSA in the Male Reproductive System

About 70% of PSA isolated from seminal plasma occurs in a single-chain form and manifests catalytic activity (16,22). This was shown by incubation of purified PSA with an excess of α_1-antichymotrypsin and subsequent gel filtration (22). Up to 30% of purified PSA manifests no enzymatic activity due to internal peptide bond cleavages C-terminal of lysine 145 and lysine 182, which results in inactive, two-chain forms of PSA (22,110). These fractions are unreactive with extracellular protease inhibitors (22).

Up to 5% of detected PSA immunoreactivity in seminal plasma occurs in a complex with PCI. Purification of PCI from seminal plasma reveals immunoreactivity corresponding to a low and high molecular form of PCI (76). A purification procedure including heparin-Sepharose and affinity chromatography with both anti-PCI and anti-PSA monoclonal antibodies was used (74). The low molecular form represents cleaved, inactive inhibitor. Amino acid sequence analysis of the high molecular form of PCI demonstrates that PSA is target for PCI. *In vitro* studies have demonstrated complex formation between purified PCI and purified PSA resulting in an SDS-stable 90 kDa complex (74). In parallel with the appearance of complexes on the SDS-PAGE, complexes between PSA and PCI are detected with immunofluorometric assays and the enzymatic activity of PSA decreases (74,77). However, the rate of complex formation is slow compared to the one reported for PCI and activated protein C, but is enhanced up to sixfold in the presence of heparin (74,77).

In freshly collected ejaculates, the rate of PCI-PSA complex formation is similar to that observed between the purified proteins. This has been shown by immunofluorometric assays and immunoblotting on SDS-PAGE (74,77).

During gel dissolution in freshly collected ejaculates, approximately 40% of immunodetected PCI becomes complexed to PSA. Although PCI is a slow inhibitor of PSA, complexes between PCI and PSA are detected at levels that correspond to an inactivation of up to 5% of the PSA-activity in the ejaculate (74). However, PSA can not be efficiently regulated by the insufficient levels of PCI present in the ejaculate, due to tenfold higher concentrations of PSA in the seminal fluid. Furthermore, the cleavage site in PCI, between arginine 357 and leucine 358, seems inconsistent with the chymotrypsin-like activity of PSA, but may nonetheless be appropriate with this substrate specificity as other chymotrypsin-like enzymes also have been shown to be inhibited by PCI (68).

Recently, fragments of PCI have been found in complex with hK2, suggesting that PCI may regulate the activity of hK2 in seminal plasma (27). The above described purification procedure for the PCI-PSA complex does not catch a potential complex between PCI and hK2 since the anti-PSA monoclonal antibody (2E9) does not crossreact with hK2 (28).

Acrosin, located at the acrosome of the sperm, is inhibited by purified PCI at a fast rate indicating that PCI may control acrosin activity *in vivo* (72). The reaction rate is 1000-fold faster than that for PSA and PCI (74). This rate of inhibition is stimulated 200-fold in the presence of heparin. In the female reproductive tract heparin-like molecules, glycosaminoglycans, are present at high concentrations (111) and may stimulate the acrosin-PCI reaction. These findings indicate that the role of PCI in seminal fluid may be to control the activity of acrosin released from damaged sperm in order to prevent unwanted degradation of proteins in the female reproductive tract (72,73).

REFERENCES

1. Riegman PH, Vlietstra RJ, Klaassen P, van der Korput J, Geurts van Kessel A, Romijn J, Trapman J. The prostate-specific antigen gene and the human glandular kallikrein-1 gene are tandemly located on chromosome 19. FEBS Lett 1989; 247: 123–126.

2. Papsidero LD, Wang MC, Valenzuela LA, Murphy GP, Chu TM. A prostate antigen in sera of prostatic cancer patients. Cancer Res 1980; 40: 2428–2432.

3. Nadji M, Tabei SZ, Castro A, Chu TM, Murphy GP, Wang MC and Morales AR. Prostate-specific antigen: An immunohistologic marker for prostatic neoplasms. Cancer 1981; 48: 1229–1232.

4. Qiu S-D, Young CY-F, Bilhartz DL, Prescott JL, Farrow GM, He WW and Tindall DJ. *In situ* hybridization of prostate-specific antigen mRNA in human prostate. J Urol 1990; 144: 1550–1556.

5. Henttu P, Lukkarinen O, Vihko P. Expression of the gene coding for human prostate-specific antigen and related hGK-1 in benign and malignant tumors of the human prostate. Int J Cancer 1990; 45: 654–660.

6. Riegman PH, Vlietstra RJ, van der Korput J, Brinkmann AO, Trapman J. The promotor of the prostate-specific antigen gene contains a functional androgen responsive element. Mol Endocrinology 1991; 5: 1921–1930.

7. Henttu P, Liao S, Vihko P. Androgens up-regulate the human prostate-specific antigen messenger ribonucleic acid (mRNA), but down-regulate the prostatic acid phosphatase mRNA in the LNCaP cell line. Endocrinology 1992; 130: 766–772.

8. Yu H, Diamandis EP, Sutherland DJ. Immunoreactive prostate-specific antigen levels in female and male breast tumors and its association with steroid hormone receptors and patient age. Clin Biochem 1994; 27: 75–79.

9. Clements J, Mukhtar A. Glandular kallikreins and prostate-specific antigen are expressed in the human endometrium. J Clin Endocrin Metab 1994, 78: 1536–1539.

10. Wang MC, Valenzuela LA, Murphy GP, Chu TM. Purification of a human prostate specific antigen. Invest Urol 1979; 17: 159–163.

11. Lilja H and Abrahamsson P-A. Three predominant proteins secreted by the human prostate gland. The Prostate 1988; 12: 29–38.

12. Bélanger A, van Halbeek H, Graves HC, Grandbois K, Stamey TA, Huang L, Poppe I, Labrie F. Molecular mass and carbohydrate structure of prostate specific antigen: Studies for establishment of an international PSA standard. The Prostate 1995; 27: 187–197.

13. Lundwall Å and Lilja H. Molecular cloning of human prostate specific antigen cDNA. FEBS Lett 1987, 214: 317–322.

14. Lundwall Å. Characterization of the gene for prostate-specific antigen, a human glandular kallikrein. Biochem Biophys Res Comm 1989; 161: 1151–1159.

15. Watt K, Lee P-J, M'Timkulu T, Chan W-P and Loor R. Human prostate-specific antigen: Structural and functional similarity with serine proteases. Proc Natl Acad Sci USA 1986; 83: 3166–3170.

16. Lilja H. A kallikrein-like serine protease in prostatic fluid cleaves the predominant seminal vesicle protein. J Clin Invest 1985; 76: 1899–1903.

17. Lilja H, Oldbring J, Rannevik G and Laurell C-B. Seminal vesicle-secreted proteins and their reactions during gelation and liquefaction of human semen. J Clin Invest 1987; 80: 281–285.

18. Lilja H, Lundwall Å. Molecular cloning of epididymal and seminal vesicular transcripts encoding a semenogelin related protein. Proc Natl Acad Sci USA 1992; 89: 4559–4563.

19. McGee R and Herr J. Human seminal vesicle-specific antigen is a substrate for prostate-specific antigen (or P-30). Biol Reprod 1988; 39: 499–510.

20. Lilja H, Laurell C-B. Liquefaction of coagulated human semen. Scand J Clin Lab Invest 1984; 44: 447–452.

21. McGee R, Herr J. Human seminal vesicle-specific antigen during semen liquefaction. Biol Reprod 1987; 37: 431–439.

22. Christensson A, Laurell C-B, Lilja H. Enzymatic activity of Prostate-specific antigen and its reactions with extracellular serine proteinase inhibitors. Eur J Biochem 1990; 194: 755–763.

23. Morris BJ. hGK-1: A kallikrein gene expressed in human prostate. Clin Exp Pharm Phys 1989; 16: 345–351.

24. Schedlich LJ, Bennetts B and Morris BJ. Primary structure of a human glandular kallikrein gene. DNA 1987; 6: 429–437.

25. Young CY-F, Andrews PE, Montgomery BT, Tindall DJ. Tissue-specific and hormonal regulation of human prostate-specific glandular kallikrein. Biochemistry 1992; 31: 818–824.

26. Chapdelaine P, Paradis G, Tremblay RR, Dubé JY. High level of expression in the prostate of a human glandular kallikrein mRNA related to prostate-specific antigen. FEBS Lett 1988; 236: 205–208.

27. Deperthes D, Chapdelaine P, Tremblay R, Brunet C, Berton J, Hébert J, Lazure C, Dube J. Isolation of prostatic kallikrein hK2, also known as hGK-1, in human seminal plasma. Biochimica et Biophysica Acta 1995; 1245: 311–316.

28. Lövgren J, Piironen T, Övermo C, Dowell B, Karp M, Pettersson K, Lilja H, Lundwall Å. Production of recombinant PSA and hK2 and analysis of their immunologic cross-reactivity. Biochem Biophys Res Commun 1995; 213: 888–895.

29. Piironen T, Lövgren J, Karp M, Eerola R, Lundwall Å, Dowell B, Lövgren T, Lilja H, Pettersson K. Immunofluorometric assay for sensitive and specific measurement of human prostatic glandular kallikrein (hK2) in serum. Clin Chem 1996; 42. In press.

30. Baba T, Watanabe K, Kashiwabara S, Arai Y. Primary structure of human proacrosin deduced from its cDNA sequence. FEBS Lett 1989; 244: 296–300.

31. Adham I, Grzeschik K, Geurts van Kessel A, Engel W. The gene encoding the human preproacrosin (ACR) maps to the q13-qter region on chromosome 22. Hum Genet 1989; 84: 59–62.

32. Keime S, Adham I, Engel W. Nucleotide sequence and exon-intron organization of the human proacrosin gene. Eur J Biochem 1990; 190: 195–200.

33. Phi-van L, Müller-Esterl W, Flörke S, Schmid M, Engel W. Proacrosin and the differentiation of the spermatozoa. Biol Reprod 1983; 29: 479–486.

34. Polakoski K, McRorie R. Boar acrosin II. Classification, inhibition, and specificity studies of a proteinase from sperm acrosomes. J Biol Chem 1973; 248: 8183–8188.

35. Schleuning W,. Fritz H. Some characteristics of highly purified boar sperm acrosin. Hoppe-Seyler's Z Physiol Chem 1974; 355: 125–130.

36. Fock-Nüzel R, Lottspeich F, Henschen A, Müller-Esterl W. Boar acrosin is a two-chain molecule. Isolation and primary structure of the light chain; homology with the pro-part of other serine proteinases. Eur J Biochem 1984; 141: 441–446.

37. Baba T, Michikawa Y, Kawakura k, Arai Y. Activation of boar proacrosin is effected by processing at both N- and C-terminal portions of the zymogen molecule. FEBS Lett 1989; 244: 132–136.

38. Baba T Kashiwabara S, Watanabe K, Itho H, Michikawa Y, Kimura K, Takada M, Fukamizu A, Arai Y. Activation and maturation mechanisms of boar acrosin zymogen based on the deduced primary structure. J Biol Chem 1989; 264: 11920–11927.

39. Rabin M, Watson M, Kidd V, Woo SLC, Breg RW, Ruddle FH. Regional location of α_1-antichymotrypsin and α_1-antitrypsin genes on human chromosome 14. Somatic Cell Mol Genet 1986; 12: 209–214.

40. Bao J, Sifers RN, Kidd VJ, Ledley FD, Woo SLC. Molecular evolution of serpins: Homologous structure of the human α_1-antichymotrypsin and α_1-antitrypsin. Biochemistry 1987; 26: 7755–7759.

41. Chandra T, Stackhouse R, Kidd VJ, Robson KJH, Woo SLC. Sequence homology between human α_1-antichymotrypsin, α_1-antitrypsin, and antithrombin III. Biochemistry 1983; 22: 5055–5061.

42. Sefton L, Kelsey G, Kearney P, Povey S, Wolfe J. A physical map of the human PI and AACT genes. Genomics 1990; 7: 382–388.

43. Hachulla E, Laine A, Hayem A. α_1-antichymotrypsin microheterogeneity in crossed immunoaffinoelectrophoresis with free concanavalin A: a useful diagnostic tool in inflammatory syndrome. Clin Chem 1988; 34: 911–915.

44. Lindmark B, Lilja H, Alm R, Eriksson S. The microheterogeneity of desialylated α_1-antichymotrypsin: the occurrence of two amino-terminal isoforms, one lacking a His-Pro dipeptide. Biochim Biophys Acta 1989; 997: 90–95.

45. Baumann U, Huber R, Bode W, Grosse D, Lesjak M, Laurell C-B. Crystal structure of cleaved human α_1-antichymotrypsin at 2·7 Å resolution and its comparison with other serpins. J Mol Biol 1991; 218: 595–606.

46. Travis J, Bowen J, Baugh R. Human α_1-antichymotrypsin: Interaction with chymotrypsin-like proteinases. Biochemistry 1978; 17: 5651–5656.

47. Laine A, Davril M, Rabaud M, Vercaigne-Marko D, Hayem A. (1985) Human serum α_1-antichymotrypsin is an inhibitor of pancreatic elastases. Eur J Biochem 1985; 151: 327–331.

48. Perlmutter DH, Dinarello CA, Punsal PI, Colten HR. Cachectin/tumor necrosis factor regulates hepatic acute-phase gene expression. J Clin Invest 1986; 78: 1349–1354.

49. Berninger RW. Alpha1-antichymotrypsin. Journal of Medicine 1985; 16: 101–128.

50. Papadimitriou CS, Stein H, Papacharalampous NX. Presence of α_1-antichymotrypsin and α_1-antitrypsin in haematopoietic and lymphoid tissue cells as revealed by the immunoperoxidase method. Pathol Res Pract 1980; 169: 287–297.

51. Kittas C, Aroni K, Matani A, Papadimitriou CS. Immunocytochemical demonstration of α_1-antitrypsin and α_1-antichymotrypsin in human gastrointestinal tract. Hepato-gastroenterol 1982; 29: 275–277.

52. Permanetter W, Meister P. Distribution of lysozyme (muramidase) and α_1-antichymotrypsin in normal and neoplastic epithelial tissues: A survey. Acta Histochem 1984; 74: 173–179.

53. Nathrath WB, Meister P. Lysozyme (muramidase) and alpha1-antichymotrypsin as immunohistochemical tumour markers. Acta Histochem suppl 1982; 25: 69–72.

54. Bergman D, Kadner S, Cruz M, Esterman A, Tahery M, Young B, Finlay T. Synthesis of alpha-1-antichymotrypsin and alpha-1-antitrypsin by human trophoblast. Pediatr Res 1993; 34: 312–317.

55. Bjartell A, Björk T, Matikainen M-T, Abrahamsson P-A, di Sant'Agnese A, Lilja H. Production of alpha-1-antichymotrypsin by PSA-containing cells of human prostate epithelium. Urology 1993; 42: 502–510.

56. Björk T, Bjartell A, Abrahamsson P-A, Hulkko S, di Sant'agnese A, Lilja H. Alpha1-antichymotrypsin production in PSA-producing cells is common in prostatic cancer but rare in benign prostatic hyperplasia. Urology 1994; 43: 427–434.

57. Schill W-B, Wallner O, Palm S, Fritz H. Kinin stimulation of spermatozoa motility and migration in cervical mucus. In: Human semen and fertility regulation in man. (ed Hafez ES) Mosby. St Louis. 1976 pp 442–451.

58. Aronsen K-F, Ekelund G, Kindmark C-O, Laurell C-B. Sequential changes of plasma proteins after surgical trauma. Scand J Lab Invest 1972; 29 suppl 124: 127–136.

59. Meijers JCM, Kanters DH, Vlooswijk RA, van Erp HE, Hessing M, Bouma BN. Inactivation of human plasma kallikrein and factor XIa by protein C inhibitor. Biochemistry 1988; 27: 4231–4237.

60. Billingsley G, Walter M, Hammond G, Cox D. Physical mapping of four serpin genes: alpha 1-antitrypsin, alpha 1-antichymotrypsin, corticosteroid-binding globulin, and protein C inhibitor, within a 280 kb region on chromosome 14q32.1. Am J Hum Genet 1993; 52: 343–353.

61. Long GL, Chandra T, Woo SLC, Davie EW, Kurachi K. Complete sequence of the cDNA for human α_1-antitrypsin and the gene for the S variant. Biochemistry 1984; 23: 4828–4837.

62. Suzuki K, Nishioka J, Hashimoto S. Protein C inhibitor. Purification from human plasma and characterization. J Biol Chem 1983; 258: 163–168.

63. Suzuki K, Deyashiki Y, Nishioka J, Kurachi K, Akira M, Yamamoto S, Hashimoto S. Characterization of a cDNA for human protein C inhibitor. A new member of the plasma serine protease inhibitor superfamily. J Biol Chem 1987; 262: 611–616.

64. Laurell M, Stenflo J. Protein C inhibitor from human plasma: Characterization of native and cleaved inhibitor and demonstration of inhibitor complexes with plasma kallikrein. Thromb Haemostas 1989; 62: 885–891.

65. Marlar RA, Griffin JH. Deficiency of protein C inhibitor in combined Factor V/VIII deficiency disease. J Clin Invest 1980; 66: 1186–1189.

66. Suzuki K, Stenflo J, Dahlbäck B, Teodorsson B. Inactivation of human coagulation Factor V by activated protein C. J Biol Chem 1983; 258: 1914–1920.

67. Esmon CT. The regulation of natural anticoagulant pathways. Science 1987; 235: 1348–1352.

68. Suzuki K, Nishioka J, Kusumoto H, Hashimoto S. Mechanism of inhibition of activated protein C by protein C inhibitor. J Biochem 1984; 95: 187–195.

69. España F, Berrettini M, Griffin JH. Purification and characterization of plasma protein C inhibitor. Thromb Res 1989; 55: 369–384.

70. Ecke S, Geiger M, Resch I, Jerabek I, Sting L, Maier M, Binder BR. Inhibition of tissue kallikrein by protein C inhibitor. Evidence for identity of protein C inhibitor with the kallikrein binding protein. J Biol Chem 1992; 267: 7048–7052.

71. Geiger M, Huber K, Wojta J, Stingl L, España F, Griffin JH, Binder BR. Complex formation between urokinase and plasma protein C inhibitor *in vitro* and *in vivo*. Blood 1989; 74: 722–728.

72. Hermans J, Jones R, Stone SR. Rapid inhibition of the sperm protease acrosin by protein C inhibitor. Biochemistry 1994; 33: 5440–5444.

73. Zheng X, Geiger M, Ecke S, Bielek E, Donner P, Eberspacher U, Schleuning W, Binder B. Inhibition of acrosin by protein C inhibitor and localization of protein C inhibitor to spermatozoa. Am J Physiol 1994; 267: C466–472.

74. Christensson A, Lilja H. Complex formation between Protein C Inhibitor and Prostate-specific antigen *in vitro* and in human semen. Eur J Biochem 1994; 220: 45–53.

75. Taylor G, Yorke S, Harding D. Glycosaminoglycan specificity of a heparin-binding peptide. Pept Res 1995; 8: 286–293.

76. Laurell M, Christensson A, Abrahamsson P-A, Stenflo J, Lilja H. Protein C inhibitor in human body fluids. Seminal plasma is rich in inhibitor antigen deriving from cells throughout the male reproductive system. J Clin Inv 1992; 89: 1094–1101.

77. España, F., Gilabert, J., Estelles, A., Romeu, A., Aznar, J. & Cabo, A. Functionally active protein C inhibitor/plasminogen activator inhibitor-3 (PCI/PAI-3) is secreted in seminal vesicles, occurs at high concentrations in human seminal plasma and complexes with prostate-specific antigen. Thromb Res 1991; 64: 309–320.

78. Kan C-C, Solomon E, Belt KT, Chain AC, Hiorns LR, Fey G. Nucleotide sequence of cDNA encoding human α_2-macroglobulin and assignment of the chromosomal locus. Proc Natl Acad Sci USA 1985; 82: 2282–2286.

79. Barrett AJ, Starkey PM. The interaction of α_2-macroglobulin with proteinases. Characteristics and specificity of the reaction, and a hypothesis concerning its molecular mechanism. Biochem J 1973; 133: 709–724.

80. Feldman SR, Gonias SL, Pizzo SV. Model of α_2-macroglobulin structure and function. Proc Natl Acad Sci USA 1985; 82: 5700–5704.

81. Sottrup-Jensen L. α-macroglobulins: Structure, shape, and mechanism of proteinase complex formation. J Biol Chem 1989; 264: 11539–11542.

82. Björk I, Fish WW. Evidence for similar conformational changes in α_2-macroglobulin on reaction with primary amines or proteolytic enzymes. Biochem J 1982; 207: 347–356.

83. Kaplan J, Nielsen ML. Analysis of macrophage surface receptors. Binding of α-macroglobulin•protease complexes to rabbit alveolar macrophages. J Biol Chem 1979; 254: 7323–7328.

84. Gliemann J, Röll Larsen T, Sottrup-Jensen L. Cell association and degradation of α_2-macroglobulin-trypsin complexes in hepatocytes and adipocytes. Biochim Biophys Acta 1983; 756: 230–237.

85. Kristensen T, Moestrup SK, Gliemann J, Bendtsen L, Sand O, Sottrup-Jensen L. Evidence that the newly cloned low-density-lipoprotein receptor related protein (LRP) is the α_2-macroglobulin receptor. FEBS Lett 1990; 276: 151–155.

86. Imber MJ, Pizzo SV. Clearance and binding of two electrophoretic fast forms of human α_2-macroglobulin. J Biol Chem 1981; 256: 8134–8139.

87. Laurell C-B, Jeppsson J-O. Protease inhibitors in plasma. In Plasma proteins. (ed Putnam FV) Academic Press, Inc, New York. 1975; pp 229–264.

88. Mosher DF, Saksela O, Vaheri A. Synthesis and secretion of α_2-macroglobulin by cultured adherent lung cells. J Clin Invest 1977; 60: 1036–1045.

89. Sottrup-Jensen L, Folkersen J, Kristensen T, Tack BF. Partial primary structure of human pregnancy zone protein: Extensive sequence homology with human α_2-macroglobulin. Proc Natl Acad Sci USA 1984; 81: 7353–7357.

90. Sand O, Folkersen J, Westergaard JG, Sottrup-Jensen L. Characterization of human pregnancy zone protein. J Biol Chem 1985; 260: 15723–15735.

91. Christensen U, Simonsen M, Harrit N, Sottrup-Jensen L. Pregnancy zone protein, a proteinase-binding macroglobulin. Interactions with proteinases and methylamine. Biochemistry 1989; 28: 9324–9331.

92. Stimson WH, Farquharson DM, Shepherd A, Andersson JM. Studies on the synthesis of pregnancy-associated α_2-glycoprotein by the liver, placenta and peripheral blood leucocyte populations. J Clin Lab Immunol 1979; 2: 235–238.

93. Folkersen J, Teisner B, Grunnet N, Grudzinskas JG, Westergaard JG, Hindersson P. Circulating levels of pregnancy zone protein: Normal range and the influence of age and gender. Clin Chim Acta 1981; 110: 139–145.

94. Catalona WJ, Smith DS, Ratliff TL, Basler JW. Detection of organ-confined prostate cancer is increased through prostate-specific antigen-based screening. JAMA 1993; 270: 948–954.

95. Stenman UH, Leinonen J, Alfthan H, Ranniko S, Tuhkanen K, Alfthan O. A complex between prostate-specific antigen and α_1-antichymotrypsin is the major form of prostate-specific antigen in serum of patients with prostatic cancer: assay of the complex improves clinical sensitivity for cancer. Cancer Res 1991; 51: 222–226.

96. Morii M, Travis J. Amino acid sequence at the reactive site of human α_1-antichymotrypsin. J Biol Chem 1983; 258: 12749–12752.

97. Pettersson K, Piironen T, Seppälä M, Liukkonen L, Christensson A, Matikainen M-T, Suonpää M, Lövgren T, Lilja H. Free and complexed prostate-specific antigen (PSA): *In vitro* stability, epitope map, and development of immunofluorometric assays for specific and sensitive detection of free PSA and PSA-α_1-antichymotrypsin complex. Clin Chem 1995, 41: 1480–1488.

98. Lilja H, Christensson A, Dahlén U, Matikainen M-T, Nilsson O, Pettersson K, Lövgren T. Prostate-Specific Antigen in Serum Occurs Predominantly in Complex with α_1-antichymotrypsin. Clin Chem 1991; 37: 1618–1625.

99. Wood WG, Sloot van der E, Böhle A. The establishment and evaluation of luminescent-labelled immunometric assays for prostate-specific antigen-α_1-antichymotrypsin complexes in serum. Eur J Clin Chem Clin Biochem 1991; 29: 787–794.

100. Christensson A, Björk T, Nilsson O, Dahlén U, Matikainen M-T, Cockett T K, Abrahamsson P-A, Lilja H. Serum prostate-specific antigen complexed to α_1-antichymotrypsin as an indicator of prostate cancer. J Urology 1993; 150: 100–105.

101. Catalona WJ, Smith DS, Wolfert RL, Wang TJ, Rittenhouse HG, Ratliff TL, Nadler RB. Evaluation of percentage of free serum prostate-specific antigen to improve specificity of prostate cancer screening. JAMA 1995, 274: 1214–1220.

102. Price A. Abrupt changes in prostate-specific antigen concentration in acute renal failure. Clin Chem 1993; 39: 161–162.

103. Pizzo SV, Mast AE, Feldman SR, Salvesen G. *In vivo* catabolism of α_1-antichymotrypsin is mediated by the serpin receptor which binds α_1-proteinase inhibitor, antithrombin III and heparin cofactor II. Biochim Biophys Acta 1988; 967: 158–162.

104. Perlmutter DH, Glover GI, Rivetna M, Schasteen CS, Fallon RJ. Identification of a serpin-enzyme complex receptor on human hepatoma cells and human monocytes. Proc Natl Acad Sci USA 1990; 87: 3753–3757.

105. van Straalen JP, Bossens MMP, de Reijke TM, Sanders GT. Biological half-life of prostate-specific antigen after radical prostatectomy. Eur J Clin Chem Biochem 1994; 32: 53–55.

106. Björk T, Abrahamsson P-A, Lilja H, Pettersson K, Cockett ATK. Rates of clearance of free and complexed forms of PSA in serum after radical prostatectomy and transurethral microwave therapy. J Urol 1995; 153: 295A (abstract 265).

107. Agha AH, Schechter E, Roy JB, Culkin DJ. Prostate specific antigen is metabolized in the liver. J Urol 1996; 155: 1332–1335.

108. Xemjonow A, Oberpennig F, Surdel W, Weining C, Brandt B, Brandau W, Hertle L, Hamm M. Prostate-specific antigen and radical prostatectomy: variation after manipulation of the prostate and consecutive elimination half-life of free and total PSA. J Urol 1996; 155: 697A (abstract 1545).

109. Luderer AA, Chen Y-T, Soriano TF, Kramp WJ, Carlson G, Cuny C, Sharp T, Smith W, Petteway J, Brawer MK. Measurement of the proportion of free to total prostate-specific antigen improves diagnostic performance of prostate-specific antigen in the diagnostic gray zone of total prostate-specific antigen. Urology 1995; 46: 187–194.

110. Zhang WM, Leinonen J, Kalkkinen N, Dowell B, Stenman U-H. Purification and characterization of different molecular forms of prostate-specific antigen in human seminal fluid. Clin Chem 1995; 41: 1567–1573.

111. Bushmeyer S, Bellin M, Brantmeier S, Boehm S, Kubajak C, Ax R. Relationships between bovine follicular fluid glycosaminoglycans and steroids. Endocrinology 1985; 117: 879–885.

17

SERPINS AND PROGRAMMED CELL DEATH

Guy S. Salvesen

The Burnham Institute
10901 North Torrey Pines Road
San Diego, California 92037

1. PROGRAMMED CELL DEATH

Cell death is a normal feature of animal development. Studies on a number of cell types, such as neurons that die during growth of young animal nervous systems and lymphocytes that die during receptor repertoire selection in adults, have shown that cells usually kill themselves by activating built in suicide mechanisms[1]. In mature organisms cell number is maintained by a balance between cell proliferation and death, and though apoptosis is often studied as a normal process in development, it is clear that it plays important roles in pathology[2,3]. In the absence of compensating responses of cell proliferation, changes in cell death rates result in either cell accumulation or cell loss. Increases in cell death are involved in AIDS, neurodegenerative disorders and ischemic injury. Decreases in cell death are involved in the pathogenesis of autoimmune diseases and viral infection but the most obvious result of the suppression of cell death is cancer. In fact, two of the genes now known to regulate apoptosis, *bcl-2* and p53, were originally identified because of their roles in neoplasia. Indeed the role played by programmed cell death in a number of cancers is central to the disease progress.

2. THE CASPASES

In 1992 two groups that had been working on the proteolytic maturation of interleukin 1β reported the identification and molecular cloning of the proteinase responsible, interleukin 1β-converting enzyme (ICE).[4,5] A year later, a group working on the genetics of programmed cell death in *Ceanhorabditis elegans* discovered that a key effector gene in the pathway, *ced-3*, exhibited important homology with ICE.[6] These seminal studies laid the groundwork for the identification of several mammalian homologs of ICE and Ced-3 that are currently under intense investigation for roles in programmed cell death. At present the family, which are now referred to as "caspases,"* consists of Ced-3, and the following ten human proteins: ICE, Ich1,[7] CPP32/Yama/apopain,[8–10] Mch3/ICELAP3/Cmh1,[11–13] Mch2,[14] ICErel-II/TX,[15,16] ICErel-III/TY,[15] ICELAP6,[17] Mch4,[18] and FLICE/Mach-1/Mch5[18–20] (Fig. 1). Originally iden-

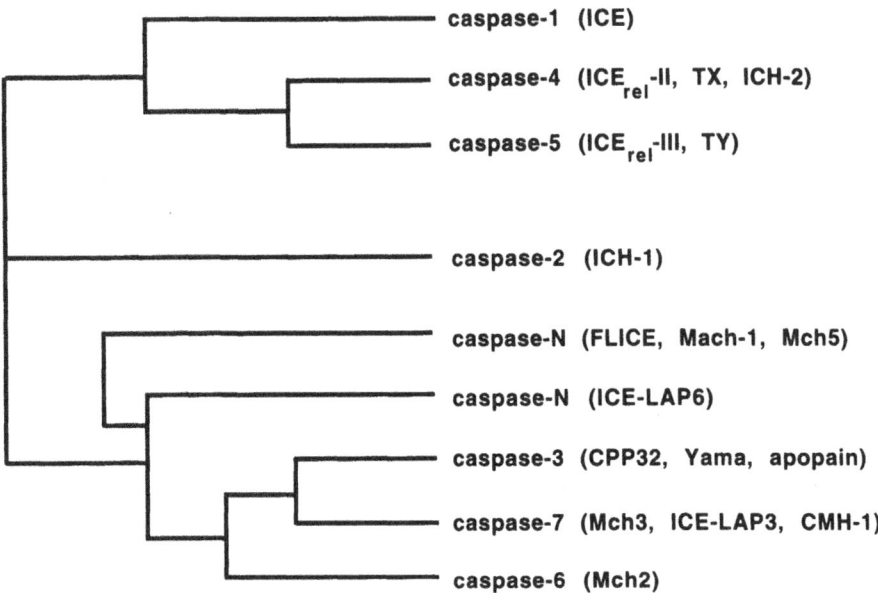

Figure 1. The human caspase family. For convenience members of the ICE/Ced3 family, based on sequence homology and conservation of critical residues, are called "caspases", following recommendations of the ICE/Ced-3 Nomenclature Committee. The name of the family should not be construed to signify too much, except that they are expected to be <u>cys</u>teine-dependent <u>asp</u>artate-specific prote<u>ases</u>. Membership of the family is based on sequence similarities in the catalytic chains. All of the caspases have now been shown to cause apoptosis when transfected into at least one cell line, and have either proteolytic activity on protein or synthetic peptide substrates. Those caspases labeled "N" in the figure have not yet been designated a number.

tified by homology screening of nucleotide databases using ICE and Ced-3 sequences, each contains characteristic conserved sequences important for the proteolytic activity of ICE and are thus suspected to be proteolytically competent. As with most proteases, the caspases require activation through proteolysis of their single chain zymogens. Each is able to cause apoptosis when overexpressed in recipient cells, though it is not clear whether this results from their intrinsic function, or just a response of cells to adventitious proteolysis. Consequently it is not clear whether all the caspases are normally involved in programmed cell death, nor whether there is a relationship between their specificities, their activities and activations.

ICE has a preference for Asp in its primary specificity pocket in both synthetic and natural substrates (including interleukin 1β and the ICE precursor itself) [5,21,22]. The crystal structure shows that this preference is attributed to one Gln and two Arg side-chains that line the pocket and contribute charge stabilization and hydrogen bonds to the P_1 β-carboxylate of the substrate[23,24]. All three residues are conserved in the caspases, thus the potential for Asp-specific proteolytic activity is present in all, indeed the caspases all have the ability to cleave substrates after Asp residues. Specificity for Asp is unusual in proteases, in fact of the hundreds of known animal proteases, the only other ones to share this specificity is granzyme B[25,26] — a serine protease that initiates apoptosis when delivered

---

* This family designation follows recommendations of the Nomenclature Committee. Publication of the recommendations is pending and any citations of the new family terminology should reference the appropriate publication, not this chapter

to susceptible target cells in models of cytolytic cell attack. Either this is an enormous coincidence, or it suggests that proteases that promote apoptosis do so by cleaving target substrates after Asp residues. Indeed there is a close biological link between granzyme B and the caspases, as we and others have recently shown[27–29]. Granzyme B is able to activate the precursor of Yama, thereby defining for this cytotoxic cell protease a role in entering target cells to initiate the pre-existing death program. It is now known that granzyme B can also activate Mch3[30], ICE-LAP6[17] and FLICE[19].

3. THE COWPOX VIRUS SERPINS

Many protease inhibitor families are found in plants and animals, and the serpins are no exception to this. However, members of the serpin family have also found a home in the orthopox viruses, giving them the widest distribution of any known protease inhibitor family (Fig. 2).

Presumably this derives from their success in adapting to different cellular environments. The orthopox viruses, which include variola, vaccinia and cowpox, contain the largest of viral genomes, in the region of 200 genes of which only a minority are required for viral replication. The majority of the gene products endow aspects of host range specificity and defend the virus against anti-viral defense measures mounted by the infected host. In cowpox virus three of these genes encode serpins: CrmA, B24R and K2L. These genes are sometimes termed the "SPI" genes in other orthopox viruses, which should not be confused with the "Spi" locus designation of mouse genes related to $\alpha_1$proteinase inhibitor or $\alpha_1$antichymotrypsin[32,33].

The function of B24R and K2L is not known. K2L is distantly related to the SERP1 serpin encoded by myxoma virus[34,35], yet they both contain an Arg in the presumptive P1 position, and consensus signal peptides suggesting they are probably secreted. Thus the function of K2L may be the same as SERP1, which acts to inhibit inflammation through inactivation of an as yet unidentified proteinase[35]. The rabbitpox virus ortholog of B24R has been reported to control the ability of infected cells to fuse, but the mechanism is unknown[36].

The biological role of CrmA[37], and its target proteinase(s) have been more closely investigated. Cowpox virus variants that contain lesions in the coding region for CrmA

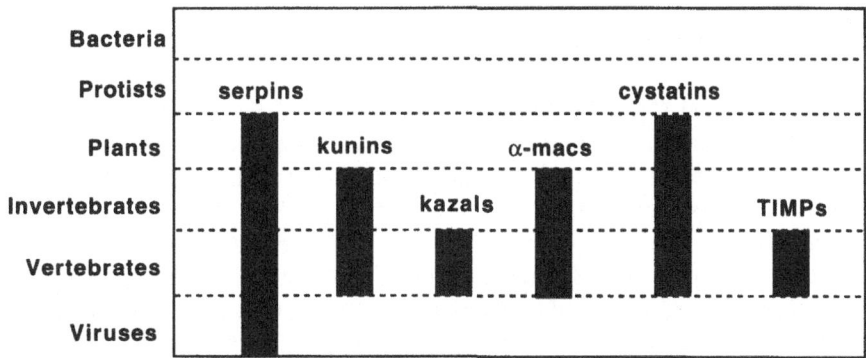

Figure 2. Distribution of protease inhibitor families. Most of the six main families of protease inhibitors found in humans have members in other orders of the biotic kingdoms[31]. The serpins contain the most diverse distribution so far known, though of course we should not rule out that members of the other families just have not been found yet in as wide a variety of life forms.

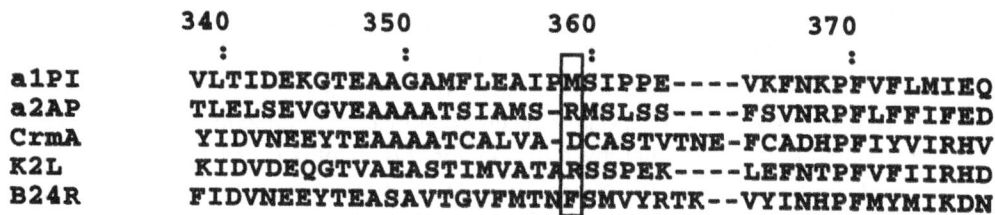

Figure 3. Alignment of the reactive site loop regions of the three cowpox virus serpins with α₁proteinase inhibitor (α₁PI) and α₂antiplasmin (α₂AP). The residue numbering is for α₁PI. The P1 residue directing the specificity of α₁PI for elastase and α₂AP for plasmin is boxed to demonstrate major presumptive specificity determinants of the viral serpins. It is on the basis of this alignment that the CrmA target was discovered.

have a characteristic phenotype when used to infect the chorioalantoid membrane of chicken eggs. This phenotype presents as white pocks, compared to red pocks of the wild type virus, and is due to massive influx of white blood cells into the pock. Because white blood cells are absent from the red pocks (which arise from CrmA-competent virus), we suggested that CrmA was able to prevent the signal that would normally attract inflammatory cells to an infected locus[38]. Moreover, since CrmA is a serpin, we argued that it was likely to inhibit the action of a proteinase component of the inflammatory signaling pathway. ICE, which had recently been discovered, fit the bill since it was responsible for activation of the pro-inflammatory mediator interleukin 1β. Moreover, ICE had been shown to cleave the precursor of interleukin 1β following an Asp residue, which matched the presumptive P1 residue of CrmA (see Fig. 3). We discovered that CrmA was able to prevent the action of ICE on the precursor of interleukin 1β, and also on a synthetic substrate spanning the ICE cleavage site[38]. Later we showed that CrmA binds ICE rapidly and tightly, indeed inhibition is among the fastest and tightest of protein–protein interactions[39].

The discovery of the inhibition of ICE by CrmA was also interesting from a mechanistic point of view since ICE is a cysteine proteinase without any relationship in primary sequence or structure to serine proteinases for which serpins were previously thought specific. We also showed that CrmA was able to function as a "normal" serpin since it also rapidly inhibits the serine proteinase granzyme B[40], a chymotrypsin family member with primary specificity for Asp.

4. INHIBITION OF CELL DEATH BY CrmA

On the basis that Ced-3 is a key regulator of apoptosis in nematodes, it was predicted that a human ortholog would carry out the same function. The problem here is that, while Ced-3 appears to be unique in the nematode, humans contain at least ten homologs, complicating the assessment of their individual functions (Fig. 1). The first demonstration that a caspase could cause death of mammalian cells was provided by Yuan and colleagues who examined the effects of ICE transfected into fibroblasts or neurons[41,42]. In both cases, ICE-mediated death was prevented by overexpression of a CrmA construct in the targeted cells. This study was undertaken in response to our earlier finding[38,39] that the serpin CrmA, encoded in the cowpox virus genome[37], is a specific inhibitor of ICE. However, such evidence is only correlative, since artificial introduction into cells of a number of proteases, including chymotrypsin, proteinase K, and trypsin, causes significant apoptosis[43]. Clearly overexpression of a protease in cells could lead to cell death, and naturally this will be rescued by CrmA since it is such a good ICE inhibitor. Targeted disruption of the ICE gene in mice abolishes the generation of mature interleukin 1β, but has no obvious defects in cell death, with the exception that one of the studies reported a decline in Fas-mediated killing of ICE[-/-] cells *ex vivo*[44,45]. This is in stark contrast to disruption of the *bcl-x* gene which results in early embryonic lethality due to massive cell death in a number of tissues[46]. Thus ICE is either redundant, or it is not involved in programmed cell death during development.

More informative are studies in which CrmA rescued cells from number of death stimuli[9,42,47–50]. This implies that CrmA can have an additional function to the one ascribed to it in limiting the generation of mature IL-1β by inhibiting ICE, and suggests that a CrmA-inhabitable protease, probably an ICE homolog but not ICE itself, is a central component of programmed cell death in several distinct systems.

5. WHAT IS THE CRMA TARGET IN THE CELL DEATH PATHWAY?

If the most well characterized caspase, ICE, plays a minor role, if any, in cell death, there is good evidence that others are more directly involved (Fig. 4). For example, Yama and Mch3 are processed rapidly following a death stimulus[17,51,52], though there is no evidence that these processed forms are actually active. Probably they are, but since the degree of activity depends on the extent of proteolytic processing for ICE[53], it is likely that

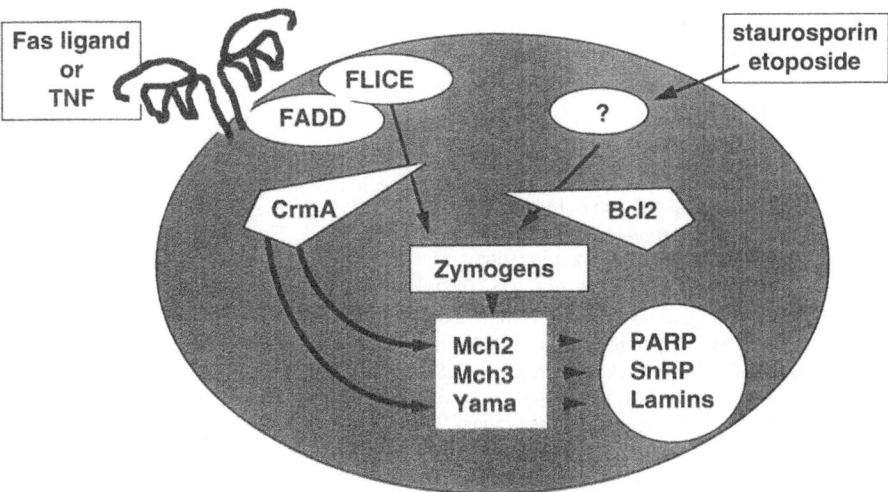

Figure 4. Two pathways to cell death. Programmed cell death can be initiated by a number of stimuli, here are shown two types. Ligation of the TNF receptor or Fas at the cell surface results in oligomerization of the receptors and recruitment of the adaptor molecule FADD to their cytoplasmic tails. FLICE binds via the adaptor complex and is thought to be activated to a proteolytically competent form that is capable of transmitting the death signal to a pathway whose result is activation of the executioners, including Yama, Mch2 and Mch3. These three caspases cleave a number of proteins, including poly (ADP) ribose polymerase (PARP), 70 kDa U1 SnRP (SnRP), and nuclear lamins. The execution phase can also be triggered by drugs such as staurosporin and etoposide that interfere with kinase-mediated signal transduction, via a pathway that is inhibitable by Bcl2, but not by CrmA. Although CrmA can inhibit Mch2 and Yama weakly (as shown by the arrows) its main target is upstream.

this will also be the case for other caspases. Mch2 has been identified as an activator of Yama[54], placing it upstream in the death pathway. There is no data of how Mch2 itself is activated. Clearly there is a need to determine the relationship between proteolytic processing (observed *in vivo* — and duplicated *in vitro*) and catalytic activity. Given the observation that Yama and Mch3 and Mch2 can act on nuclear substrates that are cleaved during apoptosis it is thought that they represent downstream effectors — executioners — of the death pathway[51].

FLICE, on the other hand, is probably an upstream component of the pathway since it is part of the protein complex that transduces the death signal following Fas ligation[19,20]. Engagement of the "death inducing signaling complex" of which FLICE is a component, results in recruitment of FLICE; in the activated complex can be found the full length precursor and the N-terminal peptide[19]. It is presumed that FLICE becomes activated through proteolysis and translocates to a cellular compartment to promote the death response.

One way to determine the key effector of cell death promoted by the CrmA target is to analyze the kinetics of inhibition of given caspases by CrmA. For example, Yama is inhibited by CrmA, but the inhibition is so weak (Ki 0.6 µM) that it represents an unlikely target *in vivo*[11]. Similarly, the inhibition of Mch3 is so weak that we have not been able to calculate an inhibition constant[11].

Another line of evidence that rules out Yama, Mch2 or Mch3 as the CrmA target comes from studies of the activation of these caspases during cell death triggered by Fas ligation or TNF receptor ligation, compared with staurosporin treatment. The latter is an inhibitor of protein kinases and results in progress to death via a pathway that is not inhibited by CrmA, yet all three triggers result in the conversion of Yama, Mch2 and Mch3[51, 55]. Since CrmA inhibits cell death triggered by engagement of Fas and the TNF receptors, the CrmA target must lie on this pathway upstream of Yama, Mch2 and Mch3. A likely candidate is FLICE.

REFERENCES

1. Raff, M.C. *Nature* **356**, 397–400 (1992).
2. Thompson, C.B. *Science* **267**, 1456–1462 (1995).
3. Nicholson, D.W. *Nature Biotechnology* **14**, 297–301 (1996).
4. Cerretti, D.P., *et al. Science* **256**, 97–100 (1992).
5. Thornberry, N.A., *et al. Nature* **356**, 768–774 (1992).
6. Yuan, J., Shaham, S., Ledoux, S., Ellis, H.M. & Horvitz, H.M. *Cell* **75**, 641–652 (1993).
7. Wang, L., Miura, M., Bergeron, L., Zhu, H. & Yuan, J. *Cell* **78**, 739–750 (1994).
8. Fernandes-Alnemri, T., Litwack, G. & Alnemri, E.S. *J. Biol. Chem.* **269**, 30761–30764 (1994).
9. Tewari, M., *et al. Cell* **81**, 801–809 (1995).
10. Nicholson, D.W., *et al. Nature* **376**, 37–43 (1995).
11. Fernandes-Alnemri, T., *et al. Cancer. Res.* **55**, 6045–52 (1995).
12. Duan, H., *et al. J Biol Chem* **271**, 1621–5 (1996).
13. Lippke, J.A., Gu, Y., Sarnecki, C., Caron, P.R. & Su, M.S. *J Biol Chem* **271**, 1825–8 (1996).
14. Fernandes-Alnemri, T., Litwack, G. & Alnemri, E.S. *Cancer Res.* **55**, 2737–2742 (1995).
15. Munday, N.A., *et al. J. Biol. Chem.* **270**, 15870–15876 (1995).
16. Faucheu, C., *et al. EMBO J.* **14**, 1914–1922 (1995).
17. Duan, H., *et al. J. Biol. Chem.* **In press**, (1996).
18. Fernandes-Alnemri, T., *et al. Proc. Natl. Acad. Sci. USA.* **In press**, (1996).
19. Muzio, M., *et al. Cell* **85**, 817–827 (1996).
20. Boldin, M.P., Goncharov, T.M., Goltsev, Y.V. & Wallach, D. *Cell* **85**, 803–815 (1996).
21. Black, R.A., Kronheim, S.R., Merriam, J.E., March, C.J. & Hopp, T.P. *J. Biol. Chem.* **264**, 5323–5326 (1989).
22. Kostura, M.J., *et al. Proc. Natl. Acad. Sci. USA* **86**, 5227–5231 (1989).
23. Wilson, K.P., *et al. Nature* **370**, 270–275 (1994).

24. Walker, N.P.C., *et al. Cell* **78**, 343–352 (1994).
25. Odake, S., *et al. Biochemistry USA* **30**, 2217–2227 (1991).
26. Poe, M., *et al. J. Biol. Chem.* **266**, 98–103 (1991).
27. Quan, L.T., *et al. Proc. Natl. Acad. Sci. USA* **93**, 1972–1976 (1996).
28. Darmon, A.J., Nicholson, D.W. & Bleackley, R.C. *Nature* **377**, 446–8 (1995).
29. Martin, S.J., *et al. EMBO J.* **15**, 2407–2416 (1996).
30. Gu, Y., *et al. J. Biol. Chem.* **271**, 10816–10820 (1996).
31. Salvesen, G. & Enghild, J.J. in *Acute Phase Proteins. Molecular Biology, Biochemistry, and Clinical Applications* (eds. Mackiewicz, A., Kushner, I. and Baumann, H.) (CRC Press, Boca Raton, Florida, 1993).
32. Hill, R.E. & Hastie, N.D. *Nature* **326**, 96–99 (1987).
33. Inglis, J.D. & Hill, R.E. *EMBO J.* **10**, 255–261 (1991).
34. Lomas, D.A., Evans, D.L., Upton, C., McFadden, G. & Carrell, R.W. *J Biol Chem* **268**, 516–521 (1993).
35. Macen, J.L., Upton, C., Nation, N. & McFadden, G. *Virology* **195**, 348–363 (1993).
36. Turner, P.C. & Moyer, R.W. *J Virol* **66**, 2076–2085 (1992).
37. Pickup, D.J., Ink, B.S., Hu, W., Ray, C.A. & Joklik, W.K. *Proc Natl Acad Sci USA* **83**, 7698–7702 (1986).
38. Ray, C.A., *et al. Cell* **69**, 597–604 (1992).
39. Komiyama, T., *et al. J. Biol. Chem.* **269**, 19331–19337 (1994).
40. Quan, L.T., Caputo, A., Bleackley, R.C., Pickup, D.J. & Salvesen, G.S. *J. Biol. Chem.* **270**, 10377–10379 (1995).
41. Miura, M., Zhu, H., Rotello, R., Hartwieg, E.A. & Yuan, J. *Cell* **75**, 653–660 (1993).
42. Gagliardini, V., *et al. Science* **263**, 826–828 (1994).
43. Williams, M.S. & Henkart, P.A. *J. Immunol.* **153**, 4247–4255 (1995).
44. Kuida, K., *et al. Science* **267**, 2000–2003 (1995).
45. Li, P., *et al. Cell* **80**, 401–411 (1995).
46. Motoyama, N., *et al. Science* **267**, 1506–1510 (1995).
47. Tewari, M. & Dixit, V.M. *J. Biol. Chem.* **270**, 3255–3260 (1995).
48. Los, M., *et al. Nature* **375**, 81–83 (1995).
49. Enari, M., Hug, H. & Nagata, S. *Nature* **375**, 78–81 (1995).
50. Boudreau, N., Sympson, C.J., Werb, Z. & Bissell, M.J. *Science* **267**, 891–893 (1995).
51. Chinnaiyan, A.M., *et al. J Biol Chem* **271**, 4573–6 (1996).
52. Schlegel, J., *et al. J Biol Chem* **271**, 1841–4 (1996).
53. Yamin, T.-T., Ayala, J.M. & Miller, D.K. *J. Biol. Chem.* **271**, 13273–13282 (1996).
54. Liu, X., Kim, C.N., Pohl, J. & Wang, X. *J. Biol. Chem.* **271**, 13371–13376 (1996).
55. Orth, K., Chinnaiyan, A.M., Garg, M., Froelich, C.J. & Dixit, V.M. *J. Biol. Chem.* **271**, 16443–16446 (1996).

18

C1 INHIBITOR

Functional Analysis of Naturally-Occurring Mutant Proteins

Alvin E. Davis III

Division of Nephrology
Children's Hospital Research Foundation
3333 Burnet Avenue
Cincinnati, Ohio 45229-3039

1. INTRODUCTION AND HISTORICAL PERSPECTIVE

C1 inhibitor, although frequently considered primarily a complement protein, is able to inactivate proteinases that participate in several different proteolytic cascades, and is essential in the regulation of activation of both the complement system and the contact system of kinin formation. It also may serve a backup function in regulation of coagulation and of fibrinolysis. C1 inhibitor, therefore, is most important in host defense and in the mediation of vascular permeability. Its name is derived from its discovery as the inactivator of the first complement component. This came about as a result of the studies of Lepow and his colleagues in the late 1950s and early 1960s; these were directed toward the isolation of C1 and its constituent subunits: C1q, C1r and C1s[1-5]. C1 inhibitor was characterized as a heat labile serum protein that inhibited the esterolytic activity of C1 and its proteolytic activity against C4 and C2. It subsequently was shown to inactivate several other plasma proteases, including kallikrein, plasmin, tissue plasminogen activator, and coagulation factors XIa and XIIa (Hageman factor)[5-10]. The fact that C1 inhibitor inactivated proteases by formation of denaturant stable equimolar complexes was first clearly shown by Harpel and Cooper with both C1s and plasmin[6]. C1 inhibitor does not appear to differ in any significant way from other serpins in its mechanism of action. The P1 and P1' residues are arginine and threonine, respectively[11].

The biologic importance of C1 inhibitor is best demonstrated by the dramatic effects resulting from its deficiency. Individuals who express a single normal C1 inhibitor gene together with either a nonfunctional or a nonexpressed allele develop the disease hereditary angioedema. Inheritance, therefore, is autosomal dominant. The causative role of C1 inhibitor deficiency in this disease was first shown in the early 1960s by Donaldson and Evans and by Landerman and colleagues[12,13].

Table 1. C1 inhibitor

- Single polypeptide chain — 478 amino acids.
- Electron microscopy — two domain structure (globular serpin domain and rod-like NH$_2$-terminal domain).
- Heavily glycosylated — 26% carbohydrate by weight.
  - 7 O-linked oligosaccharides (all in NH$_2$-terminal domain).
  - 6 N-linked oligosaccarides (3 in NH$_2$-terminal domain).
- Peptide molecular weight 52,869.
  - Peptide + carbohydrate, molecular weight 71,100.
  - SDS-PAGE, apparent molecular weight 104,000.
- Serpin domain 16–33% identity with other serpins.
- P1 residue — Arginine.
- Single gene copy on chromosome 11.2q13.

2. STRUCTURE AND COMPARISON WITH OTHER SERPINS

The first detailed structural analysis of C1 inhibitor showed that it was extremely heavily glycosylated and that it had an estimated molecular weight (by analytical ultracentrifugation) of 104,000[14] (Table 1). Additional structural studies over the subsequent years, together with information from cDNA cloning showed that C1 inhibitor consists of 478 amino acids and that it is approximately 26% carbohydrate by weight[6,15–19]. It is among the most heavily glycosylated plasma proteins. The total peptide molecular weight is 52,869, while the carbohydrate increases its molecular weight to 71,100[15,19]. The apparent molecular weight of 104,000 on SDS-PAGE or ultracentrifugation[6,14,17] is probably an anomaly related both to sequence characteristics and to the glycosylation pattern. The molecule contains six N-linked oligosaccharides and, surprisingly for a plasma protein, 7 O-linked oligosaccharides[15,19]. All of the O-linked sugars and 3 of the six N-linked sugars are within the amino terminal 100 residue non-serpin-like domain of the molecule. This constitutes the longest amino terminal non-serpin domain among the serpins and is the only one which is O-glycosylated. It is assumed that this glycosylation pattern is responsible for the unusual two domain appearance of C1 inhibitor on electron microscopy[20]. One domain is globular, while the other, presumably the amino terminal segment is rod-like. Although the molecular measurements from this study may not be precisely correct, the overall shape almost certainly is, and likely account for the lower than expected sedimentation coefficient and the abberantly high apparent molecular weight estimates on gel filtration[14,21]. C1 inhibitor is encoded by a single copy gene on chromosome 11q11-q13.1[22]. The serpin domain of C1 inhibitor reveals approximately 20–25% identity with other inhibitor serpins. It has conserved virtually all the highly conserved serpin sequence elements, and therefore clearly also must share the three dimensional structural elements that are common to the inhibitory serpins.

3. BIOLOGIC ROLE OF C1 INHIBITOR

3.1. Complement System

C1 inhibitor is the only protease inhibitor in plasma that inactivates the C1r and C1s subcomponents of the first component of complement (Table 2). This was shown by Sim and colleagues who added active radiolabelled C1r and C1s to plasma and demonstrated the formation of labelled complexes by gel filtration and SDS-PAGE[23]. All of the com-

Table 2. Physiologic inactivation of C1 inhibitor target proteases

	C1INH	α_2AP	α_2M	α_1AT	ATIII
C1r & C1s	100%	—	—	—	—
Factor XIIa	90%	3%	5%	—	2%
Factor XIIf	74%	26%	—	—	—
Kallikrein	42–84%	—	16–58%	—	—
Factor XIa	8%	—	—	68%	12%
Plasmin	—*	64–89%	7–23%	—	—
tPA	—*	—	—	—	—

*Small amounts of complexes with plasmin and tPA may be seen in some circumstances (endotoxin shock, exhaustive exercise).

plexed radioactivity was removed by immunoabsorption with antibody to C1 inhibitor. This observation was strengthened by showing that the kinetics of inhibition of C1r and C1s in serum were the same as with the purified proteins[24]. The rate of inhibition of C1r is significantly slower than that of C1s (2.8×10^3 M^{-1}sec^{-1} vs. 1.2–6.0×10^4 M^{-1}sec^{-1})[25,26]. However, *in vivo*, C1r and C1s exist primarily as components of the C1 macromolecule, which consists of one molecule of C1q and two molecules each of C1r and C1s[27]. C1 is activated by binding of C1q to the C_H2 domain of IgG (or the C_H3 domain of IgM) within an immune complex. This binding induces a conformational change in C1q which results in autoactivation of C1r from the single chain zymogen form to the two chain proteolytically active form. Activated C1r in turn similarly activates C1s.

Ordinarily, C1 inhibitor does not prevent immune complex activation of C1, but functions to inhibit C1r and C1s following their activation. C1 inhibitor, therefore, functions to limit the activity of activated C1, so that, in its presence, consumption of the C1s substrates (C4 and C2) will be limited. The interaction of C1 inhibitor with C1r and C1s leads to the dissociation of the activated C1s–C1r–C1r–C1s tetramer from C1q as two C1 inhibitor–C1r–C1s–C1 inhibitor complexes[28,29]. As with the fluid phase isolated proteases, C1s is more rapidly inhibited than is C1r. However, complex formation with C1r is the primary determinant of dissociation from C1q. This is biologically important because immune complex bound C1q, following removal of C1r and C1s, is able to interact with its receptor(s), which is (are) present on a number of cell types, particularly neutrophils and macrophages. In general, inhibition of activated C1 is more efficient with weaker C1 activators. C1 bound to larger immune complexes is much less efficiently inactivated than is C1 bound to smaller complexes. For example, C1 bound to an antibody sensitized sheep erythrocyte surface requires as much as 100 fold more C1 inhibitor for inactivation in comparison with the same amount of fluid phase activated C1[30]. With smaller immune complexes or other weak activators, C1 inhibitor can prevent C1 activation[31].

Several studies in the past had analyzed the possibility that the heavily glycosylated amino terminal domain of C1 inhibitor might play a role in its inhibitory activity[32–35]. These all suggested that this region, in general, and the carbohydrate, specifically, were not required for inhibition of free fluid phase active protease. In order to confirm these data and to examine the hypothesis that this region of the molecule might modulate access to C1r and C1s within the intact C1 macromolecule, two different truncated C1 inhibitor molecules were constructed. One of these was truncated at amino acid 76, which eliminates two of the four N-linked sites within C1 inhibitor and all 7 of the O-linked sites; the other was truncated at amino acid 100, which eliminates an additional 7 potential O-linked sites[15]. These two recombinant proteins expressed in COS cells were equivalent in activity to the normal wild type recombinant protein when tested against isolated C1r, C1s, β factor XIIa or kallikrein[36]. These studies thus confirmed the previous suggestions. However, in preliminary studies differences have been observed when assays are performed that are designed to detect differences in the ability to inhibit intact activated C1. Hemolytic assays have been performed in two ways: in the first, C1 inhibitor preparations are incubated with fluid phase activated C1, following which residual C1 is assayed using the standard hemolytic assay for C1[37]. In this assay, the activity of the truncated inhibitors is the same as that of the wild type protein. In the second assay, C1 inhibitor preparations are incubated with activated C1 that is on the surface of an antibody sensitized erythrocyte[30]. The recombinant truncated inhibitors are 2 to 3 fold more active than the recombinant wild type inhibitor when tested in this assay (unpublished data). This suggests that the amino terminal glycosylated segment may modulate access of the inhibitor to C1r and C1s within the C1 macromolecule.

3.2. Contact System

The role of C1 inhibitor in regulation of the contact system was less clearcut than its role in regulation of classical complement pathway activation. Although it had been apparent for some time that C1 inhibitor could inhibit kallikrein and factor XIIa, because other inhibitors (particularly α_2-macroglobulin) also inactivated these proteases, the relative biologic role of C1 inhibitor was unclear. In a number of studies during the 1980s this issue was clarified. In retrospect, the observation in 1962 that plasma from patients with hereditary angioedema was deficient in kallikrein inhibitory capacity was the first indication that C1 inhibitor was important in regulation of the contact system[13]. However, this was not appreciated at the time because it was not known until the following year that this deficiency was due to deficiency of C1 inhibitor. Based on a combination of kinetic analysis and quantitation of complex formation in plasma to which active kallikrein was added, different studies indicated that C1 inhibitor provided 42–84% of the plasma kallikrein inhibitory capacity[38–40] (Table 2). Similar experiments indicated that C1 inhibitor provided as much as 90% of the plasma factor XIIa and XIIf inhibitory capacity[41,42]. In the case of kallikrein and factor XIIa, α_2-macroglobulin provides the remainder of the inhibitory activity, while with factor XIIf (β factor XIIa), α_2-antiplasmin also is involved. There also is clinical and experimental evidence that C1 inhibitor, probably via regulation of contact system activation, is important in protection from endotoxin shock associated with gram negative bacterial sepsis[43–45].

These conclusions are strengthened by the observations that excessive activation of both the complement and contact systems takes place in the plasma of patients with hereditary angioedema. The immediate result of activation of the complement system is consumption of C4 and C2. Little or no consumption of other distal components takes place because the C1 activation is primarily in the fluid phase rather than on the surface of a cell or immune complex. In addition, complexes of C1r and C1s with C1 inhibitor are readily detected in the plasma of patients with hereditary angioedema. Evidence for contact system activation consists primarily of the detection of complexes of C1 inhibitor with kallikrein and factor XIIa in patients' plasma. However, we recently confirmed previous suggestions and clearly demonstrated bradykinin generation in plasma from patients with hereditary angioedema[46]. This provided supporting evidence for the importance of C1 inhibitor in regulation of activation of the contact system and, in addition, indicated that bradykinin is a major mediator of angioedema. These points are further emphasized by studies analyzing a dysfunctional mutant C1 inhibitor protein in which the mutation resulted in replacement of

Table 3. Naturally-occurring dysfunctional C1 inhibitor mutant proteins

Mutation	Secondary structure	Functional outcome
Lys 251 Deletion	Helix F–strand 3A	Multimerization Conversion to substrate
Val 432 Glu	Hinge region–P14	Conversion to substrate
Ala 434 Glu	Hinge region–P12	Conversion to substrate
Ala 436 Thr	Hinge region–P10	No protease recognition Multimerization
Ala 443 Val	Reactive center–P2	Alteration in specificity
Arg 444 His, Cys, Ser, Leu	Reactive center–P1	Alteration in specificity
Val 451 Met	Strand 1C	No protease recognition Multimerization
Phe 455 Ser	Strand 1C–strand 4B	No protease recognition Multimerization
Pro 476 Ser	Carboxyl terminus	No protease recognition Multimerization

alanine-443, which is at the P2 position, with valine[47] (Table 3). This protein is a poor inhibitor of both C1r and C1s, with rates of inhibition prolonged by approximately 8–10 fold, but recent studies indicate that it is a normal inhibitor of kallikrein and factor XIIa (unpublished data). None of the individuals with this mutation had angioedema, which again emphasizes the importance of C1 inhibitor in regulation of the contact system and the importance of the contact system in the mediation of symptoms of angioedema.

4. CONTRIBUTIONS FROM THE STUDY OF C1 INHIBITOR MUTATIONS

Analysis of mutations in the C1 inhibitor gene, particularly those that result in synthesis of dysfunctional proteins, has provided important contributions to understanding the mechanism of serpin action, in addition to contributing to knowledge relating to mechanisms of mutagenesis. The mutations that result in dysfunctional proteins may be divided into two broad groups: those that interfere with the serpin mechanism of action and those that interfere with target protease specificity. In general, mutations that disrupt the inhibitory mechanism of action interfere with the mobile domains, the movement of which are required for complex formation. These include mutations at the proximal and distal ends of the reactive center loop, and mutations that affect the stability of β sheet A. These mutations either convert the inhibitor to a substrate or result in an inhibitor that is susceptible to multimerization. Those mutations that interfere with target protease specificity usually have resulted from substitutions at the P1 residue, although one naturally occurring P2 mutant, which was discussed above, has been described. These mutations result in an inhibitor that, at least theoretically, remains capable of complex formation.

Two mutations in the proximal end of the reactive center loop of C1 inhibitor, an Ala 434→Glu (P12) and a Val 432→Glu (P14), each converted the inhibitor to a substrate[48,49] (Table 3). Each of these mutants circulated in plasma in the cleaved form and were efficiently cleaved by catalytic quantities of target protease. These mutants, therefore, although recognized by protease, were unable to form a stable complex. Furthermore, they do not express the neoepitopes that normally are expressed on cleaved C1 inhibitor following reactive center cleavage. Molecular modelling studies, based on the crystal structure of cleaved α_1-antitrypsin, suggested that the side chain of the Glu residue at P12 could not be accommodated in the pocket within sheet A into which it would have to fit during insertion of the cleaved reactive center loop. These studies, together with the crystal structures available at the time, and peptide insertion studies, provided strong support for the idea that a degree of insertion of the reactive center loop into sheet A, at least to the level of P12, was required for stable complex formation[50–53].

However, another mutant within the hinge region, Ala 436→Thr (P10), results in a non-reactive inhibitor that extensively multimerizes[49,54,55] (Table 3). This mutant did not complex with target proteases and revealed little or no cleavage, even after prolonged incubation with protease. It also was not cleaved at the P1-P1′ peptide bond by trypsin. The inhibitor was relatively thermostable, although not quite as stable as reactive center cleaved normal C1 inhibitor. In addition, the P10 mutant expressed the neoepitope that is expressed on the complexed normal inhibitor but did not express the epitope expressed on cleaved normal C1 inhibitor. This inhibitor was highly multimerized, as demonstrated by gel filtration, ultracentrifugation, non-denaturing electrophoresis and electron microscopy. This mutant expressed the "complex" neoepitope on both the monomer and multimer. This is important in interpreting the specific conformation of the monomeric mutant and the mechanism of multimerization. As discussed below, mutations in the distal reactive center loop also result in exposure of this epitope in the monomer and very likely are associated with over-insertion of the reactive center loop, presumably in a "locked" form very much like the form present in the complex with protease. It is highly likely that the P10 mutant C1 inhibitor also adopts a similar locked conformation with loop overinsertion, and that its mechanism of multimerization is similar to that hypothesized for other mutants with insertion of the reactive center loop of one molecule into either sheet A or sheet C of another molecule. This type of dimer formation has been observed directly with the crystal structure of an intact antithrombin dimer in which the reactive center loop of one molecule is inserted into sheet C of another cleaved or latent-like molecule[56,57]. However, since the Thr residue at P10 might not be predicted to fit into the available pocket within sheet A[48], it is not clear if this mutant would adopt an overinserted, locked form. It was originally suggested that this mutant polymerized via interaction between the reactive center loop of one molecule and sheet A of another. Whether this mutant polymerizes via interaction of the loop with the C or A sheet may depend upon whether insertion up to P10 is sufficient to result in the withdrawal of s1C from sheet C. The data currently available do not appear to strongly favor one model over the other.

The other group of C1 inhibitor mutants that result in multimerization are those substitutions that affect the anchoring of the distal end of the reactive center loop (Table 3). Eldering and colleagues have described three such mutants, two within s1C and one which is the third residue from the carboxyl terminus: Val 451→Met, Phe 455→Ser, and Pro 476→Ser[21]. Similar to the P10 mutation, these mutants also showed little complex formation or cleavage with target proteases. The mutants were thermostable and expressed the neoepitope in common with the complexed, but not cleaved, normal C1 inhibitor. They revealed variable amounts of multimerization, but it was clear that both the multimer and the monomer expressed the "complex" epitope. It seemed likely that these mutations, as suggested, cause release of s1C which results in overinsertion of the proximal portion of the reactive center loop in a form similar to the conformation of the inhibitor in complex with protease[21,58]. This then would allow the reactive center loop of one molecule to insert in place of s1C in sheet C of such a mutant molecule. As with the P10 mutant, direct evidence supporting this model is not yet available.

The deletion of Lys-251 in C1 inhibitor-Ta alters the sequence to create an N-glycosylation signal sequence[59,60] (Table 3). However, dysfunction appears to be a result of the deletion itself rather than the additional oligosaccharide group. This mutant is of interest because it shares characteristics with each of the above-described mutants and does not clearly fit into either category. Specifically, it did not complex with C1r, C1s or kallikrein but did inefficiently complex with β-factor XIIa. With each protease, however, 50–75% of the recombinant protein was susceptible to cleavage. It was not as thermostable as the P10 mutant or the carboxyl terminal mutants, but was more stable than the normal wild type protein. Variable proportions of the recombinant protein (20–40%) in culture supernatants were multimerized; multimer formation was not enhanced at elevated temperature. Multimers were stable in the presence of SDS, and were visible on SDS-PAGE. The multimers expressed the neoepitope that is expressed on protease complexed normal C1 inhibitor, but it was not detected on the monomeric mutant protein. Taken together, these and other data suggest that this mutation appears to produce two populations of molecules, one of which multimerizes, expresses the neopepitopes and is not recognized by target proteases. This form very likely is characterized by loop over-insertion approaching a latent-like conformation. The other form is converted primarily to a substrate. It is capable of complex formation, but only inefficiently forms a stable complex. This form exists only as a monomer, does not express the neoepitope and it is likely that it is unable to appropriately insert its reactive center loop into β sheet A.

Recently, a number of new C1 inhibitor mutations have been identified. Although the function of these proteins has not yet been analyzed, several are within regions which are likely to induce multimerization via the third mechanism that has been described: interference with the sheet A movement that is required in order to allow reactive center loop insertion during complex formation. At least three mutations are present in s6B and helix B[58], within the shutter domain described by Stein and Chothia[61]. The majority of the others we have observed are at highly conserved residues that might be expected to interfere with function or with folding, but will require structure-function analysis to define the mechanism by which they produce deficiency.

In summary, a number of mutations (>100) have now been defined within the C1 inhibitor gene, and the structural and functional consequences of many of these have been characterized. Among these mutants are examples of mutations that interfere with the mobile domains involved in the serpin conformational rearrangement that takes place during complex formation, mutations that interfere with target protease specificity and a variety of mutations that disrupt at other levels during protein synthesis or secretion. The functional analysis of C1 inhibitor mutant proteins has contributed to the current understanding of serpin function and has also helped to more completely define the biologic role of C1 inhibitor.

REFERENCES

1. Pensky, J., L. R. Levy, and I. H. Lepow. 1961. Partial purification of a serum inhibitor of C'1 esterase. *J Biol Chem.* 236:1674–1679.
2. Lepow, I. H., O. D. Ratnoff, F. S. Rosen, and L. Pillemer. 1956. Observations on a proesterase associated with partially purified first component of complement (C1). *Proc Soc Exp Biol Med.* 92:32–37.
3. Lepow, I. H., O. D. Ratnoff, and L. R. Levy. 1958. Studies on the activation of a proesterase associated with partially purified first component of human complement. *J Exp Med.* 107: 451–474.
4. Levy, L. R., and I. H. Lepow. 1959. Assay and properties of serum inhibitor of C'1 esterase. *Proc Soc Exp Biol Med.* 101:608–611.
5. Ratnoff, O. D., and I. H. Lepow. 1957. Some properties of an esterase derived from preparations of the first component of complement. *J Exp Med.* 106:327–343.

6. Harpel, P. C., and N. R. Cooper. 1975. Studies on human plasma C1-inactivator-enzyme interactions. I. Mechanisms of interaction with C1s, plasmin and trypsin. *J Clin Invest*. 55:593–604.

7. Ratnoff, O. D., J. Pensky, D. Ogston, and G. B. Naff. 1969. The inhibition of plasmin, plasma kallikrein, plasma permeability factor, and the C1'r subcomponent of complement by serum C1' esterase inhibitor. *J Exp Med*. 129:315–331.

8. Gigli, I., J. W. Mason, R. W. Colman, and K. F. Austen. 1970. Interaction of plasma kallikrein with the C1 inhibitor. *J Immunol*. 104:574–581.

9. Forbes, C. D., J. Pensky, and O. D. Ratnoff. 1970. Inhibition of activated Hageman factor and activated plasma thromboplastin antecedent by purified C1 inactivator. *J Lab Clin Med*. 76:809–815.

10. Schreiber, A. D., A. P. Kaplan, and K. F. Austen. 1973. Inhibition by C1-INH of Hageman factor fragment activation of coagulation, fibrinolysis, and kinin generation. *J Clin Invest*. 52:1402–1409.

11. Salvesen, G. S., J. J. Catanese, L. F. Kress, and J. Travis. 1985. Primary structure of the reactive site of human C1-inhibitor. *J Biol Chem*. 260:2432–6.

12. Donaldson, V. H., and R. R. Evans. 1963. A biochemical abnormality in hereditary angioneurotic edema. *Am J Med*. 35:37–44.

13. Landerman, N. S., M. E. Webster, E. L. Becker, and H. H. Ratcliffe. 1962. Hereditary angioneurotic edema. II. Deficiency of inhibitor for serum globulin permeability factor and/or plasma kallikrein. *J Allergy*. 33:330–341.

14. Haupt, H., N. Heimburger, T. Kranz, and H. G. Schwick. 1970. Ein beitrag zur isolierung und characterisierung des C1-inaktivators aus humanplasma. *Eur J Biochem*. 17:254–261.

15. Bock, S. C., K. Skriver, E. Nielsen, H. C. Thogersen, B. Wiman, V. H. Donaldson, R. L. Eddy, J. Marrinan, E. Radziejewska, R. Huber, T. B. Shows, and S. Magnussen. 1986. Human C1 inhibitor: primary structure, cDNA cloning, and chromosomal localization. *Biochemistry*. 25:4292–4301.

16. Reboul, A., G. J. Arlaud, R. B. Sim, and M. G. Colomb. 1977. A simplified procedure for the purification of C1-inactivator from human plasma. Interaction with complement subcomponents C1r and C1s. *FEBS Lett*. 79:45–50.

17. Harrison, R. A. 1983. Human C1 inhibitor: improved isolation and preliminary structural characterization. *Biochemistry*. 22:5001–5007.

18. Nilsson, T., and B. Wiman. 1982. Purification and characterization of human C1-esterase inhibitor. *Biochim Biophys Acta*. 705:271–276.

19. Perkins, S. J., K. F. Smith, S. Amatayakul, D. Ashford, T. W. Rademacher, R. A. Dwek, P. J. Lachmann, and R. A. Harrison. 1990. Two-domain structure of the native and reactive centre cleaved forms of C1 inhibitor of human complement by neutron scattering. *J Mol Biol*. 214:751–763.

20. Odermatt, E., H. Berger, and Y. Sano. 1981. Size and shape of human C1-inhibitor. *FEBS Lett*. 131:283–285.

21. Eldering, E., E. Verpy, D. Roem, T. Meo, and M. Tosi. 1995. COOH-terminal substitutions in the serpin C1 inhibitor that cause loop overinsertion and subsequent multimerization. *J Biol Chem*. 270:2579–2587.

22. Theriault, A., K. Whaley, A. R. McPhaden, E. Boyd, and J. M. Connor. 1989. Regional assignment of the human C1-inhibitor gene to 11q11-q13.1. *Human Genetics*. 84:477–479.

23. Sim, R. B., A. Reboul, G. J. Arlaud, C. L. Villiers, and M. G. Colomb. 1979. Interaction of 125-labelled complement components C1r and C1s with protease inhibitors in plasma. *FEBS Lett*. 97:111–115.

24. Ziccardi, R. J. 1981. Activation of the early components of the classical complement pathway under physiological conditions. *J Immunol*. 126:1768–1773.

25. Lennick, M., S. A. Brew, and K. C. Ingham. 1986. Kinetics of interaction of C1 inhibitor with complement C1s. *Biochemistry*. 25:3890–8.

26. Sim, R., G. Arlaud, and M. Colomb. 1980. Kinetics of reaction of human C1-inhibitor with the human complement system proteases C1r and C1s. *Biochim Biophys Acta*. 612:433–449.

27. Ziccardi, R. J., and N. R. Cooper. 1977. The subunit composition and sedimentation properties of human C1. *J Immunol*. 118:2047–2052.

28. Laurell, A. B., U. Martensson, and A. G. Sjoholm. 1976. C1 subcomponent complexes in normal and pathological sera studied by crossed immunoelectrophoresis. *Acta Pathol Microbiol Scand*. 84:455–464.

29. Ziccardi, R. J., and N. R. Cooper. 1979. Active disassembly of the first complement component C1 by C1-inhibitor. *J Immunol*. 123:788–792.

30. Tenner, A. J., and M. M. Frank. 1986. Activator-bound C1 is less susceptible to inactivation by C1 inhibition than is fluid-phase C1. *J Immunol*. 137:625–630.

31. Doekes, G., L. A. van Es, and M. R. Daha. 1983. C1 inactivator: its efficiency as a regulator of classical complement pathway activation by soluble IgG aggregates. *Immunology*. 49:215–222.

32. Minta, J. O. 1981. The role of sialic acid in the functional activity and the hepatic clearance of C1-INH. *J Immunol*. 126:245–249.

33. Prandini, M. H. 1986. Biosynthesis of complement C1 inhibitor by HepG2 cells. Reactivity of different glycosylated forms of the inhibitor with C1s. *Biochem J.* 237:93–98.

34. Reboul, A., M. H. Prandini, and M. G. Colomb. 1987. Proteolysis and deglycosylation of human C1 inhibitor. *Biochem J.* 244:117–121.

35. Patston, P. A., M. Qi, J. A. Schifferli, and M. Schapira. 1995. The effect of cleavage by a Crotalus atrox alpha-proteinase fraction on the properties of C1-inhibitor. *Toxicon.* 33:53–61.

36. Coutinho, M., K. S. Aulak, and A. E. Davis III. 1994. Functional analysis of the serpin domain of C1 inhibitor. *J Immunol.* 153:3648–3654.

37. Gigli, I., S. Ruddy, and K. F. Austen. 1968. The stoichiometric measurement of the serum inhibitor of the first component of complement by the inhibition of immune hemolysis. *J Immunol.* 100:1154–1164.

38. Schapira, M., C. F. Scott, and R. W. Colman. 1982. Contribution of plasma protease inhibitors to the inactivation of kallikrein in plasma. *J Clin Invest.* 69:462–468.

39. van der Graaf, F., J. A. Koedam, and B. N. Bouma. 1983. Inactivation of kallikrein in human plasma. *J Clin Invest.* 71:149–158.

40. Harpel, P. C., M. F. Lewin, and A. P. Kaplan. 1985. Distribution of plasma kallikrein between C1 inactivator and a2-macroglobulin in plasma utilizing a new assay for a2-macroglobulin-kallikrein complexes. *J Biol Chem.* 260:4257–4263.

41. de Agostini, A., H. R. Lijnen, R. A. Pixley, R. W. Colman, and M. Schapira. 1984. Inactivation of factor XII active fragment in normal plasma. Predominant role of C1-inhibitor. *J Clin Invest.* 73:1542–1549.

42. Pixley, R. A., M. Schapira, and R. W. Colman. 1985. The regulation of human factor XIIa by plasma proteinase inhibitors. *J Biol Chem.* 260:1723–1729.

43. Nuijens, J. H., A. J. M. Eerenberg-Belmer, C. C. M. Huijbregts, W. O. Schreuder, R. J. F. Felt-Bersma, J. J. Abbink, L. G. Thijs, and C. E. Hack. 1989. Proteolytic inactivation of plasma C1 inhibitor in sepsis. *J Clin Invest*:443–450.

44. Nuijens, J. H., C. C. M. Huijbregts, A. J. M. Eerenberg-Belmer, J. J. Abbink, R. J. M. Strack van Schijndel, R. J. M. Felt-Bersma, L. G. Thijs, and C. E. Hack. 1988. Quantification of plasma factor XIIa-C1-inhibitor and kallikrein-C1-inhibitor complexes in sepsis. *Blood.* 72:1841–1848.

45. Guerrero, R., F. Velasco, M. Rodriguez, A. Lopez, R. Rojas, M. A. Alvarez, R. Villalba, V. Rubio, A. Torres, and D. d. Castillo. 1993. Endotoxin-induced pulmonary dysfunction is prevented by C1-esterase inhibitor. *J Clin Invest.* 91:2754–2760.

46. Shoemaker, L. R., S. J. Schurman, V. H. Donaldson, A. E. Davis III. 1994. Hereditary angioneurotic edema: Characterization of plasma kinin and vascular permeability-enhancing activities. *Clin Exp Immunol.* 95:22–28.

47. Zahedi, R., J. J. Bissler, A. E. Davis III, C. Andreadis, and J. J. Wisnieske. 1995. Unique C1 inhibitor dysfunction in a kindred without angioedema. II. Identification of an Ala443-Val substitution and functional analysis of the recombinant mutant protein. *J Clin Invest.* 95:1299–1305.

48. Skriver, K., W. R. Wikkoff, P. A. Patston, F. Tausk, M. Schapira, A. P. Kaplan, and S. C. Bock. 1991. Substrate properties of C1 inhibitor Ma (alanine 434 glutamic acid). Genetic and structural evidence suggesting that the P12-region contains critical determinants of serine protease inhibitor/substrate status. *J Biol Chem.* 266:9216–9221.

49. Davis III, A. E., K. S. Aulak, R. B. Parad, H. P. Stecklein, E. Eldering, C. E. Hack, J. Kramer, R. C. Strunk, J. Bissler, and F. S. Rosen. 1992. C1 inhibitor hinge region mutations produce dysfunction by different mechanisms. *Nature Genetics.* 1:354–358.

50. Stein, P. E., A. G. Leslie, J. T. Finch, W. G. Turnell, P. J. McLaughlin, and R. W. Carrell. 1990. Crystal structure of ovalbumin as a model for the reactive centre of serpins. *Nature.* 347:99–102.

51. Loebermann, H., R. Tokuoka, J. Deisenhofer, and R. Huber. 1984. Human alpha1-proteinase inhibitor. Crystal structure analysis of two crystal modifications, molecular model and preliminary analysis of the implications for function. *J Mol Biol.* 177:531–557.

52. Carrell, R. W., D. L. Evans, and P. E. Stein. 1991. Mobile reactive centre of serpins and the control of thrombosis. *Nature.* 353:576–8.

53. Schulze, A. J., U. Baumann, S. Knof, E. Jaeger, R. Huber, and C. B. Laurell. 1990. Structural transition of alpha 1-antitrypsin by a peptide sequentially similar to beta-strand s4A. *Eur J Biochem.* 194:51–6.

54. Levy, N. J., N. Ramesh, M. Cicardi, R. A. Harrison, and A. E. Davis III. 1990. Type II hereditary angioneurotic edema that may result from a single nucleotide change in the codon for alanine-436 in the C1 inhibitor gene. *Proc Natl Acad Sci USA.* 87:265–8.

55. Aulak, K. S., E. Eldering, C. E. Hack, Y. P. T. Lubbers, R. A. Harrison, A. Mast, M. Cicardi, and A. E. Davis III. 1993. A hinge region mutation in C1-inhibitor (Ala436Thr) results in nonsubstrate-like behavior and in polymerization of the molecule. *J Biol Chem.* 268:18088–94.

56. Carrell, R. W., D. L. Evans, and P. E. Stein. 1994. Biological implications of a 3 Å structure of dimeric antithrombin. *Structure*. 2:257–270.
57. Schreuder, H. A., B. de Boer, R. Dijkema, J. Mulders, H. J. M. Theunissen, P. D. J. Grootenhuis, and W. G. J. Hol. 1994. The intact and cleaved human antithrombin III complex as a model for serpin-proteinase interactions. *Structural Biology*. 1:48–54.
58. Stein, P. E., and R. W. Carrell. 1995. What do dysfunctional serpins tell us about molecular mobility and disease? *Structural Biology*. 2:96–113.
59. Parad, R. B., J. Kramer, R. C. Strunk, F. S. Rosen, and A. E. Davis III. 1990. Dysfunctional C1 inhibitor Ta: deletion of Lys-251 results in acquisition of an N-glycosylation site. *Proc Natl Acad Sci USA*. 87:6786–6790.
60. Zahedi, R., K. S. Aulak, E. Eldering, and A. E. Davis III. 1996. Characterization of C1 inhibitor-Ta: A dysfunctional C1INH with deletion of lysine-251. *J Biol Chem.*, submitted.
61. Stein, P. E., and C. Chothia. 1991. Serpin tertiary structure transformation. *J Molecular Biology*. 221:99–102.

19

SERP-1, A POXVIRUS-ENCODED SERPIN, IS EXPRESSED AS A SECRETED GLYCOPROTEIN THAT INHIBITS THE INFLAMMATORY RESPONSE TO MYXOMA VIRUS INFECTION

Piers Nash,[1] Alexandra Lucas,[2] and Grant McFadden[*]

[1]Department of Biochemistry
[2]Division of Cardiology
Department of Medicine, University of Alberta
Edmonton, Alberta, T6G-2H7 Canada

INTRODUCTION

Many viruses, especially large DNA viruses such as the poxviruses, achieve virulence and pathogenicity by using a variety of strategies to evade or subvert the antiviral and inflammatory response of their host (Gooding, 1992; Marrack and Kappler, 1994; McFadden, 1995; Smith, 1994). In many cases, these strategies are directed against the effector cells that directly mediate natural and acquired immunity, such as monocytes/macrophages, natural killer cells, neutrophils, and B and T lymphocytes. Some viruses attempt to evade detection by interfering with major histocompatability complex antigen presentation in infected cells, effectively masking the outward signs of infection recognized by T-cells (Maudsley and Pound, 1991; McFadden and Kane, 1994). Virus encoded proteins may also be produced that actively seek to stifle the immune response by interfering with the basic mechanisms by which immune and inflammatory signals are transmitted. Such proteins include secreted mimics of host ligands/regulators (virokines) or cellular cytokine receptors (viroceptors) (McFadden *et al.*, 1995). Poxviruses employ a variety of such techniques, encoding proteins that block complement activation (Kotwal and Moss, 1988), stimulate mitogenesis (McFadden, Graham, and Opgenorth, 1995), bind tumour necrosis factor (Smith, Farrah, and Goodwin, 1994; Smith and Goodwin, 1995), interferon-γ (Mossman, Barry, and McFadden, 1995), and interleukin-1β (Alcami and Smith, 1992; Spriggs, 1994). Poxviruses can also inhibit host proteinases through the action of serpins (Turner, Musy, and Moyer, 1995).

Serpins are used in many eukaryotic tissue systems to regulate complex proteinase dependent pathways in order to maintain homeostasis (for review see Potempa, Korzus,

---

[*] To whom correspondence should be addressed. Tel: 1-519-663-3184; Fax: 1-519-663-3847.

Chemistry and Biology of Serpins, edited by Church *et al.*
Plenum Press, New York, 1997

and Travis, 1994). Examples of this include the activation of cytokines in the inflamma-tory network (Forsyth, Talbot, and Beckman, 1994; Komiyama *et al.*, 1994; Matsuda *et al.*, 1994), the complement pathway, fibrinolysis (Lijnen and Collen, 1990), thrombotic cascades (Olds *et al.*, 1994; Olson and Bjork, 1994), tissue remodeling (Smirnova *et al.*, 1994), apoptosis (Houenou *et al.*, 1995; Sarin, Adams, and Henkart, 1993), and a number of signalling pathways (Altieri, 1995; Coughlin, 1994; Nystedt *et al.*, 1994; Strickland, Kounnas, and Argraves, 1995). In most cases a few key proteinases can be modulated to regulate an entire cascade. It is not surprising, then, that certain pathogens might employ serpins in order to interrupt proteinase dependent host processes that are antagonistic to the invading organism. For example, nematodes of the *Brugia* family of filarial parasites have been reported to express serpins (Blanton, Licate, and Aman, 1994; Ghendler, Arnon, and Fishelson, 1994; Yenbutr and Scott, 1995). Of interest to virologists is that several ser-pins have been identified within the genomes of members of the poxvirus family of large DNA viruses. The fact that serpins are expressed even by viruses, the simplest of all or-ganisms, is intriguing, and testifies to the importance of proteinases in the immune re-sponse to pathogens. Among the serpins produced by members of the poxvirus family is a unique myxoma virus protein designated SERP-1, on which this review will focus.

POXVIRUSES ARE THE ONLY VIRUSES THAT HAVE BEEN SHOWN TO ENCODE SERPINS

The poxviruses comprise a family of large, complex DNA viruses that are capable of autonomous replication within the cytoplasm of infected cells. This sets them apart from the vast majority of DNA viruses which rely on the ability to parasitize the host cell replication machinery, and, consequently, require localization in the nucleus of infected cells. The pox-virus family is divided into two subfamilies, the chordopoxviridae (vertebrate poxviruses), and the entemopoxviridae (insect poxviruses) (Moss, 1990). The chordopoxviridae are fur-ther divided into seven genera, of which four, the orthopoxviruses, leporipoxviruses, avi-poxviruses, and suipoxviruses have been found by sequencing studies to encode serpins. The remainder of the poxviruses may well also encode serpins, but sequencing of the ge-nomes of many members of this family is still at a rudimentary stage. Together the pox-viruses constitute the etiologic agents of a large number of diseases ranging from benign and localized (e.g. vaccinia virus and Shope fibroma virus) to systemic and highly lethal (smallpox and myxomatosis). All poxviruses have a genome composed of a single linear double stranded DNA with cross-linked termini, which ranges in size from 130 to 375 kbp (for reviews see Buller and Palumbo, 1991; Fenner, Wittek, and Dumbell, 1989; Turner and Moyer, 1990). The central region of the genome is highly conserved, not only between strains, but even to some extent among all poxviruses. In comparison, the 20–50 kb regions of the DNA closest to the termini are subject to considerable variation among natural poxvirus isolates. The most variable region is the terminal inverted repeat (TIR) region which is duplicated at each end of the genome, and accounts for the only genes that are pre-sent in two copies. The degree to which virus genes are conserved between different poxviruses can be correlated to their function. The conserved core of the genome codes pri-marily for the so-called "housekeeping" genes that are required for the basic survival of the virus and are required for virus replication in cultured cells. The highly variable genes which map outside of the central core of the genome are generally termed non-essential because they are not required for virus propagation in tissue culture. However, many of these genes encode for what are referred to as "virulence factors" which are so named

Table 1. Summary of poxvirus-encoded serpins

Serpin nomenclature[1]	Poxvirus genera/species	Site of action	Inhibits	Comments
SERP-1/Spi-4	Leporipoxivirus/ myxoma virus	Secreted	Plasmin, tPA, uPA, Xa, thrombin, and C1s.	Virulence factor, anti-inflammatory
Spi-2/crmA	Orthopoxviruses	Cytoplasmic	ICE (Interleukin 1ß-converting enzyme)	Anti-inflammatory, anti-apoptotic
Spi-1	Orthopoxviruses	Cytoplasmic	?	Anti-apoptotic
Spi-3	Orthopoxviruses	Membrane-associated?	?	Inhibits cell fusion
Spi-7	Suipoxviruses	Cytoplasmic	?	
Spi-5	Avipoxviruses	?	?	Truncated
Spi-6	Avipoxviruses	?	?	Truncated

[1]Nomenclature described in greater detail in Turner, Musy, and Moyer, 1995.

because they confer an increased capacity for propagation within immunocompetent vertebrate hosts and thereby contribute to virus spread and pathogenesis. Such virulence factors function to allow virus infected cells to survive in complex tissues in the face of challenge by the consolidated actions of the immune and inflammatory systems of the host.

Encoded within the non-essential regions of at least four poxvirus genera are a number of proteins which conform to criteria of the serpin superfamily (Table 1). The poxvirus from which the viral serpin considered here was identified, namely myxoma virus, is a rabbit-specific pathogen of the leporipoxvirus family, that causes a systemic and lethal disease known as myxomatosis (Fenner and Meyers, 1978). Myxoma virus encodes for many known virulence factors, including the secreted serpin, designated SERP-1. SERP-1 has been characterized in terms of serine proteinase inhibitor function *in vitro* and effects on virus virulence *in vivo*, and is to date the only poxviral serpin known to be secreted from virus infected cells. Three other virus serpins (Spi-1, 2, 3) have been found to be encoded by diverse members of the orthopoxvirus genus, but all of these are believed to remain associated with the virus infected cell throughout the course of infection. SERP-1 is also the only viral serpin encoded within the TIR region of a poxvirus, and is thus present in two copies in the myxoma genome. Although other large DNA viruses, including the herpesviruses and adenoviruses, also express multiple proteins that interfere with host defenses, it is interesting to note that only poxviruses are so far known to encode serpins. The origin of poxviral serpins is still uncertain. All poxvirus genes, including the serpins, are, effectively, cDNAs and do not contain introns. This suggests that if these serpin genes are derived from host cells, they may have originally been acquired from processed, cytoplasmic mRNA transcripts, rather than directly from the host genome. This is in keeping with the entirely cytoplasmic life cycle of the virus, but it should be noted that not all poxvirus genes have obvious host counterparts and thus the origin of the SERP-1 gene remains speculative.

OTHER POXVIRAL SERPINS

The first poxvirus open reading frame with serpin-homology to be discovered was the cowpox virus 38K gene (Pickup *et al.*, 1986). This was later renamed crmA, for cytokine response modifier A (Ray *et al.*, 1992), and, in vaccinia virus, is designated as Spi-2 (Kotwal and Moss, 1989). Under the proposed nomenclature for poxviral serpins, the Spi-2 name is gaining acceptance as the genetic name for this serpin (Turner, Musy, and Moyer,

1995). Spi-2 homologs have been found in most of the orthopoxviruses, including cowpox virus (CPV), vaccinia virus (VV) strain WR, rabbitpox virus (RPV) (Ali *et al.*, 1994), and variola virus (VAR) the causative agent of smallpox (Massung *et al.*, 1993). Unlike myxoma SERP-1, and most mammalian serpins, Spi-2 is neither glycosylated nor secreted. It is synthesized early during infection, and is present as a cytosolic protein (Kettle *et al.*, 1995; Pickup *et al.*, 1986). In the avian chorioallantoic membrane model of inflammation, Spi-2 expression dramatically reduced heterophil, lymphocyte, and macrophage influx into the infected area. The mechanism of this anti-inflammatory action appears to be the ability of Spi-2 to directly inhibit interleukin-1 beta converting enzyme (ICE) *in vitro* (Ray *et al.*, 1992). This inhibition was shown to be both rapid and specific with an association rate constant of 1.7×10^7 $M^{-1}s^{-1}$, and an equilibrium constant (K_i) of less than 4×10^{-12} M (Komiyama *et al.*, 1994). The P1 residue of Spi-2, the amino acid after which the serpin is cleaved, is an aspartate for the inhibition of ICE. Interestingly, ICE is a cysteine proteinase, making Spi-2 the first example of a cross-class inhibitor (Komiyama *et al.*, 1994). Furthermore, ICE has been implicated in triggering apoptosis (Miura *et al.*, 1993), and Spi-2/crmA is capable of preventing apoptosis by a variety of inducers (e.g. (Gagliardini *et al.*, 1994). Other large DNA viruses such as baculovirus, adenovirus, and herpesvirus are also known to block apoptosis of infected cells (for extensive review see White, 1996), though not through the use of serpins. Spi-2, while conforming to the serpin homology, appears to be a cytosolic cysteine proteinase inhibitor directed against homologs of the ced-3 family of ICE-like proteinases in order to prevent apoptosis of virally infected cells.

Spi-1 is approximately 45% identical to Spi-2, although it differs at the predicted P1-P1' residues, which are Phe-Ser (Kotwal and Moss, 1989). Spi-1 is found at the right end of the genome in the orthopoxviruses CPV, RPV, VV, and VAR, but has not yet been assigned a function, other than that it increases host range by preventing virus induced apoptosis in some cell lines (Turner, Musy, and Moyer, 1995). As is the case with Spi-2, and unlike SERP-1 or mammalian serpins, Spi-1 is neither secreted nor glycosylated.

A third serpin, termed Spi-3 is found in most of the orthopoxviruses examined, including VV, RPV, CPV, VAR, and the more divergent raccoonpox virus (for review see Turner, Musy, and Moyer, 1995). Spi-3 exhibits only low homology to Spi-1 and Spi-2, and is more similar to mammalian serpins in that it is N-glycosylated. Spi-3 does not, however, appear to be secreted into the medium of virus infected cells, but its cellular localization profile has yet to be reported. It has a putative P1 arginine residue, but no virulence or anti-inflammatory phenotypes have been associated with Spi-3. The only function attributed to Spi-3 is the prevention of cell-cell fusion following infection (Law and Smith, 1992; Turner and Moyer, 1992; Zhou *et al.*, 1992). The targets and nature of Spi-3 action remain unknown, but mutation of the postulated P1-P1' site does not appear to abrogate the ability of host cells to promote fusion (Turner and Moyer, 1995).

Of the remaining orthopoxviral serpins, none has been well characterized at the biochemical level. Spi-5 and Spi-6 from fowlpox appear to be truncated serpins and are unlikely to be functional inhibitors (Turner, Musy, and Moyer, 1995). The swinepox Spi-7 protein is another cytosolic relative of Spi-1, Spi-2 and Spi-3, and may serve a similar function (Massung, Jayarama, and Moyer, 1993).

SERP-1 IS A VIRULENCE FACTOR

SERP-1 (also called Spi-4) has been found to be present in the genomes of only two poxviruses: myxoma virus (myx), and malignant rabbit fibroma virus (MRV), a recombinant between Shope fibroma virus and myxoma virus (Upton et al., 1990). In myxoma virus, the SERP-1 gene lies within the TIR sequences, and is thus present as two copies, while in MRV, the identical SERP-1 gene maps within the unique sequences adjacent to the TIR, and hence is present in a single copy only (Upton *et al.*, 1990).

Myxoma virus co-evolved with members of the *Sylvilagus sp.* family of New World rabbits in which it causes moderate, non-lethal dermal lesions. In *Oryctolagus cuniculus* (European or domestic rabbits), however, myxoma virus causes a severe and lethal disease state known as myxomatosis with mortality rates greater than 99%. Myxomatosis is characterized by extensive fulminating lesions, both external and internal, and severe generalized immunodysfunction. Infected animals generally die within two weeks from a multitude of complications, including supervening Gram negative infections of the respiratory tract. Myxoma virus is such an efficient pathogen of the European rabbit that in the 1950s it was deliberately released in Australia and Europe as a population control measure in order to reduce the numbers of feral rabbits which were causing considerable denuding of the natural vegetation and devastating the local agriculture (Fenner and Ratcliffe, 1965). Attenuated myxoma virus strains appeared quickly thereafter within the Oryctolagus populations, progressively rendering myxoma virus ineffective as a field biological control agent for Australian feral rabbits. Nevertheless, wild-type myxoma virus remains a very tractable model of viral induced immunosupression, and a wide range of immunosubversive viral proteins have now been described in this system (for review see McFadden *et al.*, 1995).

When a targeted gene disruption of the single SERP-1 gene was constructed in MRV, the resultant virus exhibited normal growth characteristics in cultured cells, but was severely attenuated in infected rabbits (Upton *et al.*, 1990). Wild type MRV-infected rabbits exhibited large tumors at the primary inoculation site; secondary tumors appeared in the ears by day 7; and progressive bacterial infections of the nasal and conjunctival mucosa were so severe that all rabbits were sacrificed on or before Day 14. The survival rate for European rabbits infected with wild type MRV is less than 1% but, in contrast, rabbits infected with the MRV-S1 deletion mutant showed a marked decrease in the severity of symptoms and 60% survived (Upton *et al.*, 1990). Similarly, the targeted gene disruption of both copies of the SERP-1 gene in myxoma virus had a comparable attenuating effect upon viral virulence, with survival rates following infection with the SERP-1 knock-out myxoma virus also exceeding 60% (Macen *et al.*, 1993). Furthermore, histological sections taken from the lesions of infected animals indicated that in the absence of SERP-1, a more effective inflammatory response occurs, allowing a more rapid resolution of the infection (Macen *et al.*, 1993). This suggests that SERP-1 contributes to viral pathogenesis by interacting with host proteins associated with virus-infected lesions that regulate the early inflammatory response to the virus infection.

SERP-1 IS SECRETED AS A STABLE N-GLYCOSYLATED PROTEIN

Transcriptional analysis of the SERP-1 gene indicates that it is expressed as a late gene, meaning that it appears late in the lytic cycle of infection following virus DNA replication (Macen *et al.*, 1993). This is in keeping with the known SERP-1 promoter sequence which is similar to the consensus late promoter described for other poxviruses, particularly vaccinia virus (Moss, 1990). SERP-1 mRNA first appears at 6 hours postinfection, and remains at high levels throughout the lytic cycle. Polyclonal antisera raised against SERP-1 detect a heterogeneous 55–60 kDa species which appears in extracellular culture supernatants starting at 8 hours post-infection, and remains at high levels as a stable soluble protein (Macen *et al.*, 1993). A recombinant vaccinia virus containing the SERP-1 open reading frame under the control of a strong, synthetic late promoter also expresses and secretes SERP-1 into the culture supernatant, and does so at levels 10 times greater than myxoma. This overexpression has allowed SERP-1 to be purified to homogeneity from supernatants of virus infected cells (Nash *et al.*, 1996). On silver stained SDS-PAGE gels, purified SERP-1 protein migrates as a diffuse 55–60 kDa band, indicating possible glycosylation. Indeed, when SERP-1 produced from the recombinant vaccinia virus vector was subjected to N-glycosidase F, the mobility of SERP-1 was reduced to the predicted size of the polypeptide (42 kDa), consistent with the size predicted from the DNA sequence of the open reading frame. Thus, SERP-1 is expressed as a secreted, soluble, N-glycosylated protein from cells infected with myxoma, MRV, or VV-SERP-1, but not from any other virus yet described.

SERP-1 INHIBITS UPA, TPA, AND PLASMIN

As noted earlier, SERP-1 remains the only virally produced serpin-homology protein that has been shown to functionally inhibit known serine proteinases. While crmA (Spi-2), an intracellular serpin expressed by vaccinia and cowpox, has been shown to inhibit the cysteine proteinase, ICE (Komiyama *et al.*, 1994), none of the other orthopoxvirus serpins have yet been shown to have actual proteinase inhibitory activity associated with them. SERP-1, on the other hand, has been tested against a panel of cellular proteinases, and some conclusions can be drawn as to the type of target proteinases that SERP-1 might inhibit in infected tissues. SERP-1 forms SDS-resistant complexes with both urokinase and tissue-type plasminogen activators, as well as plasmin, indicating that SERP-1 is capable of inhibiting these proteinases (Lomas *et al.*, 1993). Although SERP-1 is derived from a rabbit virus, these studies were performed with human proteinases indicating that the target(s) of SERP-1 are likely to be highly conserved. Short lived complexes have also been observed between SERP-1 and both thrombin and Factor Xa, and SERP-1 is fully capable of inhibiting these proteinases (Nash *et al.*, 1996). SERP-1 does not form stable complexes with, but is instead cleaved by, human neutrophil elastase (hNE), cathepsin G, bovine chymotrypsin, subtilisin, thermolysin, and trypsin. Based on the P1 arginine residue, and studies with mutant SERP-1 (see below), these enzymes likely cleave SERP-1 at sites other than the P1-P1' site, consistent with their inability to be inhibited by SERP-1. No interaction is observed between SERP-1 and Factor D, C3 convertase, or factor XII (Nash *et al.*, 1996). The association rate constants for SERP-1 interacting with human uPA, tPA, plasmin, and C1s were first calculated by an end-point inhibition assay, using unfractionated supernatants from SERP-1 expressing recombinant vaccinia virus infected cells (Lomas *et al.*, 1993). Recently, more complete data has been obtained by slow binding kinetics using purified SERP-1, and the kinetic parameters of k_{on}, K_I, and k_{off} have been determined for uPA, tPA, plasmin, thrombin, factor Xa, and C1s (Nash *et al.*, 1996). The data with purified SERP-1 protein does not support the earlier conclusion that SERP-1 inhibits C1s (Lomas *et al.*, 1993), indicating that the C1s inhibitory activity may have been due to another component in the viral supernatants, though the remaining kinetic parameters do not differ significantly between the two studies (see Table 2).

A reactive centre loop mutant of SERP-1 in which the P1-P1' residues have been changed from Arg-Asn to Ala-Ala has also been expressed and purified, using a recombinant vaccinia virus expression system (Nash *et al.*, 1996). This mutant, designated SAA,

Table 2. Kinetics data for proteinase inhibition by SERP-1

	Nash *et al.* 1996			Lomas *et al.* 1993
Enzyme	K_I (pM)	k_{ass} ($M^{-1}s^{-1}$)	k_{dis} (s^{-1})	k_{ass} ($M^{-1}s^{-1}$)
Plasmin	440	4.5×10^4	3×10^{-5}	3.4×10^4
tPA	140	5.0×10^4	7×10^{-6}	3.6×10^4
uPA	160	5.0×10^4	8×10^{-6}	4.3×10^4
Thrombin	130	2.6×10^4	2×10^{-6}	n.d.
Factor Xa	4.3×10^3	1.7×10^3	8×10^{-6}	n.d.
C1s	2×10^5	3×10^2	6×10^{-5}	1.3×10^3

neither interacts with, nor appears to inhibit, plasmin, tPA, uPA, thrombin or Factor Xa. Purified SAA is still a substrate for those enzymes that cleave wild type SERP-1 (hNE, cathepsin G, chymotrypsin, and trypsin) (Nash *et al.*, 1996). Taken together, these findings indicate that the assignment of the P1-P1′ residues is correct, and that the true target of SERP-1 inhibition in virus infected cells is likely to be a serine proteinase with a preference for cleaving after arginine residues.

Because virus dissemination is achieved from the primary replication site in the sub-dermal tissue to multiple secondary sites via lymphatic channels, myxoma virus is unlike-ly to encounter circulating plasma proteinases, and as such, it would have been surprising to find a plasma proteinase with which SERP-1 interacted with the affinity of a "true tar-get". In fact, the association rate constants are no higher than 10^5 $M^{-1}s^{-1}$ for the inhibition of human fibrinolytic enzymes by SERP-1. In addition, as has already been noted, none of the enzymes tested is rabbit in origin. Additionally, the unique P1-P1′ residues of SERP-1 (Arg-Asn) indicate a target somewhat different in nature from known serine proteinases. Given the relatively low levels of SERP-1 believed to be produced during myxoma infec-tion, we would also expect the SERP-1 target to be an extracellular or cell-surface protei-nase which is present in relatively small quantities, or present only on a particular cell type. One possibility is that the true target of SERP-1 inhibition is found as a low abun-dance proteinase associated with sentinel immune cells that function to upregulate the inflammatory signal cascade in response to viral infection in subdermal tissues.

SERP-1 IS ANTI-INFLAMMATORY IN MODELS OF RESTENOSIS AND ARTHRITIS

Following the discovery that SERP-1 had potential anti-inflammatory properties (Macen *et al.*, 1993), purified SERP-1 protein has also been tested in non-viral model sys-tems of inflammation relevant to human disease. SERP-1 is a good candidate as a novel anti-inflammatory reagent in medical applications for several reasons. The still unidenti-fied SERP-1 target proteinase appears to be intimately involved in some early stage of inflammation, and the SERP-1 protein itself is active at picogram levels in complex sub-dermal tissues that are normally highly susceptible to the elements of the inflammatory response. Furthermore, the native SERP-1 protein is believed to be poorly antigenic since no circulating antibodies directed against it can be detected in animals which survive infection by attenuated variants of myxoma virus (unpublished data). The first successful application of SERP-1 as a targeted anti-inflammatory protein was made in a model of coronary restenosis following primary balloon angioplasty in which SERP-1 was employed to reduce plaque development (Lucas *et al.*, 1996). Atherosclerotic plaque growth is one of the leading causes of morbidity and mortality in North America, and has been linked to an excessive inflammatory and thrombotic response to arterial injury. Since both the thrombotic and inflammatory cascades are regulated by serine proteinases, and as such may be regulated by serpins, the effects of SERP-1 infusion at sites of angioplasty balloon injury in a rabbit model of atherosclerosis was systematically examined. A total of 74 rabbits had either focal infusions of picogram levels of purified SERP-1 protein or systemic infusion of nanogram levels of SERP-1. The reactive centre Ala-Ala mutant of SERP-1 was used as a control. There was a dramatic reduction of atherosclerotic plaque growth 4 weeks after a single SERP-1 infusion at sites of arterial damage but not after infusion of the Ala-Ala mutant. In this rabbit model the reduction in plaque development was associated with a parallel reduction in macrophage infiltration into the balloon-damaged vascular wall that could be detected within 24 hours after SERP-1 infusion (Lucas *et al.*, 1996). These studies indicate that SERP-1 functions through its serine proteinase inhibitory activity both to reduce the level of acute macrophage influx into balloon damaged vasculature, and to diminish subsequent plaque growth at the damaged arterial sites.

Purified SERP-1 protein has also been used successfully in preliminary experiments to reduce arthritis-like inflammation in a rabbit model of rheumatoid arthritis. In an antigen-induced arthritis model of chronic inflammation, purified SERP-1 protein was administered by intra-articular injection after joint inflammation had been induced by systemic antigen stimulation. The synovia were analyzed several weeks after SERP-1 treatment, and SERP-1 was seen to reduce cellular hyperplasia, chronic inflammatory infiltration and cartilage erosion (Maksymowych *et al.*, 1996). As in the restenosis model, the dosage of SERP-1 required for effect was in the picogram to nanogram range.

While the mechanism by which SERP-1 is acting in these systems remains to be elucidated, it is nonetheless important to note the potential of using serpins or small molecule inhibitors of proteinases in such small doses to achieve dramatic clinical results.

CONCLUSIONS

Members of the poxvirus family are the first examples of viruses that encode serpins. The fact that poxviruses have evolved to utilize multiple serpins as part of their concerted anti-immune strategies underscores their importance in regulating key steps in the host's anti-viral immune reponse. The ability of SERP-1 to dampen the inflammatory response to the virus infection offers clues to the underlying role that serine proteinases may be playing in these early processes. A better understanding of the underlying mechanisms may help in the development of anti-inflammatory drugs based on serpins and small molecule proteinase inhibitors. As SERP-1 itself has now been shown to have therapeutic potential in diseases of excessive inflammatory reseponses, other proteinase inhibitors which target related pathways are also likely to be effective at low pharmacological doses. The observation that Spi-2/crmA inhibits ICE-type proteinases, and in doing so blocks apoptosis, has already contributed to our understanding of programmed cell death. As a cross-class inhibitor, Spi-2/crmA is further evidence of the diversity of roles that serpins can play to modulate the host immune responses to favor virus survival.

Viruses have co-evolved with the increasingly complex immune and cellular systems of eukaryotes. While we have been studying such matters for a mere hundred years or so, viruses have been actively engaging these same immune pathways for millenia. It is not surprising then, that we have much to learn from such seemingly simple organisms.

Work on poxviral serpins has only begun and we have yet to identify the "true" targets of SERP-1 in infected tissues and the role these targets play in the regulation of inflammation. Meanwhile, the remaining poxviral serpins remain largely uncharacterized. The tantalizing clues already uncovered suggest that there is still a great deal that viruses can teach us about the immune system in general.

ACKNOWLEDGMENTS

P.N. is funded by a Studentship, G.M. a Medical Scientist award, and A.L. a Clinical Investigatorship award from the Alberta Heritage Foundation for Medical Research. Supported by operating grants from the National Cancer Institute of Canada (G.M.), the Heart and Stroke Foundation (A.L.), and BIOGEN Corporation (G.M. and A.L.). We thank Adrain Whitty, Michele Barry, Penelope Stein, Kimberly Ellison and Marita Hobman for helpful discussions.

REFERENCES

Alcami, A., and Smith, G. L. (1992). A soluble receptor for interleukin-1ß encoded by vaccinia virus: a novel mechanism of virus modulation of the host response to infection. *Cell* **71**, 153–167.

Ali, A. N., Turner, P. C., Brooks, M. A., and Moyer, R. W. (1994). The SPI-1 gene of rabbitpox virus determines host range and is required for hemorrhagic pock formation. *Virology* **201**(1), 305–314.

Altieri, D. C. (1995). Xa receptor EPR-1. *The FASEB Journal* **9**, 860–865.

Blanton, R. E., Licate, L. S., and Aman, R. A. (1994). Characterization of a native and recombinant Schistosoma haematobium serine protease inhibitor gene product. *Molecular & Biochemical Parasitology* **63**(1), 1–11.

Buller, R. M., and Palumbo, G. J. (1991). Poxvirus pathogenesis. *Microbiological Reviews* **55**(1), 80–122.

Coughlin, S. R. (1994). Protease-activated receptors start a family. *Proceedings of the National Academy of Sciences of the United States of America* **91**, 9200–9202.

Fenner, F., and Meyers, K. (1978). Myxoma virus and myxomatosis in retrospect: The first quarter century of a new disease. *In* "Viruses and the Environment" (E. Kurstok, and K. Maramorosch, Eds.), pp. 539–570. Academic Press, New York.

Fenner, F., and Ratcliffe, F. N. (1965). "Myxomatosis." Cambridge University Press, London.

Fenner, F., Wittek, R., and Dumbell, K. R. (1989). "The orthopoxviruses." Academic, New York.

Forsyth, K. D., Talbot, V., and Beckman, I. (1994). Endothelial serpins—protectors of the vasculature? *Clinical & Experimental Immunology* **95**(2), 277–82.

Gagliardini, V., Fernandez, P.-A., Lee, R. K. K., Drexler, H. C. A., Rotello, R. J., Fishman, M. C., and Yuan, J. (1994). Prevention of vertebrate neuronal death by the *crm*A gene. *Science* **263**, 826–828.

Ghendler, Y., Arnon, R., and Fishelson, Z. (1994). Schistosoma mansoni: isolation and characterization of Smpi56, a novel serine protease inhibitor. *Experimental Parasitology* **78**(2), 121–31.

Gooding, L. R. (1992). Virus proteins that counteract host immune defenses. *Cell* **71**, 5–7.

Houenou, L. J., Turner, P. L., Li, L., Oppenheim, R. W., and Festoff, B. W. (1995). A serine protease inhibitor, protease nexin I, rescues motoneurons from naturally occurring and axotomy-induced cell death. *Proceedings of the National Academy of Sciences of the United States of America* **92**(3), 895–9.

Kettle, S., Blake, N. W., Law, K. M., and Smith, G. L. (1995). Vaccinia virus serpins B13R (SPI-2) and B22R (SPI-1) encode M(r) 38.5 and 40K, intracellular polypeptides that do not affect virus virulence in a murine intranasal model. *Virology* **206**(1), 136–47.

Komiyama, T., Ray, C. A., Pickup, D. J., Howard, A. D., Thornberry, N. A., Peterson, E. P., and Salvesen, G. (1994). Inhibition of interleukin-1 beta converting enzyme by the cowpox virus serpin CrmA. An example of cross-class inhibition. *Journal of Biological Chemistry* **269**(30), 19331–7.

Kotwal, G. J., and Moss, B. (1988). Vaccinia virus encodes a secretetory polypeptide structurally related to complement control proteins. *Nature* **335**, 176–178.

Kotwal, G. J., and Moss, B. (1989). Vaccinia virus encodes two proteins that are structurally related to members of the plasma serine protease inhibitor superfamily. *Journal of Virology* **63**(2), 600–606.

Law, K. M., and Smith, G. L. (1992). A vaccinia serine protease inhibitor which prevents virus-induced cell fusion. *Journal of General Virology* **73**(Pt 3), 549–57.

Lijnen, H. R., and Collen, D. (1990). Regulation and control of the fibrinolytic system. *In* "Serine protease and their serpin inhibitors in the nervous system" (B. W. Festoff, Ed.). Plenum Press, New York.

Lomas, D. A., Evans, D. L., Upton, C., McFadden, G., and Carrell, R. W. (1993). Inhibition of plasmin, urokinase, tissue plasminogen activator, and C1S by a myxoma virus serine proteinase inhibitor. *Journal of Biological Chemistry* **268**(1), 516–21.

Lucas, A., Liu, L., Macen, J. L., Nash, P. D., Dai, E., Stewart, M., Yan, W., Graham, K., Etches, W., Boshkov, L., Nation, P. N., Humen, D., Hobman, M., and McFadden, G. (1996). A virus-encoded serine proteinase inhibitor, SERP-1, inhibits atherosclerotic plaque development following balloon angioplasty. *Circulation*, (in press).

Macen, J. L., Upton, C., Nation, N., and McFadden, G. (1993). SERP1, a serine proteinase inhibitor encoded by myxoma virus, is a secreted glycoprotein that interferes with inflammation. *Virology* **195**(2), 348–63.

Maksymowych, W. P., Nation, N., Nash, P. D., Macen, J., Lucas, A., McFadden, G. and Russell, A. S. (1996). Amelioration of antigen-induced arthritis in rabbits treated with a secreted viral serine proteinase inhibitor. *Journal of Rheumatology* **23**(5): 878–882.

Marrack, P., and Kappler, J. (1994). Subversion of the immune system by pathogens. *Cell* **76**, 323–332.

Massung, R. F., Esposito, J. J., Liu, L.-I., Qi, J., Utterback, T. R., Knight, J. C., Aubin, L., Yuran, T. E., Parsons, J. M., Loparev, V. N., Selivanov, N. A., Cavallaro, K. F., Kerlavage, A. R., Mahy, B. W. J., and Venter, J. C. (1993). Potential virulence determinants in the terminal regions of variola smallpox virus genome. *Nature* **366**, 748–751.

Massung, R. F., Jayarama, V., and Moyer, R. W. (1993). DNA sequence analysis of conserved and unique regions of swinepox virus: identification of genetic elements supporting phenotypic observations including a novel G protein-coupled receptor homologue. *Virology* **197**, 511–528.

Matsuda, Y., Kawata, S., Nagase, T., Maeda, Y., Yamasaki, E., Kiso, S., Ishiguro, H., and Matsuzawa, Y. (1994). Interleukin-6 in transcatheter arterial embolization for patients with hepatocellular carcinoma. Effects of serine protease inhibitor. *Cancer* **73**(1), 53–7.

Maudsley, D. J., and Pound, J. D. (1991). Modulation of MHC antigen expression by viruses and oncogenes. *Immunology Today* **12**, 429–430.

McFadden, G., Ed. (1995). Viroceptors, virokines and related immune modulators encoded by DNA Viruses. Austin, Texas: R. G. Landes Co.

McFadden, G., Graham, K., Ellison, K., Barry, M., Macen, J., Schreiber, M., Mossman, K., Nash, P., Lalani, A., and Everett, H. (1995). Interruption of cytokine networks by poxviruses: lessons from myxoma virus. *Journal of Leukocyte Biology* **57**, 731–738.

McFadden, G., Graham, K., and Opgenorth, A. (1995). Poxvirus growth factors. *In* "Virokines, viroceptors and related modulators encoded by DNA viruses" (G. McFadden, Ed.). R. G. Landes Co., Austin, TX.

McFadden, G., and Kane, K. (1994). How DNA viruses perturb functional MHC expression to alter immune recognition. *Advances in Cancer Research* **63**, 117–209.

Miura, M., Zhu, H., Rotello, R., Hatweig, E. A., and Yuan, J. (1993). Induction of apoptosis in fibroplasts by IL-1ß-converting enzyme, a mammalian homolog of the *C. elegans* cell death gene ced-3. *Cell* **75**, 653–660.

Moss, B. (1990). Poxviridae and their replication. 2 ed. *In* "Virology" (B. N. Fields, and D. M. Knipe, Eds.), Vol. 2, pp. 2079–2111. 2 vols. Raven Press, Ltd., New York.

Mossman, K., Barry, M., and McFadden, G. (1995). Interferon-gamma receptors encoded by poxviruses. *In* "Virokines, viroceptors and related modulators encoded by DNA viruses" (G. McFadden, Ed.). R. G. Landes Co, Austin, TX.

Nash, P. D., Whitty, A., Macen, J. L., and McFadden, G. (1996). Kinetic characterization of wild type and a reactive centre mutant of myxoma virus Serp-1 serine proteinase inhibitor., (in preparation).

Nystedt, S., Emilsson, K., Wahlestedt, C., and Sundelin, J. (1994). Molecular cloning of a potential proteinase activated receptor. *Proceedings of the National Academy of Sciences of the United States of America* **91**, 9208–9212.

Olds, R. J., Lane, D. A., Mille, B., Chowdhury, V., and Thein, S. L. (1994). Antithrombin: the principle inhibitor of thrombin. *Seminars in Thrombosis and Hemostasis* **20**(4), 353–372.

Olson, S. T., and Bjork, I. (1994). Regulation of thrombin activity by antithrombin and heparin. *Seminars in Thrombosis and Hemostasis* **20**(4), 373–409.

Pickup, D. J., Ink, B. S., Hu, W., Ray, C. A., and Joklik, W. K. (1986). Haemorrhage in lesions caused by cowpox virus is induced by a viral protein that is related to plasma protein inhibitors of serine protease. *Proceedings of the National Academy of Sciences of the United States of America* **83**, 7698–7702.

Potempa, J., Korzus, E., and Travis, J. (1994). The serpin superfamily of proteinase inhibitors: structure, function, and regulation. *Journal of Biological Chemistry* **269**(23), 15957–60.

Ray, C. A., Black, R. A., Kronheim, S. R., Greenstreet, T. A., Sleath, P. R., Salvesen, G. S., and Pickup, D. J. (1992). Viral inhibition of inflammation: cowpox virus encodes an inhibitor of the interleukin-1 beta converting enzyme. *Cell* **69**(4), 597–604.

Sarin, A., Adams, D. H., and Henkart, P. A. (1993). Protease inhibitors selectively block T cell receptor-triggered programmed cell death in a murine T cell hybridoma and activated peripheral T cells. *Journal of Experimental Medicine* **178**(5), 1693–700.

Smirnova, I. V., Ho, G. J., Fenton, J. W. n., and Festoff, B. W. (1994). Extravascular proteolysis and the nervous system: serine protease/serpin balance. *Seminars in Thrombosis & Hemostasis* **20**(4), 426–32.

Smith, C. A., Farrah, T., and Goodwin, R. G. (1994). The TNF receptors superfamily of cellular and viral proteins: activation, costimulation and death. *Cell* **76**, 959–962.

Smith, C. A., and Goodwin, R. G. (1995). TNF receptors in the poxvirus family: biology and genetic implications. *In* "Virokines, viroceptors and related modulators encoded by DNA viruses" (G. McFadden, Ed.), pp. 29–40. R. G. Landes Co., Austin, Texas.

Smith, G. L. (1994). Virus strategies for the evasion of host response to infection. *Trends in Microbiology* **2**, 81–88.

Spriggs, M. (1994). Cytokine and cytokine receptor genes 'captured' by viruses. *Current Opinion in Immunology* **6**, 526–529.

Strickland, D. K., Kounnas, M. Z., and Argraves, S. (1995). LDL receptor-related protein: a multiligand receptor for lipoprotein and proteinase catabolism. *The FASEB Journal* **9**, 890–898.

Turner, P. C., and Moyer, R. W. (1990). The molecular pathogenesis of poxvirus. *Current Topics in Microbiology and Immunology* **163**, 125–150.

Turner, P. C., and Moyer, R. W. (1992). An orthopoxvirus serpinlike gene controls the ability of infected cells to fuse. *Journal of Virology* **66**(4), 2076–85.

Turner, P. C., and Moyer, R. W. (1995). Orthopoxvirus fusion inhibitor glycoprotein SPI-3 (open reading frame K2L) contains motifs characteristic of serine proteinase inhibitors that are not required for control of cell fusion. *Journal of Virology* **69**(10), 5978–87.

Turner, P. C., Musy, P. Y., and Moyer, R. W. (1995). Poxvirus Serpins. *In* "Viroceptors, virokines and related immune modulators encoded by DNA Viruses" (G. McFadden, Ed.), pp. 67–88. R. G. Landes Co., Austin, Texas.

Upton, C., Macen, J. L., Wishart, D. S., and McFadden, G. (1990). Myxoma virus and malignant rabbit fibroma virus encode a serpin-like protein important for virus virulence. *Virology* **179**, 618–631.

White, E. (1996). Life, death, and the persuit of apoptosis. *Genes and Development* **10**, 1–15.

Yenbutr, P., and Scott, A. L. (1995). Molecular cloning of a serine proteinase inhibitor from Brugia malayi. *Infection & Immunity* **63**(5), 1745–53.

Zhou, J., Sun, X. Y., Fernando, G. J., and Frazer, I. H. (1992). The vaccinia virus K2L gene encodes a serine protease inhibitor which inhibits cell-cell fusion. *Virology* **189**(2), 678–86.

20

DYSFUNCTIONAL VARIANTS AND THE STRUCTURAL BIOLOGY OF THE SERPINS

Robin Carrell, David Lomas, Penelope Stein, and James Whisstock

Department of Haematology
University of Cambridge
MRC Centre, Hills Road
Cambridge CB2 2QH
England

The development of our knowledge of the serpins illustrates the advantages of considering a protein superfamily as a whole. The serpins have all retained a common tertiary structure despite the individual evolution of diverse functions; for example, the homology of the plasma protease inhibitor α_1-antitrypsin is closer to that of corticosteroid binding globulin than is the homology of the two heparin-binding plasma inhibitors — antithrombin and heparin cofactor II — one to another. This retention of a well conserved structure necessarily requires the retention of strong homologies in primary and secondary structures in all the members of the family, across functions as well as species. For this reason, from the beginning, the study of the serpins has been a collective process with our understanding of the function of each member being greatly strengthened by parallel studies of other serpins. This has been particularly true of the lessons learnt from the human dysfunctional variants; one by one they have provided clues as to the normal function in individual members but when considered together with structural studies, in terms of the family as a whole, they have opened our understanding to a degree that far surpasses the contribution of more conventional approaches.

This is a bold claim but it reflects the power of natural experiments where the subjects self-select for functional abnormalities and come complete with their own unique variant. Furthermore, each variant is authentically expressed and glycosylated and usually present in multi-milligram yields. The progenitor of all these natural experiments in the serpins was the observation of Laurell and Eriksson[1] in 1963 of a genetic deficiency of plasma α_1-antitrypsin; this was associated with cumulative loss of lung elasticity and hence led to the identification of α_1-antitrypsin as a physiological inhibitor of elastase rather than of trypsin. Here we illustrate with examples from our research group over the last twenty years, the insights provided by other variants, particularly to an understanding of the inherent mobility of the molecule.

1. STATIONARY DOMAINS

1.1. Reactive Site

The first indication of the existence of the serpin family came in 1979 from the comparison[2,3] of the partial sequence of α_1-antitrypsin with the then recently completed sequence of antithrombin. We had sequenced the C-terminal third of α_1-antitrypsin in an attempt to ascertain the significance of the S (Glu264Val) and Z (Glu342Lys) mutations. Comparison of this partial sequence of α_1-antitrypsin with that of antithrombin not only showed a close general homology between the two but also the presence of a homologous reactive centre[4], positioned some 40 residues from the C-terminus of each molecule. The reactive site sequence had previously been deduced by others[5] but a series of groups had subsequently all placed the site near the amino-terminus[6]. Their experimental data were based on aminoterminal sequencing of α_1-antitrypsin after cleavage at the reactive site loop. The results in the 1970's, using the recently introduced automated Edman sequencer, consistently showed a single new amino-terminus inferring cleavage of the reactive centre had occurred near the commencement of the molecule. It was only much later realised that the glutamate at the amino-terminus of the α_1-antitrypsin readily undergoes cyclisation, blocking its measurement by the dansyl-Edman technique. Nevertheless, despite the other sequence data, the result obtained by automated sequencing was so consistent that the debate as to the placement of the site, *i.e.* — N-terminal vs C-terminal continued beyond 1980. The finding that finally put the matter beyond doubt arose from the observation of Lewis and colleagues[7] in Pittsburgh of a bleeding disorder in a patient with an electrophoretically abnormal α_1-antitrypsin. Sequencing of this variant[8] (Fig. 1) provided a series of insights into the inhibitory function of the serpins and settled the validity of the homologous alignment of the reactive centres of α_1-antitrypsin and antithrombin near their carboxy-termini.

As is so often the case with functional variants, the characterisation of the Pittsburgh mutation posed as many interesting new questions as it provided answers. The observation

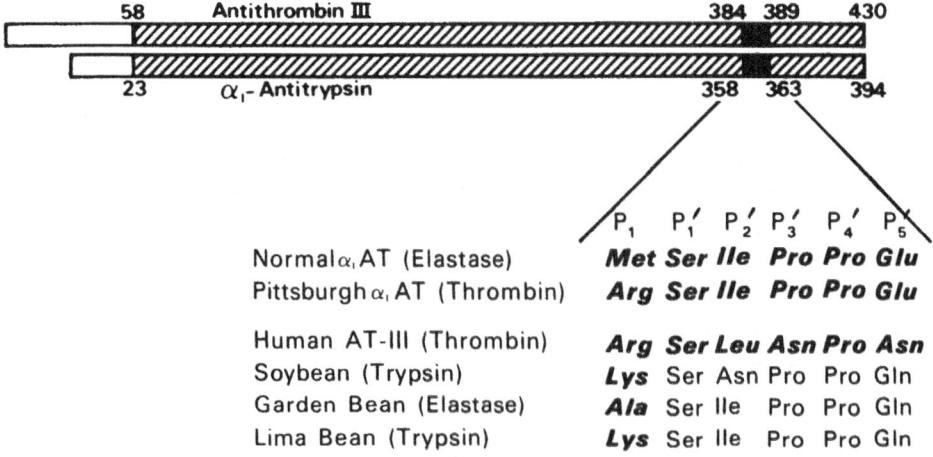

Figure 1. The Pittsburgh mutant of α_1-antitrypsin in which the mutation of the P_1 reactive centre methionine to arginine converted it to a highly effective inhibitor of thrombin. The identification of this mutant settled the placement of the reactive centre of the serpins by homology alignment and from the homologous reactive site of the Bowman-Birk bean serine protease inhibitors. (Reproduced with permission from Owen et al, New Eng. J. of Med. 1983.)

that the variant was also associated with presence of proalbumin in the plasma led to the identification of Arg358 α_1-antitrypsin as an inhibitor of the subtilisin-like proteases that cleave propeptides[9]. But a puzzling question that arose was the ability of a single substitution at the reactive centre to convert a physiological inhibitor of elastase to an inhibitor of thrombin with kinetics that approach those of heparin activated antithrombin. The conclusion[8] from this was that antithrombin circulated in a constrained conformational form that was released on the binding of heparin to give a form likely to match the unconstrained conformation of α_1-antitrypsin. An answer as to the likely conformational change involved is now provided by the new structure of the reactive centre of α_1-antitrypsin[10] (reported by Elliott *et al.* at this meeting) which shows the reactive loop of α_1-antitrypsin to be fully exposed in a canonical conformation.

1.2. Heparin Binding Site

The power of the combination of homologous structural alignments coupled with the study of human variants is illustrated by the deduction of the placement of the heparin-binding site. In 1987, at a stage when the amino-acid sequences of some 20 serpins had been completed, three of these — antithrombin, heparin cofactor II and protease nexin — were known to bind to sulphated polysaccharides. It seemed likely that these three serpins would have a common binding site formed of basic residues that would bind the sulphates of the core pentasaccharide fragment of heparin. A quick analysis of the aligned sequences showed that this was indeed so — with the uniquely conserved presence in the three heparin-binding serpins of a cluster of arginines and lysines centred on the D helix of the molecule[11,12]. The site fitted with previous biochemical data and strong support came from a series of natural mutants of antithrombin, with a decreased binding affinity for heparin and with amino-acid substitutions of the basic residues at the site[13-15].

2. MOLECULAR MOBILITY

The particular contribution of studies of serpins to biology as a whole has been as a first example of molecular mobility i.e. of the ability of an intact molecule to undergo major conformational rearrangements. Perhaps more importantly a large proportion of the pathological variants of the serpins result from a dysfunction of this structural rearrangement and hence provide a prototypic example of what is now being recognised as the new entity of conformational diseases.

The characteristic features of many of the dysfunctional variants have recently been reviewed[16,17] and are explained by an increased conformational lability — giving rise to a plasma deficiency due to incomplete intracellular processing, a partial or complete loss of inhibitory activity, and an increased plasma turnover. At the same time, that this lability of the dysfunctional variants became noticed, there was also an awareness of the functional lability of the serpins; stored or pasteurised samples were observed to lose inhibitory activity, only part of the loss being explicable by reactive loop cleavage. This spontaneous loss of activity was most dramatically seen in the PAI-1 but it could also be induced in the other inhibitory serpins by exposure to mild denaturant stress. The conclusion that these changes resulted from an insertion of the reactive centre loop into the main A-sheet of the molecule was strengthened by the deduction that the common dysfunctional abnormalities of the serpins resulted from a mutational lability of this conformational rearrangement. Subsequently it was realised that this rearrangement involved more than a single confor-

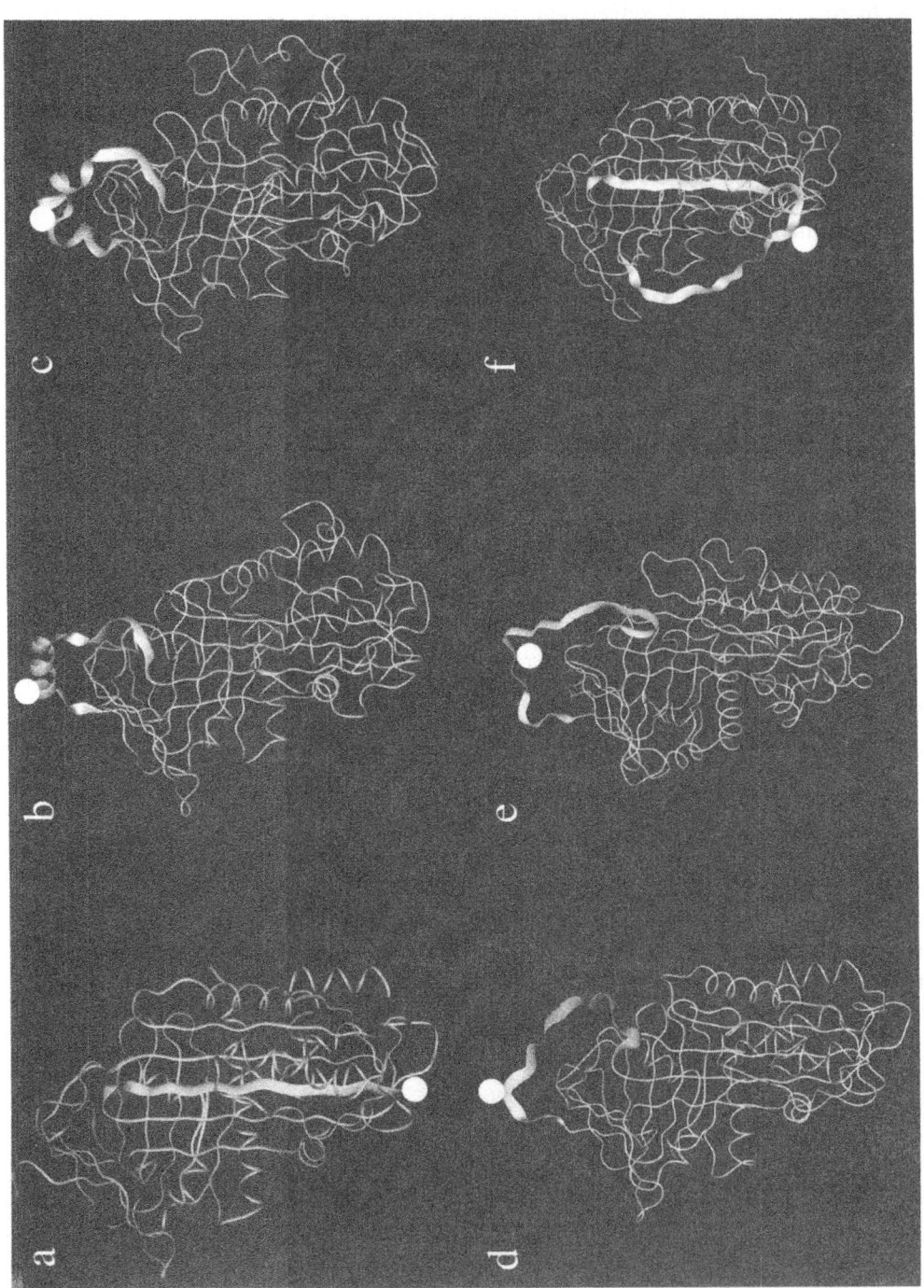

Figure 2. Mobility of the reactive loop in ribbon form a) cleaved α_1-antitrypsin,[18] b) intact ovalbumin,[36] c) intact α_1-antichymotrypsin,[50] d) intact (canonical) α_1-antitrypsin,[10] e) intact active antithrombin,[48,49] and L–(latent) antithrombin.[48] The P_1 residue is indicated by a white circle.

mational change and was accompanied by an intermolecular linkage to give polymer formation. This finding proved to be of particular significance to an understanding of disease mechanisms as the process of polymerisation is now known to be a prime cause of pathology, most notably that of the common Z variant of α_1-antitrypsin. All studies to date however emphasise the combination of conformational events that result in disease: unfolding with intracellular degradation, insertion of the loop to give inactivation, and polymerisation to give intracellular or circulatory aggregation. The insertion of the reactive loop that can be induced in inhibitory serpins and occurs spontaneously in labile variants has been described as the latent-transition, though a more correct term would be the L-transition. This acknowledges not only that the L-conformation may be irreversible but also that it may be present in subtly different structural forms.

2.1 Flexibility of Reactive Loop

The realisation that the serpins had a flexible reactive site loop first came from the crystal structure of cleaved α_1-antitrypsin determined by Robert Huber's group in Martinsried[18]. This structure (Fig. 2a) showed the complete insertion of the proximal portion of the cleaved reactive loop into the central 4th strand position of the A-sheet. This first structure posed two immediate questions:

1. Is the incorporation of the loop, either partial or complete, a necessary component of the inhibitory mechanism?
2. How do aberrations of this movement result in disease?

The study of the incorporation of the reactive loop into the A-sheet in a range of serpins was simplified by the observation of an accompanying remarkable increase in thermal stability — the S→R change[19]. The initial view that this change in stability was solely due to change from a stressed 5-stranded structure to a relaxed six-stranded structure is now known to be an over-simplification (Fig. 3). Thermal stability also depends on the readiness of polymer formation and hence on accessibility to the C-sheet as well as the A-sheet. Nevertheless the ability to undergo the S→R change is a good index of the readiness of A-sheet insertion and the failure of the change is only observed in non-inhibitory serpins such as ovalbumin and angiotensinogen[20,21]. The converse of this is not true, as numbers of non-inhibitors, including corticosteroid and thyroxine binding globulins[22], do undergo the S→R change but have no known inhibitory activity. Recently this was put further to the test by Paul Edgar in our laboratory[23] — he showed that mutation of the active centre of corticosteroid binding globulin to an optimal (Arg-Ser) sequence still did not lead to the predictable inhibitory activity against trypsin.

Overall, the strength of evidence[24–26], amply covered by others in this volume, is that loop insertion is a necessary requirement for the formation of a stable protease-inhibitor complex. Indeed, it seems that the ability to form this final 'locked' conformation is the reason for the evolutionary success of the serpins as opposed to the other families of smaller serine protease inhibitors. The related theme, developed by others in this volume, is that the effectiveness of serpins as protease inhibitors depends on not just the ability to undergo reactive loop insertion but also the rate at which this occurs. The partitioning of activity as an inhibitor, rather than as a substrate, being dependent on the stabilisation of complex by the locking movement of the reactive loop into the A-sheet.

These proposals are strongly supported by observations with natural mutations of the serpins which result in their conversion to substrates rather than inhibitors. The first

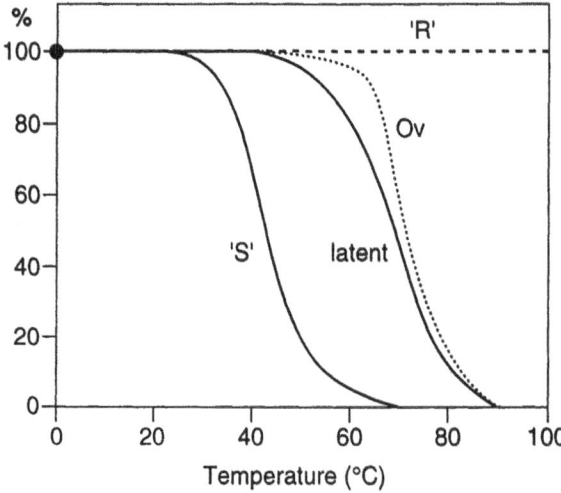

Figure 3. A diagrammatic comparison of the thermal stabilities of serpin conformations: native (S) and latent PAI-1[38] ovalbumin (Ov)[21] and cleaved α_1-antitrypsin (R)[19]. The change from the native S to the cleaved R conformation involves a change from a five-stranded (S) to six-stranded (R) A-sheet, but this does not in itself account for the change in stability as latent PAI-1 has a six-stranded sheet and ovalbumin five strands. (Reproduced with permission from Ref 33 the Brit. J. Haemat. 1996.)

examples of these were identified in variant antithrombins present in individuals with a history of thrombotic disease[27–32]. These mutations particularly involved the proximal hinge of the reactive loop formed by a sequence of small side-chain residues, mainly alanines, at P_{14}-P_{10}. At first it seemed as if mutations in this region might prevent loop insertion but this has not been confirmed by subsequent studies; these have shown that even the presence of a proline (which is not insertable into a β-sheet) does not prevent overall strand addition though it does slow and destabilise the linkage. The conclusion is that the dysfunction generally arises not from the inability to give loop insertion[33] but rather from the lack of effectiveness of this incorporation in terms of the stability and rate of formation of the six-stranded structure[17, 34].

2.2. The Latent (L) Conformation

Two different approaches were taken in 1990 to demonstrate that the intact — as opposed to cleaved — serpins had an inherent facility to incorporate a sixth strand in the A-sheet. The Martinsried[35] group showed that a homologue of the P_{14}-P_1 reactive loop peptide would readily anneal into α_1-antitrypsin with consequent loss of inhibitory activity and a typical increase in thermal stability. A different approach was taken by our group. We based our reasoning on the inhibitory serpins being in a metastable state with conversion to the stable state resulting from the movement of the reactive loop from its exposed external position to give full insertion into the A-sheet. This proposal was considerably influenced by our completion, at the beginning of 1990, of the structure of the first intact serpin — ovalbumin[36]. This high resolution structure showed a helical reactive loop with a proximal stem that was clearly situated in a position appropriate for incorporation between the partly opened strands 3 and 5 of the A-sheet (Fig. 2b). The likelihood that the intact loop of the inhibitory serpins could incorporate into the A-sheet was indicated by results from hinge mutant variants. Furthermore such incorporation would explain the then puzzling observation that another serpin — PAI-1 — spontaneously underwent a change in conformation to give the inactive latent form.

Our simplistic model of an inhibitory serpin at that time (and now!) was that of a set-mousetrap. A prediction from this model[33] is that the dormant sprung-form of the trap should be more stable than the active set-form and hence, by projection from this simple model, that the latent form of PAI-1[37] should similarly be more stable than the active form. This prediction was supported by preliminary measurements of the thermal stability of PAI-1 in the laboratory of Dr. Nuala Booth[38] and by the earlier published results of the stabilities of latent and native PAI-1 by Katagiri and colleagues[39]. To induce the stable sprung-form of a mousetrap requires vigorous shaking of the set-trap. Similarly the induction of loop insertion in the serpins should, we reasoned, occur with the molecular perturbation produced by mild denaturants or heating[33]. This turned out to be so, but it was only subsequently realised that the change in conformation to the incorporated form was also accompanied by intermolecular loop-sheet linkage with polymer formation[40]. Both the latent and polymerised forms share an identical increase in thermal stability as compared to the relatively labile inhibitory monomer.

The initial denaturing stress used to trigger this transition was by exposure to 1M guanidinium chloride but similar results, with predominant formation of polymers, were also obtained using temperature-induced denaturation[41,42]. The use of heat to stress the stability of serpins had been previously well-studied, though at the time it had not been interpretable in structural terms. This work had been carried out some 15 years earlier[43] to study the effects of pasteurisation, that is incubation for several hours at 55–60°, on plasma-fractionated α_1-antitrypsin and antithrombin. It had been noted then that this sterilising process was accompanied by the formation of high molecular weight aggregates and that this could be countered by the presence of sugars or citrate in the incubation buffers. These studies had been overlooked but coincidentally in 1991, at the time we became aware of the problem of polymerisation, we were also examining samples of commercial concentrates of antithrombin and noted that some of these contained as high as 20% of inactive, but intact monomeric antithrombin. This component is the induced six-stranded form of antithrombin, i.e. the latent or, more correctly, L-form of antithrombin.

It is now realised that the conditions that induce loop incorporation in the serpin monomer will also be accompanied by intermolecular loop-sheet linkage to give polymer formation[35,44,45]. Depending on the conditions, the proportion of the two species can be varied; if the conformational change occurs slowly as with ovalbumin[46], or very rapidly as with PAI-1, then there may be nil or negligible polymer formation. In general, however, the conformational change to the six-stranded form is accompanied by significant polymer formation whether it is induced by denaturant stress of a normal serpin or occurs spontaneously due to a hinge or other dysfunctional mutation in a natural variant[16]. This latter occurrence helps to explain some of the apparently discordant results obtained with hinge variants[34], not only between different groups but also at different times within the same laboratory. The important practical point is that such experiments should be monitored throughout by native (non-denaturing) gel electrophoresis which is the most sensitive method of detecting incidental polymer formation.

The conditions have now been standardised to prepare and isolate the L-forms of antithrombin and α_1-antitrypsin in quantitative yields, by incubation above 60°C in 0.25–1.0M citrate[41,42]. The new conformation in α_1-antitrypsin can be correctly referred to as the latent conformation, as it can be reversed by dialysis from 8M guanidinium chloride to give the active inhibitor. In antithrombin the conformation is not reversible, as the presence of disulphide bonds prevents unfolding and re-folding and the new conformation is termed L-antithrombin. Nevertheless there is firm structural evidence that this induced L-antithrombin has an equivalent conformation to that of latent PAI-1[47] The structure of the dimer of antithrombin[48] crystallised in the presence of polyethylene glycol (PEG), shows one molecule in the fully inserted conformation with the same characteristic fold as latent PAI-1. Recently, Lei Jin in our laboratory has also shown that this L-component that spontaneously forms in PEG can be replaced by heat induced L-antithrombin to give identical crystal forms.

2.3. Structural Biology, Inhibition, and Heparin Inactivation

Our concepts of the inhibitory mechanism and functional diversity of the serpins has been underpinned by a series of crystallographic structures. The alignment and homologies of the family as a whole, in a functionally meaningful sense, were dependent on the original template structure of cleaved α_1-antitrypsin[12]. The contribution from our laboratory has been to the series of intact structures (ovalbumin, active and latent antithrombin and α_1-antitrypsin, Fig. 2) that have demonstrated the range of movement that can take place around the reactive centre. This ranges from the fully exposed helical reactive loop of ovalbumin to the completely incorporated loop of L-antithrombin[48]. Together with the two intervening structures, of the fully exposed but unfolded loop in intact α_1-antitrypsin[10] and the partly inserted loop of active antithrombin[48,49], these four structures can be linked to give a video representation of the flexibility of the intact reactive loop[17]. This shows not only how movement occurs at the hinges of the loop but gives detailed representations of the movement of the A-sheet and underlying structures, that accompany the insertion of strand 4A. Now, with a 2.6Å structure of the antithrombin dimer (refined by Richard Skinner and colleagues in Cambridge), it is also possible to see the sidechain movement that occurs, during the transition from the five- to the six-stranded structures, at the core heparin-binding site on the D and A helices.

In terms of the inhibitory mechanism of the serpins three reactive loop conformations are significant: i) the quiescent conformation of the circulating inhibitor, ii) the docking conformation adopted by the inhibitor on initial association with the target protease, and iii) the locking conformation that gives the unique stabilisation of the serpin-protease complex.

It is now known that the serpins can adopt a range of quiescent conformations. The inherent flexibility of the loop was evident from the initial structure of intact ovalbumin and subsequent structures have each shown how evolution has adapted this flexibility in different ways. In α_1-antichymotrypsin, determined by the Philadelphia group,[50] the loop is in the form of a distorted helix and clearly requires reordering to provide a suitable docking conformation. This is even more so in quiescent antithrombin (Fig. 2d) which has partial insertion of the loop, with an internal orientation of the P_1 reactive centre arginine[48,49]. It was clear from other evidence that the activation of antithrombin by heparin involved a change in the conformation of the reactive loop and van Boeckel and colleagues[51] suggested that this required an expulsion of the hinge of the loop from the sheet. Evidence as to the likely structural changes is provided by our recent structure of intact α_1-antitrypsin.

The structure of α_1-antitrypsin at 2.9Å has provided an answer to a number of previous queries[10]. The conformation of the P_3-P_3' loop is in the canonical form that precisely fits the active site of serine proteases and closely matches the loop conformation of the other families of non-serpin serine proteinase inhibitors. If α_1-antitrypsin had been the first intact serpin structure to be determined then this result would not have been a surprise, but the previous series of structures had raised the expectation of a helical or disordered loop and it had been assumed that the transition to the canonical form would also involve partial loop insertion. However, the structure of α_1-antitrypsin shows that the canonical form can be adopted without such insertion and it thus provides a prototype docking conformation for the family. In the case of α_1-antitrypsin the canonical structure has been stabilised, in an evolutionary sense, by the formation of a well-structured series of hydrogen bonds linking a glutamate at P_5 to the main body of the molecule. It is likely that although the reactive site loops of the other inhibitory serpins may circulate in a variety of forms, they too will adopt a similar canonical conformation on activation or on initial recognition of their target proteinases.

The demonstration that the reactive loop can take up the canonical form, without strand insertion into the A-sheet, also favours the recent proposal by others[24–26] of a two-stage mechanism of inhibition. The canonical conformation will allow an initial rapid association with the protease with putative cleavage to give the acyl-enzyme complex, followed by the second 'locking' stage with insertion of the proximal portion of the loop into the A-sheet and distortion and stabilisation of the acyl-enzyme bond.

A comparison of the structures of the reactive loops of α_1-antitrypsin and antithrombin (Fig. 4) immediately suggests the likely conformational change induced in the reactive loop of antithrombin by heparin. The expulsion of the loop of antithrombin from the A-sheet and its adoption of a canonical-like conformation would be in keeping with both the increase in inhibitory activity and changes in protease vulnerability that occur in antithrombin on its activation by heparin. The presence in α_1-antitrypsin of a reactive loop

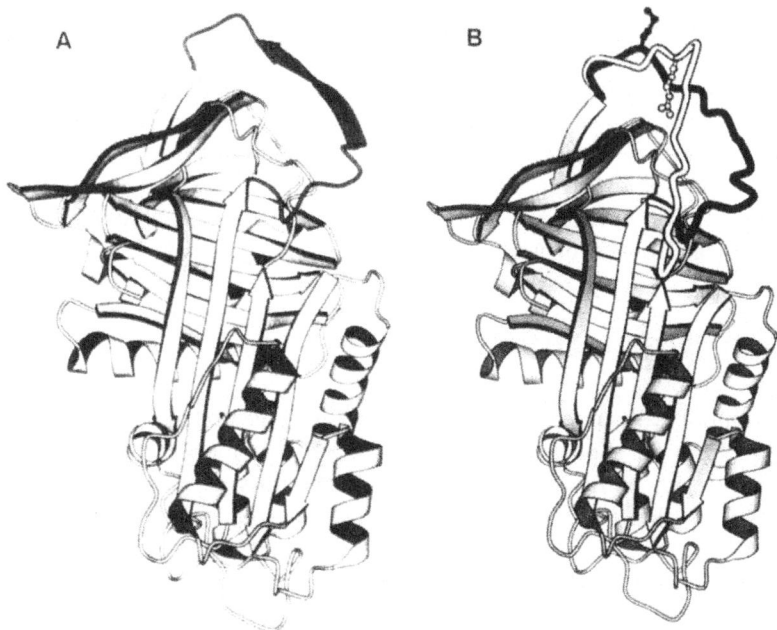

Figure 4. (A) Ribbon structure of α_1-antitrypsin[10] showing the loop in canonical form P_3-P_3' with, as a ribbon, the P_3-P_8 β-pleated strand (B) comparison of the reactive loop conformations of α_1-antitrypsin (black) and antithrombin (white) with the orientation of the respective P_1 residues; expulsion of the loop of antithrombin from the A-sheet to a conformation as in α_1-antitrypsin provides a model of the likely change induced by heparin (Modified with permission from Ref 10 Nature Structural Biology 1996.)

held, by the P_5 glutamate bonding, in an optimal inhibitory conformation explains the much earlier observations from the Pittsburgh mutant of α_1-antitrypsin[8]. The replacement of the reactive centre methionine by arginine in α_1-antitrypsin not only converts it to a specific inhibitor of thrombin but does so in a three-dimensional context equivalent to that of heparin-activated antithrombin.

2.4. β-Sheet Linkage and Polymerisation

The structure of the reactive loop in α_1-antitrypsin also clarified the likely mechanism involved in the loop-sheet polymerisation of serpins. From the time of the initial observation of polymerisation[35] it had been deduced that the linkage in both stress-induced and natural polymers[44,45] was through insertion of the loop of one molecule into the A-sheet of the next. This assumption of loop-A-sheet linkage, although persuasive, was not based on direct evidence and was put in question by the subsequent structural finding that the dimer of antithrombin was formed by a β-pleated strand linkage between the reactive loop of one molecule and the C-sheet of the next[48,49]. Other evidence for the two alternatives, i.e. for A- or C-sheet linkage, was equivocal and has been reviewed elsewhere[16,17]. Recently two lines of evidence have made it likely that A-sheet linkage is the predominant mode of polymerisation. These are 1) the clustering of natural mutations causing polymerisation at the key site of A-sheet opening and 2) the observation that the dimer of antithrombin is readily dissociable[17].

The apparently overall contradictory results have been brought into context by the findings of Salvesen and colleagues in this volume indicating that both mechanisms can occur, depending on the buffering conditions; A-sheet linkage giving stable polymer formation and C-sheet linkage giving the formation of short-chain reversible polymers as readily occurs with antithrombin. The clarification provided by the recent structure of α_1-antitrypsin is that it shows the P_3-P_8 portion of the loop to be in a exposed β-pleated conformation which convincingly models to give extended loop-A-sheet polymers. In more general terms the lesson from both the structures of α_1-antitrypsin[10] and the antithrombin dimer[48] is the readiness of the reactive loop of the serpins to form a β-pleated structure with any appropriate available β-sheet — be it an A-sheet, or C-sheet of a serpin or even the β-pleated sheets of Alzheimers' plaques[52].

CONFORMATIONAL DISEASE

The inhibitory advantages inherent in the ability to undergo the sudden conformational change from an exposed to incorporated reactive loop explains the evolutionary success of the serpin family of serine protease inhibitors. However, this change from a native metastable state to a stable dormant state also renders the serpins vulnerable to mutations that alter the triggering of this trap. The significance of this becomes apparent when the more than one hundred different mutations in the plasma serpins that result in a variety of diseases are plotted[16] on a template serpin structure (Fig. 5). Over half of these mutations are seen to cluster in the domains directly involved in the movement of the reactive loop or the opening of the A-sheet. The studies of a series of such mutants[17] provides convincing evidence of a general mechanism of conformational disease in the serpins which arises from premature and inappropriate loop incorporation.

The first examples of this were in the group of mutants of α_1-antitrypsin, epitomised by the common European Z-variant,[1] that are associated with a failure of secretion of the

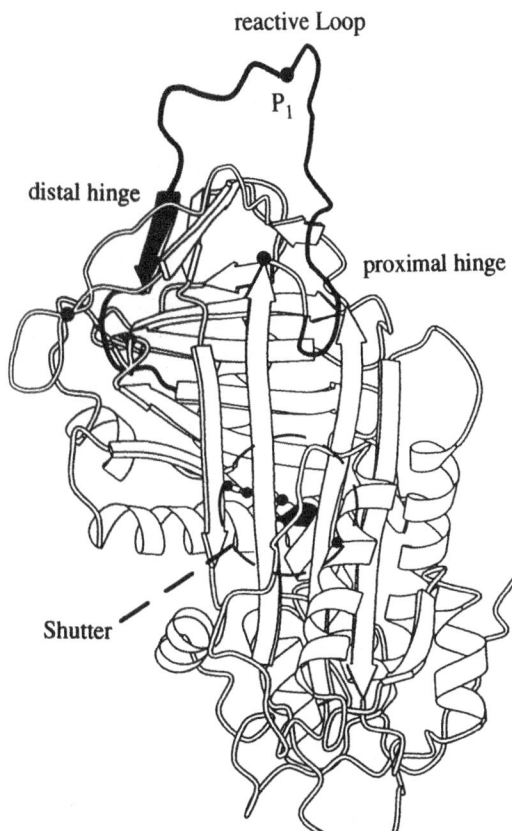

Figure 5. Mobile domains of serpins. Plotting of 100 dysfunctional serpins on a template structure shows clustering at defined functional domains (above) one half of the total being in the domains controlling mobility. The proximal and distal hinge domains were predictable but the unexpected finding was the cluster of mutations in the 'shutter' domain centred on Ser53 at the commencement of the B helix. (Reproduced with permission from Ref 17 Biological Chemistry Hoppe-Seyler 1996.)

abnormal protein due to its accumulation, at the site of synthesis, in the hepatocyte. The mutation in Z-antitrypsin, Glu342Lys, is in the proximal hinge of the reactive loop and it was shown that this results in a tendency to spontaneously form fibrils of loop-sheet polymers. Furthermore, examination of the large inclusions of Z-antitrypsin present in the hepatocytes of affected individuals showed them to be formed of tangles of such fibrils[45]. An even more convincing demonstration of this disease mechanism is seen with the rare Siiyama mutant, (Ser53Phe), with a mutation in the shutter domain directly underlying the bifurcation of the A-sheet[53]. This is associated with typical liver inclusions and when the variant is studied *in vitro* it spontaneously and completely forms long polymers.

The significance of the shutter domain in controlling the conformational transition, is demonstrated not only by the number of mutations that result in dysfunction but also by the converse experiments in which other mutations at this site result in increased stability. The Seoul group of Yu and colleagues[54] showed that the replacement of the phenylalanine 51 in the shutter domain by leucine stabilised the five-stranded structure of the A-sheet. Sidhar and Foreman of Southampton[55], combined both these observations using the Xenopus oöcyte expression system, to show that the suppressed secretion of the Siiyama mutant (Ser53Phe) was restored to normal by a second stabilising (Phe51Leu) mutation (Fig. 6). It is clear that the triggering of the conformational change is dependent on tightly defined packing in the shutter domain. As Yu and colleagues point out[56] mutations that ease the density of packing favour a five-stranded, and hence stable, conformation. The reverse provides an explanation for the conformationally labile dysfunctional variants. A good example is the strongly conserved alanine 336 in strand 5A[12]. When this shutter resi-

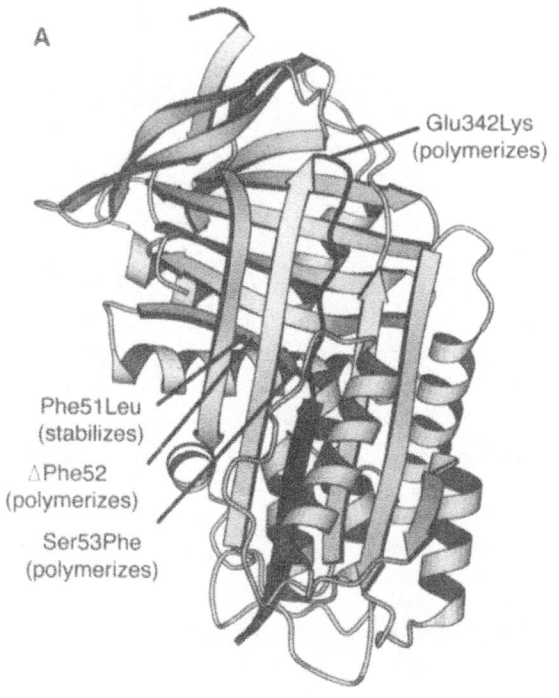

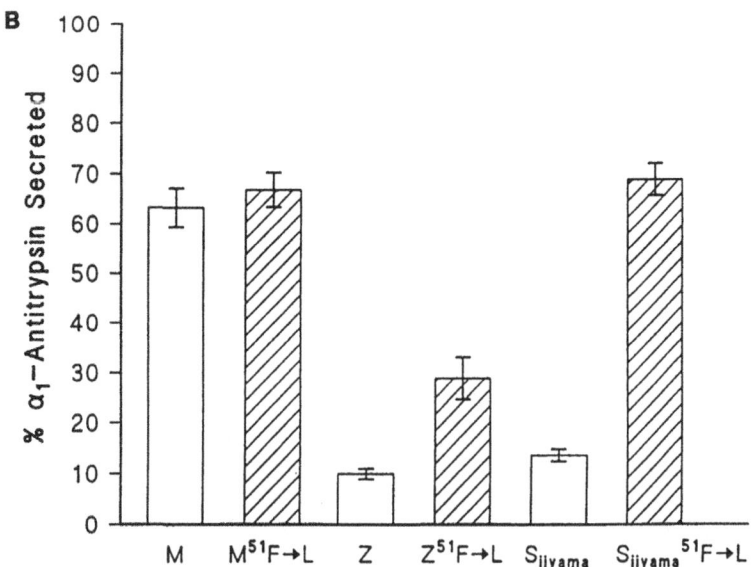

Figure 6. (A) cleaved α_1-antitrypsin illustrating mutations that cause polymerisation. The key role of the steric packing of the shutter domain (Fig. 5) is indicated by the stabilising mutation at Phe 51, adjacent to the destabilising mutations at Phe52 and Ser53 at the junction of S6B and helix B. (Reproduced with permission form Ref 10 Nature Structural Biol 1966) (B) the recombinant combination of the 52Leu and 53Phe mutations restores secretion to normal in a Xenopus oöcyte model. (Reproduced with permission from Ref 55 J. Biol. Chem. 1995.)

due is replaced by the larger threonine in an α_1-antitrypsin variant[57] it results in lability and consequent emphysema. In the naturally labile serpin PAI-1, position 336 is occupied by the even larger valine and, as predicted by this model, its recombinant replacement by alanine results in stabilisation of PAI-1 as an active inhibitor[58].

The formation of polymers is just one indication of the consequences of conformational lability. A better example of the more subtle changes involved was shown by an unstable mutant of antithrombin identified by Dr. Jean-Yvonne Borg in a family with a history of severe episodic thromboses[59]. The underlying mutation caused a marginal change in the conformational lability of the variant antithrombin such that it slowly converted in the circulation to the six-stranded L-(latent) form. However, this conformational transition was greatly accelerated by even the minor increases in temperature that occur in fever, to give a rapid loss of inhibitory activity with polymer formation[41]. The clinical consequences explain the severe episodic nature of thrombosis in affected members of this family which occurred whenever they had incidental infections resulting in fever.

Although the study of the conformational changes that occurs in variant serpins associated with disease has highlighted the formation of polymers, the consequences of inappropriate transitional changes are manifold. They may also result in misfolding of the newly-expressed nascent protein and hence in a complete or near-complete failure in secretion. The wider spectrum of phenotypes has been demonstrated by others in cell-culture models, notably by Eldering, Tosi and colleagues[60] with mutants of C1-inhibitor and Brodbeck, Sifers and their colleagues[61,62] with α_1-antitrypsin. Although the consequences of these mutations varies from total non-expression to massive intravascular polymerisation the position and nature of the underlying mutations make it clear that they all share the common defect of conformational lability. The seal to this conclusion has been the subsequent series of hyperstable recombinant mutants, identified by random selection, in PAI-1 from the laboratory of Lawrence and Ginsberg[58], and in α_1-antitrypsin from the Seoul group[56]. These recombinant mutations that result in increased stability not only map to the same domains as the natural labile mutants but in several instances, as previously exampled, involve precisely reciprocal mutations[17].

CONCLUSION

The combination of structural determinations together with studies of natural mutants has opened our understanding of the inherent molecular mobility of the serpins. It is true that "one structure is worth a thousand speculations" but a structure just by itself is static and lifeless. To bring it to life requires accompanying functional studies often involving painstaking biochemical and kinetic measurements. A complementary approach, as we have illustrated here, is through the study of the natural mutants that cause dysfunction. But it needs to be emphasised that the key natural mutants which have really opened our knowledge represent the distillate of thousands of hours of work. Our own laboratory has contributed just a proportion of these variants and yet we have screened 3000 individuals with a family history of thrombotic disease to identify 200 with abnormalities of antithrombin only a small proportion of which directly contribute to new understandings. An ancillary advantage of such studies is the information they also provide as to the associated disease processes and the steps that can be taken to alleviate them. The greatest advantage however, of accompanying structural studies with studies of dysfunctional variants, is not only that natural variants provide an answer to the questions we pose on structure-function relationships but they also provide answers to questions we wished we had had the insight to ask.

ACKNOWLEDGMENTS

Supported by the Wellcome Trust, The Lister Institute, The British Heart Foundation and European Community Grant BMH1-LT-13-1592.

REFERENCES

1. Laurell, C.-B. and Eriksson, S. (1963). The electrophoretic α_1-globulin pattern of serum in α_1-antitrypsin deficiency. Scand. J. Clin. Lab. Invest. 15: 132–140.
2. Petersen, E.E., Dudek-Wojciechowska, G., Sottrup-Jensen, L. and Magnusson, S. (1979). The primary structure of antithrombin III (heparin cofactor). Partial homology between α_1-antitrypsin and antithrombin III. In eds Collen: The physiological inhibitors of coagulation and fibrinolysis, Elsevier — North Holland Biomedical Press Amsterdam, 43–54.
3. Carrell, R., Owen, M., Brennan, S. and Vaughan, L. (1979). Carboxy terminal fragment of human α_1-antitrypsin from hydroxylamine cleavage: homology with antithrombin III. Biochem. Biophys. Res. Commun. 91: 1031–1037.
4. Boswell, D.R., Owen, M.C., Brennan, S.O., Carrell, R.W. and McLachlan, A.D. (1980). Ligand-binding properties of proalbumin Christchurch. Biochem. J. 191: 281–283.
5. Johnson, D. and Travis, J. (1978). Structural evidence for methionine at the reactive site of human α-1-proteinase inhibitor. J. Biol. Chem. 253: 7142–7144.
6. Martodam, R.R. and Liener, I.E. (1981). The interaction of α_1-antitrypsin with trypsin, chymotrypsin and human leukocyte elastase as revealed by end group analysis. Biochim. Biophys. Acta 667: 328–340.
7. Lewis, J.H., Iammarino, R.M., Spero, J.A. and Hasiba, U. (1978). Antithrombin Pittsburgh: An α_1-antitrypsin variant causing hemorrhagic disease. Blood 51: 129–137.
8. Owen, M.C., Brennan, S.O., Lewis, J.H. and Carrell, R.W. (1983). Mutation of antitrypsin to antithrombin. α_1-antitrypsin Pittsburgh (358 Met to Arg), a fatal bleeding disorder. N. Eng. J. Med. 309: 694–698.
9. Bathurst, I.C., Brennan, S.O., Carrell, R.W., Cousens, L.C., Brake, A.J. and Barr, P.J. (1986). Yeast KEX2 protease meets unique requirements for human proalbumin converting enzyme. Science 235: 348–350.
10. Elliott, P.R., Lomas, D.A., Carrell, R.W. and Abrahams, J.P. (1996). Inhibitory conformation of the reactive loop of α_1-antitrypsin. Nature Struct. Biol. 3: 676–681.
11. Carrell, R.W., Christey, P.B. and Boswell, D.R. (1987). Serpins: antithrombin and other inhibitors of coagulation and fibrinolysis; evidence from amino acid sequences. In eds Verstraets, Vermylen, Lijnen and Arnout: Thrombosis and Haemostasis Leuven University Press, 1–15.
12. Huber, R. and Carrell, R.W. (1989). Implications of the three-dimensional structure of α_1-antitrypsin for structure and function of serpins. Biochemistry 28: 8951–8966.
13. Koide, T., Odani, S., Takahashi, K., Onon, T. and Sakuragawa, N. (1984). Antithrombin III Toyama: replacement of Arginine 47 by Cysteine in hereditary abnormal antithrombin III that lacks heparin binding ability. Proc. Nat. Acad. Sci. USA 81: 289–293.
14. Peterson, C.B., Noyes, C.M., Pecon, J.M., Church, F.C. and Blackburn, M.N. (1987). Identification of a lysyl residue in antithrombin which is essential for heparin binding. J. Biol. Chem. 262: 8061–8065.
15. Borg, J.Y., Owen, M.C., Soria, C., Soria, J., Caen, J. and Carrell, R.W. (1988). Proposed heparin binding site in antithrombin based on Arginine 47. A new variant Rouen-II, 47 Arg to His. J. Clin. Invest. 81: 1292–1296.
16. Stein, P.E. and Carrell, R.W. (1995). What do dysfunctional serpins tell us about molecular mobility and disease? Nature Struct. Biol. 2: 96–113.
17. Carrell, R.W. and Stein, P.E. (1996). The biostructural pathology of the serpins: Critical function of sheet opening mechanism. Biological Chemistry Hoppe-Seyler 377: 1–17.
18. Loebermann, H., Tokuoka, R., Deisenhofer, J. and Huber, R. (1984). Human α_1-proteinase inhibitor. Crystal structure analysis of two crystal modifications, molecular model and preliminary analysis of the implications for function. J. Mol. Biol. 177: 531–556.
19. Carrell, R.W. and Owen, M.C. (1985). Plakalbumin, α_1-antitrypsin, antithrombin and the mechanism of inflammatory thrombosis. Nature 317: 730–732.
20. Gettins, P. (1989). Absence of large scale proteolytic change upon limited proteolysis of ovalbumin, the prototypic serpin. J. Biol. Chem. 264: 3781–3785.
21. Stein, P.E., Tewkesbury, D.A. and Carrell, R.W. (1989). Ovalbumin and angiotensinogen lack serpin S-R conformational change. Biochem. J. 262: 103–107.

22. Pemberton, P.A., Stein, P.E., Pepys, M.B., Potter, J.M. and Carrell, R.W. (1988). Hormone binding globulins undergo serpin conformational change in inflammation. Nature 336: 257–258.
23. Edgar, P.F. (1989). The structure and function of corticosteroid binding globulin.
24. Wilczynska, M., Fa, M., Ohlsson, P.-I. and Ny, T. (1995). The inhibition mechanism of serpins. Evidence that the mobile reactive centre loop is cleaved in the native protease-inhibitor complex. J. Biol. Chem. 270: 29652–29655.
25. Shore, J.D., Day, D.E., Francis-Chmura, A.M., Verhamme, I., Kvassman, J., Lawrence, D.A. and Ginsburg, D. (1995). A fluorescent probe study of plasminogen activator inhibitor-1: Evidence for reactive center loop insertion and its role in the inhibitory mechanism. J. Biol. Chem. 270: 5395–5398.
26. Engh, R., Huber, R., Bode, W. and Schulze, A. (1995). Divining the serpin inhibition mechanism : a suicide substrate 'springe'? Trends in Biotech. 13: 503–510.
27. Devraj-Kizuk, R., Chui, O.H.K., Prochownik, E.V., Carter, C.J., Ofosu, F.A. and Blajchman, M.A. (1988). Antithrombin III-Hamilton : a gene with a point mutation (guanine to adenine) in codon 382 causing impaired serine protease reactivity. Blood 72: 1518–1523.
28. Mohlo-Sabatier, P., Aiach, M., Gaillard, I., Fiessinger, J.J., Fischer, A.M., Chadeuf, G. and Clauser, E. (1989). Molecular characterisation of antithrombin III variants using polymerase chain reaction: identification of the ATIII Charleville as an Ala384Pro mutation. Journal of Clinical Investigation 84: 1236–1241.
29. Perry, D.J., Harper, P.L., Fairham, S., Daly, M. and Carrell, R.W. (1989). Antithrombin Cambridge, 384 Ala to Pro: a new variant identified using the polymerase chain reaction. FEBS Letts. 254: 174–176.
30. Perry, D.J., Daly, M., Harper, P.L., Tait, R.C., Price, J., Walker, I.D. and Carrell, R.W. (1991). Antithrombin Cambridge II, 384 Ala to Ser. Further evidence of the role of the reactive centre loop in the inhibitory function of the serpins. FEBS Letts. 285: 248–250.
31. Skriver, K., Wikoff, W.R., Patston, P.A., Tausk, F., Schapira, M., Kaplan, A.P. and Bock, S.C. (1991). Substrate properties of C1 inhibitor Ma (alanine 434→glutamic acid). J. Biol. Chem. 266: 9216–9221.
32. Davis III, A.E., Aulak, K., Parad, R.B., Stecklein, H.P., Eldering, E., Hack, C.E., Kramer, J., Strunk, R.C., Bissler, J. and Rosen, F.S. (1992). C1 inhibitor hinge region mutations produce dysfunction by different mechanisms. Nature Genetics 1: 354–358.
33. Carrell, R.W., Evans, D.L. and Stein, P.E. (1991). Mobile reactive centre of serpins and the control of thrombosis. Nature 353: 576–578.
34. Carrell, R.W. and Perry, D. (1996). The unhinged antithrombins. Brit. J. Haemat. 93: 253–257.
35. Schulze, A.J., Baumann, U., Knof, S., Jaeger, E., Huber, R. and Laurell, C.-B. (1990). Structural transition of α₁-antitrypsin by a peptide sequentially similar to β-strand s4A. Eur. J. Biochem. 194: 51–56.
36. Stein, P.E., Leslie, A.G.W., Finch, J.T., Turnell, W.G., McLaughlin, P.J. and Carrell, R.W. (1990). Crystal structure of ovalbumin as a model for the reactive centre of serpins. Nature 347: 99–102.
37. Hekman, C.M. and Loskutoff, D.J. (1985). Endothelial cells produce a latent inhibitor of plasminogen activators that can be activated by denaturants. J. Biol. Chem. 260: 11581–11587.
38. Sancho, E., Declerck, P.J., Price, C., Kelly, S.M. and Booth, N.A. (1995). Conformational studies on plasminogen activator inhibitor (PAI-1) in active, latent, substrate and cleaved forms. Biochemistry 34: 1064–1069.
39. Katagiri, K., Okada, K., Hattori, H. and Yano, M. (1988). Bovine endothelial cell plasminogen activator inhibitor. Purification and heat activation. Eur. J. Biochem. 176: 81–87.
40. Carrell, R.W., Evans, D.L. and Stein, P.E. (1993). Mobile reactive centre of serpins and the control of thrombosis (correction). Nature 364: 737.
41. Bruce, D., Perry, D.J., Borg, J.-Y., Carrell, R.W. and Wardell, M.R. (1994). Thromboembolic disease due to thermolabile conformational changes of antithrombin Rouen VI (187 Asn→Asp). J. Clin. Invest. 94: 2265–2274.
42. Lomas, D.A., Elliott, P.R., Chang, W.-S.W., Wardell, M.R. and Carrell, R.W. (1995). Preparation and characterisation of latent α₁-antitrypsin. J. Biol. Chem. 270: 5282–5288.
43. Busby, T.F., Atha, D.H. and Ingham, K.C. (1981). Thermal denaturation of antithrombin III. Stabilization by heparin and lyotropic anions. J. Biol. Chem. 256: 12140–12147.
44. Mast, A.E., Enghild, J.J. and Salvesen, G. (1992). Conformation of the reactive site loop of α₁-proteinase inhibitor probed by limited proteolysis. Biochemistry 31: 2720–2728.
45. Lomas, D.A., Evans, D.L., Finch, J.T. and Carrell, R.W. (1992). The mechanism of Z α₁-antitrypsin accumulation in the liver. Nature 357: 605–607.
46. Huntington, J.A., Patson, P.A. and Gettins, P.G.W. (1995). S-ovalbumin, an ovalbumin conformer with properties analogous to those of loop-inserted serpins. Protein Science 4: 613–621.
47. Mottonen, J., Strand, A., Symersky, J., Sweet, R.M., Danley, D.E., Geoghegan, K.F., Gerard, R.D. and Goldsmith, E.J. (1992). Structural basis of latency in plasminogen activator inhibitor-1. Nature 355: 270–273.
48. Carrell, R.W., Stein, P.E., Fermi, G. and Wardell, M.R. (1994). Biological implications of a 3Å structure of dimeric antithrombin. Structure 2: 257–270.

49. Schreuder, H.A., de Boer, B., Dijkema, R., Mulders, J., Theunissen, H.J.M., Grootenhuis, P.D.J. and Hol, W.G.J. (1994). The intact and cleaved human antithrombin III complex as a model for serpin-proteinase interactions. Nature Struct. Biol. 1: 48–54.

50. Wei, A., Rubin, H., Cooperman, B.S. and Christianson, D.W. (1994). Crystal structure of an uncleaved serpin reveals the conformation of an inhibitory loop. Nature Struct. Biol. 1: 251–258.

51. Van Boeckel, C.A.A., Grootenhuis, P.D.J. and Visser, A. (1994). A mechanism for heparin-induced potentiation of antithrombin III. Nature Struct. Biol. 1: 423–425.

52. Janciauskiene, S., Eriksson, S. and Wright, H.T. (1996). A specific structural interaction of Alzheimer's peptide Aβ1-42 with α_1-antichymotrypsin stimulates amyloid fibril formation. Nature Struct. Biol., in press.

53. Lomas, D.A., Finch, J.T., Seyama, K., Nukiwa, T. and Carrell, R.W. (1993). α_1-antitrypsin S_{iiyama} (Ser^{53}→Phe); further evidence for intracellular loop-sheet polymerisation. J. Biol. Chem. 268: 15333–15335.

54. Kwon, K.-S., Kim, J., Shin, H.S. and Yu, M.-H. (1994). Single amino acid substitutions of α_1-antitrypsin that confer enhancement in thermal stability. J. Biol. Chem. 269: 9627–9631.

55. Sidhar, S.K., Lomas, D.A., Carrell, R.W. and Foreman, R.C. (1995). Mutations which impede loop/sheet polymerisation enhance the secretion of human α_1-antitrypsin deficiency variants. J. Biol. Chem. 270: 8393–8396.

56. Lee, K.N., Park, S.D. and Yu, M.-H. (1996). Probing the native strain in α_1-antitrypsin. Nature Struct. Biol. 3: 497–500.

57. Holmes, M.D., Brantly, M.L., Fells, G.A. and Crystal, R.G. (1990). α_1-antitrypsin $W_{bethesda}$: molecular basis of an unusual α_1-antitrypsin deficiency variant. Biochem. Biophys. Res. Commun. 170: 1013–1020.

58. Berkenpas, M.B., Lawrence, D.A. and Ginsburg, D. (1995). Molecular evolution of plasminogen activator inhibitor-1 functional stability. EMBO J. 14: 2969–2977.

59. Perry, D.J., Marshall, C., Borg, J.-Y., Tait, R.C., Daly, M.E., Walker, I.D. and Carrell, R.W. (1995). Two novel antithrombin variants, Asn187Asp and Asn187Lys, indicate a functional role for asparagine 187. Blood Coagulation and Fibrinolysis 6: 51–54.

60. Eldering, E., Verpy, E., Roem, D., Meo, T. and Tosi, M. (1995). COOH-terminal substitutions in the serpin C1 inhibitor that cause loop overinsertion and subsequent multimerization. J. Biol. Chem. 270: 2579–2587.

61. Brodbeck, R.M. and Brown, J.L. (1992). Secretion of α-1-proteinase inhibitor requires an almost full length molecule. J. Biol. Chem. 267: 294–297.

62. Sifers, R.N., Hardick, C.P. and Woo, S.L.C. (1989). Disruption of the 290–342 salt bridge is not responsible for the secretory defect of the PiZ α_1-antitrypsin variant. J. Biol. Chem. 264: 2997–3001.

21

STRUCTURE-FUNCTION STUDIES ON PEDF

A Noninhibitory Serpin with Neurotrophic Activity

S. Patricia Becerra

Laboratory of Retinal Cell and Molecular Biology
National Eye Institute
National Institutes of Health
Building 6, room 308
6 Center Dr. MSC 2740
Bethesda, Maryland 20892-2740

Pigment epithelium-derived factor (PEDF) is a neurotrophic factor. *In vitro* it induces neuronal differentiation of retinoblastoma cells and promotes survival of cerebellar granule cell neurons. By virtue of amino acid sequence homology PEDF is a unique serpin member.

1. DISCOVERY OF PEDF

In the vertebrate eye the interphotoreceptor matrix (IPM) is the extracellular compartment between the retinal pigment epithelium (RPE) and the neural retina. The interactions between the RPE and the retina play an important role in the visual system. The biochemical components of the IPM contribute to these RPE-retina interactions. The idea that the RPE releases biochemical signals which aid in the development, differentiation, proliferation, function and survival of the neural retina is supported by studies on the effect of products secreted by cultured RPE on cells of retinal origin (1–5). It was initially shown that the conditioned media of human fetal RPE cells induced neuronal differentiation in a retinoblastoma Y-79 cell culture system (6). The Y-79 cell line is an established cell line from a human retinoblastoma tumor (7) and thought to be derived from primitive multipotential retinoblasts (8). To identify the specific factor secreted by RPE and responsible for such neurotrophic activity, proteins in RPE conditioned media were isolated after fractionation by SDS-polyacrylamide gel electrophoresis (9). One isolated protein of ~50,000-MW induced differentiation of retinoblastoma Y-79 cells when added at nanomolar levels to the culture media. Treated cells exhibited both morphological and immunochemical characteristics of neurons. Thus, the RPE-secreted product with potent neuronal differentiating activity on cells of retinal origin, was termed pigment epithelium-derived factor (PEDF).

Chemistry and Biology of Serpins, edited by Church *et al.*
Plenum Press. New York. 1997

223

```
                                                          ^^^^^^^^^^^^
AT       mpssvswgillaglcclvpvslaEDPQGDAAQKTDTSHHDQD--HPTFNKITPNLAE
huPEDF   mqalvlllcigallghsscqNPASPPEEGSPDPDSTGALVEEEDPFFKVPVNKLAAAVSN    60
bovPEDF       wt       fgr   AG--Q A   LT E      P                      58


         ^hA^^^^^^^^^^    -s6B^^^^^^hB^^^^^^^^^^^^^^^hC^^^^^^ ^hC1^^^^^^hD
AT       FAFSLYRQLAHQSNSTNIFFSPVSIATAFAMLSLGTKADTHDEILEGLNFNLTEIPEAQIHE
huPEDF   FGYDLYRVRSSMSPTTNVLLSPLSVATALSALSLGAEQRTESIIHRALYYDL--ISSPDIHG   120
bovPEDF        GE   A                                N              N     118


         ^^^^^^^^^^^    ----s2A------    ^^^^^^hE^^^^^--s1A--  ^^^^^^hF^
AT       GFQELLRTLNQPDSQLQLTTDGGLFLSEGLKLVDKFLEDVKKLYHSEAFTVNFGDTEEAKKQI
huPEDF   TYKELLDTVTAPQKNLK-SASRIVF-EKKLRIKSSFVAPLEKSYGTRPRVLT-GNPRLDLQEI   180
bovPEDF        D AS           I   R    A IP          I    S V            178


         ^^^^^^^^^^        -------s3A----     ^hF1---s4C-------s
AT       NDYVEKGTQGKIVDLVKELDRDTVFALVNYIFFKGKWERPFEVKDTEEEDFHVDQVTTVK
huPEDF   NNWVQAQMKGKLARSTKEIPDEISILLLGVAHFKGQWTKFDSRKTSLEDFYLDEERTVR   240
bovPEDF                 V  RVS     F     Y                      K      238


         3C------ ------^hF2^^----s2B------s3B--   ^^hG^^  ^^^^^hH
AT       VPMMKRLG-MFNIQHCKKLSSWVLLMKYLGNANAIFFLPDE--GKLQHLENELTHDIITK
huPEDF   VPMMSDPKAVLRYGLDSDLSCKIAQLPLTGSMSIIFFLPLKVTQNLTLIEESLTSEFIHD   300
bovPEDF        Q              Q       N        T         Q              298


         ^^^^    --s2C-----s6A---^^^hI^^^  ^hI1              -------s5A-
AT       FLENEDRRSASLHLPKLSITGTYDLKSVLGQLGITKVFSNGADLSGVTEEAPLKLSKAVHKA
huPEDF   IDRELKTVQAVLTVPKLKLSYEGEVTKSLQEMKLQSLF-DSPDFSKITGK-PIKLTQVEHRA   360
bovPEDF        I              L    V  L        A                   V      358


         ---------s4A-------- *  --s1C- ---s4B-^hI2^---s5B---^hI3
AT       VLTIDEKGTEAAGAMFLEAIPMSIPPEVKFNKPFVFLMIEQNTKSPLFMGKVVNPTQK
huPEDF   GFEWNEDGAGTTPSPGLQPAHLTFPLDYHLNQPFIFVLRDTDTGALLFIGKILDPRGP   418
bovPEDF               NS   V   R                             I           416
```

Figure 1. Alignment of amino acid sequences of PEDF and the secondary structural elements of the α_1-antitrypsin sequence. Structural elements and sequence of α_1-antitrypsin are indicated in the first and second lines (13). α-helix ^, β strand —. Conserved residues important in maintaining structural integrity are underlined. Deduced human PEDF precursor (10) is indicated in full, deduced bovine PEDF precursor (20) is given only where it differs from the human PEDF. Numbers at the right indicate PEDF positions of last residue in each line. Amino acids are given in single letter code, and mature polypeptide are given in upper case. Note that the predicted signal peptide region is 17 amino acids (10) however amino terminal residues of the mature proteins start at position 20 (17, 21). Position P1 is indicated by an asterisk leucine 382, consensus glycosylation site is double underlined.

2. SEQUENCE COMPARISON IDENTIFIES PEDF AS A SERPIN

Isolation and characterization of cDNA clones from two independent human libraries allowed the deduction of the primary structure of the PEDF protein. One PEDF cDNA was obtained from a fetal human eye library (10). Its isolation was based on the selection of sequences complementary to synthetic oligonucleotides generated from PEDF tryptic peptide sequences. This cDNA (πFS17) is 1.5kb in length and contains the initiator methionine codon and a polyadenylation signal. Its deduced polypeptide predicts a native PEDF precursor of 418 amino acid residues (Fig. 1). This predicted sequence contains a secretion signal region at its N-terminal end and a consensus glycosylation site (Asn-Leu-Thr) at residue positions 285–287. The deduced PEDF sequence is unique among those in Genbank databases. It shares primary sequence homology (about 30%) with members of the serpin superfamily. The other cDNA sequence was obtained from a subtraction library of a fetal human diploid fibroblast-like cell line WI-38 (11). Isolation of this cDNA was based on the difference in gene expression between nonproliferating (young) and late (senescent) population doubling level (PDL) WI-38 cells, and thus termed EPC-1 (early PDL cDNA-1). The deduced amino acid sequence from WI-38 cells (Accession No. M90439) was aligned and compared with that from the eye library (Accession No. M76979). The PEDF sequence in the WI-38 derived cDNA shares identity with the one derived from the eye, πFS17, but starts at Asn-60.

To demonstrate the close homology of PEDF to the serpin family, an automated assembly of protein blocks for data searching (12) was performed on human PEDF sequence. The PEDF sequence shows five scoring peptide blocks that align with more than 50 serpin sequences (Fig. 2). In addition, its linear sequence contains small insertions and deletions that are compatible with the three-dimensional serpin model, i.e., α_1-antitrypsin. The PEDF sequence has 40 of the 51 conserved serpin residues (see Fig. 1). These serpin residues are considered critical for maintaining the serpin folded structure. Table 1 lists the conserved amino acids absent in the PEDF sequence. Five of these are unique to PEDF when compared to serpin members known to date. Interestingly, all of these positions are located in close proximity in the serpin spatial models, e.g., antitrypsin, ovalbumin (13, 14). The most significant changes are in conserved positions Y297 and N186 that in PEDF are glycine residues (G323 and G209). Both positions in the serpin prototype are in sheet A (s6A and s3A) and interact with sheet B. The homologous serpin reactive site P1 is occupied by leucine. Thus, PEDF has a reactive site specific for a leucyl-enteropeptidase such as chymotrypsin-like enzymes. Altogether these observations indicate that PEDF has a strong relationship to the serpin family.

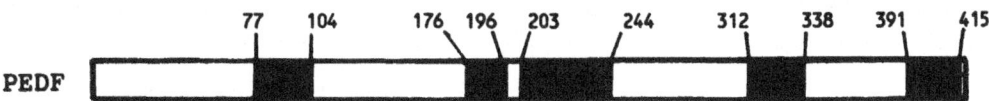

Figure 2. Serpin primary sequence homology for PEDF. Peptide segments of PEDF with high homology to members of the serpin family are shown in closed boxes. To aid in the detection and verification of protein sequence homology, the PEDF amino acid sequence was compared to database of protein blocks by an automated assembly program (12). This program is based on PROSITE 10 and SWISS_PROT-24. The alignment of the PEDF sequence with the closest to it in the BLOCKS database resulted in five segments all of serpin sequences. The score for each block was the highest given by the program, 100.00th percentile. The locations of these blocks are shown in closed boxes. Numbers above them are amino acid position.

Table 1. Comparison of PEDF sequence to conserved amino acids
in the serpin family necessary for structural integrity[a]

Structure in serpin[b]	Conserved residue[b,c]	Amino acid in PEDF[c]	Other serpins with amino acid as in PEDF
Helix B	I 57	V 85	Angiotensinogen, antiplasmin, plasminogen activator inhibitor
Helix F	T 165	M 188	None
—	I 169	L 192	None
Strand 3A	N 186	G 209	None
Strand 3A	I 188	A 211	None
Strand 6A	Y 297	G 323	None
—	L 318	F 343	Cortisol binding globulin, plasminogen activator inhibitor, RAB ORF1, tyroxine binding globulin
—	G 320	K 345	Angiotensinogen, protease nexin-1
Strand 4B	V/L 371	I 395	Cortisol binding globulin
Strand 5B	V 388	I 411	Protease nexin-1, RAB ORF1
Strand 5B	N 390	D 414	C1 inhibitor

[a]A total of 51 conserved amino acids, all other conserved amino acids are in PEDF.
[b]Structure and residues positions of human A1-antitrypsin as in reference 13.
[c]Amino acid position is from the cDNA-derived precursor polypeptide sequence (10).

3. PRODUCTION AND PROPERTIES OF RECOMBINANT PEDF

First, the PEDF cDNA in πFS17 was tested *in vitro* for production of PEDF protein. Synthetic transcripts were prepared from the full-length PEDF cDNA and tested for translation in cell-free extracts. As shown in Figure 3A, the PEDF mRNA translates a polypeptide of 46,000-MW, the expected size for the PEDF coding region. Then this cDNA was used to produce PEDF in Escherichia coli. Bacterial expression vectors were prepared

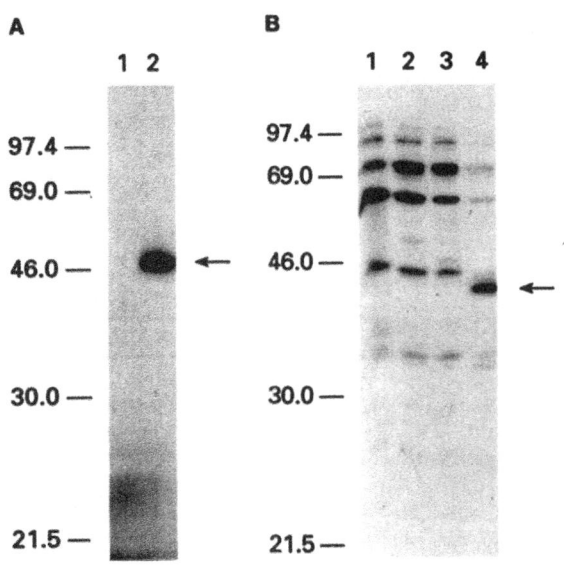

Figure 3. Recombinant PEDF. Proteins were synthesized in the presence of L-[^{35}S] methionine and then analyzed by SDS-polyacrylamide gel electrophoresis followed by autoradiography. Panel A. Cell-free translation of PEDF. PEDF mRNA was transcribed *in vitro* from πFS17 cDNA using T7 RNA polymerase and a Cap analog. Rabbit reticulocyte lysates were programmed with: lane 1, no mRNA; and lane 2, PEDF mRNA (23 µg/ml). Panel B. Bacterially-derived PEDF. PEDF bacterial expression vectors were prepared from πFS17 cDNA. Expression was induced in *E. coli* cells containing: lane 1, parent plasmid pEV-vrf1; lane 2, full length PEDF vector pRC-SH; lane, full length PEDF vector pEV-SH; and lane 4, truncated (44–418) PEDF vector pEV-BH. numbers to the left of each panel indicate migration of molecular weight markers given in kDa. The arrows point to the migration of recombinant protein.

with the full length open-reading frame Met^1-Pro^{418}, and truncated form Asp^{44}-Pro^{418} (15). Another bacterial vector with a truncated PEDF (Leu^{89}-Pro^{418}) has been independently constructed from the WI-38 derived cDNA (11). Bacterial cells bearing vectors with the full length coding region did not produce significant amounts of recombinant protein. However, transfectants with the truncated region overexpressed recombinant PEDF protein (Fig. 3B). These observations suggest that removal of the first 43 residues of PEDF enhance the production of stable recombinant protein in *E. coli*. The rPEDF (44–418) is produced at 1.3 mg per g of wet cells. Although it accumulates in inclusion bodies, rPEDF is refolded in buffers containing >2M urea. After solubilization, the rPEDF is purified by gel filtration and cation-exchange chromatography. The N-terminal sequence and size match that of the Asp^{44}-Pro^{418} expression vector. Thus this rPEDF does not contain the signal peptide region and is truncated by 23 residues from the mature protein. Note that rPEDF (44–418) contains residues at positions that aligned with alpha helix A of the serpin prototype antitrypsin (see Fig. 1). The truncated rPEDF contains one cysteine residue of the three deduced for the PEDF sequence (Cys-9, 19 and 261). Interactions based on disulfide bonds between these derived cysteine positions might play a role in solubility of the full length PEDF. Since *E. coli* does not perform a number of eukaryotic post-translational modifications, rPEDF is not acetylated, amidated or glycosylated. Thus, this bacterial expression system provides a more accessible source of human PEDF protein and with higher yields than from the fetal human RPE conditioned media.

4. RECOMBINANT PEDF IN NEUROTROPHIC STUDIES

4.1. Neuronal Differentiating Activity

The purified rPEDF (44-418) shows neurotrophic activity *in vitro* using the retinoblastoma cell system. In this assay Y-79 cells are grown in suspension at low density (15). Cells are then exposed to the rPEDF in media without serum and, after a week, attached to a poly-D-lysine substratum. Neurotrophic activity is assessed daily by phase contrast microscopy. Only Y-79 cell cultures exposed to rPEDF show significant evidence of neuronal differentiation as observed for native PEDF (Fig. 4). Untreated cells or cells treated only with the same buffer used for rPEDF do not show differentiation. Cells treated with 50 ng/ml of rPEDF show that between 50–65% of the cell aggregates have neurite-like extensions by day 3 post-attachment. The number of differentiating aggregates, the number of differentiating cells per aggregate, and the length of the neurite-like processes increase with post-attachment time. By day 5 post-attachment, 75–85% of the aggregates show signs of differentiation with neurites extending from most of their peripheral cells. rPEDF treated cells reach the maximal extend of morphological differentiation by day 7 post-attachment. These had 85–98% of cell aggregates showing abundant neurite-like extensions. At this point, two types of neuronal processes are observed: (i) single neurites, 2–3 fold longer than those observed on day 3, extending from peripheral cells of individual aggregates and (ii) much longer and thinner processes forming a branching network between neighboring cell aggregates. Beyond 10 days post-attachment, there is a decrease in the proportion of the network connections and no further growth of the single neurites. Viability of the cell aggregates is not markedly affected by rPEDF addition; it remains at 75–80%. Additions as low as 1 ng rPEDF per ml of Y-79 cell culture medium promote significant differentiation (40–60% at 7 days post-attachment). This differentiating activity has been tested in another retinoblastoma cell line, the human reti-

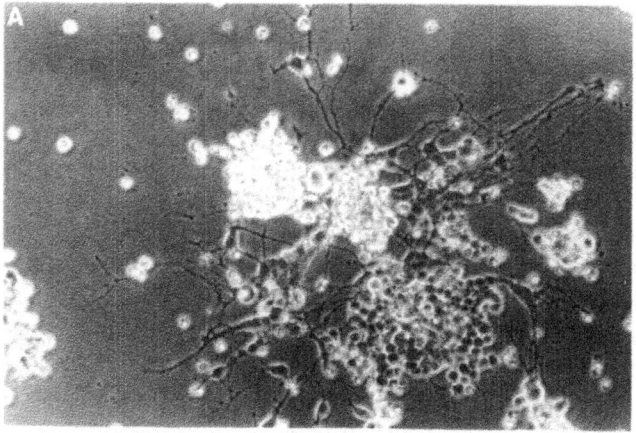

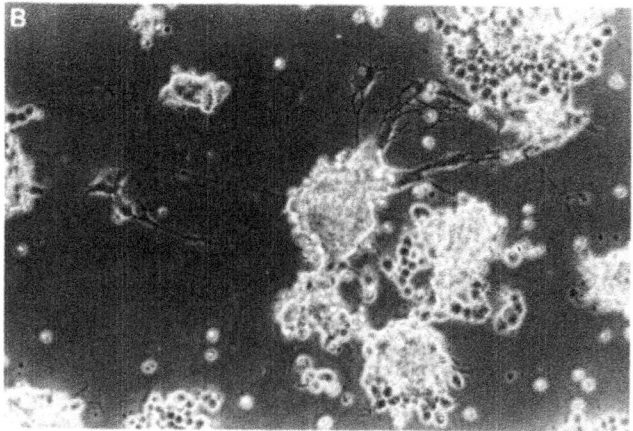

Figure 4. Neurotrophic activity of rPEDF on human retinoblastoma Y-79 cells. PEDF was added in cultures of Y-79 cells growing in suspension and without serum (15). Additions were panel A, 50 ng rPEDF per ml culture; and panel B, 100 ng native bovine PEDF per ml culture. After attachment on poly-D-lysine coated plates, morphological differentiation was monitored daily. Photographs are shown from cells on day 9 post attachment.

noblastoma Weri cell line (Fig. 5). rPEDF also induces morphological differentiation on Weri cells that resembles that of the Y-79 cells. Additions of rPEDF to the cell culture induce the formation of neurite-like processes that extend from Weri cells. The treated cells survive more than the untreated counterparts. Differentiation of rPEDF treated Weri cells is observed at 11 days post-attachment, a longer period when compared to rPEDF treated Y-79 cells.

4.2. Neuronal Survival Activity

The rPEDF also demonstrates a neurotrophic survival effect on cerebellar granule cells in culture. The neurotrophic effects of PEDF on neurons other than those of retinal derivation was investigated. A cerebellar granule cell (CGC) culture system was established for examining rPEDF (16). rPEDF increases the CGC cell number in a dose-dependent fashion as demonstrated by a viable cell counting assay, direct cell count and immunocytochemistry. A greater effect is observed in chemically defined medium than in serum containing medium, with ED_{50} in chemically defined and serum-containing medium of 30 ng/ml and 100 ng/ml respectively (Fig. 6). PEDF does not have an effect on cell proliferation or on neurite outgrowth. Thus, unfolded and refolded *in vitro* truncated rPEDF retains potent neurotrophic activity.

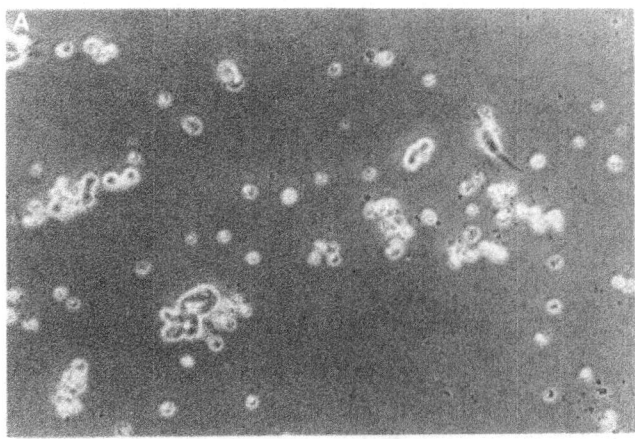

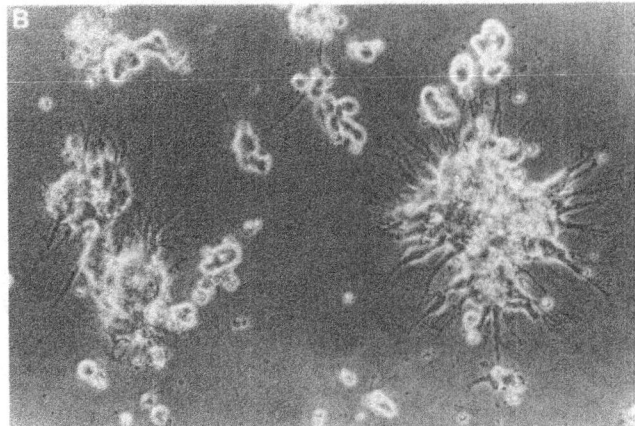

Figure 5. Neurotrophic assay of rPEDF on Weri cells. Cells were treated as for Y-79 (15). Additions were as indicated: panel A, PBS; and panel B, 50 ng rPEDF per ml culture. Photographs are from day 11 post attachment.

5. ANTI-PEDF ANTIBODIES AND BLOCKING ACTIVITY

Purified rPEDF has been used to develop a rabbit polyclonal antiserum, Ab-rPEDF, which immunoreacts in a specific, sensitive, and linear fashion with PEDF protein (17). In western transfers of rPEDF (0.5 μg) the immunoreactive signal is detected even at antiserum Ab-rPEDF dilutions as low as 1:50,000. This detection is abolished when the antiserum is preincubated with an excess of rPEDF antigen. On slot-blot transfers of increasing amounts of rPEDF protein (1–20 ng) a clear immunoreactive signal is detected with antiserum Ab-rPEDF but not with preimmune serum. Furthermore, Ab-rPEDF antiserum blocks the PEDF-mediated neurotrophic activities. As shown in figure 7, Y-79 cells treated with a mixture of PEDF and antiserum Ab-rPEDF do not exhibit neuronal differentiation. These cells show similar characteristics to those induced with mixtures in which PEDF and antiserum had been omitted.

6. IDENTIFICATION OF PEDF IN THE VERTEBRATE EYE

The presence of PEDF protein in the native eye has been investigated using Ab-rPEDF. The Ab-rPEDF specifically recognizes a 49,500-MW protein in western trans-

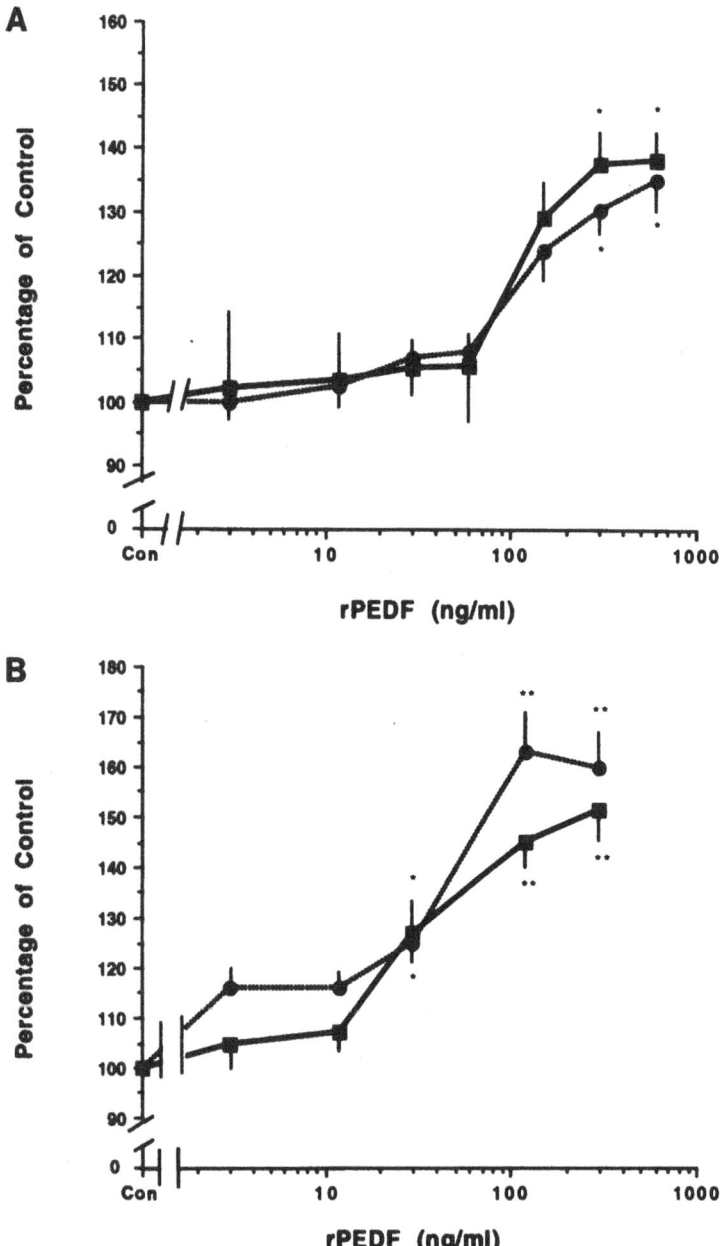

Figure 6. Neuronal survival activity on cerebellar granule cells (CGC). Dose-dependent curve for PEDF effect on cell number in (A) serum-containing and (B) chemically defined medium. Postnatal day 8 CGCs were cultured in 96-well plates. Different concentrations of rPEDF were added on DIV0, and the MTS assay (●) and cell counting (■) were carried out on DIV7. Data are mean ± SEM (bars) values (n = 6–12), expressed as a percentage of the control. statistical analysis was done by one-way ANOVA with a post hoc Scheffe F test: *$p < 0.05$, **$p < 0.005$ versus the control. [Reproduced from reference (16) with permission].

fers of an IPM wash from bovine eyes (17). PEDF is present as approximately 1% of total soluble protein of the IPM and at a concentration of 250 nM. Optimization of a purification protocol from bovine IPM wash yields highly pure PEDF with a recovery of 47%. This protocol includes ammonium sulfate saturation and cation-exchange chromatography

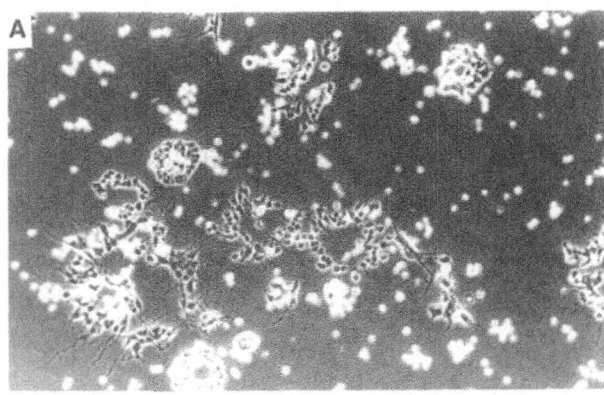

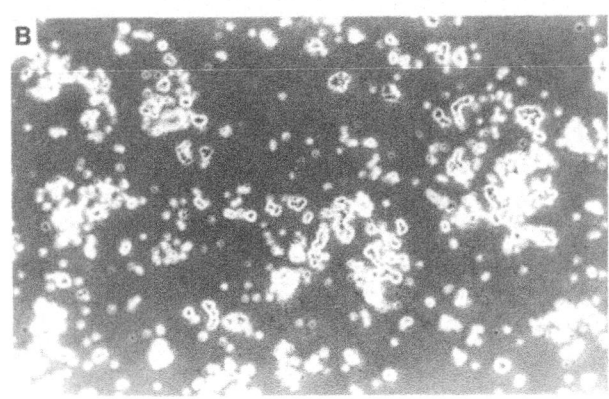

Figure 7. Antiserum Ab-rPEDF blocks the rPEDF neurotrophic activity. A total of 100 ng rPEDF was mixed without and with antiserum Ab-rPEDF (33.7 µg protein) in 25 µl containing 1% BSA in PBS. The mixtures were preincubated at 4°C for 4 hours before adding to cell cultures. Retinoblastoma Y-79 cells were treated as in figure 4 and as follows: panel A, 50 ng/ml rPEDF; and panel B, 50 ng/ml rPEDF plus 16.8 µg protein/ml antiserum Ab-rPEDF. Photographs of treated cells at day 9 post- attachment are shown.

Figure 8. Purification of PEDF from IPM. Soluble components of IPM were extracted by gentle lavage of the subretinal matrix of bovine eyes. Proteins were subjected to ammonium sulfate fractionation. PEDF was purified from the 45–80% ammonium sulfate precipitate by cation-exchange chromatography. SDS-polyacrylamide gel electrophoresis analysis of the following fractions is shown: lane 1, IPM soluble extract (11.4 µg); lane 2, soluble fraction of 45% ammonium sulfate (10.5 µg); lane 3, precipitate of 80% ammonium sulfate fractionation (9.4 µg); lane 4, purified PEDF (2.5 µg); lane 5, purified PEDF (1.4 µg); lane 6, purified PEDF (0.7 µg); and lane 7, purified PEDF (0.2 µg).

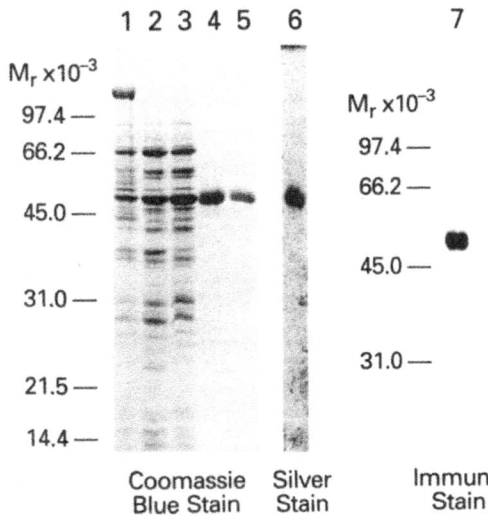

(Fig. 8). A total of 1.5 mg is purified from 1000 eyes from steers 18–20 months old of 900–1300 pounds. Furthermore, the antiserum Ab-rPEDF also blocks the IPM-mediated neurotrophic activity, indicating that PEDF is the sole IPM component that confers neurotrophic activity to the IPM (17). Western analysis indicates that the interphotoreceptor matrix (IPM) of other species, including human and avian, has PEDF in a soluble form (18). PEDF is also identified in vitreous and aqueous humors of bovine eyes (19). PEDF protein accumulates more abundantly in vitreous than in other extracellular spaces of the bovine eye and at concentrations ~30 nM. Although the relative amount of PEDF protein is <1% of total vitreal protein, it is purified to near homogeneity with yields about 7.2 μg per vitreous, i.e., 4.8-fold higher yields than from IPM (18). In a bovine aqueous humor, the PEDF concentration is ~3 nM, e.g., a total of 0.3 μg of PEDF protein per aqueous humor. Similarly, PEDF protein has been purified from vitreous of monkey eyes (S.P. Becerra, personal observations). Thus, PEDF is identified in the native eye as an extracellular protein.

7. STRUCTURAL CHARACTERISTICS OF PEDF

7.1. Size of PEDF

The size of native PEDF protein has been analyzed by 1D- and 2D-gel electrophoresis and gel filtration chromatography (17). PEDF migrates as a single protein of 49,000± 1,500 apparent molecular weight under non-reducing and reducing conditions (see Fig. 6). Under native conditions it focuses between pH 7.0–7.7. PEDF protein elutes as a single symmetrical peak by gel filtration chromatography. Its retention time is slightly behind that of ovalbumin (43,000-MW).

7.2. Amino-Terminal End of Mature PEDF Protein

The deduced PEDF sequence predicts the removal of a signal peptide of 17 residues from its amino-terminus (10, 20). Amino acid sequence determination of secreted purified protein reveals that PEDF starts at position 21 of the precursor (see Fig.1). Thus it appears that the factual signal peptide region required for protein maturation and secretion is 20 residues in length, differing in 3 residues from the predicted region. Determination of PEDF purified from vitreous does not provide sequence, implying that in this matrix the protein undergoes a post-translational modification that blocks Edman degradation, e.g., acetylation at its end-terminus.

Figure 9. Summary of sequence analysis of PEDF proteolytic products. Controlled proteolysis of bovine PEDF with several proteases showed a preferred cleavage site towards the C-end (21). PEDF sequence is represented by a line, N-end to the left. Arrows represent cleavage sites. Serpin position P1 is indicated and proteases are as follows: Glu-C: endoproteinase Glu-C, E: elastase; Subt., subtilisin; T trypsin, CT, chymotrypsin.

7.3. Deglycosylation of PEDF

The deduced sequence for PEDF also predicts an Asn-glycosylation attachment site for the mature PEDF polypeptide (see Fig. 1). Enzymatic deglycosylation with N-glycosidase F increased the migration of PEDF by ~2500-MW, as analyzed by SDS-PAGE (17). This observation demonstrates that PEDF protein contains oligosaccharides attached to internal Asn residue(s) which add 5% to its apparent molecular weight. Thus, PEDF is a secreted monomeric glycoprotein.

8. CONFORMATIONAL ANALYSIS BY CONTROLLED PROTEOLYSIS

The folded protein conformation of native PEDF has been investigated (21, 22). Controlled proteolysis of native PEDF (50 kDa) has been performed with several proteases. N-terminal sequence analysis of the products indicates a favored cleavage region located toward the C-terminal end of PEDF (Fig. 9). A chymotrypsin/PEDF reaction mixture reveals two overlapping sequences: that of the N-terminus of intact PEDF and that of an internal region starting at position P1′ (383). A circular dichroism spectrum confirms that native bovine PEDF protein contains 35% β structures as shown for the folded structure of serpins. These data indicate that PEDF protein has a globular conformation with one protease-sensitive exposed loop that contains the homologous serpin-reactive site. Thus, not only the linear but also the overall folded protein conformation of PEDF is homologous to the serpin family of proteins.

9. PEDF BEHAVES AS A NONINHIBITORY SERPIN

9.1. Search for Protease Targets of PEDF Activity

PEDF has a homologous reactive site specific for a leucyl-enteropeptidase such as the chymotrypsin-like proteases cathepsin G, chymase and chymotrypsin itself. PEDF shares this structural feature with the inhibitory serpins, α_1-antichymotrypsin (23) and heparin cofactor II (HCII) (24). Interestingly, HCII changes its target specificity to thrombin when in the presence of sulfated polysaccharides. It forms a complex with thrombin and the rate of inhibition increases >1000-fold when heparin or dermatan sulfate is present (25). In addition, protease nexin-I/glia-derived nexin (PN-1/GDN) is another serpin that mediates neurotrophic activity by inhibition of thrombin. In contrast to α_1-antichymotrypsin, HCII and PN-1/GDN, recombinant and native PEDF protein do not inhibit nor form serpin:protease complex with chymotrypsin, cathepsin G, thrombin, trypsin, elastase, with or without sulfated polysaccharides in the protease inhibition reaction assays (15, 22). Thus, PEDF and PN-1/GDN have similar neurotrophic properties but are distinct structurally and biochemically.

9.2. PEDF lacks the serpin S→R conformational change

Thermal stability curves for unmodified and cleaved PEDF reveal that PEDF lacks the S→R conformational change upon cleavage within its reactive-site loop (Fig. 10). PEDF has lower denaturation temperatures than ovalbumin differing by 20°C. The

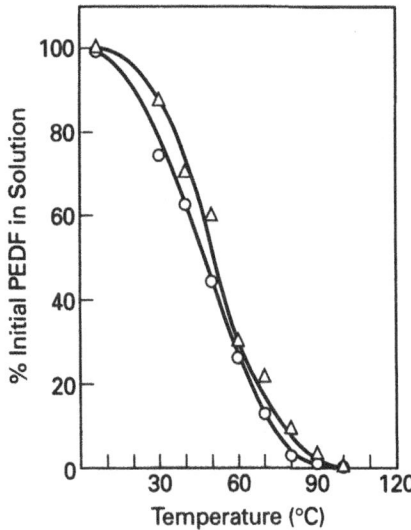

Figure 10. S→R transition assay. Aliquots of intact bovine PEDF and PEDF cleaved by trypsin at its homologous serpin reactive loop were incubated between 30°C and 100°C for 2 h as described before (25). Soluble proteins were then fractionated by centrifugation and analyzed by SDS-polyacrylamide gel electrophoresis. PEDF proteins were followed by immunoblot and quantified by their immunoreactivity against antiserum to PEDF, Ab-rPEDF. Curves in open circles are with intact PEDF and in open triangles with trypsin-treated PEDF. [Reproduced from reference (22) with permission].

denaturation temperature for PEDF is about 50–60°C, similar to the ones for the nonihibitory serpins angiotensinogen (26) and maspin (27). It is supposed that in PEDF as in ovalbumin, angiotensinogen and maspin, the S→R transition may serve no useful purpose and therefore has been lost by evolutionary change. The stabilization of inhibitory serpins upon cleavage is explained by the insertion of their A4 strand into the center of sheet A. The failure of PEDF to behave as an inhibitor may be explained by the presence of unfavorable residues in its P12-region, i.e. the A4 strand (Fig. 11). Overall, PEDF behaves like a typical nonihibitory serpin, with no demonstrable inhibitory activity and lack of the S→R conformational change.

10. SEARCH FOR A NEUROTROPHIC REGION

Limited proteolysis and recombinant DNA techniques provide means to prepare PEDF polypeptide fragments. Proteolysis has been used to prepare PEDF with a cleaved homologous reactive center. Abolishing the putative determinant for serpin inhibition does not affect the PEDF neurotrophic activity (Table 2). This observation indicates that the integrity of the serpin reactive loop is dispensable for the PEDF biological activity. To map

Table 2. Neutrophic activity of PEDF fragments

PEDF fragment	Size (kDa)	Amino acid residues	Differentiation of Y-79 cells
Native	50	20-418	+
native + Ab-rPEDF	50	20-418	−
trypsin-cleaved	46	20-382	+
subtilisin-cleaved	46	32-382	+
rPEDF	43	44-418	+
BP	28	44-269	+
BX	25	44-227	+
BA	9	44-121	+
AH	35	121-418	−

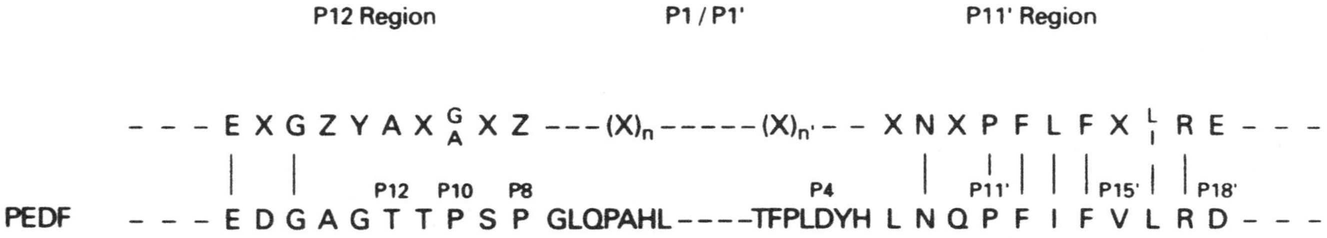

P12 Region   P1 / P1'   P11' Region

- - - E X G Z Y A X $_A^G$ X Z - - - (X)$_n$ - - - - - (X)$_{n'}$ - - X N X P F L F X^L $\vert$ R E - - -
$\qquad$ | $\quad$ | P11' $\vert$ P15' $\vert$ P18'
$\qquad$ P12 P10 P8 $\qquad\qquad$ P4

PEDF - - - E D G A G T T P S P GLQPAHL - - - - TFPLDYH L N Q P F I F V L R D - - -

Dysfunctional Serpins

P12	ATIII Hamilton A382T
P12	CIinh Ma A434E
P10	ATIII Charleville A384P
P10	CIinh Ca A436T
P10	ATIII Cambridge A384P
P10	CIinh Mo A436T
P11'	a1AT Heerlen P369L
P11'	ATIII Utah P407L
P12-P10	a2AP Enschede insertion of an additional alanine between A354 and A357

Functional Serpins

P4	a1ATChristchurch E363K
P15'	CIinh M458V
P18'	a1AT M3 E376D

Non-Inhibitory Serpins

P8	TBG (proline)
P12	AGTR (proline)
P12	AGTH (proline)
P12	CBG (threonine)

Figure 11. Amino acid sequence comparison around reactive sites for serpins and PEDF. A consensus sequence from inhibitory serpins is at the top. Three regions are represented P1/P1', P12- and P11'-regions. P12-region aligns with the secondary structural element sA4. Residue Z and Y correspond to amino acids with small side chains, e.g. T, S, V, and large charged residues, E, Q, respectively. Positions of PEDF residues shared with natural variants and noninhibitory serpins are indicated. Mutation and variations in sequence of natural serpin variants and noninhibitory serpins are indicated at the bottom. [Reproduced from reference 15 with permission].

the location involved in neurotrophic activity, PEDF expression constructs with further truncations from the carboxy-terminal end of PEDF were prepared: BP (positions 44–267), BX (positions 44–229), BA (positions 44–121) and AH (121–418) (21). The resulting expression vectors encode polypeptides of 223, 185, 77 and 298 PEDF residues, respectively. The results show that BP, BX and BA exhibit neurotrophic activity, while AH did not (Table 2). Experiments with cerebellar granule cells have also shown that PEDF polypeptide fragments lacking the exposed loop have a neurotrophic survival effect (16, 28). Thus, even when large segments from the carboxy-terminal end of PEDF are deleted, the neurotrophic activity is retained. These observations demonstrate that the amino-terminal region (BA) of PEDF confers a neurotrophic function to the protein. Consequently, the neurotrophic induction must be mediated by other than PEDF serpin inhibition.

The BA region has a sequence that is apparently unique, with the highest degree of homology to members of the serpin family (20–30%). This region is composed of putative α-helices A, B, C and part of D when aligned to antitrypsin and ovalbumin (Fig. 1). Spatial models for PEDF based on the three-dimensional structure of ovalbumin (14) show that the BA region is located to the opposite pole from the serpin exposed loop. This model is in agreement with the fact that the PEDF neurotrophic activity is not lost with cleavage of the exposed loop. Thus in the folded protein structure of PEDF the neurotrophic domain is also separated from the exposed loop.

11. SUMMARY

The neuronal differentiating and survival activities and presence of PEDF next to the neural retina support the idea that this serpin plays a neurotrophic role *in vivo*. The knowledge of the PEDF structure has provided information on the determinants for the nonihibitory and neurotrophic activities. PEDF has characteristics of a substrate rather than an inhibitor of serine proteases. Further studies are needed to identify the missing structural elements on PEDF that would confer serpin inhibitory activity. An N-terminus peptide region provides the neurotrophic function to the PEDF protein while other structural characteristics are dispensable (e.g. signal peptide, oligosaccharides on the polypeptide backbone, serpin exposed loop). During evolution PEDF might have lost its inhibitory activity and gained its neurotrophic function. Particular activities on other serpins have been reported (e.g. angiotensinogen, maspin, etc.). PEDF is an example of the separation of inhibitory and particular activities in a serpin.

REFERENCES

1. Bryan, J.A. and Campochiaro, P.A. (1986) A retinal pigment epithelial cell-derived growth factor(s). Arch.Ophthalmol. 104: 422–425.
2. Campochiaro, P.A. (1993) Cytokine production by retinal pigmented epithelial cells. Int. Rev. Cytol. 146: 75–82
3. Hewitt, A.T., Lidnsay, J.D., Carbott, D. and Adler, R. (1990) Photoreceptor survival-promoting activity in interphotoreceptor matrix preparations: characterization and purification. Exp. Eye Res. 50:79–80.
4. Gaur, V.P., Liu, Y. and Turner, J.E. (1992) RPE conditioned medium stimulates photoreceptor cell survival, neurite-outgrowth and differentiation *in vitro*. Exp. Eye Res. 54: 645–659.
5. Jaynes, C.D. and Turner, J.E. (1995) Müller cell survival and proliferation in response to medium conditioned by the retinal pigment epithelium. Brain Res. 678: 55–64.
6. Tombran-Tink J. and Johnson, L.V. (1989) Neuronal differentiation of retinoblastoma cells induced by medium conditioned by human RPE cells. Inv. Ophthalmol. Vis. Sci. 30: 1700–1707.

7. Reid, T.W., Albert, D.M., Rabson, A.S., Russell, P., Craft, J., Chu, E.W., Tralka,T.S. and Wilcox, J.L. (1974) Characteristics of an established cell line of retinoblastoma. J. Natl. Cancer Inst. 53: 347–360.

8. Kyristis, A.P., Tsokos, M., Triche, T.J. and Chader, G.J. (1984) Retinoblastoma: origin from a primitive neuroectodermal cell? Nature 307: 471–473.

9. Tombran-Tink, J., Chader, G.J. and Johnson, L.V. (1991) PEDF: a pigment epithelium-derived factor with potent neuronal differentiative activity. Exp. Eye Res. 53: 411–414.

10. Steele, F.R., Chader, G.J., Johnson, L.V. and Tombran-Tink, J. (1993) Pigment epithelium-derived factor: neurotrophic activity and identification as a member of the serine protease inhibitor gene family. Proc. Natl. Acad. Sci. USA 90: 1526–1530.

11. Pignolo, R.J., Cristofalo, V.J. and Rotenberg, M.O. (1993) Senescent WI-38 cells fail to express EPC-1, a gene induced in young cells upon entry into G_0 state. J. Biol. Chem. 268: 8949–8957.

12. Henikoff, S. and Henikoff, J.G. (1991) Automated assembly of protein blocks for database searching. Nucleic Acids Res. 19: 6565–6572.

13. Huber, R. and Carrell, R.W. (1989) Implications of the three-dimensional structure of α_1-antitrypsin for structure and function of serpin. Biochemistry 28: 8951–8966.

14. Wright, H.T., Qian, H.X. and Huber, R. (1990) Crystal structure of plakalbumin, a proteolytically nicked form of ovalbumin: its relationship to the structure of cleaved α-1-proteinase inhibitor. J. Mol. Biol. 213: 513–528.

15. Becerra, S.P., Palmer, I., Kumar, A., Steele, F., Shiloach, J., Notario, V. and Chader, G.J. (1993) Overexpression of fetal human pigment epithelium-derived factor in *Escherichia coli*: a functionally active neurotrophic factor. J. Biol. Chem. 268: 23148–23156.

16. Taniwaki, T., Becerra, S.P., Chader, G.J. and Schwartz, J.P. (1995) Pigment epithelium-derived factor is a survival factor for cerebellar granule cells in culture. J. Neurochem. 64: 2509–2517.

17. Wu, Y.Q., Notario, V., Chader, G.J. and Becerra, S.P. (1995) Identification of pigment epithelium-derived factor in the interphotoreceptor matrix of bovine eyes. Protein Expr. Purif. 6: 447–456.

18. Tombran-Tink, J., Shivaram, S.M., Chader, G.J., Johnson, L.V. and Bok, D. (1995) Expression, secretion, and age-related downregulation of pigment epithelium-derived factor, a serpin with neurotrophic activity. J. Neurosci. 15: 4992–5003.

19. Wu, Y.Q. and Becerra, S.P. (1996) Proteolytic activities directed towards pigment epithelium-derived factor (PEDF) in vitreous of bovine eyes: implications of proteolytic processing. Inv. Ophthalmol. Vis. Sci. (in press).

20. Perez-Mediavilla, L.A., Chew, C, Campochiaro, P., Zack, D.J. and Becerra, S.P. (1996) Expression of bovine PEDF: a neurotrophic noninhibitory serpin. International Symposium on the biology and chemistry of serpins. Abstract #121.

21. Stratikos, E., Alberdi, E., Gettins, P.G.W. and Becerra, S.P. (1996) Recombinant human pigment epithelium-derived factor (PEDF): overexpression by eukaryotic cells and secretion of correctly folded PEDF. International Symposium on the biology and chemistry of serpins. Abstract #127.

22. Becerra, S.P., Sagasti, A., Spinella, P. and Notario, V. (1995) Pigment epithelium-derived factor behaves like a noninhibitory serpin: Neurotrophic activity does not require the serpin reactive loop. J Biol. Chem. 270: 25992–25999.

23. Morii, M. and Travis, J. (1983) Amino acid sequence at the reactive site of human α_1-antichymotrypsin. J. Biol. Chem. 258: 12749–12752.

24. Blinder, M.A., Marasa, J.C., Reynolds, C.H., Deaven, L.L. and Tollefsen, D.M. (1988) Heparin cofactor II: cDNA sequence, chromosome localization, restriction fragment length polymorphism and expression in *Escherichia coli*. Biochemistry 27: 752–759.

25. Tollefsen, D.M., Petstka, C.A. and Monafo, W.F. (1983) Activation of heparin cofactor II by dermatan sulfate. J. Biol. Chem. 258: 6713–6716.

26. Stein, P.E., Tewksbury, D.A., and Carrell, R.W. (1989) Ovalbumin and angiotensinogen lack S→R conformational change. Biochem. J. 262: 103–107.

27. Pemberton, P.A., Wong, T.D., Gibson, H.L., Kiefer, M.C., Fitzpatrick, P.A., Sager, R., and Barr P.J. (1995) The tumor suppressor maspin does not undergo the stresseed to relax transition or inhibit trypsin-like proteases. J. Biol. Chem. 270: 15832–15837.

28. Taniwaki, T., Becerra, S.P., Chader, G.J. and Schwartz, J.P. (1994) 24th Annual Meeting of the Society for Neuroscience (Miami Beach, FL, November 13–18, Abstracts Vol. 20:1691–1692.

22

COLLIGIN, A COLLAGEN BINDING SERPIN

E. H. Ball, N. Jain,[*] and B. D. Sanwal

Department of Biochemistry
University of Western Ontario
London, Canada, N6A 5C1

1. INTRODUCTION

Colligin is a major glycoprotein in many cells and has been discovered several times by different approaches and hence given different names. Kurkinen et al. (1984) first found it as a collagen-binding protein in parietal endoderm cells and proposed the name 'colligin' but it was later detected as a differentiation related protein in myoblasts ('gp46') (Cates et al., 1984) and as a heat shock responsive protein in chick fibroblasts (HSP47) (Nagata & Yamada, 1986). Sequencing later confirmed the identity of the various proteins. Each of the routes of discovery delivered clues about the function of the protein, leading eventually to the current idea of a role as a molecular chaperone in collagen synthesis.

Collagen is the end product of an elaborate series of steps leading from nascent procollagen chains extruded into the endoplasmic reticulum (ER) to a mature trimeric molecule in the extracellular space. Figure 1 illustrates the early steps in the process and some of the proteins that may be involved. Procollagen chains have N- and C-terminal extensions (propeptides) that are eventually cleaved after secretion. In the ER, folding to form the characteristic triple helical structure, beginning at the C-terminal end, does not start until the chains are completely synthesized and proline hydroxylation has taken place (Bruckner et al., 1981; Prockop, 1990). Further covalent modifications occur as the protein traverses the endoplasmic reticulum and golgi. Interestingly, much of the procollagen that is synthesized is degraded in the endoplasmic reticulum (Berg, 1986), perhaps as the result of a quality control system. In keeping with the complexity of the process, many proteins are known to be required, from prolyl hydroxylase to peptidases.

Initially colligin was thought to be a plasma membrane protein (Cates et al., 1984; Kurkinen et al., 1984) but it was later definitively shown to be exclusively in the lumen of the endoplasmic reticulum (Hughes et al., 1987; Saga et al., 1987; Nandan et al., 1988). This organelle is principally concerned with synthesis, assembly and quality control of secreted and transmembrane proteins. Several molecular chaperones are known to reside there, including grp78 and protein disulfide isomerase (PDI). When living cells are treated with a

* Current address: Institute of Molecular and Cell Biology, 10 Kent Ridge Crescent, Singapore, 119260.

Chemistry and Biology of Serpins, edited by Church *et al.*
Plenum Press, New York, 1997

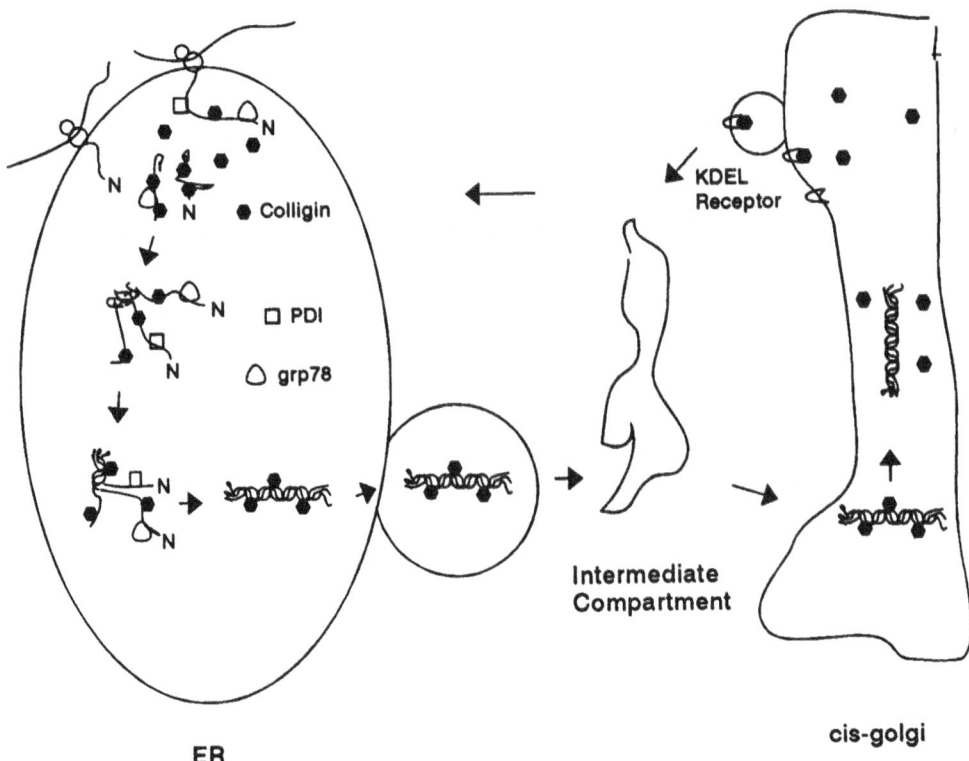

ER   **cis-golgi**

Figure 1. Early steps in procollagen synthesis. Procollagen chains interact with several other proteins during or immediately after translation including colligin, protein disulfide isomerase (PDI) and grp78.

cross-linker to stabilize interactions a complex of proteins can be immunoprecipitated with anti-colligin (Figure 2). As illustrated, grp78 and PDI are prominent constituents. In order to show procollagen, cells have been labelled with proline (panel B) as well as methionine (panel A). Thus colligin is well placed to act in the collagen assembly process.

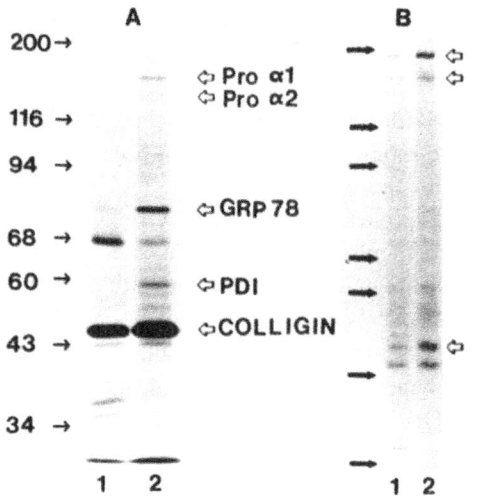

Figure 2. Binding of colligin to procollagen *in vivo*. L6 myoblasts labelled with either $\{^{35}S\}$-methionine (panel A) or $\{^{3}H\}$-proline (panel B) were lysed and immunoprecipitated using a monoclonal anti-colligin antibody. Lanes 1, without cross-linking; lanes 2, cells treated with 150μM dithio*bis*(succinimidyl propionate). Samples were run on SDS-PAGE and autoradiographed.

We have approached the characterization of colligin in several ways; most recently by a combination of molecular techniques. Starting with a monoclonal antibody generated against the rat protein, a cDNA encoding rat colligin was isolated from an L6 myoblast library and sequenced (Clarke et al., 1991). Probes generated from this cDNA allowed the isolation of the human clone from a human fibroblast library (Clarke & Sanwal, 1992). It became clear that colligin is a unique member of the serpin family located in the endoplasmic reticulum.

2. SEQUENCE OF COLLIGIN

The amino acid sequence of colligin revealed several interesting features of the protein. An N-terminal hydrophobic signal sequence specifies secretion. Retention in the ER is explained by the sequence RDEL at the C-terminal of the protein. This sequence is recognized by a receptor that retrieves proteins from the golgi or intermediate compartment and returns them to the ER (Pelham, 1989). Removal of this sequence allows secretion of the protein (Satoh et al., 1996). Membership in the serpin family of protease inhibitors is also evident from the sequence. Colligin is about 27% identical to α1-antitrypsin and shows roughly the same degree of similarity to other serpins. As an example, Figure 3 illustrates an alignment of colligin with protease nexin I (Sommer et al., 1987), another serpin that binds

```
Colligin     1               mrslllgtlcllavalaAEVKKPVEAAAPGTAEKLSSKAT
PNI          1                  mrwhlplfllasvtlpsicSHFNPL


Colligin    41  TLAEPSTGLAFSLYQAMAKDQAVENILVSPVVVASSLGLVSLGGKATTASQAKAVLSAEQ
PNI         26  SLEELGSNTGIQVFNQIVKSRPHDNIVISPHGIASVLGMLQLGADGRTKKQLAMVMRYGV
                 * *              *         ** **  .** **    **       *   * *


Colligin   101  LRDEEVHAGLGELLRSLSNSTARNVTWKLGSRLYGPSSVSFADDFVRSSKQHYNCEHSKI
PNI         86  NGVGKILKKINKAIVSKKNKDIVTVA----NAVFVKNASEIEVPFVTRNKDVFQCEVRNV
                     * *    *                        ** *     **  *    **


Colligin   161  NFPDKRSALQSINEWAAQTTDGKLPEVTKDVERTDGAL----LVNAMFFKPHWDEKFHHK
PNI        142  NFEDPASACDSINAWVKNETRDMIDNLLSP-DLIDGVLTRLVLVNAVYFKGLWKSRFQPE
                ** *  **  *** * *        *   *       ** *  **** **  *  *


Colligin   217  MVDNRGFMVTRSYTVGVTMMHRTGLYNYYDDEKEK---LQLVEMPLAHKLSSLIILMPHH
PNI        201  NTKKRTFVAADGKSYQVPMLAQLSVFRCGSTSAPNDLWYNFIELPYHGESISMLIALPTE
                      * *         * *                     * *     * * *


Colligin   274  VE-PLERLEKLLTKEQLKIWMGKHQKKAVAISLPKGVVEVTHDLQKHLAGLGLTEAIDKN
PNI        261  SSTPLSAIIPHISTKTIDSWMSIMVPKRVQVILPKFTAVAQTDLKEPLKVLGIITDMFDSS
                   **        ** * * *   ***       **  *  ** *      *


Colligin   333  KADLSRMS-GKKDLYLASVFHATAFELDTDGNPFDQDIYGREELRS-PKLFYADHPFIFL
PNI        321  KANFAKITTGSENLHVSHILQKAKIEVSEDGTKASAATTAILIARSSPPWFIVDRPFLFF
                **      *   *       *   **      ** *   * **  *


Colligin   391  VRDTQSGSLLFIGRLVRLKGDKMRDEL
PNI        381  IRHNPTGAVLFMGQINKP
                  *    *  ** *
```

Figure 3. Comparison of amino acid sequences of human colligin and human protease nexin I (PNI). Alignment performed by the program SIM (Huang & Miller, 1991) and gave 25% identity over a 375 residue overlap. Asterisks denote identical residues.

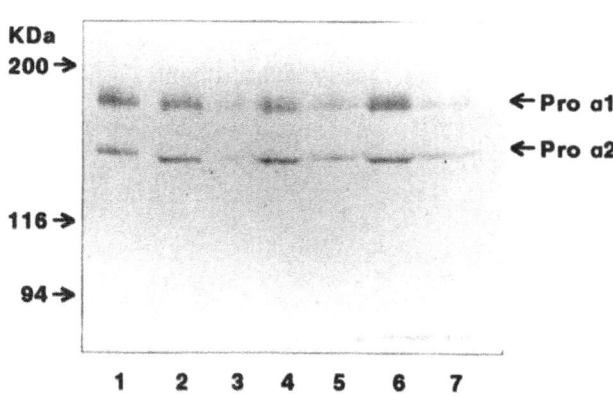

Figure 4. Effect of colligin on degradation of procollagen. $\{^{3}H\}$-procollagen was incubated with rat liver microsomes for 60 min at 37°. Samples were boiled and run on SDS-PAGE. An autoradiogram of the gel is shown. Lane 1, procollagen only; lane 2, with microsomes (without detergent); lane 3, with microsomes and detergent (complete); lane 4, complete with 1 μM recombinant colligin; lane 5, complete with 0.4 μM recombinant colligin; lane 6, complete with 1 μM non-recombinant colligin; lane 7, complete containing colligin buffer.

collagen type IV. Several key sequences such as the PFXF near the C-terminal are conserved and there are identical residues throughout a large region of similarity. The P1 or protease specificity-determining residue in the serpin family is an arginine in colligin. The sequence upstream of the P1 residue makes it unlikely that colligin acts as an inhibitory serpin, however. The large residues at P9-P11 instead of the small alanines found in inhibitors (such as protease nexin I, Fig. 3) and the proline at P13 are likely to render the protein non-inhibitory (Hopkins et al., 1993). In fact, no inhibition of several serine proteases has been detected, nor could SDS stable complexes of the protein be found (our unpublished observations). On this basis colligin appears to belong to a class of serpins that have evolved other functions, such as ovalbumin and corticosteroid-binding globulin.

3. COLLIGIN AS A COLLAGENASE INHIBITOR

Despite the suggestive evidence from the sequence that colligin was not an inhibitory serpin we set out to look for protease inhibition. A microsomal fraction from liver was able to degrade the isolated procollagen chains, in keeping with the known degradation of procollagen and other proteins in the ER (Klausner & Sitia, 1990; Bonifacino & Lippincott-Schwartz, 1991). The activity was lumenal since detergent was required to permeabilize microsomal vesicles (Figure 4) and could be inhibited by TPCK or TLCK, serine protease inhibitors (Jain et al., 1994). In this *in vitro* system we find colligin protects procollagen but not fibronectin from degradation, indicating a degree of specificity. Figure 4 shows that recombinant colligin was almost as active as non-recombinant in protecting procollagen. Only procollagen I and IV, not type III, were protected by colligin in agreement with its binding characteristics (see below). The mechanism of the inhibition seems more likely to involve shielding rather than a proteolytic inhibition, however, judging by the high concentrations of colligin (1 μM) required. Thus one aspect of colligin action may be a passive protection from proteases.

4. COLLIGIN-PROCOLLAGEN INTERACTION

The binding properties of colligin are compatible with a function in procollagen folding. The purified protein binds collagen and procollagen types I and IV, as well as denatured collagen (gelatin). Both the α1 and α2 chains of procollagen I are active (Jain et al., 1994). Thus colligin recognizes binding sites in the central triple helical part of the collagen molecule, (as well, perhaps, as in the propeptide regions (Hu et al., 1995)) and in both unfolded and folded protein. These properties suggest that colligin is bound to procollagen throughout the ER part of the journey; in fact binding to polysomal procollagen has been detected (Sauk et al., 1994; Smith et al., 1995).

ELISA type binding assays yield a Kd of 10^{-7}–10^{-8} with a slight selectivity for the proα$_2$(I) over the proα$_1$(I) chain (Jain et al., 1994). Binding stoichiometry is approximately 3 colligin/procollagen. The binding is specific for types I and IV collagen; type III does not interact. Further, binding is very sensitive to pH, decreasing as the pH decreases until near pH 6 no binding is detectable. This provides a possible release mechanism: as the pH drops in the golgi, procollagen dissociates and colligin is recycled back to the endoplasmic reticulum. Interestingly the arg-gly-asp (RGD) peptide sequence interferes with the binding (Nandan et al., 1988) (but control RGE does not): this sequence forms part of the recognition site for a variety of extracellular matrix receptors (Hynes, 1992). Procollagens do contain this sequence, so it may also be involved with colligin binding as well. No sequence similarities between colligin and receptors that might explain this common binding specificity are evident.

As a first step towards defining the procollagen binding site of colligin we have examined truncations of the protein. When proteins produced by *in vitro* translation were tested for binding to procollagen beads, a block of amino acids (104 to 169) were found to be essential for binding (E. Clarke, B. Sanwal, unpublished results). To confirm this observation, fragments of colligin were expressed as GST fusions and tested for binding (Figure 5). These experiments gave a similar result — amino acids 105 to 175 or even a subset (145 to 175) were sufficient for binding although much reduced relative to the whole protein. It is notable that this region shows a high degree of similarity (14 out of 31 residues identical) with the type IV collagen-binding protease nexin I (Figure 3). In serpin structure, much of this 70 amino acid sequence is solvent exposed (parts of sheet s2a, helix E, sheet s1A and helix F (Huber & Carrell, 1989)), indeed it forms part of the heparin-binding site in several serpins. In the case of heparin cofactor II, heparin has an allosteric effect (Sheehan et al., 1994), raising a similar possibility for colligin.

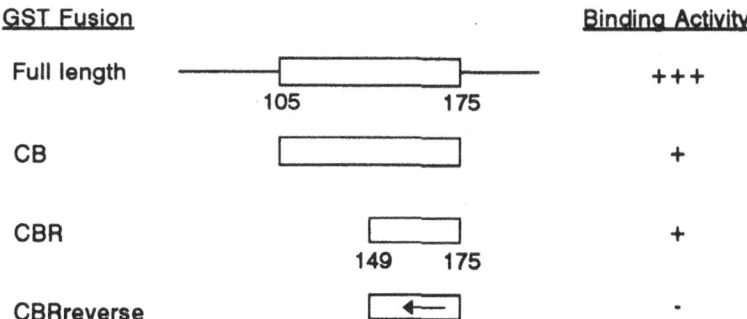

Figure 5. Binding of colligin truncations to procollagen. Plasmids encoding either whole colligin (without signal sequence) or shorter fragments as fusions with glutathione-S-transferase (GST) were constructed. The fusion proteins bound to glutathione-agarose were incubated with radiolabelled procollagen and washed to determine binding activity. CB refers to the putative collagen-binding region (amino acids 105 to 175), CBR to the right half of the region and CBRreverse to the product of a control plasmid containing the coding region in reverse orientation.

5. PARALLEL REGULATION OF COLLIGIN AND PROCOLLAGEN

Another observation that points to an intimate relationship between collagen and colligin is their parallel regulation. Circumstances that lead to increased collagen I and/orIV production also increase colligin synthesis and vice versa. Examples include the increase in colligin and type IV collagen with F9 cell differentiaton (Wang & Gudas, 1990) or the decrease of collagen and colligin in transformed cells (Nakai et al., 1990). We first noted a similar decrease in colligin and procollagen I after L6 myoblast differentiation (Cates et al., 1987). To explore this effect, growth factors that alter rates of collagen synthesis were tested (Clarke et al., 1993). TGFβ leads to large increases in extracellular matrix synthesis while EGF has the opposite effect. When L6 cells were treated with these factors, colligin protein and mRNA levels behaved exactly as did the procollagen. Thus some form of transcriptional control is acting in parallel to regulate procollagen and colligin.

It is notable, however, that an exception to this rule occurs during heat shock: colligin levels increase without a corresponding increase in procollagen (Clarke et al., 1993). In these circumstances colligin may be required immediately simply to bind existing unfolded procollagen.

6. CONCLUSIONS

While the overwhelming weight of evidence implies a role for colligin in connection with procollagen, the exact nature of colligin function is unclear. Future approaches involving gene knockouts and mutation would be appropriate to shed light on the situation. Decreased colligin levels lead to decreased procollagen secretion both in mutant cells (our observations) and after treatment with antisense oligonucleotides (Sauk et al., 1994). Thus from the evidence currently available colligin appears to act in an assisting role (like a chaperone) rather than as part of a quality control system. However, colligin would be unusual for a chaperone in that it is specific for a single type of protein and it seems to associate with both the mature and the unfolded form. Procollagen may be especially vulnerable to proteolysis because folding into the triple helix cannot begin until the chains are completely synthesized. In view of the major importance of collagen and its elaborate synthetic pathway, it would perhaps not be surprising to have a 'dedicated' chaperone, but colligin may turn out to have other functions as well.

Colligin has been a difficult protein to purify and keep in an active form, like some other serpins. Biochemical studies should be greatly facilitated by production of the protein in bacteria and this has proven possible. We find that the recombinant protein is as active in binding procollagen as the native molecule (Jain et al., 1994), despite the absence of glycosylation. This should allow detailed investigation of structure/function relationships.

In the long run a reconstituted system of procollagen and various associated proteins in cells and *in vitro* will be necessary to probe questions of colligin function. Significant steps have been made in working towards this goal and rapid progress can be anticipated in the near future.

REFERENCES

Berg, R.A. (1986). Intracellular turnover of collagen. ed. R. P. Mecham. *Regulation of matrix accumulation*. NY, Academic Press. 29–52.

Bonifacino, J.S. and Lippincott-Schwartz, J. (1991) Degradation of proteins within the endoplasmic reticulum. *Curr. Op. Cell Biol. 3*, 592–600.

Bruckner, P., Eikenberry, E.F. and Prockop, D.J. (1981) Formation of the triple helix of type I procollagen: a kinetic model based on cis-trans isomerization of peptide bonds. *Eur. J. Biochem. 118*, 607–613.

Cates, G.A., Brickenden, A.M. and Sanwal, B.D. (1984) Possible involvement of a cell surface glycoprotein in the differentiation of skeletal myoblasts. *J. Biol. Chem. 259*, 2646–2650.

Cates, G.A., Nandan, D., Brickenden, A.M. and Sanwal, B.D. (1987) Differentiation defective mutants of skeletal myoblasts altered in a gelatin-binding glycoprotein. *Biochem. Cell Biol. 65*, 767–775.

Clarke, E.P., Cates, G.A., Ball, E.H. and Sanwal, B.D. (1991) A collagen-binding protein in the endoplasmic reticulum of myoblasts exhibits relationship with serine protease inhibitors. *J. Biol. Chem. 266*, 17230–17235.

Clarke, E.P., Jain, N., Brickenden, A., Lorimer, I.A.J. and Sanwal, B.D. (1993) Parallel regulation of procollagen I and colligin, a collagen-binding protein and a member of the serine protease inhibitor family. *J. Cell Biol. 121*, 193–199.

Clarke, E.P. and Sanwal, B.D. (1992) Cloning of a human collagen-binding protein, and its homology with rat gp46, chick hsp47, and mouse J6 proteins. *Biochim. Biophys. Acta 1129*, 246–248.

Hopkins, P.C.R., Carrell, R.W. and Stone, S.R. (1993) Effects of mutations in the hinge region of serpins. *Biochem. 32*, 7650–7657.

Hu, G., Gura, T., Sabsay, B., Sauk, J., Dixit, S.N. and Veis, A. (1995) Endoplasmic reticulum protein Hsp47 binds specifically to the N-terminal globular domain of the amino-propeptide of the procollagen Iα1(I)-chain. *J. Cell. Biochem. 59*, 350–367.

Huang, X. and Miller, W. (1991) A time-efficient, linear-space local similarity algorithm. *Adv. Appl. Math. 12*, 337–357.

Huber, R. and Carrell, R.W. (1989) Implications of the three dimensional structure of α_1-antitrypsin for structure and function of serpins. *Biochem. 28*, 8951–8966.

Hughes, R.C., Taylor, A., Sage, H. and Hogan, B.L.M. (1987) Distinct patterns of glycosylation of colligin, a collagen-binding glycoprotein, and SPARC (osteonectin), a secreted Ca^{2+}-binding glycoprotein. *Eur. J. Biochem. 163*, 57–65.

Hynes, R.O. (1992) Integrins: Versatility, modulation, and signaling in cell adhesion. *Cell 69*, 11–25.

Jain, N., Brickenden, A., Lorimer, I., Ball, E.H. and Sanwal, B.D. (1994) Inhibition of procollagen I degradation in liver microsomes by the collagen binding protein colligin. *Arch. Bioch. Biophys. 314*, 23–30.

Jain, N., Brickenden, A., Lorimer, I., Ball, E.H. and Sanwal, B.D. (1994) Interaction of procollagen I and other collagens with colligin. *Biochem. J. 304*, 61–68.

Klausner, R.D. and Sitia, R. (1990) Protein degradation in the endoplasmic reticulum. *Cell 62*, 611–614.

Kurkinen, M., Taylor, A., Garrels, J.I. and Hogan, B.L.M. (1984) Cell surface-associated proteins which bind native type IV collagen or gelatin. *J. Biol. Chem. 259*, 5915–5922.

Nagata, K. and Yamada, K.M. (1986) Phosphorylation and transformation sensitivity of a major collagen-binding protein of fibroblasts. *J. Biol. Chem. 261*, 7531–7536.

Nakai, A., Hirayoshi, K. and Nagata, K. (1990) Transformation of BALB/3T3 cells by simian virus 40 causes a decreased synthesis of a collagen-binding heat-shock protein (hsp47). *J. Biol. Chem. 265*, 992–999.

Nandan, D., Cates, G.A., Ball, E.H. and Sanwal, B.D. (1988) A collagen-binding protein involved in the differentiation of myoblasts recognizes the arg-gly-asp sequence. *Exptl. Cell Res. 179*, 289–297.

Pelham, H.R.B. (1989) Control of protein exit from the endoplasmic reticulum. *Ann. Rev. Cell Biol. 5*, 1–24.

Prockop, D.J. (1990) Mutations that alter the primary structure of type I collagen. *J. Biol. Chem. 265*, 15349–15352.

Saga, S., Nagata, K., Chen, W.T. and Yamada, K.M. (1987) pH-dependent function, purification, and intracellular location of a major collagen-binding glycoprotein. *J. Cell Biol. 105*, 517–527.

Satoh, M., Hirayoshi, K., Yokota, S., Hosokawa, N. and Nagata, K. (1996) Intracellular interaction of collagen-specific stress protein HSP47 with newly synthesisized procollagen. *J. Cell Biol. 133*, 469–483.

Sauk, J.J., Smith, T., Norris, K. and Ferreira, L. (1994) Hsp47 and the translation-translocation machinery cooperate in the production of α1(I) chains of type I procollagen. *J. Biol. Chem. 269*, 3941–3946.

Sheehan, J.P., Sadler, J.E. and Tollefsen, D.M. (1994) Heparin cofactor II is regulated allosterically and not primarily by template effects. *J. Biol. Chem. 269*, 32747–32751.

Smith, T., Ferreira, L.R., Hebert, C., Norris, K. and Sauk, J.J. (1995) Hsp47 and cyclophilin B traverse the endoplasmic reticulum with procollagen into pre-golgi intermediate vesicles. *J. Biol. Chem. 270*, 18323–18328.

Sommer, J., Nick, H., Gloor, S.M., Meier, R., Rovelli, G.F., Monard, D. and Hofsteenge, J. (1987) cDNA sequence coding for a rat glia-derived nexin and its homology to members of the serpin family. *Biochem. 26*, 6407–6410.

Wang, S.-Y. and Gudas, L.J. (1990) A retinoic acid-inducible mRNA from F9 teratocarcinoma cells encodes a novel protease inhibitor homologue. *J. Biol. Chem. 265*, 15818–15822.

ABSTRACTS

COAGULATION, NEUROBIOLOGY, AND CANCER

1. EXPRESSION AND CHARACTERIZATION OF THE SERPINS SCCA AND LEUPIN

Ruth C. Barnes and D. Margaret Worrall

Department of Biochemistry, University College Dublin, Belfield, Dublin 4, Ireland

The squamous cell carcinoma antigen (SCCA) is a tumour antigen associated with uterine and cervical carcinomas and is a member of the ov-serpin branch of the serpin superfamily (1). Serum levels of SCCA correlate with the degree of metastasis and a role for SCCA in tumour development has been proposed. "Cross-class" inhibition of the cysteine proteases papain and cathepsin L by SCCA has been demonstrated (2). We have recently discovered a novel serpin, leupin, which is high in sequence identity (92%) to SCCA but which contains differences in the reactive loop sequence including the P1 leucine residue, suggesting it may inhibit a different target protease (3). The leupin gene was independently sequenced and located to chromosome 18q by others and has also been designated SCCA2 (4). The high degree of sequence similarity between the two serpins means that previous protease inhibition and SCCA expression studies are likely to have detected both SCCA and leupin.

In this study we have designed specific nucleotide primers to detect and distinguish the expression of SCCA and leupin in human tissues. RT-PCR was carried out from total RNA isolated from a range human tissues and cell lines.Expression of leupin was found in uterine carcinoma cell lines which also express SCCA. Both genes were cloned into the expression vector pRSET and the recombinant proteins expressed and purified. Purified proteins were independently tested for their ability to inhibit proteases including chymotrypsin, cathepsin L and papain.

REFERENCES

1. Suminami, Y, Kishi, F., Sekiguchi, K. & Kato, H. (1991) Biochem. Biophys. Res. Comm.181,51–58.
2. Takeda, A., Yamamoto,T., Nakamura, Y., Takahashi, T, and Hibino, T, (1995) FEBS Lett, 359, 78–80.
3. Barnes, R.C. and Worrall, D.M. (1995) FEBS Lett., 373, 61–65.
4. Schneider, S.S., Schick, C., Fish, K.E., Miller, E., Pena, J.C., Treter, S.D., Hui, S.M. and Silverman, G.A. (1995), P.N.A.S., 92, 3147–3151.

2. ROLE OF THROMBIN'S 60-INSERTION LOOP IN ACTIVE SITE AVAILABILITY TO SERPINS, HIRUDIN AND THE THROMBIN RECEPTOR

Susannah J. Bauman,[*] Alireza R. Rezaie,[¶] Charles T. Esmon,[¶] Dougald M. Monroe,[*] Maureane Hoffman,[†] and Frank C. Church[*]

[*]Departments of Pathology and Medicine, University of North Carolina, Chapel Hill, North Carolina 27599, [¶]Oklahoma Medical Research Foundation, Oklahoma City, Oklahoma 73104, and [†]Department of Pathology, Duke University, Durham Veterans Affairs Medical Center, Durham, North Carolina 27705

α-Thrombin is a trypsin-like serine protease involved in coagulation and wound healing. The 60-insertion loop (L60–Y60a–P60b–P60c–W60d–D60e–K60f–N60g–F60h–T60i–E61) is a region of thrombin that influences macromolecular substrate access to the active site. The activities of five recombinant thrombin mutants, W60dA, W60dF, E192Q, W60dA/E192Q, and W60dF/E192Q, were compared with various inhibitors and substrates to define interactions with thrombin's binding pocket. Recent studies by Le Bonniec *et al.* (*J. Biol. Chem.* 268: 19055–61, 1993), show that the 60-loop influences access to the active site cleft. This may be a reason for thrombin's narrow specificity as compared to the highly homologous trypsin. Rydel *et al.* (*J. Molec. Biol.* 221, 583–601: 1991), show that based on a crystal structure of the thrombin-hirudin complex, hirudin interacts with multiple residues in the 60-loop (Y60a, P60c, W60d, and K60f) and around the binding pocket (E192). The goal of this study was to compare thrombin mutants and expose any differences in the interaction of thrombin and macromolecular substances. Rates of inhibition by two serine protease inhibitors (serpins), antithrombin (AT) and heparin cofactor II (HCII), and the leech anticoagulant, hirudin, were measured. The ability of the thrombin mutants to activate the tethered ligand thrombin receptor in platelets was monitored by calcium flux.

In the absence of glycosaminoglycan, inhibition rates of AT and HCII are slightly decreased (2–3 fold less) with the conservative mutation W60dF. A more drastic mutation, W60dA, has 20 and 6.9 fold less inhibition with AT and HCII, respectively. E192Q, a mutation opposite the 60-loop in the binding pocket, has inhibition rates similar to wild-type thrombin. When E192Q is paired with the mutations at W60d, serpin inhibition rates become more like wild-type thrombin. In the case of W60dA/E192Q, the inhibition rates with AT and HCII are only 3 and 5 fold less, respectively, compared to wild-type thrombin.

Likewise, in the presence of glycosaminoglycan, W60dF, E192Q and W60dF/E192Q have AT and HCII inhibition rates comparable to wild-type thrombin. In the presence of heparin, W60dA has a 8.5 and 6.1 fold decrease in inhibition rates of AT and HCII, respectively. In the presence of dermatan sulfate, the same mutant has a 12.5 fold decrease in inhibition rate with HCII. With heparin, the double mutant W60dA/E192Q returns to only 2.1 and 1.5 fold less inhibitory activity with AT and HCII, respectively. While with dermatan sulfate, W60dA/E192Q has rate constants only 2.4 fold less than wild-type thrombin.

Inhibition by the leech anticoagulant hirudin showed different trends. All mutants had K_i values similar to wild-type thrombin except E192Q whose inhibition pattern changed from slow tight to non-competitive inhibition. The range of K_i values of all mutants and wild-type thrombin except E192Q is from 1 to 7 pM. The E192Q mutant has a K_i value of 57 pM which is approximately 19 fold higher than wild-type thrombin.

Activation of platelets via the tethered ligand thrombin receptor was monitored by the uptake of calcium. The ED_{50} values for all mutants and wild-type thrombin fell into a range of 1 to 8 nM.

These results suggest that W60d is important for thrombin's interaction with serpins, in particular AT, but not with hirudin or the tethered ligand thrombin receptor. In addition, E192 plays a role in thrombin's interaction with hirudin and the E192Q mutation compensates for changes at W60d during serpin inhibition. However, the E192Q mutation has no effect on tethered ligand thrombin receptor activation in platelets. In conclusion, W60d and E192 influence some macromolecular substances, but not all. Furthermore, the combination of mutations at these positions appears to compensate for a loss of activity with only a single mutation.

3. NOVEL ANTITHROMBIN VARIANTS ASSOCIATED WITH THROMBOSIS

H. L. Fitton,* I. D. Walker,[+] L. Jones,[∅] K. Brown,* and P. Coughlin*

*University of Cambridge Department of Haematology, MRC Centre, Hills Road, Cambridge, CB2 2QH, U.K., [+]Department of Haematology, 3rd Floor, Macewen Building, Glasgow Royal Infirmary, Castle Street, Glasgow, G4 0SF, U.K., and [∅]Epsom General Hospital, Dorking Road, Epsom, Surrey, KT18 7EG, U.K.

Antithrombin is a critical regulator of thrombin and other serine proteases of the coagulation pathway. Deficiency or dysfunction of antithrombin is associated with venous thromboembolism. Mutations causing antithrombin dysfunction have been useful in identifying important areas of the antithrombin structure. Those with altered heparin affinity have helped localise the heparin binding site to the D-helix and beginning of the A-helix. Several mutations that affect the inhibitory activity of antithrombin are present at or close to the reactive site, Arg 393–Ser 394 (P_1–P_1'). A number of other mutations have pleiotropic effects, demonstrating reduced heparin affinity and inhibitory activity.

We have identified five mutations in antithrombin by direct sequencing of exons amplified using PCR. Four of these were associated with familial thrombophilia. Of these, three caused a type 1 antithrombin deficiency and the other a type II deficiency. The significance of the last mutation is uncertain but it may be a polymorphic variant.

The type 1 mutations are in exon 2, exon 3b and exon 4. The first of these is a missense mutation causing substitution of Tyr→STOP at position-15 within the secretion signal sequence. The second mutation is an in-frame deletion resulting in the loss of Ile 186. This is a highly conserved residue in the serpin superfamily and results in the disruption of the F-helix. The last is a missense mutation resulting in the substitution Cys→Ser at position 247. This disrupts the disulphide bond with Cys 430 leaving a free cysteine residue and the C-terminus unconstrained. The type II mutation was associated with a borderline antithrombin antigen level (75%) and heparin cofactor activity (79%). The mutation, in exon 3a, resulted in the substitution of SER162 by Asn. This residue is sited in the E-helix and the replacement of the buried side chain of serine by the larger asparagine side chain will predictably cause structural perturbation. The last example, Val 415→Asn, was an apparently incidental finding as a follow up investigation of a nephrotic patient. Although one other member of the family also had the mutation there was no linked history of thrombotic disease. Valine 415 is an external residue on the turn of strand 4 to strand 5 of the B-sheet. Structurally its replacement should have minimal effect, consistent with the observed lack of clinical consequences.

4. INACTIVATION OF THROMBIN BY ANTITHROMBIN IS ACCOMPANIED BY INACTIVATION OF REGULATORY EXOSITE I

Paul E. Bock,[‡] Steven T. Olson,[§] and Ingemar Björk[¶]

[‡]Department of Pathology, Vanderbilt University School of Medicine, Nashville, Tennessee, [§]Center for Molecular Biology of Oral Diseases, University of Illinois, Chicago, Illinois, and [¶]Department of Veterinary Medical Chemistry, Swedish University of Agricultural Sciences, Uppsala, Sweden

Regulatory exosite I of the blood clotting proteinase, thrombin, plays an important role in directing the activity of the enzyme toward certain specific inhibitors and physiological substrates, and in mediating its interactions with regulatory proteins of the hemostatic system. A fluorescent derivative of the COOH-terminal dodecapeptide of hirudin containing fluorescein at the amino-terminus ([5-F]Hir$^{54-65}$) was synthesized and its specific binding to exosite I was used to probe changes in the function of the regulatory site accompanying thrombin inactivation by its physiological serpin inhibitor, antithrombin. Characterization of [5-F]Hir$^{54-65}$ binding to human α-thrombin measured by a decrease in the fluorescein fluorescence showed binding to 0.94±0.04 sites with a dissociation constant of 26±2 nM, and a maximum fluorescence change of −25±1%. In contrast, the stable thrombin–antithrombin complex (T–AT) decreased the fluorescence of [5-F]Hir$^{54-65}$ by <3% at concentrations up to 3 μM, indicating no detectable binding. Thrombin that had been active-site-blocked with D-Phe-Pro-Arg-CH$_2$Cl bound [5-F]Hir$^{54-65}$ with the same affinity as native thrombin. The T-AT complex at a concentration of 4 μM did not have a significant competitive effect on binding of the fluorescent peptide to active-site-blocked thrombin, indicating a >100-fold decrease in affinity of thrombin in the T–AT complex for the peptide. Analysis of changes in [5-F]Hir$^{54-65}$ fluorescence accompanying reactions of thrombin with antithrombin showed that the loss of binding sites for the peptide occurred with the same time course as the loss of thrombin catalytic activity. The bimolecular rate constants for enzyme inactivation and peptide dissociation decreased in parallel with increasing concentration of [5-F]Hir$^{54-65}$, with a maximum effect of 1.7-fold, and with an inhibition constant (16±13 nM) indistinguishable from the dissociation constant determined for peptide binding by fluorescence titration. Similar results were obtained in additional studies with the unlabeled hirudin peptide (Hir$^{54-65}$), which indicated that the affinity of thrombin for the peptide was decreased in the T-AT complex, and that binding of the peptide to thrombin resulted in a 2.2-fold decrease in the bimolecular rate constant. The partial inhibition of the reactions by [5-F]Hir$^{54-65}$ and Hir$^{54-65}$ was not due to increased partitioning of antithrombin to the cleaved form of the inhibitor, as shown by the unchanged reaction stoichiometry in the presence of the peptides. Results of these studies support a mechanism in which free thrombin and the thrombin-Hir$^{54-65}$ complex associate with antithrombin and undergo formation of the T–AT complex at modestly different rates, with dissociation of the peptide from exosite I occurring subsequent to rate-limiting formation of the T–AT complex. This mechanism suggests that antithrombin inactivation of thrombin bound in exosite I-mediated complexes with regulatory proteins, such as thrombomodulin and fibrin, may be accompanied by essentially irreversible release of the stable T–AT complex due to exosite I inactivation. This mechanism may function to clear inactivated thrombin from complexes with regulatory proteins without prior dissociation of these complexes, and thus serve to localize the activity of thrombin physiologically.

5. PROBING SERPIN REACTIVE LOOP CONFORMATIONS BY PROTEOLYTIC CLEAVAGE

Wun-Shaing W. Chang,* Mark R. Wardell, David A. Lomas, and Robin W. Carrell

Department of Haematology, University of Cambridge, MRC Centre, Hills Road,
    Cambridge CB2 2QH, U.K.

Several crystal structures of intact members of the serpin superfamily have recently been solved but the relationship of their reactive loop conformations to those of circulating forms remains unclear. Here we examine reactive loop conformational changes of antitrypsin and antithrombin using limited proteolysis and binary complex formation with synthetic homologous reactive loop peptides. Proteolysis at the P10–P9, P8–P7 and P7–P6 bonds of antitrypsin was distorted by binary complex formation. The P1'–P2' bond in antithrombin was more accessible to proteolysis following binary complex formation whereas cleavage at the P4–P3 bond was variably altered by synthetic peptide insertion. The proteolytic accessibility of the reactive-site P1–P1' bond of antitrypsin and antithrombin binary complexes was identical to that of the native form and no cleavage was observed in the hinge region (P15–P10) of either protein whether native or as binary complexes. These results fit with the proposal that the hydrophobic reactive loop of serpins adopts a modified helical conformation in the circulation with the hinge region being partially incorporated into the A-sheet. This loop can be displaced by peptides and induced to adopt a new conformation similar to the three-turn helix of ovalbumin. Both the native and binary complexed forms of antithrombin showed a greatly increased proteolytic sensitivity in the presence of heparin, indicating that heparin either induces a conformational change in the local structure of the helical reactive loop or facilitates the approximation of enzyme and inhibitor.

6. ARGININE 200 OF HEPARIN COFACTOR II STABILIZES BOTH TERNARY COMPLEX FORMATION AND INTRAMOLECULAR INTERACTION WITH ITS ACIDIC DOMAIN

Angelina V. Ciaccia and Frank C. Church

Departments of Pharmacology and Pathology, School of Medicine, University of North Carolina at Chapel Hill, Chapel Hill, North Carolina

Heparin cofactor II (HCII) is a serine proteinase inhibitor (serpin) that inhibits the coagulation proteinase thrombin in a manner that is greatly accelerated in the presence of glycosaminoglycans, such as heparin or dermatan sulfate. It is thought that binding of gly-cosaminoglycans to the HCII D-helix displaces the amino terminal HCII "acidic domain", which then interacts with anion binding exosite 1 (ABE-1) of thrombin and further stabilizes the bimolecular complex. The role of HCII residue Arg 200 in thrombin inhibition and glycosaminoglycan binding was investigated. R200 is in strand 2 of sheet A, which is immediately after the D-helix. A comparison of glycosaminoglycan-binding serpins shows that HCII is the only serpin with a basic residue at this position. HCII mutants containing substitutions of either Ala or Glu at Arg 200 were generated by site-directed mutagenesis of the wildtype HCII cDNA. The rHCII mutants were expressed in High-Five insect cells, partially purified, and characterized in comparison to wt-rHCII.

Heparin and dermatan sulfate template curves were performed to determine if R200 is involved in glycosaminoglycan-accelerated thrombin inhibition. In the presence of dermatan sulfate, the mutants were identical to wt-rHCII. The maximal inhibition rates (k_2, $M^{-1}min^{-1}$) were 1.3×10^9 (wt-rHCII), 1.4×10^9 (R200A), and 1.2×10^9 (R200E), and the optimal dermatan sulfate concentration was about 800 µg/mL for all proteins. The maximal k_2 values with heparin were 4.0×10^8 (wt-rHCII), 5.1×10^8 (R200A), and 4.8×10^8 (R200E), but both rHCII mutants required higher heparin concentrations for optimal inhibition (500 µg/mL for mutants $vs.$ 200 µg/mL for wt-rHCII). Heparin-Sepharose salt elution profiles were performed in order to determine the relative affinity of the proteins for heparin. There were slight differences in the NaCl concentration required to elute the proteins (250–300 mM for wt-rHCII, 230–260 mM for R200A, and 220–260 mM for R200E). When heparin template curves were performed with R93,97,101A-thrombin, which does not bind heparin or form a ternary complex with heparin-HCII, the difference in the heparin optimum was eradicated (wt-rHCII and both R200 mutants required 1000 µg/mL for optimal activity). The k_2 values were 2.8×10^8 (wt-rHCII), 5.6×10^8 (R200A), and 5.7×10^8 (R200E). These data suggest that although R200 is not part of the primary heparin binding site of HCII, it might be on the "path" of heparin as it bridges HCII and thrombin.

In the absence of glycosaminoglycan, there was a significant difference in thrombin inhibition by wt-rHCII as compared to the mutants. R200A and R200E were 3- and 5.6-fold more active than wt-rHCII, respectively. The k_2 values were 2.2×10^4 (wt-rHCII), 6.6×10^4 (R200A), 1.2×10^5 (R200E). When assayed with γ-thrombin, whose ABE-1 domain has been partially destroyed by limited proteolysis, the inhibition rates of the mutants were significantly decreased to about 2-fold greater than wt-rHCII. The k_2 values were 1.3×10^4 (wt-rHCII), 2.8×10^4 (R200A) and 2.9×10^4 (R200E). In addition, hirugen, which binds to

ABE-1 of thrombin, was more efficacious in reducing the inhibition rates of the mutants than of wt-rHCII (R200E > R200A > wt-rHCII). Thus, blockage of the acidic domain- ABE-1 interaction more greatly interferes with thrombin inhibition by the mutants than by wt-rHCII. These data suggest that the increased progressive anti-thrombin activity of the R200 mutants is due to increased acidic domain interaction with ABE-1. Based on these data, we hypothesize that in the absence of glycosaminoglycan, Arg 200 stabilizes acidic domain binding to HCII by acting as an "anchor" that attracts the negatively charged acidic domain. Changing Arg 200 to an Ala or Glu would favor the equilibrium towards acidic domain binding to ABE-1, resulting in increased thrombin inhibition.

7. A MONOCLONAL ANTIBODY AGAINST THROMBIN EXOSITE II SLOWS INHIBITION BY HEPARIN/ANTITHROMBIN BUT NOT BY DERMATAN SULFATE/HEPARIN COFACTOR II

Niall S. Colwell, Morey A. Blinder, Jayne C. Marasa, and Douglas M. Tollefsen

Division of Hematology, Department of Internal Medicine, Washington University, St. Louis, Missouri

Previous studies with thrombin mutants suggests that rapid inhibition by heparin/antithrombin (AT) involves a template mechanism in which the glycosaminoglycan chain interacts with anion-binding exosite II of thrombin. By contrast, mutations in exosite II do not affect the rate of inhibition of thrombin by dermatan sulfate/heparin cofactor II (HCII) (J.P. Sheehan et al., J. Biol. Chem. 269: 32747, 1994). In the current study, we have isolated an IgG-kappa monoclonal antibody that binds to thrombin exosite II from a patient with multiple myeloma and have determined the effects of this antibody on inhibition of thrombin by AT and HCII. The antibody, purified by chromatography on protein A-Sepharose and thrombin-Sepharose, immunoprecipitated ^{125}I-labeled human alpha-, beta-, and gamma-thrombin, but not prothrombin, other vitamin K-dependent coagulation factors, or fibrinogen. We used a panel of surface mutants (M. Tsiang et al., J. Biol. Chem. 270: 16854, 1995) to define the epitope on thrombin to which the antibody binds. Wild-type and mutant prothrombins were transiently expressed in Cos-7 cells, activated to thrombin by *Echis carinatus* venom, and immunoprecipitated by protein A-Sepharose beads coated with the monoclonal IgG. The antibody bound wild-type thrombin and 52 of the 54 expressed thrombin mutants, but did not bind to the single amino acid mutant R245A or to the triple mutant K252A/D255A/Q256A. R245 and K252 occur in close proximity to one another in exosite II, and both residues are required for binding to heparin or dermatan sulfate. The purified antibody bound to thrombin with an apparent K_d of 20 nM (3 µg/ml) and decreased the rate of hydrolysis of the chromogenic substrate tosyl-Gly-Pro-Arg-ρ-nitroanilide by 54% at saturation. At a concentration of 80 µg/ml, the antibody decreased the second-order rate constant for inhibition of thrombin by heparin/AT 118-fold (from 1.3×10^8 to 0.011×10^8 M^{-1} min^{-1}). By contrast, the antibody decreased the rate constant for inhibition of thrombin by heparin/HCII only 2-fold (from 1.5×10^8 to $0.76 \times 10^8 M^{-1} min^{-1}$) and had no effect on inhibition by dermatan sulfate/HCII. These results provide further evidence for the lack of involvement of the glycosamino glycan-binding site of thrombin (exosite II) in the mechanism of inhibition by dermatan sulfate/HCII.

8. HEPARIN BRIDGES ANION BINDING EXOSITE 2 OF THROMBIN WITH BOTH THE D-HELIX OF HEPARIN COFACTOR II AND THE H-HELIX OF PROTEIN C INHIBITOR

Scott T. Cooper,[‡] Alireza R. Rezaie,[¶] Charles T. Esmon,[¶] and Frank C. Church[‡]

[‡]Departments of Pathology and Medicine, The University of North Carolina at Chapel Hill, School of Medicine, Chapel Hill, North Carolina 27599, and [¶]Cardiovascular Biology Research Program, Oklahoma Medical Research Foundation, Oklahoma City, Oklahoma 73104

Therapeutically, coagulation is most often regulated by the administration of heparin, accelerating the inhibition of thrombin by the plasma serine protease inhibitor (serpin) antithrombin (AT). The inhibition of thrombin by two other serpins, heparin cofactor II (HCII) and protein C inhibitor (PCI), is also stimulated by heparin. Heparin can act as a "template", forming a trimolecular complex with thrombin and the serpin, and thus accelerating the reaction. All three of these heparin stimulated inhibitors of thrombin exhibit different characteristics in their reaction with thrombin. AT undergoes a conformational change upon binding to heparin primarily through a specific pentasaccharide, which increases its activity with proteases. Heparin spans the D-helix of AT and anion binding exosite 2 (ABE2) of thrombin further accelerating the reaction. Like AT, the heparin-binding site in HCII is centered in the D-helix. In addition, the D-helix is involved in dermatan sulfate binding, a reaction apparently unique to HCII. The amino-terminal 100 residues of HCII contains two "hirudin-like" acidic sequences which are thought to bind intramolecularly to the D-helix. Binding of heparin to the D-helix displaces the acidic domains and catalyzes their interaction with anion binding exosite 1 (ABE1) of thrombin, thus dramatically increasing the rate of thrombin inhibition. PCI is unique in that its primary heparin binding site is located on the H-helix instead of the D-helix as in AT, HCII, plasminogen activator inhibitor-1 and protease nexin-1. Another positively charged helix located at the amino terminus of PCI, the A+ helix, may also contribute to heparin binding. PCI does not go through any measurable conformational change upon binding to heparin, and compared to AT and HCII it displays the least stimulation by heparin. This study was undertaken to compare the inhibition properties of AT, HCII and PCI in the absence and presence of glycosaminoglycans using a mutant thrombin which has reduced heparin binding properties.

Mutagenesis of thrombin residues Arg_{93}, Arg_{97} and Arg_{101} (in anion binding exosite 2) to Ala decreases heparin affinity and heparin-stimulated AT inhibition of thrombin. In the absence of heparin, PCI inhibited the mutant thrombin better than wild-type thrombin. However, heparin enhanced PCI inhibition of the mutant thrombin only 1.7 fold, compared with 40-fold enhancement with wild-type thrombin. Both wild-type and mutant thrombin had similar activities with HCII in the absence of glycosaminoglycan. HCII achieved the same maximum activity in the presence of heparin with the mutant thrombin; however, the optimum heparin concentration for this reaction was 20 times higher than that for the reaction with wild-type thrombin, indicative of a decrease in heparin affinity. Thrombin inhibition in the presence of dermatan sulfate was not changed in the mutant thrombin, suggesting that dermatan sulfate does not accelerate the HCII reaction with thrombin via a template mechanism or that heparin and dermatan sulfate bind to separate sites of thrombin. Molecular models of docked complexes between thrombin and each of the serpins AT, HCII and PCI reveal that heparin can bind to ABE2 of thrombin and either the D- or H-helix of the serpin. These studies indicate that ABE2 of thrombin has a critical role in the heparin catalysis of reactions between thrombin and AT and PCI, but it is not essential for the reaction with HCII. For HCII this implies that HCII either employs other residues in thrombin to accelerate the reaction, or a template mechanism is not essential for catalysis. For PCI this is somewhat surprising as the heparin binding domains of AT and HCII are located on the D helix, whereas in PCI heparin interacts with the H helix.

9. LOCALIZATION OF HEPARIN-BINDING SERPINS IN NORMAL HUMAN OVARY AND OVARIAN CANCER

Cory A. Dunnick,* Maureane Hoffman,[#] and Frank C. Church*

*University of North Carolina School of Medicine, Chapel Hill, North Carolina 27599, and
[#]Duke University Medical Center, Durham, North Carolina 27705

Ovarian cancer is the fourth leading cause of cancer death in U.S. women with approximately 20,000 new cases per year and greater than 12,000 deaths. Although ovarian cancer represents only 6% of cancers in females, these neoplasms account for a disproportionate number of fatal cancers because they cannot be detected early. Furthermore, risk factors for ovarian cancer are much less clear than those for other gynecological tumors. Despite the large morbidity and mortality associated with ovarian cancer, its pathogenesis and etiology are not well understood.

An immunohistochemistry study was performed to see if any heparin-binding serine proteinase inhibitors (serpins) were found in normal human ovary and ovarian cancer. The serpins studied were protein C inhibitor (PCI), antithrombin (ATIII) and heparin cofactor II (HCII). PCI [also called plasminogen activator inhibitor-3 (PAI-3)] inhibits thrombin and activated protein C in blood coagulation, but it also inhibits urokinase-type plasminogen activator and tissue-type plasminogen activator (and other serine proteases) which are important during tumor cell metastasis and cell migration processes. ATIII and HCII are much more focused in their inhibition pattern in that they primarily regulate proteases in blood coagulation.

By Northern blot analysis, PCI mRNA was detected in ovary (we have not yet completed the Northerns for ATIII and HCII mRNA in this tissue). In normal ovary, PCI was most prominent in the cytoplasm of epithelial cells immediately adjacent to the follicle. PCI was also strongly positive in endothelial cells and in the cytoplasm of a few cells scattered throughout the ovary which may represent mast cells. The stroma stained for PCI in a reticular pattern. We studied three adenocarcinomas of the ovary, and all three showed decreased to absent staining for PCI except in necrotic areas which are probably staining nonspecifically. A granulosa cell cancer specimen (stromal cell tumor) showed PCI staining that was equivalent or only slightly less than normal ovary. ATIII staining was strong and almost diffuse in the normal ovary. Like PCI, the follicular epithelium was most prominently positive. However, unlike PCI, ATIII seemed to have a layered appearance around the follicle. The theca interna and cells directly adjacent to the follicle were strongly positive while the theca externa was relatively spared. ATIII stained in the stroma and endothelium. ATIII remained strongly positive in the adenocarcinoma and granulosa cell tumors of the ovary. HCII weakly stained the follicular epithelium, vascular endothelium, stromal and granular cells presumed to be mast cells in normal ovary. HCII was diminished in adenocarcinoma and absent in the granulosa cell tumor of the ovary.

These results show that there is a dramatic change in the staining pattern for PCI in epithelial cell cancers of the ovary compared to normal tissue. This is in contrast to that found for ATIII and HCII which were present in normal ovary and ovarian cancer, but were not drastically different in the neoplasms. Our Northern blot data show that PCI is produced in normal ovary and not just deposited from the serum. PCI could be diminished or absent from ovarian adenocarcinomas because of dedifferentiation of cancer cells or it could actually participate in the carcinogenesis of these cells. Since PCI is capable of regulating proteases involved in degradation of the extracellular matrix, the loss of PCI production could be related to the invasion of ovarian tumors. Whether PCI could be considered a tumor suppressor gene awaits a molecular biology probe of PCI mRNA in ovarian cancers and additional immunohistochemistry studies. Further consideration of PCI is warranted because the risk factors for ovarian cancer are not well defined, and ovarian cancer continues to be the leading cause of death from gynecological cancers.

10. INHIBITION OF SERINE PROTEASES BY REACTIVE SITE MUTANTS OF PROTEIN C INHIBITOR

Marc G. L. M. Elisen,[1] Frank C. Church,[2] Bonno N. Bouma,[1] and Joost C. M. Meijers[1]

[1]Department of Haematology, University Hospital, Utrecht, the Netherlands, and
[2]Department of Pathology and the Center for Thrombosis and Hemostasis, University
   of North Carolina School of Medicine, Chapel Hill, North Carolina

Protein C inhibitor (PCI) is a relatively non-specific inhibitor and is a member of the serine protease inhibitor (serpin) superfamily. It can be further classified as a heparin-binding serpin like antithrombin III and heparin cofactor II. Heparin and some other glycosaminoglycans act as a template thereby increasing the rate of inhibition for PCI. Protein C inhibitor can inhibit various serine proteases in haemostasis and reproduction. However, its target enzyme is still unknown. Reactive site mutants of PCI were used to define target protease specify. Elucidation of the specificity of PCI might provide clues towards its function *in vivo*. PCI has an Arg at the P3' position, similar to fibrinogen. This residue may interact with Glu39 and Glu192 of thrombin and Glu192 of activated protein

C (APC). We have studied the importance of both P3 and P3′ residues of PCI in protease recognition especially towards APC and thrombin. Factor Xa, factor XIa, kallikrein and acrosin were also used as target proteases to characterize the specifity of PCI in the region P3–P2–P1–P1′–P2′–P3′. For these purposes a set of PCI reactive site mutants was used: P3 Thr352Arg; P3 Thr352Lys; P2 Phe353Pro; P2 Phe353Gly; P1′ Ser355Val; P3′ Arg-357Asp; P3′ Arg357Ala and a double mutant P3 Thr352Glu–P3′ Arg357Asp. The kinetic parameters kass. and Ki were determined using slow-binding kinetics. Comparison of target protease specifity of PCI P3′ mutants with wild type PCI did not result in dramatic changes in kinetic parameters for the proteases mentioned above. On the other hand we observed that substitution at the P3 position (Thr352Arg) of PCI resulted in a more potent inhibitor by increasing the kass. by a factor 2–4 for the inhibition of APC, factor Xa, factor XIa and kallikrein. Surprisingly, Thr352Lys substitution resulted in a decreased association constant for these serine proteases compared to wild-type PCI. For the inhibition of acrosin, the P3 position of PCI also plays an important role towards substrate recognition. A PCI double mutant P3 Thr352Glu and P3′ Arg357Asp resulted in an 8-fold increase of kass. while the Arg357Asp mutant did not alter the kinetic parameters. Since mutations of Thr352 into Arg and Lys resulted in a mild decrease in kass. and Ki, we believe that substitution of Thr352 into a negatively charged residue can result in a more potent inhibitor of acrosin. We were not able to demonstrate a role of Arg357 at P3′ in target specifity for thrombin and APC. However, we can conclude that the P3 position of PCI plays an important role in target protease recognition.

11. INFLUENCE OF MOLECULAR WEIGHT AND ION STRENGTH ON HEPARIN STIMULATION OF THE INHIBITION OF ACTIVATED PROTEIN C AND OTHER ENZYMES BY HUMAN PROTEIN C INHIBITOR

Francisco España,[1] Jaime Sánchez-Cuenca,[1] Amparo Estellés,[1] Juan Gilabert,[2] and Justo Aznar[3]

[1]Research Center and [2]Departments of Obstetrics and Gynecology and [3]Clinical Pathology, "La Fe" University Hospital, 46009 Valencia, Spain

Protein C inhibitor is a serine protease inhibitor (serpin) that inhibits procoagulant and profibrinolytic enzymes (España & Griffin, Blood 70 (S1): 401, 1987; España et al, Thromb. Res. 55: 369, 1989). Since all its inhibition reactions are heparin-dependent, unfractionated heparins and low-molecular-weight- (LMW-) heparins were studied with respect to their ability to potentiate the inhibition of fibrinolytic and coagulation factors by protein C inhibitor. The inhibition of activated protein C, urokinase plasminogen activator, tissue plasminogen activator, thrombin, factor Xa, factor XIa and plasma kallikrein by protein C inhibitor was found to be dependent on the size of the polysaccharide. In general, maximal stimulation was obtained with unfractionated heparins, except in the case of plasma kallikrein. Differences in heparin stimulation were more pronounced for thrombin, activated protein C, urokinase and tissue plasminogen activators and factor XIa, whereas inactivation of factor Xa by protein C inhibitor was less dependent on the presence of heparin, and plasma kallikrein showed higher potentiation with LMW-heparin than with unfractionated heparin. The second-order rate constants for enzyme inhibition by protein C inhibitor were strongly dependent on the ion strength and, in general, with ion strength above 0.15 the heparin stimulation of the inhibition reactions was drastically reduced. These results may explain the wide discrepancies in the literature as to the effect of heparin on the stimulation of enzyme inhibition by protein C inhibitor. They also show that LMW-heparins stimulate the inhibition of the fibrinolytic enzymes activated protein C, urokinase and tissue plasminogen activators less efficiently, which could contribute to their profibrinolytic effect. FIS 96/1129.

12. REGULATION OF THE TISSUE FACTOR PATHWAY BY SERPINS AND OTHER INHIBITORS

Mirella Ezban and Ulla Hedner

Novo Nordisk A/S, Vessel Wall Biology-Hemostasis, Hagedornsvej 1 DK-2820 Gentofte, Denmark

The Factor VIIa (fVIIa)/Tissue Factor (TF) complex is the principal initiator of the coagulation cascade. The complex catalyses the conversion of Factor IX and Factor X into the active enzymes Factor IXa and Factor Xa, respectively, resulting ultimately in thrombin generation.

An important aspect of coagulation is its strict regulation by the proteinase inhibitors Tissue Factor Pathway Inhibitor (TFPI) and Antithrombin (ATIII). FVIIa attains significant catalytic activity only in complex with TF, and the complex is inhibited by both plasma inhibitors. The inhibition with ATIII takes place only in the presence of heparin/heparan sulfate and may be too slow to be of physiological importance. The inhibition by TFPI requires the presence of fXa. The fact that fVIIa is inhibited when complexed to its cofactor, TF, is in sharp contrast to the mode of action by which other vitamin K-dependent coagulation proteins, fIXa and fXa are neutralized. These proteins are relatively inaccessible to inhibition in their reactive tenase and prothrombinase complexes (fIXa/Factor VIII/phospholipid and fXa/Factor V/phospholipid, respectively) but are effectively neutralized when freed to the circulation. Thus, TFPI would appear to control initiation of the coagulation cascade whereas ATIII regulates the later steps of the coagulation reactions.

Interference with the fVIIa/TF complex formation represents an alternative way of controlling initiation of coagulation. We have also studied the effect of inhibiting TF-induced coagulation by means of an active site inhibited recombinant fVIIa (fVIIai), prepared by binding of a tripeptidechloromethylketone to human recombinant fVIIa. In several in vitro binding assays it was shown that rfVIIai competes efficiently with fVIIa for binding to TF. In different animal models of arterial injury it has been demonstrated that rfVIIai is able to efficiently reduce thrombus formation without causing increased bleeding. We will present data which compares these three inhibitors in their regulation of the tissue factor pathway.

13. STRUCTURE-BASED DESIGN OF SYNTHETIC HEPARIN ANALOGUES

Peter D. J. Grootenhuis and Constant A. A. van Boeckel

NV Organon, P.O. Box 20, 5430 BH Oss, The Netherlands

It has been known for over a decade that a unique, well-defined pentasaccharide domain of heparin is essential for the activation of antithrombin (AT) III. This pentasaccharide domain of heparin has been used as a lead to new synthetic antithrombotics and it has been shown that various well-defined pentasaccharides can be synthesized that solely stimulate the AT III-mediated factor Xa inhibition. In order to accelerate thrombin inactivation, oligo-saccharides which comprise about 18 saccharides in addition to containing the unique pentasaccharide, are required. We have tried to assist the design of new compounds by providing a molecular basis for the specific interactions between AT III, heparin and the coagulation proteases. The resulting 3D-models, in combination with structural and binding data that recently became available, lead us to propose a molecular mechanism for the heparin-induced potentiation of AT III. In addition, the models inspired the design and synthesis of novel glycoconjugates with both thrombin and factor Xa inhibitory activities. Interestingly, the inhibitory profile of such synthetically feasible compounds can be tuned in a rational way, leading to novel anticoagulants with unprecedented characteristics.

REFERENCES

1. C.A.A. van Boeckel & M.Petitou Angew.Chem.Int.Ed.Engl. 32(1993)1671.
2. P.D.J.Grootenhuis & C.A.A. van Boeckel J.Am.Chem.Soc. 113(1991)2743.
3. C.A.A. van Boeckel, P.D.J.Grootenhuis & A.Visser Nat.Struct.Biol. 1(1994)423.
4. P.D.J.Grootenhuis et al. Nat.Struct.Biol. 2 (1995)736.

14. MOLECULAR REGULATION OF PROTEASE NEXIN-1

Denis C. Guttridge and Dennis D. Cunningham

Department of Microbiology and Molecular Genetics, University of California, Irvine, Irvine, California 92717

Protease Nexin-1 (PN-1) or Glia-Derived Nexin is a potent inhibitor of serine proteases in the extracellular environment. It is abundantly expressed in the nervous system where it is thought to participate in local injury and repair processes. Tissue culture studies have identified several injury-related factors that regulate PN-1 expression. In particular, cytokines interleukin-1 beta and tumor necrosis factor alpha increase PN-1 secretion (Vaughan and Cunningham, 1993), while angiotensin II and dexamethasone function separately to down regulate PN-1 expression (Guttridge et al., 1993; Bleuel et al., 1995). Elucidating the regulatory mechanisms of these factors should provide a better understanding of PN-1 function during development and wound repair. Previous evidence showed that regulation by some of these factors is mediated at the transcriptional level. Although some information has been obtained regarding the PN-1 gene structure, relatively little is known about the cis and trans-acting factors that regulate its transcriptional activity. Utilizing the human version of the PN-1 promoter we have identified an activation domain which maps to positions −199 to −45 relative to the start site of transcription. This domain is highly G/C rich and contains multiple putative Sp1 binding sites. Electrophoretic mobility shift assays and DNase I footprinting revealed multiple Sp1 binding in two regions of the activation domain. Cotransfection experiments into the Sp1-deficient Drosophila SL2 cell line also showed Sp1 activates PN-1 promoter activity in a dose-dependent fashion via the Sp1 binding sites. However, our results with nuclear extract from a PN-1 expressing cell line indicates that other unidentified proteins also bind within the activation domain upstream of the Sp1 binding sites. Our analysis thus demonstrates that Sp1 is a major regulator of PN-1 transcription, but other transcription factors are most likely involved, and function either independently or cooperatively with Sp1 to stimulate PN-1 transcription.

REFERENCES

Bleuel et al., (1995) J. Neurosci. 15, 750–761.
Guttridge et al., (1993) J. Biol. Chem. 268, 18966–18974.
Vaughan and Cunningham (1993) J. Biol. Chem. 268, 3720–3727.

15. HEPARIN FACILITATES DISSOCIATION OF COMPLEXES BETWEEN THROMBIN AND A REACTIVE SITE MUTANT (L444R) OF HEPARIN COFACTOR II

Jin-hua Han, Vivianna M. D. Van Deerlin, and Douglas M. Tollefsen

Division of Hematology, Department of Internal Medicine, Washington University, St. Louis, Missouri

Heparin cofactor II (HCII) is a human serpin that inhibits thrombin and chymotrypsin. Heparin and dermatan sulfate increase the rate of inhibition of thrombin by HCII≥1000-fold. We previously reported that the mutation L444R at the P1 position of the reactive site of HCII increases the rate of inhibition of thrombin ~100-fold in the absence of a glycosaminoglycan (V.M. Derechin, M.A. Blinder, and D.M. Tollefsen, J. Biol. Chem. 265: 5623, 1990). We now find that inhibition of thrombin by this mutant is reversible in the presence of heparin but not dermatan sulfate. Thrombin (7 nM) is rapidly inhibited by HCII(L444R) (260 nM) in the presence of heparin (50 µg/ml), but ~50% of the thrombin activity subsequently reappears with a $t_{1/2}$ of ~30 min. By contrast, thrombin activity remains undetectable for ≥ 75 h when dermatan sulfate is substituted for heparin. At higher HCII(L444R): thrombin ratios, the onset of dissociation caused by heparin is delayed and the final plateau of thrombin activity is decreased. Electrophoretic analysis shows proteolysis of excess HCII(L444R) during the lag phase. Addition of heparin at longer intervals after formation of the thrombin-HCII(L444R) complex causes a progressive decrease in the thrombin plateau. Thus, in the absence of heparin the complex is slowly converted to an irreversibly inhibited form. Addition of the hirudin C-terminal peptide to trap thrombin as it is being released from HCII(L444R) in the presence of heparin results in nearly full recovery of thrombin amidolytic activity, which is consistent with the hypothesis that irreversible inhibition depends on proteolysis of the complex by a trace amount of free thrombin. In summary, heparin facilitates dissociation of the thrombin-HCII(L444R) complex to yield a proteolytically cleaved form of the inhibitor and active protease. Our data indicate that the P1 residue determines not only the protease specificity of HCII, as reflected by the rate of complex formation, but also affects the stability of the resulting complex.

16. MECHANISM OF HEPARIN ACTIVATION OF ANTITHROMBIN: EVIDENCE FOR REACTIVE CENTER LOOP PRE-INSERTION WITH EXPULSION UPON HEPARIN BINDING

James A. Huntington, Bingqi Fan, Steven T. Olson, and Peter G. W. Gettins

Department of Biochemistry, University of Illinois at Chicago

Heparin binds tightly to antithrombin in a two-step mechanism, the second of which is a conformational change that is necessary for enhancing the rate at which antithrombin inhibits factor Xa, but not thrombin or trypsin. It has been proposed [van Boeckel et al. (1994) *Structural Biology* 1, 423–425], based on recent x-ray structures of heterodimeric forms of antithrombin, that hinge region residues up to P14 are also inserted into β-sheet A in the native solution structure and that heparin binding results in their displacement from the β-sheet, thereby altering the conformation of the reactive center and making it a better target for factor Xa binding. We have previously shown that heparin binding does change the conformation of the P1 residue [Gettins et al (1993) *Biochemistry 32*, 8385–8389]. Here we have examined the properties of a recombinant P14 variant of antithrombin (S380W) to address the question of whether the proposed mechanism of heparin activation by loop displacement is correct. We found that the fluorescence emission spectrum of the P14 tryptophan showed a 17nm red shift upon heparin binding consistent with a change in environment from hydrophobic to polar, as would be expected from displacement of the P14 residue from β-sheet A. The variant also had a seven-fold increased K_a for heparin binding at I 0.6, suggesting that the mutation had altered the thermodynamics of the heparin-induced conformational change. Stoichiometries of inhibition

(S.I.) of trypsin, thrombin and factor Xa were all increased by the P14 substitution, to 26, 12 and 29 respectively. Such changes in SI are in keeping with the branched pathway, suicide substrate inhibition mechanism of serpins and the location of the mutation in the hinge region. Using these S.I. values, we found that corrected second-order rate constants for proteinase inhibition in the absence of heparin were the same as for wild-type antithrombin for inhibition of thrombin and trypsin, but were enhanced 2-fold for inhibition of factor Xa, suggesting that the P14 variant represents a partially-activated form of antithrombin that has a reactive center conformation between that of wild-type antithrombin and heparin-activated antithrombin. Heparin binding resulted in normal enhanced rate constants for inhibition factor Xa, showing that the final heparin-bound reactive center conformation was the same as for activated wild-type antithrombin. Taken together, our findings are consistent with a variant that binds and is activated by heparin, and interacts with proteinases by the same mechanisms as wild-type antithrombin, but does so with changed values that are interpretable in terms of a P14 residue that is already inserted into β-sheet A in native antithrombin, but is displaced upon heparin binding. These conclusions strongly support the proposed mechanism of heparin-activation of antithrombin by reactive center loop displacement.

Supported by grant HL49234.

17. LIMULUS COAGULATION CASCADE IS EFFECTIVELY REGULATED BY THREE TYPES OF ENDOGENOUS SERPINS

Shun-ichiro Kawabata, Yoshiki Miura, Kishan Lal Agarwala, Yuka Kuroki, and Sadaaki Iwanaga

Department of Biology, Faculty of Science, Kyushu University 33, Fukuoka 812–81, Japan

Hemocytes called granulocytes or amebocytes in the hemolymph of horseshoe crab (limulus) contain an intracellular clotting system, which are extremely sensitive to bacterial endotoxin, lipopolysaccharide (LPS). Based on analysis of electron microscopy, large and small granules are present in the hemocyte. The clotting system is composed of at least four serine protease zymogens including factors C, B, G, and proclotting enzyme and a clottable protein, coagulogen. These clotting factors are all localized in the large granules of the hemocytes. When the hemocytes make contact with Gram-negative bacteria or LPS, they begin to degranulate, and the clotting factors secreted initiate hemolymph coagulation. This response is thought to be important for the host defense in engulfing invading microbes, in addition to preventing the leakage of hemolymph. All the serine protease zymogens are activated by limited proteolysis. Factors C and G are biosensors of the clotting reaction, which are autocatalytically activated by LPS and (1,3)-β-D-glucans, respectively. The formation of an active form of factor C (factor $\overline{C}$ or factor G (factor $\overline{G}$) results in the activation of proclotting enzyme and the resulting clotting enzyme catalyzes the conversion of coagulogen to insoluble coagulin gel. Factor $\overline{C}$ activates proclotting enzyme through the activation of factor B, whereas factor $\overline{G}$ directly activates proclotting enzyme.

In our ongoing studies on the molecular mechanism of the limulus clotting pathway, we found recently two types of coagulation inhibitors, named limulus intracellular coagulation inhibitor type-1 (LICI-1) with an apparent $Mr. = 48,000$ (Miura, Y., Kawabata, S., and Iwanaga, S. (1994) *J. Biol. Chem.* **269**, 542–547) and LICI-2 with an apparent $Mr. = 42,000$ (Miura, Y., Kawabata, S., Wakamiya, Y., Nakamura, T., and Iwanaga, S. (1995) *J. Biol. Chem.* **270**, 558–565). These inhibitors are functionally and structurally close to the mammalian plasma serpins. The LICI-1 specifically inhibits factor $\overline{C}$ ($k_I = 2.5 \times 10^6$ M^{-1} S^{-1}), whereas LICI-2 inhibits not only factor $\overline{C}$ ($k_I = 7.1 \times 10^4$ M^{-1} S^{-1}) but also clotting enzyme ($k_I = 4.3 \times 10^5$ M^{-1} S^{-1}). During these studies, we found a third serpin, designated LICI-3, which strongly inhibits factor $\overline{G}$ ($k_I = 3.9 \times 10^5$ M^{-1} S^{-1}). Glycosaminoglycans have no apparent effect on LICI-3 activity as well as LICI-1 and LICI-2, a finding clearly different from mammalian serpins, such as antithrombin III, heparin cofactor II, protease nexin I, and plasminogen activator inhibitor-1. LICI-3 with an apparent $Mr. = 53,000$ (392 amino acid residues) shows significant sequence identities to LICI-1 (46%) and LICI-2 (34%). It also shares sequence identities to other members of the serpin superfamily including mammalian and insect serpins, such as human monocyte/neutrophil elastase inhibitor (35%), human plasminogen activator inhibitor type 2 (29%), human antithrombin III (29%), antitrypsin (30%) from *Bombyx mori*, and elastase inhibitor from *Manduca sexta* (29%). LICI-3 contains a putative reactive site, –Arg–Ser–, distinct from that of LICI-2 (–Lys–Ser–) but the same as LICI-1. Component sugar analysis suggests that the content of sugars in LICI-3 is three times higher than those of LICI-1 and LICI-2, both of which contain one sugar chain, and therefore, the three potential sites in LICI-3 are probably all glycosylated by the similar sugar chain present in LICI-1 or LICI-2. No sialic acid is present in LICIs. Expression of LICI-3 mRNA is only detected in hemocytes, but not in heart, brain, stomach, intestine, coxal gland, and skeletal muscle. Although mammalian plasma serpins are mainly expressed in liver, mRNAs for LICIs are not detected in hepatopancreas with functions analogous to the liver and pancreas of vertebrates. The immunoblotting of large and small granules with antiserum against LICI-3 suggests that LICI-3 is stored specifically in large granules, as in the case of LICI-1 and LICI-2. LICI-3 as well as other two serpins is released by adding calcium ionophore A23187. Therefore, in limulus, all of the serine protease zymogens for coagulation and the three types of serpins with different specificities are co-localized in the same granules and are secreted in response to external stimuli, thereby indicating a more effective coagulation and regulation at local lesions.

18. ANALYSIS OF INTRACELLULAR MECHANISM OF ANTITHROMBIN III DEFICIENCY AND CHARACTERIZATION OF RECOMBINANT MUTANTS

Takehiko Koide, Hiroko Shirotani, and Fuminori Tokunaga

Department of Life Science, Faculty of Science, Himeji Institute of Technology, Harima Science Park City, Hyogo 678–12, Japan

Inherited antithrombin III (ATIII) deficiency is frequently associated with a familial venous thromboembolic disease. Among various mutations of ATIII hitherto characterized, mutants causing type I (classical) and pleiotropic effects-type deficiencies remain to be studied on their intracellular events. Pleiotropic effects-type mutants show the impaired interactions with thrombin and heparin, coupled with a reduced plasma concentration. To examine the intracellular mechanism of ATIII deficiency, we expressed two secretion defect mutants, ΔGlu (deletion of Glu313) and P→Stop (Pro429→Stop codon), and three well-known pleiotropic effects-type mutants, Oslo (Ala404→Thr), Kyoto (Arg406→Met), and Utah (Pro407→Leu), and studied the secretion rate and thrombin-AT III (TAT) complex formation ability, comparing with those of wild type (Wt), two heparin-binding defect mutants, Toyama (Arg47→Cys) and Rouen II (Arg47→Ser). Pulse-chase experiments using a pool of stable BHK cells showed that Oslo and Kyoto mutants gave the secretion rate similar to those of Wt, Toyama, and Rouen II, and after a 2 h-chase, almost all the labeled ATIII was recovered from culture media. Utah mutant, however, showed 50% decrease of secreted antigen, and ΔGlu and P→Stop mutants were secreted little (<10%) into media after an 8h-chase. The decrease in the total amount of the radioactivity suggested the intracellular degradation of these three mutants. In contrast that a part of the intracellular molecules of Wt ATIII were shown to be endoglycosidase H resistant after 1 h-chase, all the intracellular molecules of ΔGlu mutant were sensitive to endoglycosidase H throughout the chase period, suggesting that the intracellular ΔGlu mutant had high mannose-type sugar chains and localized mainly in the endoplasmic reticulum. The intracellular ΔGlu mutant was shown to associate selectively with 100 kDa and 80 kDa proteins in the endoplasmic reticulum, suggesting their function as molecular chaperons. The ability of three pleiotropic effects-type mutants to form a TAT complex was in the decreasing order of Kyoto, Utah and Oslo mutants. Furthermore, the TAT complex formations of Kyoto and Utah were enhanced in the presence of heparin, although neither one of these mutants showed as good ability as Wt. Oslo mutant, however, showed no ability to form a TAT complex in the presence or absence of heparin. From these results, we conclude that each of three mutants, ATIII Oslo, Kyoto and Utah, although they are grouped as "pleiotropic effects-type mutants", have different properties in secretion, intracellular degradation and functional activities.

19. DETERMINING THE SPECIFIC ROLES OF LYS 114, LYS 125, LYS 136 AND LYS 139 IN HEPARIN BINDING AND ACTIVATION OF HUMAN ATIII PRODUCED IN INSECT CELLS

Steven J. Kridel and Daniel J. Knauer

Department of Developmental and Cell Biology, University of California, Irvine, Irvine, California 92717

The binding of heparin to ATIII, and the subsequent activation of antithrombotic activity, has been biochemically mapped to a region in and adjacent to the D-helix. This region is rich in positively charged amino acids that are believed to form coordinate bonds with the sulfate groups in heparin sulfate. Two naturally occurring genetic variants of ATIII with amino acid substitutions at Arg 47 and Arg 129, have demonstrated the key role of these residues in heparin binding and activation. Chemical modification studies have suggested the involvement of several other residues in this region, including Lys 114, Lys 125, Lys 136 and Lys 139. With the exception of Lys 125, none of the residues have been evaluated independently. While these studies have been very informative, and provided the basis for the biochemical localization of the heparin binding site, each of the residues must be evaluated individually to determine their contribution in heparin binding and activation. In the present studies variant forms of ATIII were expressed with specific point mutations that changed Lysine to Glutamine at residues 114, 125, 136 and 139 in a Baculovirus driven expression sysytem. The variant proteins were purified by immuno-affinity chromatography and analyzed for their ability to inhibit thrombin in both the absence and presence of heparin. Direct binding studies using ^{125}I-Flouresceinamine-heparin were employed to determine binding constants of the variant proteins for heparin. The recombinant native form of ATIII expressed by the insect cells was found to inhibit thrombin in the absence of heparin with a second order rate constant of 1.45×10^5 M^{-1} min^{-1}. This value is in good agreement with published second order rate constants for ATIII purified from human plasma. The K114Q (Lys 114 changed to Gln) variant had a comparable second order rate constant of 1.13×10^5 $M^{-1}min^{-1}$, indicating that the reactive center was not affected by the introduced mutation. In the presence of heparin, however, the K114Q variant displayed a severely impaired heparin cofactor activity, while the recombinant native ATIII had heparin cofactor activity identical to human plasma ATIII. This observation was supported by direct binding studies that yielded a heparin binding constant of $K_d \sim 6$ nM for recombinant native ATIII, while the binding of heparin to the K114Q variant was so poor that a K_d could not be determined. Similar results were found for the K139Q variant. In contrast, the K125Q variant had only about a 50% reduction in heparin binding and a concurrent increase in K_d to approximately 9.5 nM. Although previous studies using chemical modification had suggested a role for Lys 136, surprisingly the K136Q variant had a heparin binding constant identical to recombinant native ATIII, indicating that this residue is not involved in heparin binding or activation. These results support the development of a preliminary heparin binding model, in which there is an initial interaction of heparin with the positively charged face of the D-helix, followed by a transition to a stabilized binding state that requires Lys residues 114 and 139 and has a 300-fold higher affinity.

20. ROLE OF THE PROPOSED SEC RECEPTOR RECOGNITION SITE IN BINDING AND INTERNALIZATION OF THROMBIN-HEPARIN COFACTOR II COMPLEXES BY HEPATOCYTES

Hisato Maekawa and Douglas M. Tollefsen

Division of Hematology, Department of Internal Medicine, Washington University, St.
    Louis, Missouri

Serpin-enzyme complexes (SEC) bind to a receptor on the surface of hepatocytes which mediates their endocytosis and lysosomal degradation. Joslin *et al.* (J. Biol. Chem.

266: 11282, 1991) have proposed that a conserved pentapeptide sequence near the C-terminal end of the serpin (e.g., FVFLM in α1-antitrypsin, FLVFI in antithrombin, and FLFLI in heparin cofactor II) mediates binding to the SEC receptor. In experiments with synthetic peptides, they found that substitution of alanine at the fourth or fifth position in this sequence reduced the affinity of binding to HepG2 cells, whereas the presence of alanine at any of the other positions had little or no effect. To test the hypothesis that FLFLI (positions 456–460) in heparin cofactor II (HCII) is involved in binding and uptake of the thrombin-HCII complex by HepG2 cells, we prepared five recombinant HCII variants, each with one of the following mutations: F456A, L457A, F458A, L459A, or I460A. All of the variants were expressed in *E. coli*, purified, and found to inhibit thrombin with normal kinetics both in the presence or absence of a glycosaminoglycan. Radiolabeled complexes were formed by incubating ^{125}I-thrombin with HCII and were purified by heparin-agarose chromatography. At 4°C, the ^{125}I-thrombin-HCII(native) complex binds reversibly to 6.2×10^4 sites per HepG2 cell with a K_d of 19 nM. Binding is inhibited by excess unlabeled thrombin-HCII, thrombin-antithrombin, or elastase-α1-antitrypsin, but not by the free serpin or enzyme, as previously reported. Binding is also inhibited by complexes of thrombin with each of the HCII variants. A competitive binding study with various concentrations of unlabeled thrombin-HCII(native) or thrombin-HCII(I460A) indicated that the two complexes have identical affinity for HepG2 cells. At 37 °C, the ^{125}I-thrombin-HCII(native) complex undergoes time-dependent uptake and lysosomal degradation by HepG2 cells as assayed by release of [^{125}I]iodotyrosine into the medium. Uptake and degradation were inhibited ~50% by excess unlabeled thrombin-HCII and ~75% by 200 μM chloroquine. Complexes of ^{125}I-thrombin with each of the five HCII variants undergo uptake and degradation at the same rate as the complex with native HCII. Our data suggest that the putative SEC receptor recognition site of HCII identified by Joslin *et al.* is not involved in the equilibrium binding, internalization, and degradation of thrombin-HCII complexes by HepG2 cells.

21. SEQUENCE ANALYSIS OF CDNAS ENCODING PAI-1, UPA, AND UPAR FROM OVARIAN CANCER CELL LINES

B. Türkmen, V. Magdolen, P. Trommler,[*] S. Creutzburg, H. Graeff, and M. Schmitt

Klinische Forschergruppe (DFG) der Frauenklinik der TU München, Ismaninger Str. 22, 81675 München, Germany, and [*]Dermatologische und Poliklinik der LMU München

Evidence has accumulated that urokinase-type plasminogen activator (uPA), its inhibitor (PAI-1) and its receptor (uPAR) are involved in tumor invasion and metastasis (for a review see: *Schmitt et al. 1995, J. Obstet. Gynaecol. 21:151*). We analyzed whether the DNA sequence encoding these factors are altered in ovarian cancer cells. We isolated total RNA from three different cell lines (OVCAR-3, OV-MZ-19 and OV-MZ-6) which strongly differ in their expression pattern for uPA, PAI-1, and uPAR (*Will et al. 1994, Int. J. Oncol. 5:753*). Full-length cDNA fragments for uPA, PAI-1, and uPAR were generated by RT-PCR and the DNA sequences were determined. We did not find any indication for alternatively spliced uPA, PAI-1 or uPAR RNAs in the different cell lines.

Compared to published sequences a total of 8 differences were identified. 5 deviations concerned the wobble position of a codon, thus not changing the amino acid sequence of the protein.

- The uPAR cDNAs of all three cell lines encoded the wild-type sequence previously published (*Roldan et al. 1990, EMBO J. 9:467; Casey et al. 1994, Blood 84:1151*).
- In the PAI-1 cDNAs derived from OVCAR-3 and OV-MZ-6 cells, the triplett for residue 15 was found to encode threonine (which agrees with the PAI-1 sequence published by *Andreasen et al. 1986, FEBS Lett. 209:1; Strandberg et al. 1988, Eur. J. Biochem. 176:609*). In contrast, in other PAI-1 cDNA or genomic sequences a triplett encoding alanine was determined (*Ginsburg et al. 1986, J. Clin. Invest. 78:1673; Pannekoek et al. 1986, EMBO J. 5:2539; Bosma et al. 1988, J. Biol. Chem. 263:9129*).
- We failed to isolate PAI-1 specific cDNA fragments from OV-MZ-19 cells. Since we were also unable to detect PAI-1 antigen secreted by these cells using a very sensitive ELISA (detection limit: 0.05 ng/ml culture medium), OV-MZ-19 cells most probably do not express PAI-1.
- In the uPA-encoding cDNA derived from OV-MZ-6 cells (but not in the uPA-cDNAs from OVCAR-3 and OV-MZ-19) a mutation in codon 121 was identified resulting in a proline to leucine exchange (for the wild-type sequence see *e.g. Riccio et al. 1985, Nucl. Acids Res. 13:2759*). This mutation is located in the so-called kringle domain of uPA.

At present, sequence analyses of the coding regions of PAI-I and uPA using DNA isolated from tumor tissue are under way in order to investigate whether the alanine/threonine variants in the signal sequence of PAI-1 and the Pro121Leu exchange in uPA are also detectable in native tumor tissue.

22. COMPARISON OF THE ANTICOAGULANT ACTION OF SULFATED AND PHOSPHORYLATED POLYSACCHARIDES

Erica McBride, Efath Muneer, Genia Harris, and V. M. Doctor

Prairie View A&M University, Prairie View, Texas 77446

Studies were conducted to introduce sulfate groups on polysaccharides or increase sulfate groups of naturally occurring sulfated polysaccharides by sulfation using chloro-sulfonic acid-pyridine complex. The compounds sulfated were oat spelts xylan (OSX), fucoidan, carrageenan, and chondroitin sulfates A and C. Fucoidan, carrageenan and OSX were phosphorylated using methane sulfonic acid-phosphorus pentoxide mixture. The anticoagulant properties were measured by determination of prothrombin time of pooled normal human plasma with or without addition of sulfated or phosphorylated compounds. The results showed that phosphorylation significantly increased the molecular weights of the polysaccharides by forming phosphodiester and diphosphodiester bonds between chains as shown by ^{31}P-nmr spectroscopy. In general the anticoagulant properties of the

268 Abstracts

sulfated polysaccharide were directly related to the % of sulfate groups while all of the three phosphorylated polysaccharides showed increases in the anticoagulant properties which were directly related to the increase in the molecular weight (> 300K Daltons) during phosphorylation. The mechanism of action of oat spelts xylan sulfate (OSXS) or oat spelts xylan phosphate (OSXP) was studied using ^{125}I-thrombin and human plasma. The results showed that at lower concentrations of OSXS or OSXP the complexation of ^{125}I-thrombin with heparin cofactor II (HC-II) was enhanced, while at higher concentrations of the compound the complexation with both antithrombin-III (AT-III) and HC-II was enhanced. (Supported by NIH-MBRS 08094).

23. OSMOTIC STRESS, AN EXPERIMENTAL TOOL FOR STRUCTURE-FUNCTION ANALYSES OF SERPINS. COAGULATION FACTOR XA INHIBITION BY ANTITHROMBIN III

Maria P. McGee and Hoa Teuschler

Bowman Gray School of Medicine, Medicine Department, Winston-Salem, North Carolina

The osmotic stress technique is based on controlled removal of water from hydration shells of molecules reacting in solution. Osmotic stress is induced with inert co-solutes that are excluded from the volumes occupied by hydration shells. Osmotic stress either stabilizes or destabilizes configurations of reaction intermediates that are respectively, dehydrated or hydrated relative to initial reactants. The increase in water activity that is generated by the excluded co-solutes displaces the equilibrium position of activated intermediates. This shift in equilibrium is reflected in the free energy of activation, ΔG^*. The component of ΔG^* due to water transfer is obtained from the reaction rate coefficient, k_{obs}, measured at different osmotic stress levels. The volume of water transferred is given by the differential coefficient, $\Delta G^*/D\Pi$, where $\Delta \Pi$ is the difference in pressure between excluded and non-excluded volumes.

In the present studies, the technique of osmotic stress was applied to study hydration/dehydration reactions during inhibition of coagulation factor Xa by anti-thrombin III. Osmotic stress changes, DP, were induced with Polyetyleneglycol, MW~6000, and rates of factor Xa inhibition were measured with specific chromogenic substrate. Reactions were assembled with purified human anti-thrombin III and factor Xa in the presence of heparin, at pH~7.2, and ionic strength, I~0.15. In addition to $\Delta \Pi$, other variables in these experiments were reactants' concentrations, temperature, and heparin size.

Rate coefficients of factor Xa inhibition increased with osmotic stress. The osmotic effect was independent of heparin concentration and size. Rate enhancement was observed only when the molar ratio of anti-thrombin III to factor Xa was >1. (This observation indicates that the reaction step sensitive to osmotic stress is monomolecular and is preceded by formation of the anti-thrombin/factor Xa complex). The reaction rate and the osmotic stress effects were independent of the temperature at least for the experimental interval 20–33°C. Thermodynamic activation parameters were consistent with rate enhancement due to a more favorable entropy of activation under osmotic stress conditions. Plots of $\Delta G^*/\Delta\Pi$ were bi-phasic with slopes of 1638 ± 200 and 197 ± 1 cal/mol/atm. These values correspond to 3765 and 452 mol of H_2O transferred to bulk phase per mol of factor Xa inhibited.

Results are interpreted to be consistent with a model in which the rate limiting step(s) for the inhibitory reaction is a change in protein conformation stabilized by dehydration of at least two distinct protein surfaces in the anti-thrombin III/factor Xa complex. The findings in these studies are also relevant to the physiology and pathology of hemostasis. Results suggest that inhibition of factor Xa by anti-thrombin III can be modulated by localized changes in colloidosmotic pressure during tissue injury.

24. INVOLVEMENT OF ARGININE 132 AND LYSINE 133 IN HEPARIN BINDING TO AND ACTIVATION OF ANTITHROMBIN

Jennifer L. Meagher, James A. Huntington, Bingqi Fan, and Peter G. W. Gettins

Department of Biochemistry, University of Illinois at Chicago

Allosteric activation of antithrombin by heparin requires a mechanism for transmission of the effects of heparin binding from the heparin binding site to the reactive center loop region, some distance removed. It has been proposed [van Boeckel et al. (1994) *Structural Biology* 1, 423–425] that the transmission mechanism involves extension of the D-helix of antithrombin as a consequence of neutralization of adjacent positive charges in the heparin binding site. Such D-helix extension would involve contraction of the previously-extended polypeptide, which is connected to strand 2 of β-sheet A, and could thus couple changes in the heparin binding site to changes involving β-sheet A and hence the reactive center. The positively charged residues proposed to be involved include arginine 132 and lysine 133, for which there has previously been no direct evidence for involvement. We have examined here the properties of two recombinant variants of antithrombin, R132M and K133M, in which arginine and lysine, respectively, have been changed to the uncharged, but approximately isosteric methionine, to determine whether these residues are involved in heparin binding, and whether removal of the positive charge results in a conformation that partially mimics the heparin-bound conformation, as might be expected from the proposed model. We found that the K_a for long-chain high affinity heparin was reduced ~25-fold for R132M and ~30-fold for K133M, indicating involvement of each residue in heparin binding. However, the reduction in affinity for high affinity pentasaccharide was very much smaller for each variant, suggesting a location for R132 and K133 peripheral to the main pentasaccharide binding site. The rates of inhibition of factor Xa by both variants were indistinguishable from those of wild-type antithrombin, and the accelerations produced by heparin binding were normal. We conclude that neither arginine 132 or lysine 133 plays an important role in binding of heparin pentasaccharide nor do they mediate the transmission of conformational change from the heparin binding site to the reactive center region. These residues do, however, contribute to binding of longer chain heparin species.

Supported by grant HL49234.

25. LOCALIZATION OF PROTEIN C INHIBITOR IN HUMAN PROSTATIC CARCINOMA

Laura L. Neese,* Scott T. Cooper,* Tracy P. Jackson,* Mike N. DiCuccio,¶ Maureane Hoffman,¶ Christopher W. Gregory,* Frank S. French,* and Frank C. Church*

*University of North Carolina School of Medicine, Chapel Hill, North Carolina 27599, and ¶Duke University Medical Center, Durham, North Carolina 27705

Prostate tumor cell invasion and metastasis involves numerous concerted interactions between both cellular and stromal components, including proteinases and their interactions with proteinase inhibitors. Protein C Inhibitor (PCI), also called plasminogen activator inhibitor-3 (PAI-3), is a member of the serine proteinase inhibitor (serpin) superfamily. PCI inhibits proteinases found in the prostate such as thrombin, activated protein C, urokinase-type plasminogen activator, and prostate specific antigen which are important during blood coagulation, wound repair, and cell migration processes. By Northern blot analysis, PCI mRNA was found as expected in many human tissues including the liver, kidney, testis, and prostate, but unexpectedly in the spleen, pancreas, and ovary. PCI mRNA was also detected in paired hyperplastic and malignant samples from five men with prostate cancer (CaP). PCI was localized to normal human prostatic epithelial cells by immunohistochemistry. There appeared to be an increase in PCI immunoreactivity in both hyperplastic and malignant tissue (scored with Gleason grades of III and IV). The increase in PCI staining in malignant tissue was evident in surrounding stroma, possibly due to breakdown of the tubulo-acinar glands' structural integrity. Furthermore, a CaP cell line, PC-3, was found to express PCI *de novo*. These results indicate that PCI is present in normal, hyperplastic, and neoplastic human prostate.

A human CaP xenograft model is also being used to further investigate the role of PCI in CaP. This androgen-dependent xenograft, CWR22, is propagated in nude mice. Frozen sections of tumor were stained for PCI antigen. PCI was detected throughout the tumor, but not in every epithelial cell. Mice with CWR22 tumors were then orchiectomized and sacrificed on consecutive days. Sections stained for PCI one to six days post-castration had an apparent increase in the amount of PCI. PCI expression by the xenograft tumor model evaluated by immunoblot [monoclonal antibody to PCI (G4–2)] of tumor tissue extract indicated a band consistent for PCI which also increased following orchiechtomy. These results suggest that PCI is present in this human CaP model and its expression is altered by androgens.

Three types of androgen regulated prostatic tumor cells are proposed to exist in human tumors and the xenograft model: androgen-responsive, non-responsive, and -sensitive. The data presented here suggest that PCI may be expressed either in androgen non-responsive or in androgen-sensitive tumor cells. A transcription factor search of the promoter region of PCI found putative GRE, MRE, NFkβ, and ERE sites, in addition to sites of c/EBP, hsp70, AP-1, and p53. Some of these sites code for transcription factors which are androgen regulated. Future work will focus on the influence of androgen on PCI expression in the xenograft model of CaP and to understand the mechanism of PCI regulation mediated by the 5′-flanking region of the PCI gene. The role of PCI in normal and malignant prostate is not clear. Since PCI inhibits prostate-related proteinases, it may have a role in regulating the "proteolytic cascade" which facilitates prostatic metastasis.

26. ONCOIMMUNIN-L: A SERPIN FROM HUMAN TUMOR CELLS WITH DIRECT LYMPHOID IMMUNOMODULATORY ACTIVITY

Beverly Z. Packard and Akira Komoriya

OncoImmunin, Inc., 336 Paint Branch Drive, College Park, Maryland 20742;
    E-mail: oncoim@access.digex.net

The idea that the tumor environment contains immunocytes which, if properly amplified and/or activated, are potential antitumor agents has served as the basis for various immunotherapeutic clinical trials. Although many studies have been aimed toward understanding the initiation of an immune response as being a result of direct tumor cell-immunocyte contact, the possibility of tumor cells providing soluble factors with stimulatory activity for immunocytes has received less attention. The objective of this work is to define the biochemical nature of the tumor environment such that the soluble molecules which carry stimulatory activity from tumor cells to immunocytes become known (Proc. Natl. Acad. Sci. 87:4058–4062 (1990)). We have previously characterized a 36 kDa tumor cell-derived protein with sequence homologies to lactate dehydrogenase, type M which stimulates differentiation of a myeloid precursor (J. Biol. Chem. 268:4058–4062 (1993); BBA 1222:159–163 (1994)), More recently, we have identified, purified, and partially sequenced a T-lymphocyte mitogen from the same serum-free medium conditioned by an epithelial tumor cell line (BBA 1269:41–50 (1995)). Surprisingly, the molecule is a ca. 45 kDa protein which, by amino acid sequence analysis of seven tryptic fragments (ca. 30% of total expected sequence) and immunoreactivity, can be classified as a member of the serpin superfamily with significant similarity to human monocyte/neutrophil elastase inhibitor. Two other serpins, i.e., squamous cell carcinoma antigen (SCCA) and maspin, have also been associated with cancer. However, none has previously been identified to have direct immunomodulatory activity such as the lymphocyte mitogenic activity of Oncoimmunin-L. Using cells which have shown clinical efficacy as targets, i.e., human tumor infiltrating lymphocytes (TILs), this work establishes a direct immunotherapeutic link between serpins and cancer.

27. MASPIN: EXTRACELLULAR, MEMBRANE BOUND, OR CYTOPLASMIC SERPIN?

Philip A Pemberton, Daniel T Wong, Michael C Kiefer, Ian C Bathurst, Shijie Sheng, Ruth Sager*, and Philip J Barr

LXR Biotechnology Inc., Richmond, California 94804, and *Division of Cancer Genetics, Dana Farber Cancer Institute, Boston, Massachusetts 02115

The tumor suppressor Maspin (mammary specific serpin) is a component of normal human epithelial tissues present in, but not limited to, the breast, intestine, prostate, testis, and thymus (1,2). Maspin is not found in malignant breast carcinomas and is therefore thought to be downregulated as mammary tumors progress from a benign to malignant phenotype. The addition of exogenous recombinant maspin to malignant mammary tumor cell lines, or transfecting them with maspin cDNA, markedly suppresses their invasiveness and motility *in vitro* and growth and invasiveness *in vivo* ((1,3). These results clearly demonstrate that maspin can function extracellularly. Immunostaining studies using anti-peptide antibodies have indicated that maspin localizes mainly in the pericellular space with small amounts present in the cytoplasm (1). These results suggest that maspin, although lacking a secretory signal peptide, is secreted into the extracellular matrix. To confirm and extend these findings we have used two complementary approaches. 1) Subcellular fractionation of cultured human mammary epithelial cells followed by maspin localization by immunoblotting. 2) Comparative molecular analysis of the domain, also present in maspin, that is responsible for secretion of the related serpins ovalbumin (OVAL) and plasminogen activator inhibitor 2 (PAI-2).

REFERENCES

1. Zhou et al., Science, 263: 526–529, 1994.
2. Pemberton et al., unpublished data.
3. Sheng et al., J. Biol. Chem., 269: 30988–30993, 1994.

28. SERPIN INTERACTION WITH A ZYMOGEN: A POTENTIAL REGULATORY ROLE IN CELL SURFACE PROTEOLYSIS

B. S. Schwartz[1] and F. España[2]

[1]University of Wisconsin, Madison, Wisconsin, U.S.A.; [2]Hospital Universitari "La Fe", Valencia, Spain

Plasminogen activation by urokinase has been identified as the initial step in cell surface proteolysis required for cellular invasion. Recent data suggests that single chain urokinase (scu-PA, the "zymogen") is the relevant enzyme for initiating cell surface plasminogen activation. Scu-PA possesses a small, but measurable amount of enzymatic activity in solution, and this activity is greatly augmented upon binding to the cell surface urokinase receptor (u-PAR). We present evidence that the serpin, PAI-3, plays a role in maintaining a scu-PA in an inactive state prior to binding to u-PAR. PAI-3 purified from human plasma inhibited the plasminogen activating activity of solution phase scu-PA, and of the noncleavable variant, glu_{158}-scu-PA in a concentration dependent manner. Scu-PA formed non-SDS stable complexes with PAI-3, such that ^{125}I-two chain urokinase (tcu-PA, the activated enzyme) could not interact with the PAI-3. Although reversible, these scu-PA•PAI-3 complexes were appreciably more stable than scu-PA complexes with PAI-1 or PAI-2. Scu-PA bound to monocyte or U-937 cell surface u-PA receptors (u-PAR) did not form these complexes with PAI-3, whereas receptor bound tcu-PA formed the usual SDS stable complexes with PAI-3. Preformed scu-PA•PAI-3 complexes added to u-PAR bearing cells dissociated more rapidly than did complexes in the absence of cells. Scu-PA bound to cellular u-PAR, and PAI-3 remained in solution; both molecules retained full functional activity. The extent of this dissociation was dependent on the number of u-PAR added to the system, and was demonstrated by ^{125}I-scu-PA binding to cells, cell surface scu-PA activity, and soluble PAI-3 complexing to ^{125}I-tcu-PA. The rapid dissociation of scu-PA•PAI-3 complexes was u-PAR dependent, as it was prevented in a concentration dependent manner by the receptor binding amino terminal fragment of u-PA, but not by the 33 KD catalytic domain. Hence, a model is proposed wherein scu-PA is held in a proteolytically silent state in complex with PAI-3, insuring no inappropriate plasminogen activation. The expression of cell surface u-PAR favors dissociation of this complex and allows for the initiation of PA by scu-PA bound to u-PAR. These u-PAR bound scu-PA molecules are poorly inhibited by PAI-3, whereas u-PAR bound tcu-PA is efficiently inhibited. Hence, the reversible serpin-zymogen interaction allows for regulated initiation of proteolysis, and the essentially irreversible serpin- enzyme interaction serves to down regulate the system.

29. EXCHANGE MUTAGENESIS OF ANTITHROMBIN AND HEPARIN COFACTOR II

William P. Sheffield, Varsha Bhakta, Michael A. Cunningham, and Morris A. Blajchman

Department of Pathology, McMaster University, and Canadian Red Cross Society, Hamilton, Ontario, Canada

Inhibition of fluid-phase thrombin in plasma is primarily mediated via two serpins: antithrombin (AT) and heparin cofactor II (HCII). Both proteins are glycoproteins, with molecular weights of 58 and 66 kDA, respectively. While AT inhibits all the known procoagulant serine proteases, HCII is thrombin-specific; while the action of both inhibitors is accelerated by heparin, another glycosaminoglycan, dermatan sulphate, specifically accelerates the action of HCII. In order to gain insight into the molecular reasons for these differences, we have used PCR and site-directed mutagenesis to construct recombinant serpins which combine portions of the rabbit AT and HCII polypeptides. We have previously demonstrated that expression of these proteins in a cell-free system yields functional molecules capable of forming SDS-stable complexes with cognate proteases. Exchange of aligned codons 368–433 (human AT numbering, P25–P39'), followed by *in vitro* transcription and *in vitro* translation in rabbit reticulocyte lysate, yielded two chimeric polypeptides, designated H/A (HCII 1–397 fused to AT 368–433) and A/H (AT 1–369 fused to HCII 401–461). Both proteins had lost all antithrombin (anti-IIa) or anti-factor (anti-Xa) activity. Next a more conservative exchange was performed, limited to the tetrapeptide P4-P1 of each serpin (IAGR, AT; FMPL, HCII). The resulting polypeptides, AT(P4P1)H, and H(P4P1)AT, retained 10% anti-IIa activity, but did not demonstrate detectable anti-Xa activity, even in the presence of heparin. Since others have shown that human HCII L(P1)R acquires anti-Xa activity, and demonstrates increased anti-IIa activity, the results indicate that one or more of the AT residues in the P4-P2 tripeptide are incompatible with the remainder of the HCII reactive center loop. Reduction in anti-IIa activity of AT(P4P1)H suggests that the FMP tripeptide of HC-II is similarly incompatible. Recombinant AT I(P4)F, produced by site-directed mutagenesis and expressed in the same cell-free system, showed unimpaired anti-IIa activity, as did AT G(P2)P, in a previous study. These results suggest that the P3 substitution is responsible for the decreased inhibitory action of AT(P4P1)H. Experiments are in progress to generate single amino acid mutations to resolve these issues. In this regard, we constructed P1 Met mutants of both thrombin-inhibitory serpins. HCII-L(P1)M retained antithrombin activity, while AT-R(P1)M did not; neither P1 mutant acquired anti-elastase activity. These results underscore our findings of different sensitivities of HCII and AT inhibitory activities to reactive center loop mutations.

30. MASPIN, A NOVEL SERPIN, INHIBITS THE MOTILITY AND INVASIVENESS OF MAMMARY AND PROSTATIC TUMOR CELLS IN CULTURE

Shijie Sheng,* Juliana Carey,* Mary J. C. Hendrix,[±] Elisabeth A. Seftor,[±] Lauren Dias,* and Ruth Sager,*[¶]

*Division of Cancer Genetics, Dana-Farber Cancer Institute, Harvard Medical School, 44 Binney Street, Boston, Massachusetts 02115; [±] Pediatric Research Institute, Cardinal Glennon Children's Hospital and St. Louis University, School of Medicine, 3662 Park Avenue, St. Louis, Missouri 63110

Maspin, a tumor suppressor, exerts its inhibitory effect on the invasion of breast carcinoma cell lines in culture in part by inhibiting cell motility. The inhibition of motility per se was demonstrated by video microscopy. In addition, three prostatic cancer cell lines were inhibited by recombinant maspin both in invasion and motility. Endogenous maspin is localized both in the cytoplasm and on the cell surface. The recombinant maspin, on the other hand, binds specifically to cell surface as demonstrated by immunofluorescent staining and functional studies. The inhibitory functions of recombinant maspin protein and maspin re-expressed by transfectants were undistinguishable. Thus maspin, either exogenous or re-expressed by cells, can re-activate the processes that inhibit invasion and motility of tumor cells.

31. HEPARIN COFACTOR II ACTIVITY IS STIMULATED BY ARTERIAL SMOOTH MUSCLE CELL HEPARAN SULFATE PROTEOGLYCANS

Rebecca A. Shirk*, Frank C. Church[†], and William D. Wagner*

*Dept. of Comparative Medicine, Bowman Gray School of Medicine of WFU, Winston-Salem, North Carolina and [†]Dept. of Pathology and Laboratory Medicine and the Center for Thrombosis and Hemostasis, Univ. of North Carolina, Chapel Hill, North Carolina

Heparin cofactor II (HCII), a potent thrombin inhibitor in the presence of the glycosaminoglycans heparin and dermatan sulfate, is postulated to be an extravascular thrombin inhibitor that is stimulated physiologically by dermatan sulfate proteoglycans. To understand how the mitogenic and chemoattractant activity of thrombin may be down-regulated within the artery wall, cultured monkey (*Macaca fascicularis*) aorta smooth muscle cell (SMC) monolayers, underlying extracellular matrix material, and purified proteoglycans were tested for their ability to accelerate thrombin inhibition by HCII. In a modified chromogenic activity assay, early confluent SMC monolayers increased thrombin-HCII inhibition rates 2–4 fold compared to reactions in cell-free control wells (*e.g.*, 9.4×10^4 *vs.* 2.2×10^4 M^{-1} min^{-1}, with and without SMC, respectively). Extracellular matrix obtained by SMC monolayer removal also accelerated the thrombin-HCII inhibition reaction 3–5 fold. Rate increases were abolished by Polybrene or protamine sulfate. Pretreatment of SMC monolayers with heparitinase I (and of extracellular matrix with nitrous acid) to degrade heparan sulfate blocked the thrombin-HCII inhibition rate increase. In contrast, pretreatment with chondroitinase ABC to digest dermatan sulfate and chondroitin sulfate chains had no effect. These results suggested that the accelerated thrombin-HCII reaction is mediated by a SMC-associated heparan sulfate proteoglycan (HSPG). To confirm HSPG activity with HCII, "pericellular" (cell surface and extracellular matrix- derived) SMC HSPG was purified and fractionated by charge on DEAE-Sephacel by step elution with 0.35 M NaCl (eluted ~20% of total HSPG) and 1M NaCl (~80% of total). At 1 µg/ml hexuronic acid, high charge HSPG stimulated a 7-fold thrombin-HCII inhibition rate increase relative to reactions without proteoglycan, while low charge HSPG induced a 2-fold rate increase. In comparison, an 18-fold rate increase was observed with 1 µg/ml dermatan sulfate proteoglycan purified from SMC culture media. These results indicate that SMC HSPG could contribute significantly to intramural thrombin inhibition by HCII.

Changes in artery wall proteoglycan content and glycosaminoglycan chain structure occur during the progression of atherosclerosis. Because unregulated thrombin activity may contribute to atherogenesis, we hypothesized that SMC HSPG in atherosclerotic lesion has reduced activity in proteoglycan-accelerated thrombin-HCII inhibition reactions. Human aorta HSPG was purified from normal intima-media tissue and atherosclerotic lesion types II, IV, V, and VI obtained from four autopsies. Following extraction, HSPG was isolated on DEAE-Sephacel by elution at 0.35M NaCl. At 1 μg/ml hexuronic acid, all HSPG samples accelerated the thrombin-HCII inhibition rate 2–5-fold relative to reactions in the absence of proteoglycan (3.1×10^4 M^{-1} min^{-1}). In three of the sets of HSPG samples, lesion HSPG elicited a slightly larger inhibition rate than normal aorta HSPG (1.3–2.3 fold higher). HSPG from the fourth autopsy showed the reverse pattern, with normal HSPG stimulating a higher inhibition rate compared to lesion HSPG. These results confirm that human arterial HSPG stimulates HCII activity and indicate that this potential remains present in the atherosclerotic lesion. [Supported by HL-07115, HL-25161, HL-45848, and HL32656.]

32. A SYNTHETIC ANALOG OF THYMOSIN ALPHA-1 IS ANION-BINDING EXOSITE-DIRECTED THROMBIN INHIBITOR

Svetlana Strukova, Tamara Dugina, Alexandra Samal and Maria Smirnova

Department of Human and Animal Physiology, Faculty of Biology, Lomonosov Moscow State University, Moscow, Russia; Faculty of Physics, Minsk State University, Minsk, Bellorussia

Thrombin was shown to interact with the N-terminal domain of heparin cofactor II (HCII) (VanDeerlin V., Tollefsen D., 1991). This acidic domain of HCII resembles the C-terminal fragment of hirudin, which interacts with anion-binding exosite I of thrombin. The C-terminal sequence of the thrombin inhibitor thymosin alpha-1 has the high degree of local similarity to the C-terminal portion of hirudin. Interaction of the synthetic peptide EEAEN, a thymosin alpha-1 24–28 analog, with alpha-thrombin was studied by inhibition assays. Peptide EEAEN was synthesized by solid-phase method. The binding of peptide EEAEN to alpha-thrombin was demonstrated by the finding that 1. alpha-thrombin-induced polymerization of fibrinogen was inhibited by peptide EEAEN (0.01–0.1 nM) because the time of protofibril formation and the rate if its lateral assosiation were prolonged; 2. alpha-thrombin-induced human platelet aggregation also was inhibited by peptide at concentrations of 0.01–1.0 nM. Inhibition of thrombin activity by peptide was smaller or not observed at concentrations higher than 1.0 nM. Thus, the interaction of the C-terminal site of thymosin alpha-1 with alpha-thrombin was similar to the reaction of the whole chain of thymosin alpha-1 with thrombin. The rate constants for inhibition of thrombin by thymosin alpha-1 and its acidic domain in the presence and the absence of glycosaminoglycans were determined. These results suggest that peptide EEAEN of thymosin alpha-1 contributes to binding of thrombin through the thrombin anion-binding exosite 1 as in cases of C-terminal part of hirudin and N-terminal domain of HCII.

33. A STRUCTURAL COMPARISON OF NATIVE AND LATENT ANTITHROMBIN: IMPLICATIONS FOR HEPARIN BINDING AND THE S TO R TRANSITION

James Whisstock, Richard Skinner, Robin Carrell and Arthur Lesk

Department of Haematology, University of Cambridge, MRC Centre, Hills Road, Cambridge CB2 2QH, U.K.

A detailed analysis of the recently solved 2.6 Å structure of antithrombin indicates the changes that take place during the transition from the native to the latent form. The comparison between two serpin conformations of the same sequence allows us to see movements which are due to changes in conformational state and not sequence variation. Seven fragments move independently during the transition, the largest being the A, C and H helices, the B-sheet and part of the A and C-sheet. Other fragments are: hB, hC and hI with s5A and s6A; s2C and s3C; part of s2A and s3A; hE and part of s1A; hF and part of s1A s2A and s3A and lastly hD. Where a region is common to more than one fragment it can be described as a hinge region (of which there are several).

Of particular interest is the D-helix where the key residues for binding the heparin pentasaccharide are located. The D-helix and its associated loops form a separate fragment. A backbone fit of residues 115–131 gave a r.m.s. deviation of 0.618 Å. Inspection of the fit profile of the helices indicated that in the native form the helix was kinked to give a conformation appropriate for binding of the pentasaccharide, whilst in the latent form the helix had straightened out and another half turn had been added. The effect of this conformational change is to move the backbone atoms of the pentasaccharide binding residues and thus alter the charge distribution which is compatible with the observed release of the pentasaccharide.

To identify which areas of the antithrombin were responsible for this change we superimposed the native and latent structures on the largest fragment and transposed the D-helix of native into the latent structure and vice versa. We then looked to see where steric clashes were occurring. When the latent D-helix is transposed into the native structure we found that the most serious clashes were between the D-helix and the B and E helices. There were also clashes between the D-helix and Arg 47 which is on the amino-terminal end of the A-helix and has been implicated in pentasaccharide binding. However, if the native D-helix is transposed into the latent structure then there are relatively few clashes. In both experiments the amino-terminus was ignored as the incompleteness and high temperature factors of this area makes it difficult to analyse the changes in detail. The amino-terminus is joined to the D-helix by means of a disulphide bond and thus it is likely that movement in the D-helix will result in some amino-terminal movement. Both the B-helix and the E-helix are part of other fragments and undergo significant movement during the S to R transition and therefore we conclude that these movements can allow the D-helix to adopt a new conformation in the latent form.

We have analysed the other changes that take place during the S to R transition using transposition techniques (as described above) and detailed examination of the changes in packing. This has enabled us to create a kinetic video depicting fragment movement during the S to R transition. The video shows a general overview of the fragment movements and then a specific look at the movements in the shutter region and the D-helix.

34. HEPARIN, NOT DERMATAN SULFATE, PROMOTES THE "SUBSTRATE-LIKE" BEHAVIOR OF L444R RECOMBINANT HEPARIN COFACTOR II IN THROMBIN INHIBITION

Annemieke J. Willemze,* Angelina V. Ciaccia,[#] and Frank C. Church[#]

*University of Leiden, Leiden, The Netherlands and [#]Departments of Pharmacology and Pathology, University of North Carolina at Chapel Hill, Chapel Hill, North Carolina 27599

Heparin cofactor II (HCII), a member of the "serpin" family of serine protease inhibitors, is a 65,600-Da plasma glycoprotein that inhibits the coagulation proteinase thrombin at a rate that increases in the presence of glycosaminoglycans. HCII has Leu444–Ser445 in the reactive site bond, which is unusual for a thrombin-specific inhibitor. The rate of thrombin inhibition is stimulated ~17000-fold by heparin and ~36000-fold by dermatan sulfate. In this study, we have further characterized the effects of substituting an Arginine for Leu444 in recombinant HCII (L444R-rHCII), previously expressed by Derechin and Tollefsen (1990, J. Biol. Chem, **265**: 5623–5628). L444R-rHCII was expressed in High-Five insect cells and partially purified by sequential heparin- and Q-Sepharose batch adsorption. In the absence of a glycosaminoglycan, the thrombin inhibition rate (k_2, M^{-1} min^{-1}) for L444R-rHCII (1.3×10^6) was 56-fold higher than for wildtype-rHCII (2.4×10^4). Dermatan sulfate accelerated thrombin inhibition by both, but the maximal rate constant was 3.5-fold lower for L444R-rHCII (2.4×10^8) than for wt-rHCII (8.7×10^8). In the presence of heparin, L444R-rHCII (7.5×10^7) exhibited a more pronounced reduction in inhibition rate (5.5-fold lower) as compared to wt-rHCII (4.1×10^8).

To determine why L444R-rHCII, which contains a reactive site loop sequence preferred by thrombin, is a poorer thrombin inhibitor than wt-rHCII in the presence of glycosaminoglycans, we determined the stoichiometry of inhibition (SI). The tendency of a serpin to behave as a substrate is reflected in its SI value, the number of inhibitor molecules consumed before a bimolecular complex forms. An inverse correlation between SI and k_2 has been reported for most serpin-proteinase reactions. In the absence of glycosaminoglycan, approximately 10.3 molecules of L444R-rHCII were cleaved by thrombin before a stable complex was formed. Addition of dermatan sulfate to the reaction decreased the SI value to 3.6 molecules of L444R-rHCII per thrombin molecule. Interestingly, when heparin was added, the SI value increased to approximately 30 molecules. Wildtype-rHCII, in contrast, has an SI value of 1.8 in the presence of heparin. These results indicate that heparin promotes a more "substrate-like" behavior of L444R-rHCII, whereas dermatan sulfate promotes a more "inhibitor-like" behavior of the mutant.

To study the influence of glycosaminoglycan-bridging and acidic domain-anion-binding exosite 1 (ABE-1) interactions on the substrate vs. inhibitor properties of L444R-rHCII, SI-assays were performed using two thrombin variants: R93,97,101A-thrombin, a recombinant thrombin with changes in anion-binding exosite 2 (ABE-2), lacks heparin-binding ability and ternary complex formation; and γ-thrombin lacks part of ABE-1 due to limited proteolysis. The SI values of L444R-rHCII with thrombin, R93,97,101A-thrombin and γ-thrombin in the absence of glycosaminoglycan do not differ significantly (10.3, 9.5 and 10.7, respectively) in contrast to the different values obtained in their presence. While the SI value for thrombin in the presence of heparin is ~30, this decreases notably to 3.3 for R93,97,101A-thrombin. With dermatan sulfate, which is unable to promote ternary complex formation, the SI values of L444R-rHCII with thrombin and R93,97,101A-thrombin are comparable (about 3). SI values obtained using γ-thrombin show a similar, but less dramatic trend (10.7 in the absence of glycosaminoglycan, 15.4 in the presence of heparin, and 8.6 in the presence of dermatan sulfate). These results imply that ternary complex formation, mediated by heparin, increases L444R-rHCII inactivation by thrombin. Overall, these results strongly suggest different effects of heparin and dermatan sulfate in terms of the inhibitor:substrate equilibrium for the L444R-rHCII reaction with thrombin.

35. MONOMERIC AND POLYMERIC FIBRIN INHIBIT THE ANTITHROMBIN-MEDIATED ANTICOAGULANT ACTIVITY OF HEPARIN BUT NOT OF ORGARAN (ORG 10172)

Adrian Zammit and Joan Dawes

Thrombosis Research Group and CRC for Biopharmaceutical Research, The Garvan Institute of Medical Research, Darlinghurst, NSW 2010, Australia

In certain clinical situations characterised by the growth of existing thrombi the efficacy of unfractionated heparin is severely affected due to its inability to catalyse the inactivation of thrombin bound to soluble and insoluble fibrin. However, recent evidence from several laboratory and clinical studies has indicated that in such circumstances the low molecular weight (LMW) heparinoid Orgaran may be a superior antithrombotic agent than unfractionated and LMW heparins.

In the present investigation we compared the ability of fibrin II monomer and polymer to inhibit the anti-thrombin and anti-factor Xa activities catalysed by the antithrombin-binding fraction of unfractionated heparin (high-affinity heparin) and the LMW heparin Fragmin (high-affinity LMW heparin) with those catalysed by the antithrombin-binding fraction of Orgaran (Org 10849). Fibrin II monomer strongly inhibited the rates of thrombin and factor Xa inactivation catalysed by high-affinity heparin and to a lesser extent by high-affinity LMW heparin, but had very little effect on the anti-thrombin and anti-factor Xa activities of Org 10849. In the presence of 3 µM fibrin II monomer the rates of thrombin inactivation catalysed by high-affinity heparin and high-affinity LMW heparin were inhibited by approximately 77% and 29% respectively whereas that catalysed by Org 10849 was completely unaffected. Fibrin II monomer at similar concentrations inhibited the anti-factor Xa activities of high-affinity heparin and high-affinity LMW heparin by approximately 83% and 16% respectively, but did not modulate the anti-factor Xa activity of Org 10849. Half-maximal inhibition of the anti-thrombin and anti-factor Xa activities of high-affinity heparin occurred at 0.45 μM and 0.3 μM fibrin II monomer respectively. Fibrin II monomer (3 μM) also inhibited the uncatalysed rates of thrombin and factor Xa inactivation via antithrombin by 70% and 25% respectively. Fibrin clots formed by the polymerisation of 2 μM fibrinogen with 0.4 nM thrombin in the presence of approximately 67 nM of the various glycosaminoglycans bound 32% of added high-affinity heparin but only 10% and 4% of high-affinity LMW heparin and Org 10849 respectively. Taken together these results indicate that Orgaran is more effective in inhibiting fibrin-associated thrombin and factor Xa than either unfractionated or LMW heparin and may provide the basis for the clinically observed efficacy of Orgaran.

This work was supported by Organon International.

FIBRINOLYSIS, DEVELOPMENT, AND REPRODUCTION

36. INTERSPECIES INTERACTIONS OF PLASMINOGEN, UROKINASE AND PLASMINOGEN ACTIVATOR-INHIBITOR TYPE 2

John L. Andrews

Biotech Australia Pty Limited, Roseville N.S.W. Australia

The plasminogen/plasminogen activator system is but one of many zymogen activation cascades that amplify catalytic processes. The enzyme produced in this system, plasmin, plays the central role in fibrinolysis and is a significant component of other proteolytic events in both health and disease, e.g. embryonic development and extracellular matrix degradation. Plasmin activity can be inhibited by the serpin α_2-antiplasmin, while the plasminogen activators, tissue type (tPA) and urokinase (uPA), are also inactivated by serpins: plasminogen activator inhibitors types 1 and 2 (PAI-1 and PAI-2).

Animal models are often used to study the efficacy of a proposed therapeutic agent on a disease process. Assumptions are made in the use of these models, particularly, that the introduced molecule will interact with its heterologous target similarly to its natural target. We aimed to assess some of these assumptions in the uPA/PAI-2 system.

The interactions of human and rodent uPAs with recombinant PAI-2s of human and rodent origin were examined. Activation of homologous and heterologous plasminogens was examined with both human and rodent uPAs. Assays for uPA amidase activity were performed in 96-well microplates with 4-nitroaniline or 7-amino-4-methylcoumarin as the measured hydrolysis products. Plasmin was similarly measured as amidase activity or, for higher sensitivity, esterase activity. Data was analysed using equations for tight binding inhibition kinetics. Inhibition constants and association rate constants were determined for each combination using synthetic peptide substrates for uPA. Where possible, interactions were also assessed with plasminogen as substrate.

37. ELEVATED PAI-1 AND D-DIMER LINKED TO INTRAVENTRICULAR HEMORRHAGE IN PRETERM INFANTS

James P. Chen and Vichien Lorch

University of Tennessee Medical Center, Knoxville, Tennessee 37920

Intraventricular hemorrhage (IVH) is one of the most commonly encountered perinatal injuries to brain of preterm infants. Previous studies suggest that the fibrinolytic system of the human infant is not fully developed at the time of birth, and that the balance between coagulation and fibrinolysis favors fibrin formation in the newborn (Suarez *et al.* Neonatal hemostasis: current concepts and relevance of molecular markers of hemostasis. In: *Developmental and Neonatal Hematology*. Raven Press, 1988:57–86). This study examined PAI-1 activity and antigen, plasma levels of fibrinogen, plasminogen and antithrombin III (ATIII) in 11 full-term and 30 preterm infants, and related the findings to the incidence of IVH. Platelet counts were determined in premature infants at birth.

Means comparisons of PAI-1 activity indicate that on postnatal day 1 the difference between the full-term and preterm infants with IVH and without IVH was highly significant ($p=0.0018$). In particular, the preterm with IVH was different to both full-term and preterm without IVH ($p<0.05$). On day 1, the difference between PAI-1 antigens of the full-term and preterm infants with IVH and without IVH was significant ($p=0.0045$). Specifically, the preterm with IVH was different to both the full-term and preterm without IVH. However, on days 5–11, the difference among the three groups of infants was not significant for both PAI-1 activity and antigen. Plasma concentrations of fibrinogen and plasminogen were compared between the three groups of full-term and preterm infants. No difference could be observed in these values by ANOVA. Plasma concentrations of ATIII were significantly higher in full-term infants than in preterm infants on both day 1 and days 5–11 ($p<0.00001$). The difference between the platelet counts of preterm infants with IVH and without IVH was not significant ($p>0.05$). Elevation of D-dimer crosslinked fibrin degradation products (XDP) (0.2–1.6 μg/ml), determined by the Simpli-RED D-dimer test, correlated in 4 out of 5 premature infants with the diagnosis of IVH by ultrasonography. No elevation of D-dimer XDP (<0.2 μg/ml) was observed in premature infants without IVH (11/12) and full-term infants (6/6).

In conclusion, a hypercoagulable state, as indicated by a rise in D-dimer XDP, may be initiated by some types of trauma to fragile blood vessels of the preterm infants who develop IVH. The hypercoagulability seen in the preterm infant with IVH is more likely induced by the causes that are multifactorial in nature. These contributory factors possibly include hypoxia, changes in blood flow and blood pressure, and acidosis. This hypercoagulability is further exacerbated by the increased release of PAI-1 leading to suppressed fibrinolysis.

38. LOCALIZATION AND IDENTIFICATION OF FUNCTIONALLY IMPORTANT DOMAINS IN PLASMINOGEN ACTIVATOR INHIBITOR-1

Paul J. Declerck, Ann Gils, Kathleen Aertgeerts,* and Camiel De Ranter*

Laboratory for Pharmaceutical Biology and Phytopharmacology and *Laboratory for
    Analytical Chemistry and Medicinal Physicochemistry, Faculty of Pharmaceutical
    Sciences, K.U.Leuven, Belgium

Plasminogen activator inhibitor-1 (PAI-1) is a unique member of the serpins and plays an important role in fibrinolysis. PAI-1 can adopt different conformations: an *active* but labile conformation that converts spontaneously to a *latent* conformation. In addition, a *substrate* conformation has been identified. To further elucidate the structure-function relationship in PAI-1 we have characterized various mutants and we have solved the three-dimensional structure of a cleaved (at P1–P1') variant of human PAI-1(PAI-1–P12, Ala to Pro at position P12).

Nine mutants were obtained in which the amino acid at position P5, P6, P7, P8, P10, P12, P14, P15 and P16, respectively, was substituted with a proline.The purified mutants were subjected to a detailed functional, conformational and heat denaturation analysis. This characterization revealed that substitution at positions P5, P6, P7, or P16 yielded PAI-1 variants with inhibitory properties (63 to 129% compared to that observed for wild-type PAI-1) but with significantly increased substrate properties (2- to 4-fold vs wild-type, $p<0.001$). Importantly, for all of these mutants a substrate-reactive form, devoid of inhibitory activity could be isolated. Substitution at position P8 or P10 yielded PAI-1 variants exhibiting predominantly substrate characteristics, whereas substitution at position P12, P14 or P15 yielded PAI-1 variants with exclusively stable substrate properties. Comparative plasmin digestion of the various conformations of PAI-1 and PAI-1 mutants indicated the presence of a unique plasmin cleavage site exclusively present in latent forms. Amino-terminal amino acid sequence analysis revealed that this cleavage site comprised the Lys191-Ser192 bond. Based on the comparison between the three-dimensional structures of latent PAI-1 and that of the cleaved PAI-1 variant it was demonstrated that this plasmin cleavage site, exclusively susceptible in latent PAI-1, is localized indeed in a region of the molecule that constitutes a major difference between latent and reactive forms of PAI-1.

In conclusion, introduction of a proline between position P5 to P16 yields PAI-1 variants with increased substrate properties. In contrast to what has been suggested for other serpins, the functional behaviour of PAI-1 is mainly determined by its initial conformation prior to interaction with its target proteinase rather than during interaction.

39. CHARACTERISATION OF THE COMPLEX OF PAI-1 WITH TPA BY MASS SPECTROMETRY AND SIZE-EXCLUSION CHROMATOGRAPHY

Johanna Deinum,[1]* Karl-Erik Karlsson,[1] Petter Björquist,[1] Thord Johansson,[2] Jan-Olof Andersson,[2] Mona Byström,[2] Lennart Hansson[2] and Mats Strömqvist[2]

[1]Preclinical R & D, Astra Hässle AB, S-431 83 Mölndal; [2]Astra Hässle AB, Tvistevägen 48, S-901 87 Umeå, Sweden; *E.mail:Johanna.Deinum@hassle.se.astra.com

Glycosylated human plasminogen activator inhibitor type 1 (PAI-1), produced in Chinese hamster ovary (CHO) cells, showed a variety of compounds with different molecular weights when subjected to electrospray mass spectrometry (ES-MS), owing to the heterogeneity of the carbohydrate chains. However, non-glycosylated human PAI-1, produced in *E. coli,* gave rise to a prominent species with a molecular weight of 42 774, consistent with the amino acid sequence. A non-glycosylated mutant of the protease domain (B-chain) of tissue-type plasminogen activator (tPA) produced in C 127 cells, had a molecular weight of 28 168. Full-length, glycosylated, tPA showed a large heterogeneity in molecular mass. For a mass study, a complex composed of non-glycosylated PAI-1 and non-glycosylated B-chain was formed. The complex, isolated by size-exclusion chromatography was remarkably stable at neutral pH at room temperature. The mass spectrum of the complex provided two main species, a peptide with a mass of 3803 and a dominating species of 67 133. These masses are consistent with a complex where PAI-1 is cleaved at the P1-P1' position. A trace of a species with a molecular mass of 70 942 was also found, corresponding to the complex with full-length uncleaved PAI-1. Separation of the cleaved peptide, corresponding to the hydrophobic C-terminal 33 amino acid residues of PAI-1 from the complex, was achieved by size-exclusion chromatography in the presence of 30% acetonitrile. To assess whether this is specific to the tPA B-chain alone, experiments with the complex of full-length, glycosylated tPA and glycosylated PAI-1 were also performed, and it was possible to demonstrate the release of the same 33 amino acid residues C-terminal PAI-1 peptide by chromatography, mass spectrometry, as well as by SDS-PAGE.

Thus, in the complex between tPA and PAI-1, the proteins are held together by a tight covalent bond, but the C-terminal cleaved peptide of PAI-1 is only bound to the complex by hydrophobic forces.

40. PROTEIN C INHIBITOR IS A POTENT INHIBITOR OF ACROSIN AND IS LOCALIZED ON THE ACROSOMAL HEAD OF HUMAN SPERMATOZOA

Marc G. L. M. Elisen,[1] Roel J. van Kooij,[2] Martijn A. Nolte,[1] Bonno N. Bouma,[1] and Joost C. M. Meijers[1]

Departments of Haematology[1] and Obstetrics and Gyneacology[2], University Hospital
   PO Box 85500, 3508 GA Utrecht, the Netherlands

Protein C inhibitor (PCI) is a heparin binding plasma serine protease inhibitor, which was originally identified as an inhibitor of activated protein C. PCI has a broad enzyme specificity, inhibiting several enzymes in haemostasis. Recently it has been described that proteases of the reproductive system such as acrosin, prostate specific antigen and tissue kallikrein can also be effectively inhibited by PCI. The concentration of PCI in seminal plasma is 50 times higher compared to the concentration in blood plasma (4 μM vs. 80 nM). However, the role of PCI in vivo is still unknown. In this study we determined inhibition of activated protein C and thrombin, two potential physiological target proteases in haemostasis, by PCI and compared the results with the parameters obtained for the inhibition of acrosin. The second order rate inactivation constant kass. was determined using slow binding kinetics. Direct comparison of the kinetic arameters for the three possible target protease clearly show that acrosin can be very effectively inhibited by PCI with kass. values of 5.7×10^5 and 1.6×10^7 $M^{-1}s^{-1}$ in the absence and presence of heparin. The inhibitory effect of PCI for all three serine proteases is stimulated by heparin, which acts as a template for both protease and inhibitor. Combination of this observation with the high concentration of PCI in seminal plasma suggests that the physiological function of PCI could be the inhibition of acrosin. The stimulating effect of heparin andalso other glycosaminoglycans which acts as a template for both protease and inhibitor suggests the possibility of a surface-bound reaction. Therefore, experiments were performed to localize PCI on spermatozoa. Semen samples from normospermic donors were used. Immunofluorescence data show that PCI is localized on the acrosomal head. This raised questions since the very potent inhibition of acrosin would prevent its function in vivo. Therefore we repeated the labeling experiment with acrosome-reacted spermatozoa, which showed no staining of the acrosomal head but also that PCI-antigen was strictly localized at the equatorial segment. Control experiments performed with the antibody which was preincubated with saturating amounts of PCI gave no specific staining of regions of the spermatozoa. In conclusion, PCI is a potent inhibitor of acrosin, and its localization on the acrosomal head of the spermatozoa suggests that PCI can protect the acrosomal head against premature proteolytical degradation.

41. INHIBITION OF HUMAN SPERM-HAMSTER OOCYTE BINDING AND PENETRATION BY PROTEIN C INHIBITOR

Francisco España,[1] Jaime Sánchez-Cuenca,[1] Pedro J. Fernández,[2] Alberto Romeu,[2] Juan Gilabert,[2] Amparo Estellés,[1] and Charles H. Muller[3]

[1]Research Center and [2]Department of Obstetrics and Gynecology, "La Fe" University Hospital, 46009 Valencia, Spain, and [3]University of Washington School of Medicine, Seattle, Washington 98195

Protein C inhibitor is a heparin-dependent inhibitor that belongs to the protein superfamily of serine protease inhibitors (serpins) and is present in plasma at about 0.08 µmol/L. It inhibits activated protein C and other plasma coagulation and fibrinolytic enzymes. Previously, we described the presence of high levels of protein C inhibitor in human semen (3.1 µmol/L) (España et al., Thromb. Res. 64: 309, 1991), where it has multiple potential roles in human reproduction as a result of its interaction with prostate specific antigen, plasminogen activators and tissue kallikrein (España et al., Thromb. Haemost. 70: 989, 1993; España et al., Eur. J. Biochem. 234: 641, 1995). Here, we show that protein C inhibitor is present in amniotic and follicular fluid at about 0.05 and 0.1 µmol/L, respectively, and that purified human protein C inhibitor as well as two synthetic peptides derived from its sequence inhibited both binding and penetration of zona-free hamster oocytes by human sperm. The binding inhibition by protein C inhibitor was dose dependent, with 50% inhibition at 0.037 µmol/L of inhibitor (45±17 sperm/egg versus 105±23 in control experiments). Protein C inhibitor also reduced the percentage of hamster eggs penetrated by human sperm (7%±3% fertilized eggs versus 12%±5% in control experiments). A polyclonal antibody against protein C inhibitor partially abolished the serpin inhibition of the binding and penetration of hamster oocytes by human sperm. We conclude that protein C inhibitor may be involved in human reproduction at several steps, during transportation of sperm through the uterus and Fallopian tube as well as in the fertilization process. FIS 96/1129.

42. AN INTRACELLULAR SERPIN FROM INSECT BLOOD CELLS

Hong Gan, Haobo Jiang, and Michael R. Kanost

Department of Biochemistry, Kansas State University, Manhattan, Kansas 66506

Serine proteinase inhibitors from the serpin superfamily have recently been identified as hemolymph proteins from several groups of arthropods, including horseshoe crabs, crayfish, and lepidopteran insects. In the tobacco hornworm, Manduca sexta, a family of serpins present in the plasma fraction of hemolymph were found to be products of a single gene, Manduca serpin gene-1. We have recently identified a second gene in this insect,

Manduca serpin gene-2. A serpin-2 cDNA was isolated from a Manduca day 3 fifth instar larval hemocyte library. This cDNA lacked a portion of the 5' end. The missing 5' end cDNA was cloned by a RACE (rapid amplification of cDNA ends) PCR technique. The full length cDNA is 1.5 kb long and encodes a protein of 381 amino acid residues. Sequence comparisons reveal approximately 27%–38% identity, and 51%–58% similarity between serpin-2 and other invertebrate serpins, with greatest similarity to Bombyx mori (silkworm) antichymotrypsin-2. The incomplete serpin-2 cDNA was cloned into expression vector H6PQE60 and expressed as a truncated protein in E. coli. The recombinant serpin-2 was used as antigen to generate a rabbit polyclonal antibody antiserum to Manduca serpin-2. The antiserum recognizes a 43 kDa protein present in hemocytes but absent from the plasma fraction of Manduca larval hemolymph. Immunofluorescence microscopy indicates that serpin-2 is present in cytoplasmic granules of the granular hemocytes. Manduca serpin-2 was purified from larval hemocytes by anion exchange chromatography, Con A lectin affinity chromatography, and HPLC gel filtration chromatography, using western blotting to assay column fractions. Experiments are in progress to characterize the inhibitory activity of serpin-2. The intracellular location of serpin-2 suggests a function for serpin-2 different from the plasma serpin-1 proteins in insect hemolymph.

43. THE USE OF FLUORESCENT PROBES TO CHARACTERIZE CONFORMATIONAL CHANGES IN THE INTERACTION BETWEEN VITRONECTIN AND PAI-1

Angelia Gibson,[1] Baburaj Kunnumal,[1] Duane E. Day,[2] Ingrid Verhamme,[2] Joseph D. Shore,[2] and Cynthia B. Peterson[1]

[1]Department of Biochemistry and Cellular and Molecular Biology, The University of Tennessee, Knoxville, Tennessee and [2]Department of Biochemical Research Henry Ford Health System, Detroit, Michigan

Plasminogen activator inhibitor-1 (PAI-1), the primary inhibitor of tissue-type plasminogen activator and urokinase, has been observed to readily convert to a latent form. The conversion from active to latent PAI-1 is attributed to insertion of the reactive center loop into the central beta sheet. Interaction with vitronectin stabilizes the active conformation of PAI-1, so that the rate of conversion to the latent form is greatly decreased. Binding of vitronectin to PAI-1 would be expected to induce changes at the reactive center loop, but concomitant structural changes in vitronectin induced by PAI-1 binding have not been characterized. There is disagreement regarding the localization of the PAI-1 binding site in vitronectin, with some studies reporting binding to the N-terminal, somatomedin B domain and others to the heparin binding region near the C-terminus. Fluorescent probes incorporated into specific sites on PAI-1 or vitronectin have been exploited to quantitate binding and monitor structural changes in both proteins. Mutant forms of PAI-1 were designed with cysteines substituted for residues at positions P1' and P9 of the reactive center loop. Labeling of the unique cysteines in PAI-1 with NBD provided environmentally sensitive probes at the reactive center. Binding of vitronectin quenched the fluorescence of NBD-labeled P1′ PAI-1 by 15%, and titrations revealed a tight association between the proteins with a Kd less than or equal to 100 nM. In contrast, association with vitronectin did not have a significant effect on the fluorescent emission of the labeled P9 derivative. These results suggest that the scissile bond of PAI-1 becomes more solvent exposed upon interaction with vitronectin, while the N-terminal portion of the reactive loop does not experience a significant change in its environment. To investigate further the interaction between vitronectin and PAI-1, a fluorescently labeled vitronectin was prepared using an arginine selective coumarin derivative. The fluorescent probe was incorporated into the highly basic region near the C-terminus of vitronectin, abolishing heparin binding to the protein. Titration of PAI-1 binding to vitronectin, monitored by the 17% quench in the fluorescence of the coumarin probe, gave a binding constant comparable to the Kd obtained using NBD-labeled PAI-1. The observation that PAI-1 binds to a labeled vitronectin devoid of heparin binding, argues for separate binding sites on vitronectin. Furthermore, these data indicate that the reactive center of PAI-1 becomes more solvent exposed, and the heparin binding site on vitronectin is altered, when the two proteins associate to form a stable complex.

44. THE THREE DIMENSIONAL STRUCTURE OF ACTIVE SERPIN K FROM *MANDUCA SEXTA*

Elizabeth J. Goldsmith,[1] Jinping Li,[1] Zhulun Wang,[1] and Mike Kanost[2]

[1]The Department of Biochemistry, The University of Texas Southwestern Medical Center at Dallas, 5323 Harry Hines Blvd., Dallas, Texas, 75235-9050, [2]Kansas State University, Manhattan, Kansas 66506; Correspondence: Betsy@howie.swmed.edu

To resolve the problem of the conformation of the reactive center loop and the degree of formation of sheet A in free active serpins, we have carried out crystallization trials on a variety of active serpin molecules. Crystallization and determination of the structure of serpin K from *Manduca sexta* (H. Jiang, Y. Wang, and M.R. Kanost (1993) *J. Biol. Chem* 269 55–58) was carried out. Serpin K was expressed in *E. coli* with a His$_6$ tag at the N-terminus. Crystals grown in 28% PEG 6000, 1.2 M NH$_4$CL, 0.1 M MES, pH 6.0 at 21°C are 0.6 mm × 0.4 mm x 0.3 mm and diffract to 2.2 Å resolution. The space group of the serpin K crystals is C2, with cell constants a=129., b=43., c=76., β=116°. The inhibitor within the crystals is intact and active. The structure was solved by molecular replacement, with ovalbumin, latent PA-1, and cleaved antitrypsin used as search models. The best molecular replacement solution was obtained with a model based on ovalbumin truncated to polyalanine, with correlation coefficient of 54%, although several of the search models gave solutions. The structure, refined against 2.7 Å data will be presented, and compared with the active conformations of antithrombin and antichymotrypsin.

45. RAT OVARIAN GRANULOSA CELLS SYNTHESIZE HEPARAN SULFATE PROTEOGLYCANS THAT BIND AND ACTIVATE THE SERPIN ANTITHROMBIN III UNDER HORMONAL CONTROL

G. Hosseini and A. De Agostini

Sterility Clinics, Geneva University Hospital, Switzerland

The development of the ovarian follicle and its breakdown at ovulation are driven by hormonal stimulations and require extensive tissue remodelling. At the time of ovulation several serine proteases of the plasminogen activator and coagulation cascades are activated in the ovary and the heparin-activated serpins plasminogen-activator-inhibitor-1, protease nexin-1 and antithrombin III (AT) are present in the follicular environment. Ovulation is an inflammatory process involving extravascular coagulation, yet follicular fluid does not clot before expulsion of the oocyte from the broken follicle, suggesting the presence of an anticoagulant system in this locale. Granulosa cells, together with the oocyte, form the inner follicular compartment that remains avascular until ovulation and heparin-like activity has been described in follicular fluid and in ovarian granulosa cell extracts.

We have now demonstrated that rat granulosa cells in primary culture synthesize considerable amounts of anticoagulant heparan sulfate proteoglycans (aHSPGs) that bind and activate AT. Granulosa cell aHSPGs are enriched in 3-O-sulfated disaccharides which are the markers of the specific AT-binding site of heparin and they represent 6.5% of the total HS chains, a value comparable to that found for endothelial cells (1–10%). Complex formation between granulosa cell aHS and AT was shown by affinity coelectrophoresis and the biologic activity of granulosa cell aHS was demonstrated by the acceleration of formation of thrombin-AT complexes on SDS-PAGE.

The effects of stimulations by the gonadotropin FSH on granulosa cell aHSPGs was investigated using ^{125}I-AT binding assays. The results showed that FSH induces the release of aHSPGs from the cell surface toward the culture media. Finally, we have localized aHSPGs in rat follicles by ^{125}I-AT autoradiography on ovary cryosections. aHSPGs were detected on the entire granulosa cell layers of antral follicles and the intensity of the labelling increased with the size of the follicles.

Taken together, these data demonstrate that aHSPGs are synthesized in the avascular compartment formed by ovarian granulosa cells under gonadotropin control. aHSPGs could contribute to the control of proteolytic activities in the follicle by providing activators of serpins at the surface of granulosa cells and in follicular fluid. This could serve to maintain fluidity in the environment of the oocyte until its expulsion in the ovarian bursa.

46. INTRACELLULAR BINDING SITES FOR PAI-2 RECOGNIZE THE EXON 2 ENCODED SEQUENCE

Poul. H. Jensen,[1] Jørgen Gliemann,[1] Thomas G Jensen,[2] Barbara Müeller,[3] Walther E. Laug,[4] and Blake Pepinsky[5]

[1]Department of Medical Biochemistry and [2]Human Genetics, University of Aarhus, Denmark, [3]The Scripps Research Clinic, La Jolla, [4]Childrens Hospital Los Angeles, Los Angeles, California, [5]Biogen, Cambridge, Massachusetts

The intracellular PAI-2 (i-PAI-2) belongs to the "ovalbumin-like" serine proteinase inhibitor (SERPIN) family, whose members might possess novel functions, e.g. the inhibition of cysteine proteinases by the SERPIN, squamous cell carcinoma antigen. Structure/function relations of i-PAI-2 has not yet been described, even though it might possess important intracellular functions as a cytoprotective molecule against apoptogenic stimuli.

In order to identify putative intracellular PAI-2 binding molecules we performed affinity chromatography by immobilized recombinant human PAI-2 using a crude placental cytosol fraction as the source of putative ligands. A mixture of proteins, eluted from the column by 1 M NaCl, bound ^{125}I-PAI-2 with a K_d about 1–50 nM independent of Ca^{2+}-ions. PAI-2 bound to these proteins were still inhibitory toward urokinase.The binding was entirely dependent on PAI-2 amino acid residues 66–99, corresponding to the main piece of the exon 2 encoded sequence, as demonstrated by the absent binding of a PAI-2 deletion mutant. Furthermore, a synthetic (aa 66–99) peptide contained full binding activity. A major protein in the eluted mixture were lipocortin-1/annexin-1, a member of an anti-inflammatory superfamily of Ca^{2+}, actin and phospholipid binding proteins. Using purified lipocortins we demonstrated Ca^{2+}-independent ^{125}I-PAI-2 binding to lipocortins 1, 2, 4, 5, but not to lipocortins 3 and 6. The binding of PAI-2 to the lipocortins points to a convergence of antifibrinolytic and anti-inflammatory pathways of still unknown significance.

To study the organization of intracellular-PAI-2 (i-PAI-2), we took advantage of the high intracellular expression in a PAI-2-transfected human melanoma cell line. By means of immunofluorescense confocal laser scanning microscopy using a commercial PAI-2 specific MAB, we observed i-PAI-2 to be organized in a reticular pattern in resting cells, not colocalizing with actin, vimentin, the Golgi or ER organelles. Metabolic labeling of this cell line followed by crosslinking of the whole cells by a cleavable crosslinker, immunoprecipitation and SDS-PAGE revealed a approximately 100 kDa and a 70 kDa protein that coprecipitated with PAI-2 when using either a polyclonal and a monoclonal antibody. The identity of these proteins are currently being pursued.

47. CHARACTERIZATION OF TWELVE SERPIN REACTIVE SITE LOOP VARIANTS FROM MANDUCA SEXTA SERPIN GENE-1

Haobo Jiang and Michael R. Kanost

Department of Biochemistry, Kansas State University, Manhattan, Kansas 66506

We have discovered that a serpin gene from an insect, Manduca sexta, encodes, through alternate exon use, a family of serpin proteins. Manduca serpin gene-1 is composed of ten exons, the ninth of which has 12 alternate forms between exons 8 and 10. Consequently, the serpin proteins have a constant N-terminal 336 residues encoded by exons 1 through 8 and a variable C-terminal 39–46 residues encoded by one of the exon 9 alternates. The variable region includes the reactive site loop that determines selectivity for inhibiting a specific type of serine proteinase. All twelve serpin-1 variant cDNAs have been expressed in E. coli and purified by affinity chromatography. Some of these recombinant proteins can inhibit serine proteinases of different specificity, with association rate constants comparable to human serpins. Serpin-1A with a predicted P1 Arg inhibits both trypsin and plasmin. Serpin-1B (P1=Ala) and serpin-1F (predicted P1=Val) are inhibitors of porcine pancreatic elastase and human neutrophil elastase. Serpins-1H, -1K and -1Z, all with a Tyr residue at P1 position, inhibit chymotrypsin and cathepsin G. Serpin 1I (predicted P1=Leu) is an inhibitor of both elastase and chymotrypsin. An evolutionary history of the exon 9 region of Manduca serpin gene-1 is proposed, based on computer-assisted analysis of the DNA and deduced amino acid sequences. It is apparent that duplication and subsequent sequence divergence of one or more copies of exon 9 has occured multiple times, probably through unequal crossing over. Evolutionary trees derived from alignments of the exon 9 sequences are consistent with the following hypothesis. The earliest duplication gave rise to exons 9A and 9Z, and these were again duplicated to produce exons 9B and 9K. The remaining exon 9 variants, all with a four-codon insertion at the carboxyl terminal end of the reactive site loop, may have diverged from exon 9K. The most recent duplications involve exons 9E, 9F, 9G, and 9H, all near the center of the exon 9 region. Evolutionary sequence divergence of these exons has led to the ability to synthesize serpins with twelve different reactive site sequences from a single gene, through use of mutually exclusive alternate exon splicing. From serpin sequences obtained by the group of Sasaki from the silkworm, Bombyx mori, it is apparent that similar alternate exon use occurs at least in another lepidopteran insect. It will of interest to determine how widespread this type of serpin gene structure may be in other insects and arthropods.

48. CONFORMATIONAL CHANGES OF β-STRAND 5A AND THE REACTIVE CENTRE LOOP ACCOMPANY TEMPERATURE-DEPENDENT INHIBITOR-SUBSTRATE TRANSITION OF PLASMINOGEN ACTIVATOR INHIBITOR-1

Lars Kjøller, Pia M. Martensen, Lars Sottrup-Jensen, Just Justesen, Kees W. Rodenburg, and Peter A. Andreasen

Department of Molecular Biology, University of Aarhus, Denmark

We have studied conformational changes of type-1 plasminogen activator inhibitor (PAI-1) during a temperature-dependent transition between inhibitor and substrate forms by measuring susceptibility of the molecule to non-target proteinases. When incubated at 0°C instead of the normally used 37°C, a ten-fold decrease in the specific inhibitory activity of active PAI-1 was observed. Accordingly, PAI-1 was recovered in a reactive centre-cleaved form from incubations with the target proteinase urokinase-type plasminogen activator (u-PA) at 0°C, but not at 37°C. It thus behaved as a substrate for u-PA at the lower, but not at the higher temperature. Active PAI-1 was exposed to a variety of non-target proteinases, including elastase, papain, thermolysin, trypsin, and V8 proteinase, at each of the two temperatures. It was found that specific peptide bonds in the reactive centre loop (RCL) and strand 5 in β-sheet A (s5A) had a temperature-dependent proteolytic susceptibility, while the P_{17}–P_{16} (E332–S333) bond, forming the hinge between s5A and the RCL, showed indistinguishable susceptibility to proteolysis by V8 proteinase at 0°C and 37°C. In latent and reactive centre-cleaved PAI-1, all the bonds were resistant to proteolysis at the higher as well as the lower temperature. An anti-PAI-1 monoclonal antibody stabilized the inhibitory activity of PAI-1 and prevented reactive centre cleavage at 0°C, and thus prevented conversion to the substrate form. Concomitantly, it caused specific changes in protelytic susceptibility of the RCL and s5A, but it did not affect cleavage of the P_{17}–P_{16} bond by V8 proteinase. Our observations suggest that temperature dependent conformational changes of s5A and the RCL determine temperature-dependent partitioning of the serpin between inhibitor and substrate pathways. They also suggest that the RCL of PAI-1 appears to be fully extracted from β-sheet A in the inhibitory as well as in the substrate form, favoring an *induced conformational state* model as explanation for the requirement for partial insertion of the RCL into β-sheet A for inhibitory activity.

49. SERPIN INHIBITION IS CHARACTERIZED BY FAST COVALENT BOND FORMATION

Jan-Olov Kvassman, Ingrid Verhamme, and Joseph D. Shore

Henry Ford Health Sciences Center, Detroit, Michigan

The experimental approach to the serpin inhibitory mechanism was threefold. First the rate of binding of a chromogenic substrate (S2222), a Kunitz inhibitor (PTI) and a serpin (PAI-1) to the active site of β-trypsin were compared over the pH-range 4–11. Association of the substrate and the serpin, but not the Kunitz inhibitor, required the active conformation of trypsin, triggered by protonation of Ile-16. Protonation of His-57 at the active site reduced the rate of association of all three reactants. However, while productive association of the substrate and inhibitory binding of the serpin was found to depend ultimately on deprotonation of His-57, this was not the case for the Kunitz inhibitor.— *Because the two ionizations that regulate the catalytic activity of trypsin also regulate inhibitory binding of PAI-1 to this enzyme, this binding must involve covalent bond formation mediated by the catalytic apparatus of the enzyme. Obviously, this is not the case with PTI.*

Secondly, we developed a method for quenching of the serpin reaction after incubation times as short as 5 ms. This technique, followed by analysis of the quenched mixtures by SDS-PAGE, was applied to measure the rate of formation of SDS-stable complexes in the reaction of PAI-1 with tPA and β-trypsin. Both reaction exhibited hyperbolic saturation kinetics. With tPA the limiting rate was 2 s^{-1} and the saturation constant (K_M) < 0.5 μM. The corresponding values for trypsin were 28 s^{-1} and 5 μM, respectively. — *These results are in obvious conflict with the suggested mechanism for serpin inhibition where tight reversible binding is followed by a slow stabilizing isomerization. The initial binding may be tight but is rapidly followed by covalent bond formation. The different limiting rates observed with tPA and trypsin are consistent with the difference in general catalytic efficiency between the two enzymes.*

Thirdly, we studied the kinetics of loop-insertion in the reaction of PAI-1 with β-trypsin. For this purpose we used a P9-Cys mutant PAI-1, with a fluorescent probe (NBD) bound to the cysteine. The fluorescence of this probe has been shown to increase in response to insertion of the P9-region of the reactive loop of PAI-1 into β-sheet A (Shore *et al.* 1995 *J.Biol.Chem.*, **270**, 5398). The rate of the fluorescence change, caused by trypsin binding, increased hyperbolically with the concentration of the enzyme, approaching a limiting value of 116 s^{-1} with a K_M of 32 μM. At this concentration, and above, the fluorescence progress curves exhibited a distinct lag phase, reflecting the accumulation of a "Michaelis" complex prior to loop-insertion. — *With both trypsin and tPA hyperbolic saturation kinetics is observed for the reaction with PAI-1, whether this is measured as formation of SDS-stable complexes, using the wild type inhibitor, or as loop-insertion, using the labeled mutant form. The kinetic parameters associated with the latter reaction are, however, higher by approximately the same factor for both enzymes. This finding, and the fact that covalent bond formation, measured with the rapid-quench technique, is not faster with the labeled mutant PAI-1 than with the wild-type inhibitor, suggest that loop-insertion precedes covalent bond formation in the course of PAI-1 binding to trypsin.*

50. PEPSIN INHIBITORY ACTIVITY OF THE UTERINE SERPINS

Nagappan Mathialagan,[a] Thomas R. Hansen,[b] and R. Michael Roberts[a]

[a] Department of Animal Sciences, University of Missouri-Columbia, Columbia, Missouri 65211; [b]Department of Animal Sciences, University of Wyoming, Laramie, Wyoming 82071

Among the major products secreted by the uteri of cattle, sheep and pigs during pregnancy or in response to progesterone are basic glycoproteins with amino acid sequences that clearly place them in the serpin family of proteins. The inferred amino acid sequences for bovine uterine serpin (boUS-1) and ovine uterine serpin (ovUS-1) exhibit about 72% sequence identity to each other but only about 50% and 56% identity respectively to two distinct porcine uterine serpins (poUS-1 and poUS-2). Despite these differences in primary structure, the uterine serpins possess a fairly well conserved reactive center loop regions containing "KVP" and "VVKVP" motifs which are present in the propeptide regions of pepsinogens. Therefore, we tested the hypothesis that US might bind and inhibit aspartic proteinases rather than serine proteinases. Immobilized pepsin A selectively bound ovUS from sheep uterine secretions. In addition, ovUS inhibited the pepsin A and pepsin C (chymosin) but not cathepsin D or E proteolytic activities in a concentration-dependent manner. Thus, ovUS-1 is the first specific inhibitor for aspartic proteinases to be identified in vertebrates and provides another example of a serpin with "crossover" activity. To determine whether there was a "natural" target protein for ovUS in the pregnant uterus, medium from cultured placental explants was passed through a ovUS affinity column. Western blot analysis of the eluted proteins showed that pregnancy-associated glycoproteins (PAG), which are members of aspartic proteinase family, released by the trophoblast layer of sheep placenta may be the natural target partners for the uterine serpins. These data suggest that the function of the US in pregnancy is to sequester PAG secreted by the conceptus. Why the mother would find it necessary to sequester PAG by complexing them with a serpin is unclear, but it may reflect her need to limit the activities of the trophoblast during pregnancy.

Supported by USDA grants 95-37203-2035 (NM), 89-37240-4586 (RMR) and NIH grant HD21980 (RMR).

51. ANALYSIS OF SERPIN INHIBITORY FUNCTION BY MUTAGENESIS OF OVALBUMIN AND GENERATION OF CHIMERIC OVALBUMIN/PAI-2 FUSION PROTEINS

Barry J. McCarthy and D. Margaret Worrall

Dept of Biochemistry, University College Dublin, Belfield, Dublin 4, Ireland

Ovalbumin is a non-inhibitory serpin which lacks the ability to undergo the S→R transition or conformational change. Amino acid residues in the hinge region (P11–P14) of ovalbumin and other non-inhibitory serpins differ from the concensus sequence in this region of inhibitory serpins. Site directed mutagenesis using PCR overlap extension was performed on these residues in ovalbumin to create a mutant with three amino acid changes, R340T, V342A and V343A. However analysis of the mutant recombinant ovalbumin with the consensus residues failed to show inhibitory activity or increased instability, indicating that the hinge region alone is not responsible for lack of inhibition.

A series of three fusion proteins were then constructed by replacing varying C-terminal regions of ovalbumin with the corresponding region of the inhibitory ov-serpin-PAI-2 in order to further analyse serpin inhibitory function. Fusion proteins F1 and F2 contained approximately 16% and 35% PAI-2 respectively. This resulted in the replacing of structural features such as the reactive site loop, hinge region and ss sheet strands 5A and 6A. However both fusion proteins showed no inhibitory activity with the PAI-2 target protease urokinase (uPA). The third chimeric fusion protein constructed (F3) contained 65% PAI-2 and did demonstrate inhibition of uPA. Structural differences between the inactive F2 and active F3 include the replacement of ss sheet 3A and helix F of ovalbumin with those of PAI-2, suggesting that these regions may have a key role in serpin molecular flexibility and inhibition.

52. STUDIES ON THE INTERACTION OF FUCOIDAN WITH PROTEINS OF FIBRINOLYSIS OR COAGULATION

Roxanne Minix and V. M. Doctor

Prairie View A&M University, Prairie View, Texas 77446

Fucoidan [sulfated poly(L-fucopyranose)] is reported to possess anticoagulant activity *in vitro* and *in vivo* (Springer *et al.* Proc. Soc. Exp. Biol. Med. *94*, 404, 1957) and its antithrombic activity was reported to be mediated by heparin cofactor II (Church *et al.* J. Biol. Chem. 264, 3618, 1989). Our laboratory recently reported (Thrombosis Research. *79*, 237, 1995) that fucoidan enhanced the activation of glutamic plasminogen (Glu-Plg) by two-chain tissue plasminogen activator (t-PA) or by LMwt-Urokinase. In the present report the interactions of fucoidan with Glu-Plg, t-PA, Urokinase, thrombin or antithrombin-III (AT-III) were investigated using fucoidan-sepharose affinity chromatography. The hydroxyl groups of fucoidan were converted to epoxy derivative using epichlorohydrin and aminated using ammonia. The aminated fucoidan was covalently attached to Sepharose-4B by using the cyanogen bromide method. The concentrations of Glu-Plg, t-PA or urokinase bound to fucoidan-sepharose were determined by measuring their levels in the eluates by chromogenic assays for plasmin formation using H–D–Val–L–Leu–L–Lys–pNa(S-2251). On the other hand the levels of thrombin or AT-III eluted were measured by using chromogenic substrate, H–D–Phe–Pip–Arg–pNa(S-2238). The results showed: 1) a high degree of affinity between fucoidan-sepharose and Glu-Plg, urokinase or thrombin, 2) t-PA did not bind with fucoidan-sepharose and 3)the inhibitory activity of AT-III was enhanced after passing through the column. The double reciprocal plot for the interaction of Glu-Plg and urokinase showed a significantly higher affinity between the two in presence of fucoidan. This may suggest that a ternary complex may be formed between fucoidan, Glu-Plg and urokinase. On the other hand, the effect of fucoidan on the activation of Glu-Plg by t-PA was characteristic of an allosteric effect since the velocity of the reactions increased slowly at low substrate concentrations and then rose rapidly at higher levels. Fucoidan inhibited thrombin directly and this inhibition was enhanced by AT-III. The results of affinity chromatography showed a significant binding between thrombin and fucoidan-sepharose, while AT-III passed through the column with altered reactivity which resulted in the enhancement of its antithrombic activity. [Supported by NIH-MBRS 08094]

53. PLASMINOGEN ACTIVATOR INHIBITOR-1 (PAI-1) INTERACTION WITH TISSUE-PLASMINOGEN ACTIVATOR (TPA) INVOLVES A TIGHT BINDING "EXO-SITE"

Shabazz A. Muhammad, Steingrimur Stefansson, Mitchell B. Berkenpas,* and Daniel A. Lawrence

American Red Cross Holland Laboratory, 15601 Crabbs Branch Way, Rockville, Maryland 20855; *Pharmacia and Upjohn Inc., 301 Henrietta Street, Kalamazoo, Michigan 49007

Plasminogen Activator Inhibitor 1 (PAI-1), a member of the serine protease inhibitor (serpin) super family, is the primary inhibitor of plasminogen activators (PAs) *in vivo*, inhibiting both tPA and uPA with second order rate constants $> 10^7$ $M^{-1}s^{-1}$. As such, PAI-1 plays a major role in regulating vascular fibrinolysis, with elevated levels of PAI-1 in plasma correlating with increased incidence of cardiovascular disease. Functionally, inhibitory serpins are thought to interact with target proteases through their reactive center loop sequences, with the primary contact occurring via the P1 bait residue which associates with the S1 site of the enzyme. However, several studies have suggested the existence of additional sites of interaction between serpins and their target proteases, that are distant from the S1-P1 contact. Of these so called "exo-sites" the association of the carboxyl-terminal domain of α_2-antiplasmin with a lysine binding site on plasmin is probably the best known. Several reports have also suggested the presence of an exo-site on PAI-1 that interacts with tPA. These include kinetic analysis (Hekman *et al.*, *Arch of Biochem. and Biophys.* 1988, **262**: 199–210; Chmielewska *et al.*, *Biochem J.* 1988; **251**:327–332), as well as functional studies of tPA and PAI-1 mutants (Madison *et al.*, *Nature* 1989; **339**:721–724 & *J. Biol. Chem.*; 1990; **265**:21423–21426; Lawrence et al., *J. Biol. Chem.*, 1990; **265**:20293–20301). In order to examine the possibility that such a specific exo-site(s) exists on PAI-1, and if so, to characterize its (their) interaction with tPA we have used site-directed mutagenesis to convert the PAI-1 Arg residue to Ala. This change removes the P1-S1 interaction, and thus enables us to examine only binding resulting from weaker secondary interactions. The mutant PAI-1 has been purified to homogeneity and shows no detectable inhibitory activity against either uPA or tPA. Furthermore, is not cleaved by these enzymes as a substrate. Together, these data suggest that the mutant PAI-1 does not interact with the active site of PAs, consistent with the importance of the P1-S1 contact for inhibitory activity. However, direct binding studies of this mutant to immobilized tPA indicate that it binds specifically and saturably with a sub-μM K_d. In contrast, only insignificant binding is seen with uPA. Together, these data suggest that PAI-1 and tPA interact through one or more exo-sites(s), and that this interaction is specific for tPA. In an attempt to identify this site(s) we have developed a random mutagenesis strategy based on the binding of P1 Ala PAI-1 to tPA. Briefly, error prone PCR was used to construct a library of random P1 Ala PAI-1 mutants in the λExLox phage vector. The mutant phage library consists of $\sim 2 \times 10^6$ independent clones with an error rate of ~ 4 mutations per clone. Experiments are currently under way to identify clones that have lost the ability to bind tightly to tPA but retain their affinity for vitronectin. In this way we hope to identify specific residues critical for this secondary interaction with tPA, and to clarify their importance for PAI-1 inhbition of tPA, as well as to more generally understand the influence of exo-site contacts on serpin-protease inhibition.

54. ROLE OF THE CATALYTIC SERINE IN THE INTERACTIONS OF SERINE PROTEINASES WITH PROTEIN INHIBITORS OF THE SERPIN FAMILY

S. T. Olson,* R. Swanson,* P. Patston,* P. E. Bock,[+] J. Kvassman,[§] J. D. Shore,[§] D. Lawrence,[#] D. Ginsburg,[†] and I. Björk[¶]

*University of Illinois, Chicago, Illinois, [§]Henry Ford Hospital, Detroit, Michigan, [+]Vanderbilt University, Nashville, Tennessee, [#]American Red Cross, Rockville, Maryland, [†]University of Michigan, Ann Arbor, Michigan, and [¶]Swedish University of Agricultural Sciences, Uppsala, Sweden

To determine the role of the catalytic serine residue of serine proteinases in the mechanism by which protein inhibitors of the serpin family form high-affinity complexes with such enzymes, the interactions of a number of serpins with β-trypsin and the inactive, catalytic serine-modified derivative, β-anhydrotrypsin were compared. Kinetic analyses showed that anhydrotrypsin had little or no ability to compete with trypsin for binding to equimolar levels (10 nM) of the serpins, α_1-proteinase inhibitor (α_1PI), antithrombin (AT),

AT-heparin complex, and plasminogen activator inhibitor-1 (PAI-1) when the anhydroenzyme was present at up to a 100-fold molar excess over the active enzyme. By contrast, trypsin binding to the Kunitz inhibitors, bovine pancreatic trypsin inhibitor or soybean trypsin inhibitor, was blocked by equimolar anhydro-trypsin. Equilibrium binding of the serpins to trypsin and anhydrotrypsin, monitored by displacement of the fluorescent probe, p-aminobenzamidine, from the enzyme active site, confirmed that serpins bound weakly to anhydrotrypsin (K_Is ranging from 0.2 to >10 µM) but bound tightly and stoichiometrically to trypsin (K_Is << 0.1 µM). In contrast, the Kunitz inhibitors stoichiometrically bound with high-affinity to both trypsin and anhydrotrypsin (K_Is << 0.1 µM). Quantitation of K_I values for serpin-trypsin complex formation from kinetic analyses of on-rate and off-rate constants revealed that the affinities of serpins for trypsin were minimally 4 to 6 orders of magnitude greater than their affinities for anhydrotrypsin, indicating that the proteinase catalytic serine is critical for forming high-affinity complexes with serpins. The ability of anhydrotrypsin to induce the serpin conformational change involved in forming stable complexes with proteinases was examined by studying the interactions of trypsin and anhydrotrypsin with a P9 Ser→Cys PAI-1 variant labeled at the P9 residue with a nitrobenzofurazan fluorescent probe. Anhydrotrypsin binding to the labeled PAI-1 failed to induce the 6.7-fold fluorescence enhancement and 17 nm spectral shift induced by trypsin binding which has been shown to characterize the serpin conformational change (Shore, J. D., Day, D. E., Francis-Chmura, A. M., Verhamme, I., Kvassman, J., Lawrence, D. A., and Ginsburg, D. (1995) J. Biol. Chem. 270, 5395–5398). Moreover, anhydrotrypsin binding to wild-type PAI-1 reduced the rate at which the inhibitor spontaneously underwent the serpin conformational change to yield latent inhibitor by 1.8-fold. A previous report that chemical modification of the catalytic serine of trypsin by dichloroisocoumarin (DCI) did not block the ability of trypsin to form an SDS-stable complex with α_1PI (Christensen, S., Valnickova, Z., Thøgersen, I. B., Pizzo, S. V., Nielsen, H. R., Roepstorff, P., and Enghild, J. J. (1995) J. Biol. Chem. 270, 14859–14862) was reexamined by kinetic studies of the reactions of α_1PI with active and inactive enzymes. The amount of enzyme-serpin complex formed in the DCI-enzyme reaction was quantitatively accounted for by the amount of active enzyme regenerated from inactive enzyme by spontaneous hydrolysis of the inactivating group. Moreover, complex formation was blocked in the presence of D-Phe-Phe-Arg-chloromethyl ketone. These observations indicate that the observed complex originated from the serpin reacting with active enzyme rather than with inactive DCI-enzyme. Together, these results demonstrate that the catalytic serine contributes a major fraction of the binding energy to serpin-trypsin interactions and is required for inducing the serpin conformational change which accompanies enzyme binding. Such results support a suicide substrate model for serpin action in which a covalent interaction involving the catalytic serine of the proteinase and the serpin reactive bond triggers the serpin conformational change which leads to the trapping of proteinase in a high-affinity complex.

55. DIFFERENTIAL MECHANISMS FOR THE SECRETION AND INTRACELLULAR ACCUMULATION OF PLASMINOGEN ACTIVATOR INHIBITOR 2 (PAI-2) IN HUMAN PERIPHERAL BLOOD MONOCYTES

Helen M. Ritchie and Nuala A. Booth

Department of Molecular and Cell Biology, University of Aberdeen, Aberdeen AB9 1AS, U.K.

PAI-2 is a member of an emerging group of intracellular serpins, whose function(s) and target(s) within the cell are as yet unknown. PAI-2 is synthesized by many different cell types and also by transformed cell lines, and PAI-2 is found mainly as an intracellular protein. Peripheral blood monocytes are a major physiological source of PAI-2. Monocytes can be activated, by various agents, to secrete PAI-2 into the culture medium and also to accumulate PAI-2 within the cell. Here, we have investigated the secretion and intracellular accumulation of PAI-2 in isolated human peripheral blood monocytes.

Synthesis of PAI-2 by peripheral blood monocytes was stimulated by thrombin (1U/ml) or by lipopolysaccharide (LPS, 10 ng/ml). These agents had a similar effect on PAI-2 and up-regulated synthesis of secreted and intracellular PAI-2; intracellular PAI-2 was the major pool of PAI-2. An upregulation in PAI-2 mRNA, when compared to 18 S ribosomal RNA, was evident by 4 h. Secretion of PAI-2 by the endoplasmic-reticulum (ER)–Golgi dependent pathway was studied by incorporation of a specific inhibitor, brefeldin A (0.5 µg/ml). Brefeldin A completely inhibited the accumulation of intracellular PAI-2, which represented the majority of PAI-2, over a 24–96 h time period. Secretion of PAI-2 was initially inhibited by brefeldin, but by 48 h PAI-2 was secreted into the culture medium at levels that were similar to monocytes cultured in medium alone. TNFα was measured in the culture supernatants, as an example of a protein secreted via the ER-Golgi pathway. TNFα was greatly inhibited by brefeldin A over the entire time course. Tunicamycin (1 µg/ml), an inhibitor of N-linked glycosylation, was also incorporated into our culture system. Tunicamycin had a similar effect on PAI-2 to that of brefeldin A; tunicamycin completely inhibited the intracellular accumulation of PAI-2, but not the secretion of PAI-2 in the culture supernatant. Lactate dehydrogenase (LDH) was measured in the monocyte culture supernatants, and it was found that neither brefeldin of tunicamycin were having a cytotoxic effect towards the monocytes. We also assayed total protein in the monocyte cell lysates, as a measure of cell number, and found no great difference between brefeldin/tunicamycin treated and unstimulated cells. These results with brefeldin and tunicamycin have been repeated in monocytes isolated from at least three separate donors.

The molecular mass species of PAI-2 were investigated by metabolic labelling of monocytes, using ^{35}S-methionine, followed by immunoprecipitation and SDS-PAGE. Monocytes were grown in methionine-free medium, pulsed with ^{35}S-methionine and chased with cold medium. PAI-2 was immunoprecipitated from the culture supernatant and cell lysate using a monoclonal antibody, and samples were analysed by SDS-PAGE. We found that the major species of PAI-2 that was present in both supernatants and cell lysates was a 43 kDa species. Minor bands at 60 kDa and approximately 200 kDa were also present. It is likely that this 43 kDa species represents the unglycosylated species of PAI-2. Immunoprecipitation of supernatants and cell lysates with pre-immune mouse IgG served as the appropriate negative control. We also localized intracellular PAI-2 in mono-cytes using a monoclonal antibody to PAI-2, followed by FITC-labelled anti-mouse IgG. PAI-2 was found at the plasma membrane, which suggests distinct targeting of PAI-2. This was in contrast to U937 cells where PAI-2 was detected throughout the cytoplasm.

These data show that secretion of PAI-2 by peripheral blood monocytes can occur via an ER-Golgi independent pathway. The accumulation of intracellular PAI-2 is dependent on the ER-Golgi pathway and is also dependent on glycosylation. It is not yet known whether this phenomenon is a direct effect on PAI-2 or an indirect effect on some cofactor that may be involved in the accumulation of PAI-2 in monocytes. We found that a 43 kDa species of PAI-2 was the major form in both supernatants and cell lysates, but other molecular mass species were present. The distinct membrane localization of PAI-2 in monocytes suggests a potential membrane function for intracellular PAI-2.

56. MOLECULAR BASIS FOR ANTIBODY EFFECTS ON PLASMINOGEN ACTIVATOR INHIBITOR-1 ACTIVITY

J. D. Shore, S. Debrock, D. E. Day, I. Verhamme, and P. J. Declerck

Henry Ford Health System, Detroit, Michigan, and Faculty of Pharmaceutical Sciences, Catholic University of Leuven, Belgium

We have studied two monoclonal antibodies: MA-33B8, which inactivates PAI-1 without converting it to a substrate, and MA-55F4, which converts the PAI-1 from an nhibitor to a substrate. With MA-33B8, addition of 0.3 μM antibody to 0.2 μM P9 Cys NBD labeled PAI-1 resulted in a 4.6-fold enhancement of the NBD fluorescence and a 13 nm blue shift, similar to the active to latent transition in our previously reported studies. The kinetics of this process were first order, with a half life of approximately one minute at 25°C and pH 7.4. The end product after antibody binding was inactive as an inhibitor and not cleavable by a catalytic concentration of elastase, which does not cleave latent PAI-1. A P9 Cys NBD P14 Arg double mutant, which is a substrate and does not exhibit spontaneous loop insertion, showed no fluorescence change after addition of antibody, suggesting that the observed fluorescence increase is not due to binding in the vicinity of P9. The k_{cat}/K_m for hydrolysis of the P14 Arg mutant was not affected by MA-33B8. Our conclusion is that binding of MA-33B8 causes a rapid conversion of PAI-1 to a latent form, in which the reactive center loop (RCL) is inserted into β sheet A. In contrast, binding of MA-55F4, which has been shown to convert PAI-1 to a substrate form, to P9 Cys NBD PAI-1 did not cause fluorescence enhancement. However, tPA addition prior or subsequent to antibody caused a 5.2-fold enhancement and 17 nm blue shift, compared with a 3.3-fold enhancement and 13 nm blue shift due to tPA alone. Addition of MA-55F4 to the P9 Cys NBD P14 Arg double mutant caused no fluorescence enhancement, indicating that it was not binding at or near the P9 residue. Our present interpretation of these results is that MA-55F4 prevents inhibition of tPA by PAI-1 through perturbation of the structure such that during or after cleavage and insertion of the RCL the P9 is in a more hydrophobic environment than in the absence of antibody. In conclusion, the observed different biochemical mechanisms by which monoclonal antibodies can impair PAI-1 activity are further substantiated at a more detailed molecular level.

57. REACTION KINETICS OF WILD TYPE PAI-1 AND ITS SUBSTRATE FORMS WITH TARGET SERINE PROTEINASES

I. M. Verhamme,* J.-O. Kvassman,* H. R. Halvorson,* D. A. Lawrence,[¶] D. Ginsburg,[§] and J. D. Shore*

*Henry Ford Health Sciences Center, Detroit, Michigan, [¶]American Red Cross Holland Laboratory, Rockville, Maryland, [§]Howard Hughes Medical Institute, University of Michigan, Ann Arbor, Michigan

Wild type plasminogen activator inhibitor-1 (PAI-1) inhibits the serine proteinases tPA, tc-uPA, β-trypsin, and, in the presence of heparin, also α-thrombin, whereas Thr333Arg PAI-1 and PAI-1 complexed with the P14-P7 PAI-1 octapeptide are substrates for these enzymes. The early mechanistic steps governing reversible binding, and substrate or inhibitor cleavage with acyl-enzyme formation were compared by studying the kinetics of proteinase inactivation by wild type PAI-1, and hydrolysis of substrate PAI-1. It was assumed that the mechanisms have similar steps preceding covalent bond formation. The k_{cat}/K_m values for hydrolysis of substrate PAI-1 by these enzymes were 3 to 30-fold smaller than the respective second order inactivation rate constants k* for wild type PAI-1. The pseudo-first order rate constant of tc-tPA inactivation by PAI-1, measured by either competitive chromogenic substrate hydrolysis or p-aminobenzamidine displacement, increased linearly with increasing [PAI-1], up to 150 s^{-1} at 12 μM PAI-1, the experimental upper limit for accurate measurement of probe displacement. Although this linear dependence could be consistent with a one step, reversible tight binding mechanism, the serpin-enzyme mechanism is known to have a second step, involving serpin cleavage and acyl-enzyme formation. Both methods for determining the inactivation rate monitor a decrease in free enzyme, and describe reversible complex formation, which in this case was non-saturable. K_d's of 3 to 6 nM have been reported for PAI-1 and active-site modified tPA and uPA (*H.R. Lijnen, B. Van Hoef, D. Collen, J. Biol. Chem. 266, 4041–4044, 1991*). These are true dissociation constants since the mechanism was arrested at the reversible step. If the binding constant for PAI-1 and native tPA is similar, our data suggest that the Michaelis complex does not accumulate up to 12 μM PAI-1. Hence the acylation step k_2 is faster than the dissociation rate constant for reversible complex. The differences in k_{cat}/K_m and k* reflect differences in the magnitude of the microconstants determining the early steps in both mechanisms. The important structural difference of substrate PAI forms with the inhibitory PAI is their inability to pre-insert the reactive center loop. The decreased second order rate constant for substrate PAI binding may be due to a decrease in k_1, or in k_{-1}/k_2. The former would suggest pre-insertion in the free inhibitor, the latter an induced pre-insertion in the Michaelis complex. All the enzymes had small values of apparent k_{cat} for hydrolysis of substrate PAI-1 (0.06–1.6 s^{-1}), suggesting a common rate-limiting step for substrate turnover. K_m values were small (20 to 90 nM) with all the tPA forms including mini-tPA, so the putative exosite must be near the proteinase domain. The k_{cat} for substrate PAI hydrolysis by tc-tPA (0.06–0.09 s^{-1}) is of the same order of magnitude as that for fibrin-mediated plasminogen activation. Deacylation might be partly rate-limiting in the substrate reaction, because rapid chemical quenching of the Thr333Arg PAI–uPA reaction, at 1 μM enzyme and 5 μM substrate, followed by SDS gel electrophoresis of the reaction mixtures, demonstrated the existence of a transient acyl- enzyme intermediate.

58. HIGH PAI-1 LEVELS AND FAILURE OF UROKINASE TO INDUCE A SYSTEMIC FIBRINOLYTIC STATE IN A PATIENT WITH ANTIPHOSPHOLIPID SYNDROME AND HEPARIN INDUCED THROMBOCYTOPENIA

H. C. Whinna, H. Hwang, C. L. Fligner, M. H. Wener, and W. L Chandler,

Departments of Pathology and Laboratory Medicine, University of Washington, Seattle, Washington 98195

A patient was unsuccessfully treated with urokinase for multiple thromboses secondary to antiphospholipid syndrome (APS) and possibly heparin induced thrombocytopenia and thrombosis syndrome (HITTS). We hypothesize that the urokinase infusion failed to induce a systemic fibrinolysis (as measured by a 2–4 fold increase in thrombin time) due to extraordinarily high plasminogen activator inhibitor 1 levels (PAI-1) of ≥1000 pmol/L (50 AU/ml).

A 38-year old female with a 6-year history of systemic lupus erythematosus was admitted to the intensive care unit at the University of Washington Medical Center with mental status changes, increasing shortness of breath, and worsening of her severe Raynaud's phenomenon. The patient also had a history of APS with positive anticardiolipin antibody and lupus inhibitor. Heparin was started due to concern of a pulmonary embolus contributing to her shortness of breath. Over two days the patient's platelet count dropped from 181,000/ul to 54,000/ul and her hands became increasingly cyanotic. Heparin was discontinued secondary to a presumptive diagnosis of HITTS and the patient's platelet count gradually increased to 225,000/ul over the next 3 days. Doppler studies showed bilateral occlusion of radial arteries and occlusion of right ulnar artery. Post-mortem measurement of plasminogen, PAI-1 and alpha-2-antiplasmin activity in a sample taken prior to the start of the urokinase infusion showed levels of 92%, ≥1000 pmol/L (50 AU/ml), and 144%, respectively. Urokinase infusion was started on hospital day 2 at a rate of #197#5 pmol/L/sec and two bolus doses of 55,400 pmol were given approximately 6 and 18 hours after the start of the urokinase infusion. Over the next two days the infusion rate was increased stepwise up to a maximum of #197#11 pmol/L/sec, however the patient's thrombin time never reached a therapeutic level of 2–4 fold normal (maximum increase was 1.5 fold). Plasminogen, PAI-1, and alpha-2-antiplasmin levels were measured during the infusion and found to be 15%, 16 pmol/L (0.8 AU/ml) and 44% respectively. Two units of fresh frozen plasma were administered after these measurements in an attempt to replenish plasminogen levels. However, the thrombin time remained below the therapeutic level and on hospital day 5 the urokinase infusion was stopped. Over the next three days the patient developed renal failure and her respiratory status worsened. Plasmapheresis was performed in an attempt to remove any antiphospholipid antibodies contributing to her continued thrombosis. Although the patients anticardiolipin antibody and antinuclear antibody titers became negative, her condition declined and she developed multi-organ system failure. The family chose to withdraw all but comfort measures per the patient's expressed wishes prior to admission and the patient expired on hospital day ten.

An autopsy was performed and showed multiple thromboemboli involving small and large arteries of the lungs and kidneys as well as peripheral vasculature and skin. Both platelet rich and erythrocyte rich clots were observed and the age range of thromboemboli were consistent with both recent events and thromboemboli predating this hospital admission. The patient died of multiple organ system failure due to multiple thromboses secondary to APS and possibly HITTS. We believe the urokinase infusion failed to promote a systemic fibrinolytic state because of the patient's extraordinarily high levels of PAI-1 (most likely increased due to an acute inflammatory response). The infusion rate and bolus doses of urokinase were never sufficient to overwhelm the available PAI-1 and allow for systemic plasminogen activation. In addition, the patient also had alpha-2-antiplasmin activity remaining during the infusion and elimination of this activity has been reported necessary for systemic fibrinolysis. Although it is unclear that successful promotion of a systemic fibrinolytic state would have altered the patient's disease course, physicians attempting to promote systemic fibrinolysis should be aware that increased levels of PAI-1, such as those seen in acute inflammation, may necessitate the use of higher than normal urokinase or other plasminogen activator dosing regimens.

59. TISSUE LOCALIZATION AND REGULATION OF PROTEIN C INHIBITOR (PCI) GENE EXPRESSION IN MICE

J. J. Xu,[1,3] M. Zechmeister-Machhart,[1] P. Uhrin,[1] E. Bielek,[2] B. R. Binder,[1] and M. Geiger[1]

[1]Lab. Clin. Exp. Physiol., and [2]Dept. Histol. and Embryol., Univ. Vienna, Austria; [3]Dept. Cardiac Surg., 2nd Affiliated Hosp., Jiangxi Med. College, P.R. China

Human PCI is a non-specific serpin that inhibits several proteases of the coagulation and fibrinolytic systems as well as tissue kallikrein and sperm protease acrosin. It is synthesized in many organs including the liver, pancreas, and testis, and is present in plasma and several other body fluids and secretions. The physiological role of PCI has not yet been defined. Recently, we have cloned and sequenced the mouse PCI gene (M. Zechmeister-Machart, et al., manuscript submitted). This enabled us to study PCI gene expression in murine tissues using mouse PCI cDNA and cRNA probes. By Northern blot analysis, mouse PCI mRNA was exclusively found in the reproductive tract (testis, seminal vesicle, ovary); all other organs analyzed, including the liver, were negative for PCI mRNA, indicating that in the mouse — unlike in men — PCI is not a plasma protein. To determine which cells of the reproductive tract synthesize PCI, cellular location was assessed by in situ hybridization of mouse testis, seminal vesicle and ovary sections. In testis, PCI mRNA was present in the spermatogonia layer and in Leydig cells, whereas Sertoli cells and peritubular myoid cells were negative. Specific hybridization signals were also observed in epithelial cells of seminal vesicles. These results are consistent with the immunohistological localization of human PCI (Laurell et al, J. Clin. Invest. 89: 1094, 1992). In the mouse ovary, stroma cells of the medulla and around the follicles were positive for PCI mRNA. No PCI expression was detected in theca or granulosa cells. Since little is known concerning the regulation of PCI biosynthesis and since the 5' flanking region of the mouse PCI gene contains 2 sterol regulatory elements, we studied the regulation of mouse PCI gene expression by steriod hormones in vivo. In mature male mice castration led to an increased accumulation of PCI mRNA in seminal vesicles, which was reversed upon administration of testosterone. After injection of human chorionic gonadotropin (hCG) into normal and adult female mice, PCI mRNA levels decreased in testis, seminal vesicles and ovary, respectively. In castrated male mice, injection of hCG had no effect on seminal vesicle PCI mRNA. Furthermore, a decrease in ovarian PCI mRNA levels was observed after injection of either progesterone or 17-estradiol into immature (7–10 days old) female mice. These findings suggest that the transcription of the mouse PCI gene is directly down regulated by gonadal steriods. The exclusive localization and sex hormone controlled expression of PCI in the reproductive tract strongly suggests that in the mouse this serpin is involved in the regulation of processes related to reproduction.

60. MOLECULAR CLONING OF THE GENE ENCODING MOUSE PROTEIN C INHIBITOR: ORGANIZATION OF THE GENE AND TISSUE SPECIFIC EXPRESSION

M. Zechmeister-Machhart, P. Hufnagl, J. Xu, P. Uhrin, I. Korschinek, B. R. Binder, and M. Geiger

Lab. Clin. Exp. Physiol., University Vienna, Austria

The serpin protein C inhibitor (PCI) has originally been described in human plasma as an inhibitor of the anticoagulant serine protease activated protein C (Marlar & Griffin, J. Clin. Invest. 66: 1186, 1980). In the meantime we and others have shown that PCI inhibits a variety of other plasma and tissue serine proteases (e.g. enzymes involved in coagulation and fibrinolysis, tissue kallikrein, prostate specific antigen, and the sperm protease acrosin). Furthermore, PCI is present not only in plasma but in many other body fluids and secretions, and PCI synthesis has been shown in several human organs. The physiological function of PCI, however, has not been defined yet. Therefore, in order to investiate the biological role of PCI we cloned and analyzed the gene encoding mouse PCI and studied its tissue specific expression. The mouse PCI gene (transcription start site to polyadenylation site) has a length of ~5kb. Like the human counterpart, it is composed of 5 exons and 4 introns; the exon/intron boundaries are highly conserved. The mouse PCI gene encodes a pre-polypeptide of 405 amino acids, which shares the highest degree of homology (63% identity) with human PCI. The putative reactive site is identical to that of human PCI from P5 to P3′, suggesting a similar protease specificity. Also the putative heparin binding sites and the "hinge" region are highly homologous in mouse and human PCI. One might therefore speculate that mouse and human PCI have a similar biological function. However, studies analyzing tissue distribution of mouse and human PCI mRNA by Northern blotting revealed that in the mouse PCI mRNA was exclusively found in the reproductive system (testis, seminal vesicle, ovary), while human PCI is synthesized in many organs. In men plasma PCI is thought to be derived from the liver. Therefore, the absence of PCI mRNA from mouse liver, which was confirmed by RNase protection assays, additionally suggests that in the mouse PCI is not a plasma protein. Consistently, the analysis of the 5′ flanking regions of mouse and human PCI revealed that the mouse PCI gene contained a testis specific element and two sterol regulatory elements not present in human PCI. The human PCI gene, on the other hand, contained 5 liver specific elements, while mouse PCI contained only one. Based on these data, we suggest that PCI — although highly conserved in its functional protein domains — fulfills different physiological functions in mice and men.

INFLAMMATION AND NONINHIBITOR SERPINS

61. PEDF-BINDING ACTIVITY TO RETINOBLASTOMA CELLS

Elena Alberdi and S. Patricia Becerra

National Eye Institute, NIH Bethesda, Maryland 20892

Pigment epithelium-derived factor (PEDF) is a member of the serpin family with neurotrophic activity. However, PEDF belongs to the non-inhibitory subgroup of serpins and its neurotrophic activity does not require inhibition of proteases. PEDF induces neuronal survival on cerebellar granule cell neurons and neurite-outgrowth on retinoblastoma cells. These PEDF-induced neurotrophic activities are blocked by antiserum specific for PEDF, Ab-rPEDF. In order to identify a binding activity for PEDF we have studied the interactions of PEDF to retinoblastoma cell surfaces. Binding activity was determined in retinoblastoma cells treated as for the biological neurite outgrowth assay. Retinoblastoma Y-79 cell cultures were grown in suspension at 1.25×10^5 cell per ml of medium without serum. Radiolabelled I^{125}-PEDF was added to the suspension cultures and incubated at 4°C for 60 min. Free ligand was separated from bound PEDF by filtration through glass fibre filters under vacuum. Increasing amounts of unlabelled PEDF were used to compete with labelled proteins. Saturation binding curves indicated that PEDF-binding to retinoblastoma cells reaches a plateau at 15nM PEDF with a linear relationship up to 10nM PEDF. We have used antiserum Ab-rPEDF and ovalbumin to compete for the I^{125}-PEDF binding to retinoblastoma cells. PEDF binding in the presence of Ab-rPEDF, excess of cold PEDF (15 nM) or ovalbumin (15 nM) was 30%, 28% and 90%, respectively, of the assays without these competitors. These results indicate that PEDF binds specifically to retinoblastoma cells. Plasma membrane extracts were prepared by differential centrifugation and a detergent soluble fraction prepared from these with 0.5% CHAPS™ in a buffer containing 1 mM $CaCl_2$. PEDF binding activity was observed for the membrane extracts as well as the detergent fraction. We have prepared a PEDF-affinity column and used it to identify the PEDF-binding protein in the detergent soluble fraction. After affinity column chromatography we have identified a protein with an apparent molecular weight of 80,000 that has PEDF binding properties. Our results are consistent with the identification of a retinoblastoma cell surface receptor for PEDF.

62. PLASMINOGEN ACTIVATOR INHIBITOR TYPE-2 AS AN INHIBITOR OF CELL DEATH

Toni M. Antalis, Joanne L. Dickinson, and Karen Donnan

Cellular Oncology Laboratory, Queensland Institute of Medical Research, Brisbane, 4029
  Australia

The signalling of cell death involves multiple processes including the action of serine and cysteine proteinases. Plasminogen activator inhibitor type-2 (PAI-2) is unusual among serpins in that it is targeted to both extracellular and intracellular locations. PAI-2 is an efficient inhibitor of urokinase-type plasminogen activator (uPA) and functions to regulate extracellular uPA-initiated proteolytic events involved in tissue remodelling, inflammation and tumor invasion and metastasis. Our recent data demonstrates a role for intracellular PAI-2 in the regulation of cell death and apoptosis. The precise mechanism by which PAI-2 inhibits apoptosis has not yet been established, but appears to involve a signalling pathway triggered by a select group of 'death' signals. Stable expression of PAI-2 was shown to protect HeLa cells from tumour necrosis factor-α (TNF) induced cytolysis in a dose dependent manner. The protective effect of PAI-2 appears to be unrelated to its function as an inhibitor of uPA but is likely to involve an intracellular Arg-protease. PAI-2 expression also confers resistance to cellular lysis by selected viruses, indicating that the pathway affected by PAI-2 may regulate a host-response to virus infection. Investigation of the mechanism by which PAI-2 confers these effects reveals altered expression of several key inflammatory factors, suggesting that expression of PAI-2 may interfere with a central regulatory pathway.

63. PARADOXIC RESPONSE OF SPI 2.1 GENE EXPRESSION TO IL-6: SPI 2 GENE RESPONSES TO INFLAMMATION ARE NOT IL-6 SPECIFIC

Susan A. Berry,[1] Pearl L. Bergad,[1] Howard C. Towle,[2] and Sarah Jane Schwarzenberg[1]

[1]Departments of Pediatrics and [2]Biochemistry, University of Minnesota, Minneapolis,
  Minnesota

The hepatic acute phase response (APR) is a reproducible alteration in a specific set of proteins synthesized by the liver. Tissue injury produces a cascade of cytokine mediators which, in combination with other factors (including glucocorticoids), alters APR gene expression following an inflammatory stimulus. Interleukin (IL)-6, a major component of this cascade, activates latent cytoplasmic transcription factors (STAT proteins). Serum content of the APR proteins may increase or decrease following induction of the APR. The rat serine protease inhibitor (Spi) 2 gene family is an excellent model for study of the APR as it includes both positively (Spi 2.2) and negatively (Spi 2.1) regulated acute phase reactants. Because the first ~ 350 bp of the promoters of these genes differ only at the location of sites which have the characteristics of STAT binding sites, we hypothesized that the hepatic APR following inflammation induced by turpentine injection would include rapid activation of STAT proteins and that the differences in Spi 2.1 and 2.2 gene expression would result from alterations in activation of STAT proteins.

Hepatic nuclei from adult male rats treated with s.q. turpentine were isolated at varying time points after treatment and nuclear protein extracts were prepared. Binding of STAT proteins was examined using labeled DNA oglionucleotides containing known STAT binding sites in electromobility shift assays (EMSA). The optimized high affinity sis-inducible element (GTCAGACATTTCCCGTAAATCGTCGA) was used to assay for Stat1 and Stat3 and the gamma-activated site from the -casein promoter was used to assay for Stat5 (GGACTTCTTGGAATTAAGGGA). Identity of the various STAT proteins observed in these extracts was assessed by addition of antibodies to specific known STAT proteins with observation of altered mobility in EMSA. Only Stat3 and not Stats 1 or 5 is activated after turpentine inflammation. This response is present within 2 hr of turpentine treatment and is not altered by pituitary ablation.

It has been speculated that much of the APR is mediated via IL-6 action. IL-6 in turn activates Stat3, and Stat3 was the STAT protein activated after turpentine stimulation. Because the inflammatory response induced by turpentine injection results in rapid and specific activation of Stat3, but not other STAT proteins, we speculated that differences in Spi 2.1 and 2.2 gene expression after induction of the APR would be due to differential responses to IL-6. To test this in a system most like the animal model used, the effects of IL-6 treatment on gene expression directed by the Spi 2.1 and 2.2 promoters were examined in primary hepatocytes in short term culture. Primary hepatocytes had growth hormone (GH), 500 ng/ml, IL-6, 5 ng/ml, or both hormones added for 18h after transfection with either pSpi2.1(−275to+85)CAT or pSpi2.2(−167to+85)CAT. GH had no effect on Spi 2.2 promoter-directed CAT expression while IL-6 resulted in a 35 fold increase in CAT. In contrast, GH and IL-6 had essentially equal effects in increasing Spi 2.1 promoter-directed CAT gene expression. Thus, IL-6 has a paradoxic effect on Spi 2.1 gene promoter as the physiological response of Spi 2.1 after induction of the APR is down regulation. We conclude that although turpentine injection results in rapid and specific activation of Stat3, that this STAT response to IL-6 is not sufficient to direct the differential responses of Spi 2.1 and 2.2 in response to induction of the APR.

64. CHARACTERISATION OF PULMONARY AND PLASMA SERPINS IN THE HORSE USING POLYCLONAL AND MONOCLONAL ANTIBODIES AGAINST EQUINE ALPHA-1-PROTEINASE INHIBITOR

Mark Dagleish, Elspeth Milne, Cheryl Scudamore, Padraic Dixon, Sybil McAleese, Alan Pemberton, and Hugh Miller

Royal (Dick) School of Veterinary Studies, University of Edinburgh,
   Veterinary Field Station, Easter Bush, Roslin, Midlothian, EH25 9R6, Scotland;
   Email: 8514358@Lab∅.vet.ed.ac.uk

The equine alpha-1-proteinase inhibitor (API) serpin family is, unlike the single gene encoding human API, encoded by the closely linked genes Spi1, 2 & 3 and 22 haplotypes have been recognized to date, 8 of which are common in Thoroughbred horses (F, G, I, L, N, S, T & U). The fourth, Spi4, is only rarely expressed (5/22 haplotypes). A slight downward shift (approx. 1.5kDa) in the molecular weight of Spi1, 2 and 3 occurs in bronchoalveolar lavage fluid (BALF) when compared with plasma derived Spi proteins and there is an apparent partial depletion of Spi3 from BALF from normal lungs and from lungs of horses with chronic obstructive pulmonary disease. The molecular weight shift was not related to any loss of inhibitory activity nor could it be attributed to bacterial infection. These pulmonary changes may, therefore, result from loss of carbohydrate moieties on the Spi proteins or, alternatively, to significant local production of Spi1 and 2 in the airways, where molecular weight differences may reflect different glycosylation pathways. The aim of this study, was therefore, to use polyclonal and monoclonal antibodies to identify potential sites of synthesis/degradation of Spi in the liver and lungs. Western blots of purified Spi1, 2 and 3 from native polyacrylamide gel electrophoresis (PAGE) show specific staining of Spi1 and 2 when probed with a mouse monoclonal antibody (Mab) raised against Spi1. A similar Western blot of serum from the 8 common Thoroughbred haplotypes, this time incorporating blue dextran into the stacking gel to retard albumin migration, reveal similar results except that haplotypes F and N are very poorly recognised and a dimeric form of API is visualised in the other 6 haplotypes. A similar pattern is seen when Western blots from sodium dodecyl sulphate (SDS) PAGE are examined in that haplotypes F and N are poorly recognised. Indirect ELISA using Mab of Spi proteins from haplotypes U and N give positive results for Spi1 and 2 for U, very weak for Spi1 and 2 from N and negative for Spi3 in both haplotypes. A polyclonal, affinity purified, ovine anti API antibody recognised Spi1, 2 and 3 in all 8 common haplotypes on Western blots and, in immunohistochemical studies, revealed intense cytoplasmic staining of hepatocytes. Pulmonary alveolar epithelium was also positive but bronchiolar epithelial cells were negative with this antibody. Cells with foamy cytoplasm typical of activated macrophages, isolated from bronchoalveolar lavage, contained many intensely stained granules. Whilst it is clear that the liver is a major source of Spi proteins in the horse, the present results do not exclude the possibility that there is significant production of Spi1 and 2 in the lung. In future studies, the *in vitro* synthesis of Spi1 and 2 by cells from the airways will be investigated by metabolic labelling and immunoprecipitation.

65. HETEROLOGOUS EXPRESSION OF THREE INHIBITORY PLANT SERPINS

Søren W. Dahl,[1,2] Søren K. Rasmussen,[2] and Jørn Hejgaard[1].

[1]Department of Biochemistry and Nutrition, Technical University of Denmark, DK-2800 Lyngby, Denmark; [2]Plant Genetics, Environmental and Technology Department, Risø National Laboratory, DK-4000 Roskilde, Denmark

Numerous serpins from mammals, insects and vira are known but only three plant serpins have been identified. The barley serpins BSZ4 (Hejgaard, 1982) and BSZ7 (Lundgard & Svensson, 1989) and the wheat serpin WSZ1 (Rosenkrands et al., 1994) have been purified from grain and characterized with respect to molecular and immunological properties. Previously, only WSZ1 has been shown to form SDS-stable complexes with a serine proteinase (chymotrypsin).

cDNAs encoding BSZ4 (unpublished variant of gene cloned by Brandt et al., 1990), BSZ7 (unpublished) and WSZ1 (Rasmussen et al., 1996) and a gene encoding a fourth serpin BSZx (Rasmussen, 1993) have been cloned and sequenced. The putative P_1 residues of the inhibitors are M (BSZ4), Q (WSZ1) and R (BSZx and BSZ7). BSZ4, BSZx and WSZ1^T were selected for heterologous expression on assumption of distinct inhibitory properties. The cDNA clones were inserted directly into pET3d expression vector (Studier et al., 1990) and equipped with a six-histidine affinity tag at the N-terminus. The genomic clone encoding BSZx contained an intron which was removed by overlap extension PCR before insertion into pET3d. The constructs were transformed into E.coli BL21 (DE3) pLysS and the serpins were expressed as soluble and insoluble protein. Purification of the soluble serpins to homogeneity was accomplished by metal chelate chromatography on a Ni-NTA column, gelfiltration on a G-100 column and anion exchange on a Resource Q column. Amino acid sequencing of the products resulted in the common N-terminal sequence MHHHHHHMA... No trace of the sequences following the P_1-P_1' cleavage sites could be identified verifying that the serpins had been isolated with uncleaved reactive centre loops.

The ability of recombinant BSZ4, BSZx and WSZ1 to inhibit proteinases of the trypsin family was analyzed and for the first time it has been demonstrated that plant serpins may exhibit distinct inhibitory properties and are able to form SDS-stable complexes with their target proteinases(Dahl et al., 1996).

Hejgaard, J. (1982). *Physiol. Plant.*, 54, 174–182; Lundgard, R. & Svensson, B. (1989) *Carlsberg Res. Commun.*, 54, 173–180; Rosenkrands, I., Hejgaard, J., Rasmussen, S.K. & Bjorn, S.E. (1994). *FEBS Lett.*, 343, 75–80; Rasmussen, S.K. (1993). *Biochim. Biophys. Acta*, 1172, 151–154; Brandt, A., Svendsen, I. & Hejgaard, J. (1990). *Eur. J. Biochem.*, 194, 499–505; Rasmussen, S.K., Dahl, S.W., Norgaard, A. & Hejgaard, J. (1996). *Plant Mol. Biol.*, 30,673–677. Studier, F.W., Rosenberg, A.H., Dunn, J.J., and Dubendorff, J.W. (1990) *Methods Enzymol.*, 185, 60–89; Dahl, S.W., Rasmussen, S.K. & Heijgaard, J. (1996), submitted.

66. CONFORMATIONAL CHANGE OF THE REACTIVE CENTRE IN C1 INHIBITOR DUE TO A DELETION IN THE NH$_2$-TERMINAL DOMAIN

E. Eldering,* D. Roem, and C. E. Hack

Centr Lab of the Red Cross Blood Transfusion Service, Amsterdam, Netherlands

Genetic deficiency of the serpin C1 inhibitor causes hereditary angioedema. During our effort to trace the molecular causes of this disease in 22 families, a novel type of mutation, consisting of an in-frame deletion of 165 basepairs, was found in exon 3. Exon 3 of the C1 inhibitor gene codes for the entire NH$_2$-terminal domain, unique among the serpins and rich in carbohydrate and *Glx–Pro–Thr–Thr* repeats, of which the functional significance is unclear. The mutation was copied into an expression vector for COS cells. The internal deletion of 55 residues (Asp^{62}–Thr^{115}), although eliminating Cys^{102} and Cys^{108} and thereby disrupting the two disulfide bonds in C1 inhibitor, did only partially prevent secretion. This permitted analysis of the structural consequences of the deletion, which does not extend into the serpin domain. The deletion mutant was non-functional and existed in a conformation which is adopted by the normal protein only upon complex- formation with a protease, as demonstrated by intrinsic exposure of a neo-epitope and by increased thermostability. In addition, both sucrose gradient density centrifugation and a single-antibody ELISA showed that the mutant protein existed in multimeric form. These small multimers were antigenically different from the large heat-induced multimers of the normal protein, but non-reducing SDS-PAGE excluded the presence of disulfide-linked dimers. We conclude that the deletion mutant has a conformation similar to that encountered frequently when single conserved residues in the serpin domain of C1 inhibitor are substituted. This "multimerization phenotype" results from an increased tendency of the reactive centre loop to insert in the central β-sheet, and/or rotation of strand 1C. The results with this particular mutant suggest that such conformational changes at the reactive centre loop may also be triggered by the distant NH$_2$-terminal region.

Funded by the Netherlands Foundation for Preventive Medicine grant no 28-2274.

67. THE S MUTATION OF α₁-ANTITRYPSIN (^{264}Glu→Val) DISTORTS THE PACKING OF RESIDUES ADJACENT TO THE A β-SHEET AND FAVOURS THE FORMATION OF LOOP SHEET POLYMERS

Peter R. Elliott,[*†] Penelope E. Stein,[†] Diana Bilton,[¶] Robin W. Carrell,[†] and David A. Lomas[*†¶]

Departments of [*]Medicine and [†]Haematology, University of Cambridge, MRC Centre, Cambridge, CB2 2QH, UK and [¶]Chest Medical Unit, Papworth Hospital, Papworth Everard, Cambridge, CB3 8RE, U.K.

Alpha₁-antitrypsin is the archetypal member of the serine proteinase inhibitor or serpin superfamily and the most abundant circulating proteinase inhibitor. The commonest deficiency variant is the S allele which occurs in 7% of northern Europeans and nearly 19% of the population from Spanish descent. The S allele is characterised by a Glu to Val mutation at position 264 and results in plasma antitrypsin levels that are 60% of the normal M allele; this is not associated with clinical pathology. The more severe Z (^{342}Glu→ Lys), Siiyama (^{53}Ser→Phe) and Mmalton (^{52}Phe deleted) deficiency variants of antitrypsin result in misfolding of the protein and the spontaneous formation of polymers between the reactive centre loop of one molecule and a β-pleated sheet of a second. These loop-sheet polymers tangle as intracellular inclusions within the hepatocyte, predispose to juvenile cirrhosis and result in antitrypsin levels that are only 10–15% of normal.

The mechanism by which the S mutation results in plasma deficiency is unclear and has been attributed to increased clearance, aberrant protein folding and intracellular degradation. We now show that purified plasma S antitrypsin forms loop sheet polymers during incubation under physiological conditions at 37°C and 41°C. These polymers form more readily than those of M, but less readily than Z, antitrypsin which is compatible with the mild clinical phenotype. The predisposition to form polymers accounts for the thermolability of the S variant at 50°C but when heated at higher temperatures (>60°C) the S polymers were consistently shorter than those of M antitrypsin. Unlike Z antitrypsin, the S mutation had no effect on the conformation of the reactive site loop when assessed by limited proteolysis with *Staph aureus* V8 proteinase (which cleaves at P4-P5), papain (P6–P7 and P1–P1′) or papaya proteinase IV (P9-P10). Similarly S antitrypsin was able to form SDS-stable complexes with bovine α-chymotrypsin and had essentially normal association and dissociation rate constants with neutrophil elastase (k_{ass} 2.8 ± 0.02 × 10⁷ M⁻¹s⁻¹ and K_i < 5 pM compared to 5.3 ± 0.06 × 10⁷ M⁻¹s⁻¹ and <5 pM for M antitrypsin) and bovine α-chymotrypsin (k_{ass} 3.4 ± 0.02 × 10⁶ M⁻¹s⁻¹ and K_i<5 pM compared to 2.6 ± 0.02 × 10⁶ M⁻¹s⁻¹ and 9.2 ± 2.7 pM for M antitrypsin). The S mutation did not interfere with the configuration of the A β-pleated sheet as assessed by the incorporation of the exogenous antithrombin reactive loop peptide BC11 and the insertion of the endogenous reactive loop to form the cleaved or latent species. Moreover the S mutation did not perturb the tertiary structure of the antitrypsin molecule as assessed by transverse urea gradient gel electrophoresis and folding/unfolding transitions following denaturation with guanidine hydrochloride.

These data suggest that the deficiency of S antitrypsin results from a slow transition to form dimers and short chain polymers that result in a mild reduction in secretion from the hepatocyte and an increased rate of turnover in the plasma. The crystal structure of S antitrypsin shows that the S mutation breaks the ^{264}Glu–^{38}Tyr salt bridge which would predictably perturb the ^{52}Phe–^{53}Ser region at the commencement of the B helix and hence favour polymerisation as occurs more dramatically with the Siiyama and Mmalton mutants.

68. β-PLEATED CONFORMATION OF THE REACTIVE LOOP OF α₁-ANTITRYPSIN

Peter R. Elliott,*† David A. Lomas,*† Robin W. Carrell,* and Jan-Pieter Abrahams¥

Department of *Haematology and †Medicine, University of Cambridge and ¥Laboratory of Molecular Biology, MRC Centre, Hills Road, Cambridge, CB2 2QH, U.K.

We have determined the crystal structure of a mutant of antitrypsin in which the loop adopts a novel conformation with direct mechanistic and biological implications. The 51Phe→Leu mutation has been shown to render antitrypsin thermostable (Kwon et al, 1994) and was subsequently shown to increase the secretion of the Z and Siiyama deficiency variants in a Xenopus oocyte expression system. This mutant was cloned into the pTERMAT vector with expression and then purification to homogeneity. The resulting protein was 80% active as an inhibitor of bovine α–chymotrypsin, was more resistant to temperature induced loop-sheet polymerisation, accepted a reactive loop peptide less readily than the wildtype protein and showed different reactive loop cleavage profiles following limited proteolysis with papaya proteinase IV. The mutation had no effect on association or dissociation rate kinetics with human neutrophil elastase and bovine α–chymotrypsin, the formation of SDS stable complexes, the S to R transition and unfolding profiles in urea and guanidine hydrochloride. Crystals (space group C2) were formed in Tris/acetate buffer with 22–26% PEG 4000 and gave diffraction to 2.9A. The structure was solved by molecular replacement using the co-ordinates of ovalbumin, reactive loop cleaved antitrypsin and active antithrombin and, at this stage of refinement, has an R factor of 20% and a free R factor of 30%. There was good density for all regions of the protein apart from P11–P14 of the reactive loop. The structure showed a 5 membered β-sheet, but instead of a helix, the reactive loop was held above the molecule as a β-pleated strand. This loop was closer to the body of the protein than in antithrombin and was anchored by salt bridging of P5 glutamate to a well-defined pocket of 3 lysine and 1 arginine residues at the pole of the molecule beneath the s3A–s4C junction. The facility of formation of a a β-pleated conformation fits with the readiness of the serpins to form intermolecular loop-sheet linkages. Initial modelling favours a P1-P8/A-sheet linkage between serpins but the finding also explains the bonding of seprins to other β-sheet structures, notably the β-amyloid of Alzheimer's disease. The most profound implication however is for the mechanism of inhibition. Although the conformation of the loop is partly influenced by contact with the s3C-4C loop of adjacent molecules, the P5 salt bridge indicates its likely physiological significance. This conclusion is greatly strengthened by the precision of docking of the reactive site with the active site of chymotrypsin, superimposably with that of a canonical inhibitor.

69. SIGNIFICANCE OF SECONDARY STRUCTURE PREDICTIONS ON THE REACTIVE CENTER LOOP REGION OF SERPINS. A MODEL FOR THE FOLDING OF SERPINS INTO A METASTABLE STATE

Peter G. W. Gettins and Philip A. Patston

University of Illinois at Chicago

To address how serpins might fold so as to adopt the mechanistically-required metastable conformation we have compared the predicted secondary structures of the reactive center loops (RCL) of 29 serpins with those of the equivalent regions of other non-serpin protein proteinase inhibitors. The serpins include both inhibitory and non-inhibitory examples. The non-serpin inhibitors include members of the Kazal, Kunitz, Bowman-Birk and potato families as well the bait regions of several α_2-macroglobulins. Whereas the RCLs of non-serpin inhibitors are predicted to be loop or β-strand, those of inhibitory serpins are strongly predicted to be α-helical. However, non-inhibitory serpins (angiotensinogen, corticosteroid binding globulin, maspin, ovalbumin, pigment epithelium-derived factor and thyroxine binding globulin), which also adopt the metastable conformation, show no consistent preference for α-helix. We propose that the RCL primary structure plays little role in determining the metastable serpin conformation. Instead we hypothesize that preference for the metastable state results from the incorporation of part of the RCL into β-sheet C, which as a consequence precludes incorporation of the RCL into β-sheet A to give the most stable conformation. Consequently the RCL must be exposed and by default will adopt the most stable conformation in this particular context, which is likely to be an α-helix, irrespective of the primary structure. Thus the observed correlation between inhibitory properties in serpins and prediction of α-helix in the RCL is unlikely to be connected with the adoption of a metastable structure. Instead it may reflect a mechanistic need for alanine residues between positions P12 and P9 to produce efficient functioning as an inhibitor, which incidentally results in strong prediction for α-helix.

Supported by grants HL49234 (PGWG) and HL49242 (PAP).

70. ISOLATION OF ALPHA₁-ANTICHYMOTRYPSIN (ACT) ISOFORMS AS INHIBITORS OF PROCESSING ENZYMES REQUIRED FOR THE BIOSYNTHESIS OF PEPTIDE HORMONES AND NEUROTRANSMITTERS

Vivian Y. H. Hook and Nikolaos Tezapsidis*

Department of Medicine, University of Calif., San Diego, California 92103–8227, *Dept. of Biochemistry, Uniformed Services Univ., Bethesda, Maryland 20814; (*Present address, Dept. of Psychiatry, Mount Sinai School of Medicine, New York, New York 10029)

Protease inhibitor regulation of prohormone processing enzymes may provide a mechanism for observed differences in the extent of prohormone processing. We have previously reported characterization of an alpha₁-antichymotrypsin-like (ACT-like) protease inhibitor from bovine pituitary that potently inhibits the processing enzyme 'prohormone thiol protease' (PTP) (Hook et al., *J. Biol. Chem.* 268, 29570–20577). Inhibition involves complex formation of the cysteine protease PTP and the serpin ACT based on parallel P_1 specificities. Further studies in this report indicate that ACT-like proteins are colocalized with PTP in secretory vesicles of bovine adrenal medulla (chromaffin granules), as well as in pituitary. In chromaffin granules, two forms of ACT with molecular weights of 56 and 70 kDa were identified immunologically by anti-ACT western blotting. Sucrose density gradient analyses of catecholamine-containing chromaffin granules indicated the colocalization of ACT in the secretory vesicle fraction. The ACT-like proteins were purified from chromaffin granule lysate by concanavalin A-agarose and ion-exchange Mono-Q FPLC column. The 70 and 56 kDa ACT-like proteins were tested for inhibition of the processing enzymes purified from the same chromaffin granules: 'prohormone thiol protease' (PTP), PC1/3, and PC2 (PC=prohormone convertase, subtilisin-like proteases). PTP was inhibited by both 56 and 70 kDa ACTs at µM concentrations. When the 56 and 70 kDa ACTs were tested at the same concentration (µM) with the PC enzymes, the 56 kDa ACT inhibited PC1/3 and PC2, but the 70 kDa ACT had no effect. The presence of endogenous protease inhibitors of processing enzymes is consistent with the limited processing of proenkephalin in these granules; approximately 90% of (Met)enkephalin exists as incompletely processed high molecular weight intermediates. Future definition of the role of each ACT isoform in regulating propeptide processing will be important for understanding regulatory mechanisms involved in peptide hormone and neurotransmitter biosynthesis.

71. A NOVEL NEUROENDOCRINE-SPECIFIC ALPHA$_1$-ANTICHYMOTRYPSIN HOMOLOGUE REVEALED BY MOLECULAR CLONING: POSSIBLE ROLE IN PROHORMONE PROCESSING

Shin-Rong Hwang and Vivian Y. H. Hook

Department of Medicine, University of Calif., San Diego, California 92103-8227

Alpha$_1$-antichymotrypsin (ACT)-like proteins have been demonstrated by this laboratory to inhibit prohormone processing proteases localized to secretory vesicles of bovine pituitary and adrenal medulla (Hook et al., *J. Biol. Chem.* 268, 20570–20577, 1993; Azaryan et al., *J. Biol. Chem.* 270, 8201–8208, 1995). ACT is a potent inhibitor of the proenkephalin processing enzyme known as 'prohormone thiol protease' (PTP), and also inhibits the subtilisin-like processing proteases PC1/3 and PC2. To determine whether neuroendocrine isoforms of ACT exist, sequence comparison of ACT cDNAs from bovine liver (Hwang et al., *Proc. Natl. Acad. Sci.* 91, 9579–9583, 1994) with bovine adrenal medulla and pituitary neuroendocrine ACT cDNAs was achieved by molecular cloning. The deduced amino acid sequences indicated that the adrenal medulla and pituitary neuroendocrine ACTs differ from two bovine liver isoforms of ACT. RT-PCR detection of the adrenal medulla ACT mRNA indicated its expression in neuroendocrine, but not in liver, tissues. Most importantly, the bovine ACT isoforms all contain a homologous reactive site loop (RSL) that is characteristic of serpins. It is observed that the predicted arginine as P$_1$ residue of bovine adrenal medulla and pituitary ACTs parallels the known cleavage specificity of the prohormone processing enzymes — PTP, PC1/3, and PC2 — for basic residues. We are currently cloning the full-length neuroendocrine-specific adrenal medulla ACT cDNA. Expression of the cloned ACT cDNA and examination of the predicted inhibitory activities against prohormone processing enzymes will elucidate a possible role for ACT-like protease inhibitors in regulating prohormone processing.

72. A SPECIFIC STRUCTURAL INTERACTION OF ALZHEIMER'S PEPTIDE ABETA (1-42) WITH ALPHA-1-ANTICHYMOTRYPSIN

Sabina Janciauskiene,* Sten Eriksson,* and H. Tonie Wright[@]

* Dept. of Medicine, University Hospital of Malmö, Lund University, Malmö 20502, Sweden and [@]Dept. of Biochemistry, Virginia Commonwealth University, Richmond, Virginia 23298-0614

Formation in vitro of the amyloid fibrils characteristic of Alzheimer's disease (AD) from the AB peptide is a spontaneous self-assembly process. In vivo, amyloid fibrils are associated with a number of other macromolecules, including alpha-1-antichymotrpsin, which have been shown to affect fibril formation in vitro and to bind ABeta peptide with sequence specificity. We have studied the effects on fibril formation of both intact, proteolytically cleaved, and peptide-inserted ACT at low molar ratio and find that only intact ACT stimulates fibril formation. We have also determined that ABeta stabilizes ACT to denaturation, while neither the amino terminal ABeta(1–17) nor the carboxyl ABeta (28–42) peptide fragments of ABeta had a stabilizing effect. ACT, like alll inhibitor serpins has an expandable beta-sheet, sA, into which the post-proteolysis reactive site strand and exogenous peptides insert with attendant increase in stability to denaturation. A structural model of ABeta inserted into the major beta-sheet of ACT is stereochemically favorable and suggests a mechanism by which ACT stimulates fibril formation.

73. BINDING AND ENDOCYTOSIS OF U-PA/PAI-1 COMPLEX AND OTHER SERINE PROTEINASE/SERPIN COMPLEXES BY VERY LOW DENSITY LIPOPROTEIN RECEPTOR AND α_2-MACROGLOBULIN RECEPTOR/LOW DENSITY LIPOPROTEIN RECEPTOR-RELATED PROTEIN

Aneta Kasza,[1,2] Helle H. Petersen,[1] Anna C. Wiborg Simonsen,[1] Kazuhiro Oka,[3] Lars Ellgaard,[3] Lawrence Chan,[3] Christian W. Heegaard,[1] and Peter A. Andreasen[1]

[1]Department of Molecular Biology, University of Aarhus, Denmark; [2]Institute of Molecular Biology, Jagiellonian University, Kraków, Poland; [3]Departments of Cell Biology and Medicine, Baylor College of Medicine, Houston, Texas

Very low density lipoprotein receptor (VLDLR) and α_2-macroglobulin receptor/low density lipoprotein receptor-related protein (α_2MR/LRP) are multifunctional endocytosis receptors of the low density lipopotein receptor family. Both have been shown to bind u-PA/PAI-1 complex with a K_d around 1 nM, and both are able to mediate endocytosis and degradation of the complex by cultured cells. We have now studied the specificity of binding and endocytosis by VLDLR and α_2MR/LRP among a variety of serine proteinase/serpin complexes, including various combinations of the serine proteinases u-PA, LMW-u-PA, t-PA, plasmin, thrombin, horse neutrophil elastase, cathepsin G, and plasma kallikrein with the serpins PAI-1, PAI-2, horse neutrophil elastase inhibitor, protein C inhibitor, C1-inhibitor, α_2-antiplasmin, α_1-proteinase inhibitor, α_1-antichymotrypsin, and antithrombin III. Binding was estimated by ligand blotting analysis. Endocytosis was estimated by measuring receptor-associated-protein (RAP)-sensitive degradation of radiolabelled complexes by CHO cells transfected with VLDLR cDNA and by COS-1 cells, which have a high endogenous expression of α_2MR/LRP. We found that the binding specificity of the two receptors is quite restricted. The receptors bind some, but not all, combinations of plasminogen activators and thrombin with PAI-1, protein C inhibitor, and antithrombin III, while complexes of many serine proteinases with their primary inhibitor, i.e. plasmin/α_2-antiplasmin complex, did *not* bind. The binding results show that both the serine proteinase and the serpin moiety contribute to the binding specificity. The binding specificities of VLDLR and α_2MR/LRP are overlapping, but not identical. These results suggest that VLDLR and α_2MR/LRP may have different biological functions by having different binding specificities as well as by being expressed by different cell types. The functions of the two receptors may include clearance of not only plasminogen activator/inhibitor complexes, but also a few other serine proteinase/serpin complexes.

74. PROBING THE SHAPE OF A1PI CONFORMERS BY INTRINSIC FLUORESCENCE

Henryk Koloczek,[#] Agnieszka Banbula,[#] Guy Salvesen,[*] and Jan Potempa[$]

[#]Biochemistry Department, University of Agriculture and [$]Institute of Molecular Biology, Jagiellonian University, Krakow, Poland; [*]La Jolla Cancer Research Foundation, San Diego, California

Various conformational forms of the archetypal serpin human a1-proteinase inhibitor (a1PI), including ordered polymers, active and inactive monomers and heterogeneous aggregates have been produced by refolding from mild denaturing conditions. These forms presumably originate by different folding pathways during renaturation, under the influence of the A- and C-sheets of the molecule. Since a1PI contains only two Trp residues, at positions 194 and 238, it is amenable to fluorescence quenching resolved spectra and red-edge excitation measurements of the Trp environment. Thus it is possible to define the conformation of the various forms based on the observed fluorescent properties of each of the Trp residues measured under a range of conditions. We show that denaturation in GuHCl followed by renaturation leads to the formation of polymers that contain solvent exposed Trp238, which we interpret as ordered head-to-tail polymers (A-sheet polymers). However, denaturation in citrate leads to shorter polymers where some of the Trp238 residues are not solvent accessible, which we interpret as polymers capped by head-to-head interactions via the C-sheet. The latter treatment also generates monomers thought to represent a latent form, but in which the environment of Trp238 is occluded by ionized groups. These data indicate that the folding pathway of a1PI, and presumably other serpins, is sensitive to solvent composition that affects the affinity of the reactive site loop for the A-sheet or the C-sheet.

75. COMPARISON OF SERPINS WITH THE STANDARD MECHANISM INHIBITORS

Tomoko Komiyama,[1] Hanne Grøn,[1] Philip A. Pemberton,[2] and Guy S. Salvesen[3]

[1]Duke University Medical Center, Durham, North Carolina 27710, [2]LXR Biotechnology 1401 Marina Way South, Richmond California 94804, [3]Burnham Institute, 10901 North Torrey Pines Road, San Diego California 92037

Some serine proteinase inhibitors form complexes with their target enzymes that are readily resolved by X-ray diffraction. In these cases the inhibitors are seen to operate by forming a tight, almost lock-and-key, complex that does not involve cleavage of the inhibitor.

Though many of the aspects of proteinase inhibition by serpins are shared by other proteinase inhibitor families, some characteristics are different enough to imply a distinct mechanism. Two unique characters of serpin inhibition are: 1) *Partitioning*, in which the total number of serpin molecules required to inhibit one molecule of proteinase varies from one to infinity. This is a consequence of the serpin acting as a substrate for some proteinases. 2) *SDS-stable complex* formation, where the stability of the complex has been considered as evidence of covalent bond formation between $O\gamma$ of the catalytic Ser of the proteinase and P_1–P_1' peptide bond of serpin. These characteristics have been interpreted by some authors as evidence for scission of the P_1–P_1' peptide bond during inhibition. This contrasts to the known inhibitory mechanism of the other inhibitor families.

Since structure-based information is lacking for the inhibitory active form and the complexed form of serpins, the inhibitory mechanism of serpin is not well defined yet. We have examined the serpin mechanism by comparing the interaction of standard mechanism inhibitors and $\alpha_1$proteinase inhibitor (α_1PI) with subtilisin *Carlsberg* and chymotrypsin. Under appropriate conditions we observe SDS-stable complex formation between the standard mechanism inhibitors and the subtilisin that resembles the interaction of this proteinase with serpins. We discuss these findings in terms of an inhibitory mechanism for serpins that does not require covalent reactions or peptide bond scission: a mechanism essentially identical to the standard mechanism.

76. MECHANISMS OF CELLULAR-MEDIATED CATABOLISM OF SERPIN ENZYME COMPLEXES

Maria Z. Kounnas,[1] Frank C. Church,[2] W. Scott Argraves,[1] and Dudley K. Strickland[1]

[1]Holland Laboratory, Department of Biochemistry, American Red Cross, 15601 Crabbs Branch Way, Rockville, Maryland 20855 and [2]Department of Pathology and Medicine and Center for Thrombosis and Hemostasis, University of North Carolina School of Medicine, Chapel Hill, North Carolina 27599

The inhibition of proteinase activity by members of the serine proteinase inhibitor (serpin) family is a critical regulatory mechanism for a variety of biological processes. Once formed, the serpin enzyme complexes (SECs) are removed from the circulation by an hepatic receptor. The present study suggests that this receptor is very likely the low density lipoprotein receptor-related protein (LRP), a prominent liver receptor. *In vitro* binding studies revealed that antithrombin III (ATIII):thrombin, heparin cofactor II (HCII):thrombin, and α_1-antitrypsin (α_1AT):trypsin bound to purified LRP, and their binding was inhibited by the 39 kDa receptor associated protein (RAP), an antagonist of LRP-ligand binding activity. In contrast, native or modified forms of the inhibitors were unable to bind to LRP. Mouse embryonic fibroblasts, which express LRP, mediates the cellular internalization leading to degradation of these SECs, while mouse fibroblasts genetically deficient in LRP showed no capacity to internalize and degrade these complexes. SECs were also degraded by HepG2 cells, and this process was inhibited by LRP-antibodies, RAP, and chloroquine. The cellular-mediated uptake and degradation was specific for SECs; native or modified forms of the inhibitors were not internalized and degraded. Finally, *in vivo* clearance studies in rats demonstrated that RAP inhibited the clearance of ATIII:[125]I-thrombin complexes from the circulation. Together, these results indicate that LRP functions as a liver receptor responsible for the plasma clearance of SECs.

77. LOCALIZATION OF HEPARIN COFACTOR II IN INJURED HUMAN SKIN: A POTENTIAL ROLE IN WOUND HEALING

Kim L. M. Loh,* Jason Longchamps,* Frank C. Church,[#] and Maureane Hoffman*

*Pathology and Laboratory Medicine Service, Duke University and Durham Veterans Affairs Medical Center, Durham, North Carolina and [#]Department of Pathology and Lab Medicine, University of North Carolina at Chapel Hill School of Medicine

The physiologic function of the serine protease inhibitor **(serpin)** heparin cofactor II **(HCII)** is unknown. HCII is an effective inhibitor of thrombin. Thrombin has both procoagulant and cytokine-like activities. Thrombin formed at a site of injury might play a role in inflammatory and wound healing responses long after the coagulation process has been completed. We have hypothesized that HCII functions as an extravascular inhibitor of thrombin. If this hypothesis is correct, we would expect HCII to be present at a site of injury.

To begin to test our hypothesis, we examined the immunohistochemical localization of HCII in human skin, and compared it to that of another heparin-binding serpin, antithrombin III **(ATIII)**. HCII was not detected in normal skin. Diffuse HCII and ATIII staining was seen in areas of hemorrhage. In healing or recently healed wounds, ATIII was found primarily in association with dermal mast cells, while HCII was associated with mononuclear phagocytes. These cells also stained weakly for thrombin. Exogenously added thrombin bound to the same macrophage-like cells, and binding was enhanced by added heparin or dermatan sulfate. In addition, we found that blood monocytes isolated from healthy donors also stained for HCII.

We conclude that HCII is present in association with mononuclear phagocytes in wounded skin. This localization was unlikely to have resulted from uptake of plasma proteins deposited in the tissue by hemorrhage, since ATIII was not present in the same cells. We have not demonstrated a functional role for HCII in the host response to injury. However, we have shown that HCII is localized to sites of healing, and might play a role in the regulation of wound healing.

78. POSSIBLE INVOLVEMENT OF A CBG-LIKE SERPIN IN *IN VIVO* MODULATION OF THE ACROSOME REACTION

W. Miska, P. Baltes, R. Sanchez,* R. Henkel, and W.-B. Schill

Dept. of Dermatol. & Androl., Justus-Liebig University, Giessen, Germany;
   *Faculty of Medicine, University Front., Temuco, Chile

The acrosome reaction (AR) is one of the prerequisites for successful mammalian fertilization. The AR is a modified exocytotic process, in which the outer acrosomal membrane fuses with the plasma membrane of the spermatozoon, resulting in its complete loss and release of the acrosomal content. Enzymes thereby released are thought to play a role in the penetration of spermatozoa through the outer oocyte investments. Human follicular fluid (hFF) contains a component which is able to induce the AR in spermatozoa. Previous experiments performed here have yielded several indications on the protein character of the acrosome reaction-inducing substance (ARIS), and according to the current literature, steroids also seem to play an important role in AR-induction. Our current findings show that hFF, which has been stripped of proteins and/or steroids (protease-, DCC-treatment) cannot induce the AR. However, after the removal of steroids, the AR-inducing activity of hFF can be restored by exogenous progesterone, but only in the presence of intact protein. Furthermore, in gel filtration experiments with 3_H-progesterone-labelled hFF an elution of the radioactive signal in the high molecular weight range, corresponding to bound progesterone, was found. Based on these results we suggest that the effect of ARIS is a synergistic action of progesterone and a progesterone-binding protein. The protein has been shown to be immunologically identical to the corticosteroid-binding globulin (CBG), which has already been described and serves as a transport protein for progesterone and cortisole in the plasma. Our findings also show that the CBG-progesterone complex is able to induce the AR in nanomolar concentrations, where neither progesterone nor the protein alone is able to bring about an effect. In the culture medium of human cumulus oophorus cells, AR-inducing activity could be detected. The immunological as well as the radiochemical investigations strongly indicate that human cumulus cells actively express and secrete a CBG-like progesterone-binding protein.

CONCLUSIONS

CBG and thereby ARIS is a member of the superfamily of serine protease inhibitors (SERPINs). The SERPINs perform numerous regulatory functions by means of their inhibitory effects. In the case of transport proteins, to which CBG belongs, they have lost their inhibitory properties, but have retained the typical structure and functional characteristics of SERPINS. CBG is able to bind and transport small molecules and then to release them at specific targets after undergoing SERPIN-specific conformation changes in the CBG molecule through the action of a serine protease. On the basis of this the following mechanism is proposed for the role of the CBG-progesterone complex in the induction of the AR: Progesterone is transported by CBG to the spermatozoon. CBG is cleaved there by a plasma membrane-bound protease, leading to conformational changes within the protein. This causes a local release of progesterone, which possibly stimulates a Ca^{2+}-channel, thereby leading to an influx of Ca^{2+}, and finally to the induction of the AR.

79. Spi7, A SERPIN-LIKE PROTEIN FROM SWINEPOX VIRUS (SPV) WITH REACTIVE SITE LOOP HOMOLOGY TO THE COWPOX VIRUS (CPV) SERPIN *CRMA*, DOES NOT FUNCTIONALLY SUBSTITUTE FOR *CRMA*

Pierre Y. Musy and Richard W. Moyer

Department of Molecular Genetics and Microbiology, University of Florida, Gainesville, Florida

SPV is a member of the Poxviridae and the sole representative of the genus *Suipoxvirus*. It is found worldwide and causes a relatively benign exanthem. Unlike many other poxviruses, SPV has a very narrow species specificity and is capable of infecting only in pigs. Routine sequencing of the HindIII K fragment of the SPV genome revealed an open reading frame which encodes a protein with strong homology to inhibitory members of the serpin superfamily and so was named Spi7. Analysis of Spi7 expression in infected cells revealed that like crmA, it is neither secreted nor glycosylated. Unlike all other known poxviral serpins, however, it is synthesized both early and late during the viral life cycle. Disruption of the SPV Spi7 gene with a β-galactosidase marker has shown it to be non-essential in tissue culture. The resulting virus, SPVΔSpi7, was subjectively found to be mildly attenuated in weanling pigs.

Overall, Spi7 is most homologous to squamous cell carcinoma antigen, an inhibitor of the cysteine proteinase cathepsin L, and leukocyte elastase inhibitor, being roughly 40% identical to both at the amino acid level. The predicted reactive site loop, however, is most like that of crmA. Comparison of these two proteins shows that of the 13 amino acid residues from P14 threonine to the cleavage site of crmA, eight are shared by Spi7. Included in this are three residues, valine/alanine/aspartate, which immediately precede the cleavage site and are predicted to be crucial in determining proteinase specificity. Thus, several experiments were done to test the possibility that in spite of the lack of over-all homology between Spi7 and crmA, the homology of the two proteins in the reactive site loop would render them functionally similar. First, the crmA protein is required for the production of red pocks when CPV is grown on the chorioallantoic membrane (CAM) of embryonated chicken eggs. This is presumably due to its ability to inhibit interleukin-1β converting enzyme (ICE) and therefore prevent an influx of heterophils in response to the elaboration of this pro-inflammatory cytokine. A recombinant CPV strain was constructed which expresses Spi7 but not crmA. Like the parental CPV(crmA strain, this recombinant made white pocks on the CAM, indicating that expression of Spi7 did not complement the crmA mutation. Second, it has been shown that purified crmA or its orthologue Spi2 from rabbitpox virus is able to prevent the in vitro cleavage of poly(ADP-ribose) polymerase (PARP) by extracts of cells which have been committed to undergo apoptosis. Spi7 was expressed as an amino-terminally histidine tagged derivative in the vaccinia virus T7 ex-pression system and purified by affinity chromatography. This protein was unable to pre-vent PARP cleavage or the nuclear fragmentation and chromatin condensation that occur during apoptosis in this in vitro system, once again revealing a lack of functional identity between these proteins. Third, it has been reported that crmA, in addition to its ability to inhibit ICE and other ICE-like enzymes, is also capable of inhibiting granzyme B, a serine proteinase released by cytotoxic T cells and which shares with ICE both a preference for aspartate residues at the P1 position and a role in apoptosis. Therefore, purified Spi2 or Spi7 was added to extracts containing active granzyme B, and residual granzyme B activ-ity was measured by the addition of a chromogenic tripeptide thiobenzyl ester substrate. The experiment showed that Spi2 but not Spi7 was able to inhibit granzyme B. Together, these results indicate that regions outside of the putative specificity-determining region within the reactive site loop may play a role in the inhibitory specificity of crmA/Spi2, both against ICE and ICE-like proteinases as well as granzyme B. To further explore this possibility, two hybrid proteins were made. In the first, the putative P1′ serine residue of Spi7 was replaced with the cysteine found in crmA. In the second, the carboxyl-terminal 51 amino acids of Spi7, beginning with the predicted P18 residue, were removed and re-placed with the corresponding region of crmA. These proteins will be tested in both the *in vitro* apoptosis and granzyme B assays. In addition, these genes have been inserted into CPVΔcrmA and so will be tested on the CAM for the ability to restore the red pock pheno-type. These results should begin to shed some light on the specificity-determining regions of crmA and conversely on the inability of Spi7 to functionally complement crmA.

80. INHIBITION KINETICS OF WILD-TYPE AND MUTANT SERP-1 FROM MYXOMA VIRUS

Piers Nash,[†] Adrian Whitty,[§] and Grant McFadden[†]

[†]Dept. of Biochemistry, U. of Alberta, Edmonton, Alberta, Canada, T6G 2H7; [§]BIOGEN, 14 Cambridge Centre, Boston, Massachusetts 02142

SERP-1 is a secreted serine proteinase inhibitor encoded by myxoma virus and malignant rabbit fibroma virus(1). It has been shown to be required for viral virulence(2), and in addition, the purified protein has anti-inflammatory properties in a model of restenosis(3). We have produced SERP-1 in a recombinant vaccinia virus system, and purified the protein to homogeneity from viral supernatants by FPLC. The purified protein was characterized as a proteinase inhibitor by gel mobility shift assays and slow binding kinetics. SERP-1 forms complexes with, and inhibits plasmin (K_I = 440 pM, k_{ass} = 4.5 $\times$ 10^4 $M^{-1}s^{-1}$), tissue type plasminogen activator (K_I = 140 pM, k_{ass} = 5 $\times$ 10^4 $M^{-1}s^{-1}$), and urokinase type plasminogen activator (K_I = 160 pM, k_{ass} = 5 $\times$ 10^4 $M^{-1}s^{-1}$). SERP-1 was observed to form a complex with factor Xa, and a short lived complex with thrombin. SERP-1 did not form stable complexes with, but was cleaved by, human neutrophil elastase (hNE), cathepsin G, bovine chymotrypsin, subtilisin, thermolysin, and trypsin. No interaction was observed between SERP-1 and Factor D, C3 convertase, and factor XII.

A reactive center mutant of SERP-1 was constructed in which the P1–P1' residues were changed from arginine-asparagine to alanine-alanine. Purified mutant SERP-1 (SAA) failed to form stable complexes with any of the target proteases, but was cleaved by those enzymes which cleaved wild type SERP-1 (hNE, cathepsin G, chymotrypsin, and trypsin). Taken together, these findings indicate that the assignment of the P1–P1' residues is correct, and that the true target of SERP-1 is likely to be a serine proteinase with a preference for cleaving at basic residues.

REFERENCES

1. Upton, C., J. L. Macen, D. S. Wishart, and G. McFadden. 1990. Myxoma virus and malignant rabbit fibroma virus encode a serpin-like protein important for virus virulence. Virology. 179:618–631.
2. Macen, J. L., C. Upton, N. Nation, and G. McFadden. 1993. SERP1, a serine proteinase inhibitor encoded by myxoma virus, is a secreted glycoprotein that interferes with inflammation. Virology. 195:348–63.
3. Lucas, A., L. Liu, J. L. Macen, P. D. Nash, E. Dai, M. Stewart, W. Yan, K. Graham, W. Etches, L. Boshkov, P. N. Nation, D. Humen, M. Hobman, and G. McFadden. 1996. A virus-encoded serine proteinase inhibitor, SERP-1, inhibits atherosclerotic plaque development following balloon angioplasty. Circulation (in press).

81. SPI6, A NEW MOUSE SERPIN THAT IS RELATED TO MOUSE SPI3 AND HUMAN PI-6

Lisa Ooms, Jiuru Sun and Phillip Bird

Department of Medicine, Monash Medical School, Box Hill Hospital, Box Hill 3128, Australia

The majority of serpins are involved in the regulation of extracellular proteinases, however at least two serpins appear to function intracellularly: CrmA is a viral Asp-serpin involved in the regulation of apotosis and cytokine processing; and PI-6 is an endogenous Arg-serpin of unknown physiological function. The properties and distribution of PI-6 (1–3) suggest the existence of other endogenous intracellular serpins so we have surveyed a number of cDNA libraries for PI-6 related sequences. To this end we have isolated and characterized a 1.8 kb cDNA from a mouse embryo library that encodes a protein 56% identical to mouse SPI3 (the homologue of human PI-6). The gene for this new serpin has been designated Spi6 and mapping experiments show that it co-localizes with Spi3. Analysis of the distribution of SPI6 by Northern blotting shows that the mRNA is present in heart, spleen, lung and kidney. Two transcripts of approximately 2.4 kb and 4.4 kb are present in equivalent amounts in these tissues. Like SPI3/PI-6 and CrmA, SPI6 lacks a conventional signal sequence suggesting that it is also an intracellular serpin. Alignment of the SPI6 amino acid sequence with SPI3 and other serpins predicts that the P1–P1′ residues are Cys–Cys. The structure of the putative reactive center hinge region predicts that SPI6 is an inhibitory serpin. Experiments are in progress to identify interacting proteinases.

REFERENCES

1. Coughlin P, Sun J, Cerruti L, Salem H and Bird P. (1993). Cloning and molecular characterization of a novel human intracellular serine proteinase inhibitor. Proc. Natl. Acad. Sci. (USA) 90, 9417–9421.
2. Sun J, Rose J and Bird P (1995). Gene structure, chromosomal localization, and expression of the murine homologue of human proteinase inhibitor 6 (PI-6) suggests divergence of PI-6 from the ovalbumin serpins. J. Biol. Chem. 270, 16089–16096.
3. Scott FL, Coughlin P, Bird C, Cerruti L, Hayman JA and Bird P (1996). Proteinase inhibitor 6 cannot be secreted, which suggests it is a new type of cellular serpin. J. Biol. Chem. in press.

82. A NEW CLASS OF PROFLUORESCENT REAGENTS FOR THE DETECTION OF SERINE PROTEASES IN BIOLOGICAL SAMPLES

Beverly Z. Packard, Dmitri D. Toptygin, Ludwig Brand, and Akira Komoriya

OncoImmunin, Inc., 336 Paint Branch Drive, College Park, Maryland 20742 and Biology Department, Johns Hopkins University, Baltimore, Maryland 21218

Although the name serpin derives from the function most commonly associated with this group of molecules, their compact folded globular structure has permitted evolution into multiple conformational states which allow for a diversity of function. Specifically, it is the structure of the sequence variable reactive site loop, a stretch of ca. 15 amino acids, which has generated most attention in the past few years. Here, X-ray crystallographic data have defined elements comprising the four serpin structural classes, i.e., non-inhbiitory, cleaved, active, and latent. Moreover, recent structural data have clarified mechanisms of activation and dysfunction thereof, suggesting future designs for therapeutic protease inhibitors as well as substrates for detection and diagnosis.

As a first step in the design of useful reagents to assess the physiological roles of proteases and serpins in biological samples at the single cell level, we started with the amino acid sequence comprising the active site loop of one serpin, i.e., α_1-antitrypsin; a modified polypeptide substrate for the serine protease elastase was then synthesized. Included in the peptide design were reactive groups on both sides of the cleavage site for conjugation with fluorescent dyes. Additionally, in order to minimize interference from biological samples, we used dyes with both excitation and emission in the visible wavelength range. Upon addition of elastase, increase in fluoresence intensity due to cleavage of the doubly-labeled peptide was monitored as a measure of protease activity. Hence, this new class of reagents can be employed not only for characterizing protease activities in solution samples but also for histologic samples.

Although data from previous studies have shown that placement of fluorophores on both sides of a protease cleavage site can result in a protease detection reagent, the use of fluorescence resonance energy transfer (FRET) as a descriptor has limited the usefulness of this type of reagent, e.g., in fluorescence microscopy. In contrast, incorporation of our discovery of an alternative mechanism into the substrate design permits detection of serine protease activities in physiological samples such as blood, whole cells, and en bloc sections. Detection of protease activity in physiological settings, compared with identification of protease molecules, as is the case when antibodies are used for detection, will allow identification of physiologically significant protease activities.

83. SHEEP MAST CELL PROTEINASE CAUSES FIBRINOGENOLYSIS *IN VITRO*, IN THE PRESENCE OF EXCESS SHEEP PLASMA PROTEINASE INHIBITORS

Alan D. Pemberton, John F. Huntley,[†] and Hugh R. P. Miller

Department of Veterinary Clinical Studies, University of Edinburgh Veterinary Field Station, Easter Bush, Roslin, Midlothian, United Kingdom; [†]Moredun Research Institute, 408 Gilmerton Road, Edinburgh, Midlothian, United Kingdom

Chymotrypsin-like serine proteinases (chymases) are known to be expressed by subsets of mast cells in a number of mammalian species, including man, dog, mouse, rat and sheep. Sheep mast cell proteinase (SMCP) has been characterized as a chymase and immunolocalized in all gut and lung mast cells, and in a small percentage of skin mast cells. Chymases are released by mast cell degranulation during allergic inflammatory responses, but their target substrates are not well recognized. The ability of the native plasma serpin α_1-proteinase inhibitor (α_1PI) to control the activity of mast cell chymases has previously been studied, and found to vary from effective inhibition with rat mast cell proteinase-II, to the very slow inhibition of human skin mast cell chymase.

In this study, two serpins resembling α_1PI were purified from sheep serum, along with α_2-macroglobulin (α_2M). The identities of the inhibitors were verified by N-terminal amino acid sequencing and their interactions with SMCP were studied. The α_1PI-like protein of M_r 56,000 formed a 1:1 complex with SMCP which was stable to reducing SDS-PAGE conditions, but the rate of complex formation was slow (k_{ass} Å 10^3 $M^{-1}s^{-1}$). SMCP also formed a complex with α_2M, at a rate comparable with α_1PI. The higher M_r α_1PI did not inhibit SMCP, but was a very efficient inhibitor of trypsin. When excess serum was added to SMCP, residual amidolytic activity was detected, due to the entrapment of a significant proportion of SMCP by α_2M. A consequence of slow inhibition of SMCP by native plasma inhibitors is the opportunity for the enzyme to interact with potential target substrates in plasma. When plasma was added to SMCP, fibrinogen was observed to be cleaved, despite the presence of excess proteinase inhibitors. Thus, SMCP may prove to have a role in fibrinogenolysis and/or fibrinolysis.

84. EXPRESSION OF BOVINE PEDF: A NEUROTROPHIC NONINHIBITORY SERPIN

L. Alberto Perez-Mediavilla, Christina Chew,* Peter Campochiaro,* Donald J. Zack,* and S. Patricia Becerra

National Eye Institute, NIH, Bethesda, Maryland 20892 and *Department of Ophthalmology, Johns Hopkins Medical School, Baltimore, Maryland 21287

Pigment epithelium-derived factor (PEDF) was initially reported as a neurotrophic factor secreted by fetal human retinal pigment epithelium (RPE) cells in culture. PEDF protein promotes neuronal survival on cerebellar granule cell neurons and neurite outgrowth on retinoblastoma cells. PEDF has been identified as a serpin member by sequence homology. Recently, PEDF protein has been identified in bovine eyes in the interphotoreceptor matrix (IPM), the extracellular matrix between the RPE and the neural retina. The bovine protein is neurotrophic and behaves like a noninhibitory serpin. The source and complete sequence of the bovine IPM-derived protein are unknown. We have isolated RNA from RPE and retina tissues from adult bovine eyes. A total of 20 mg RNA from each sample was analyzed by northern blot using a 1.5 kb human PEDF cDNA as probe. A single 1.5 kb PEDF transcript was detected only for the RPE-derived RNA and not for the retina-RNA. We have isolated and cloned a bovine PEDF cDNA from an RPE cDNA library and determined its nucleotide sequence. The longest open reading frame in the PEDF cDNA had a coding capacity of 416 amino acid residues. The derived bovine sequence shared 88% identity with the known human sequence. The bovine cDNA fragment was placed under the control of a T7 RNA promoter. RNA was synthesized from the linearized PEDF cDNA plasmid using T7 RNA polymerase and the CAP analog 5'7MeGppp5'. Rabbit reticulocyte lysates programmed with bovine PEDF transcripts synthesized a unique translation product of 45,000-Mr that immunoprecipitated with polyclonal antiserum specific to PEDF. Thus our bovine PEDF cDNA clone is functional for expression of bovine PEDF protein in eukaryotes. However, when compared to human transcripts, the amount of PEDF protein synthesized from these transcripts was 4-fold lower for bovine than human PEDF. We prepared a PEDF expression vector containing a hybrid coding sequence composed of 90% C-terminus of bovine and 10% N-terminus of human PEDF. The hybrid PEDF transcripts were prepared and assayed for translation in rabbit reticulocyte lysates. The amount of product from the hybrid PEDF transcript was 4-fold higher than from the bovine PEDF. These results are consistent with the observation that the human PEDF sequence provides a more favorable context for translational initiation, ANNATGC, than the bovine context, GNNATGC. Our data indicate that bovine PEDF transcripts are preferentially expressed in RPE, suggesting that in the bovine eye the RPE acts as the source of the PEDF protein that accumulates in the IPM. However, in eukaryotic expression systems, bovine PEDF transcripts have lower translational efficiencies than human transcripts in eukaryotic systems.

85. LOSS OF THE LOW-FIELD PROTON RESONANCE SIGNAL OF THE HIS57-ASP102 HYDROGEN BOND IN A SERPIN/PROTEASE COMPLEX

Michael I. Plotnick, Leland Mayne, Norman M. Schechter, and Harvey Rubin

Pulmonary and Critical Care Division, Department of Medicine, and Departments
of Biochemistry and Biophysics, Dermatology, and Microbiology, University
of Pennsylvania, Philadelphia, Pennsylvania 19104

How serine protease inhibitors of the serpin family inhibit their target enzymes is not completely understood. Structural and biochemical studies have suggested that serpins utilize a mechanism that is distinct from the lock and key type binding proposed for most small protein protease inhibitors (standard mechanism inhibitors). ^{1}H-NMR spectroscopy was used in the present study to demonstrate a fundamental difference in the atomic environment of the His57–Nd1 proton of α-chymotrypsin in complex with serpins when compared to uncomplexed enzyme and enzyme in complex with standard mechanism inhibitors. A low-field resonance signal at 13–18ppm has been demonstrated for the His57–Nd1 proton in virtually every serine protease studied, consistent with proposals that the hydrogen bond between Asp102–Od2 and His57–Nd1 plays an important role in serine protease catalytic activity. The present work demonstrates that the low-field signal of the His57–Nd1 proton of α-chymotrypsin is observed in complexes (32–72kDa) with the standard mechanism inhibitors bovine pancreatic trypsin inhibitor and kunitz-type soybean trypsin inhibitor. However, the low-field signal is absent from complexes (70kDa) of α-chymotrypsin with a recombinant variant of α-1-antichymotrypsin (rACT–P3P3′). This implies that the Asp102–His57 hydrogen bond is distorted or disrupted in a serpin/protease complex. The distortion of the His57–Asp102 hydrogen bond in a serpin/protease complex may result in a marked reduction in the catalytic efficiency of the protease and explain the stability of covalent complexes that break down on time scales of hours or days.

86. PANCREATIC PROTEIN C INHIBITOR

Klaus-P. Radtke, José A. Fernández,* Judith S. Greengard,* Thomas Böttger,[$]
David J. Loskutoff,* John H. Griffin,* and I. Scharrer

University of Frankfurt, Dept. of Internal Medicine, Frankfurt, FRG.; [$]University of Mainz,
Dept. of Surgery, Mainz, FRG; *The Scripps Research Institute, La Jolla, California

Protein C inhibitor (PCI), a member of the serpin family, was originally identified as a physiologic regulator of the antithrombotic serine protease, activated protein C. Further characterization of the inhibitor showed that plasma PCI also inhibits other plasma proteases including urokinase, tissue plasminogen activator, thrombin, factor XIa, factor Xa, and plasma and tissue kallikreins. Recently PCI was also found in extra-vascular systems. High concentrations of PCI detected in human seminal plasma suggest a role for PCI in human fertility. Significant concentrations of PCI message located in proximal tubular kidney cells as well as PCI antigen in their primary lysosomes suggest an intracellular function for PCI in this environment. *In vitro* experiments showed that target proteases for PCI also include pancreatic enzymes chymotrypsin, trypsin and elastase. However, no physiologic significance for these reactions has been reported so far. Since vascular thrombosis and systemic hypercoagulable states are known complications of acute pancreatitis and other pancreatic diseases, we asked whether PCI may be involved in proteolytic regulation in the pancreas. In this study we present evidence that PCI is present in human pancreas. Reverse transcription of total pancreas RNA followed by two rounds of PCR showed the presence of PCI message in total pancreas RNA, and a resulting PCI cDNA was identical with PCI cDNA from human liver. This observation was confirmed by *in situ* hybridization experiments with paraffin sections of normal human pancreas tissue. Using ^{35}S labeled antisense RNA probes gave a strong signal, indicating the presence of PCI RNA was located in the exocrine areas of pancreas tissue. The signal was concentrated in the pancreatic acinus cells which store and secrete the zymogens of the pancreatic target enzymes chymotrypsin, trypsin, and elastase. Islets of Langerhans did not stain positive for PCI RNA. To determine if pancreas PCI RNA was expressed, pancreatic fluid was analyzed by SDS-PAGE and immunoblotting. Using monospecific antibodies directed against human plasma PCI, a 57,000 MW protein band was observed which comigrated with purified human plasma PCI. Our results show that pancreatic cells contain a significant concentration of PCI message. This message is localized in the exocrine areas of pancreas tissue. We suggest that PCI antigen found in pancreatic fluid is likely to originate in the pancreas. The role of pancreatic PCI is unknown at present. However, colocalization of PCI and zymogens of its target proteases in the exocrine tissue of pancreas indicate that PCI may be part of the inhibitor potential which protects pancreatic tissue from auto degradation. These inhibitors normally prevent the release of active pancreatic proteases into the vasculature or microcirculation where destabilization of the coagulation balance and subsequent thrombus formation could occur. Another possible role for pancreatic PCI could involve regulation of pancreatic kallikrein, a serine protease which participates in the generation of vasoactive kinins which in turn are directly involved in the development of shock during pancreatitis and pancreatic necrosis.

87. ACUTE PHASE INDUCTION IN RABBIT ACTIVATES AN INDUCIBLE MEMBER OF α_1-ANTIPROTEINASE FAMILY

Bimal K. Ray, Xiong Gao, and Alpana Ray

Department of Veterinary Pathobiology, University of Missouri, Columbia, Missouri 65211

Studies on the acute phase inducible α_1-antirypsin gene in rabbit have revealed a new member of this antiproteinase family, designated as α_1-AT S2. Sequence analysis has revealed that this protein belongs to a serine protease inhibitor (serpin) superfamily and functional assay has demonstrated that it is capable of blocking function of several proteases. This gene has been found to be highly induced in the liver during many inflammatory episodes. Induction of this gene under inflammatory condition suggests a regulatory role of this unique serpin in the pathogenesis of diseases associated with inflammatory conditions. In search of the induction mechanism, we have investigated the promoter region of this gene to look for the structural elements responsive to different inflammatory agents such as bacterial lipopolysaccharide (LPS) and proinflammatory cytokines e.g., IL-1 and IL-6. Using gel electrophorectic mobility shift assay, transient transfection and mutagenesis studies we have located several promoter elements that are capable of binding to transcription factors activated during inflammation. These elements also drive transcription of a reporter gene indicating their potential role in the induction process. Expression of α_1-AT S2 is significantly induced in the blood monocyte following exposure of the cells to LPS. Gel electrophoretic mobility shift and transfection assays have revealed a macrophage-specific enhancer in the distal promoter and located between −2438 and −1990 basepairs upstream of the transcription initiation site. Transcription factor, NF-κB has been found to bind to this distal enhancer and drive inducible expression of the gene. Two members of the NF-κB family, RelA and NFKB1, are involved in the binding to this distal enhancer and promoting transcription. Selective mutation of this enhancer virtually abolished the LPS-induced expression in the monocyte cells indicating a major role of the NF-κB/RelA family of transcription factors in the expression of α_1-AT S2 in macrophage.

88. DIVERSE EFFECTS OF pH ON THE INHIBITION OF HUMAN CHYMASE BY SERPINS

Norman M. Schechter, Michael Plotnick, Michele Walter, Barry S. Cooperman, and Harvey Rubin

Departments of Dermatology, Medicine, and Chemistry, University of Pennsylvania, Philadelphia, Pennsylvania 19104

Inhibition of serine proteases by serpins may proceed through an acylation reaction resembling that observed in serine protease hydrolysis of peptide bonds. Current studies on certain serpin-protease pairs indicate a pH dependence for inhibition similar to that associated with catalytic turnover. In this study the effect of pH on the interaction of human chymase, a mast cell protease, with serpins is investigated. Human chymase is inhibited by α_1-antichymotrypsin (ACT) and α_1-proteinase inhibitor (PI). Both reactions occur with a high apparent stoichiometry of inhibition (SI) that is attributed to a hydrolytic or substrate-like reaction inactivating the inhibitor at a rate competitive with inhibition. Although the mechanism producing the partitioning of the reaction is uncertain, a suicide substrate model of reaction appears to apply at least in a descriptive sense. Lowering the pH from 8 to 5.5 in the reaction of chymase with ACT reduced the SI from 4 to near 1, consistent with a significant effect of pH on the substrate aspect of the overall reaction. Reactive site variants of ACT (L358) also were analyzed. The SI values for L358W, L358M, and L358F, which are 1.5, 2 and 6.5 respectively, also decreased toward a value near 1 at lower pHs. In contrast to these results, the reaction of chymase with PI demonstrated an increase in the SI from 6 to 15 over the same range of pH. Inhibition rate constants (k_I) for the reactions also were measured. Rate constants were determined under pseudo-first order conditions where the inhibitor concentration was at least 10 fold higher than the product of the SI × $[E]_0$. ACT and reactive site variants demonstrated a relatively small change in k_I at pH 8.0 and 5.5; in some cases the k_I value modestly increased at pH 5.5. Assuming a suicide substrate model of inhibition, multiplication of k_I × SI to give a rate constant (k_{Ia}) adjusted for partitioning resulted in k_{Ia} values 2–3 fold lower at pH 5.5 than 8.0 for all variants. Similar measurements with PI demonstrated a much more dramatic effect of pH on k_I; approximately a 100 fold reduction in k_I was obtained by lowering the pH from 8.0 to 5.5. Applying a suicide substrate model to this reaction resulted in a 50 fold decrease in k_{Ia} due to pH. The higher SI and lower k_I for the chymase-PI reaction at pH 5.5 was not due to unfolded inhibitor as measured by complete retention of inhibitory activity toward chymotrypsin in parallel studies. We speculate that when there is a large effect of pH on k_I, the inhibitory pathway competitive with the hydrolytic reaction is significantly affected and reduced to a greater extent than the hydrolytic pathway, thereby producing an increased SI as observed in the chymase-PI reaction. When the pH effect on k_I is relatively small, the rate of reaction along the entire k_I pathway is changed little while the rate of the hydrolytic pathway decreases significantly. This produces a decrease in the SI due to pH as observed for the reactions of chymase with ACT and ACT variants. The reasons for the difference in the effect of pH on the k_I (and k_{Ia}) for the chymase-ACT (and variants) reaction compared to the chymase-PI reaction are presently unclear.

89. DIVERGENT TISSUE DISTRIBUTION OF HOMOLOGOUS RAT SERINE PROTEASE INHIBITORS

Sarah Jane Schwarzenberg,* Sherani Amarasinghe,*[&] and David C. LaPorte[&]

*Departments of Pediatrics and [&]Biochemistry, University of Minnesota, Minneapolis, Minnesota 55455

The serine protease inhibitor (serpin) family of proteins perform a wide variety physiologic functions. We have examined three highly homologous rat serpins, the Spi 2.1, 2.2 and 2.3 genes. These genes exhibit divergent regulation in a variety of physiologic circumstances. Expression of Spi 2.1 and 2.3 is dependent on growth hormone while expression of Spi 2.2 is not. Spi 2.2 mRNA is an acute phase reactant, while Spi 2.1 and 2.3 mRNA decrease with inflammation. In the current work, we examined the distribution of the protein products of the Spi genes. To distinguish among these proteins, polyclonal antisera were prepared which recognized the reactive centers which are unique to each protein. Although the Spi 2.1, 2.2, and 2.3 proteins have very similar amino acid sequences, Western blot analyses demonstrated that each serpin has a unique tissue distribution. Spi 2.1 protein was present in several tissues, Spi 2.2 protein was found predominantly in the lung, and Spi 2.3 protein was found in pancreas, liver, and kidney. Tissue Spi 2.3 protein was not glycosylated, raising the possibility that this protein functions intracellularly. Cellular localization of these serpins was investigated in the rat kidney by immunohistochemistry. Spi 2.1 protein was most prominent in or below the brush borders of tubules in both cortex and medulla. Spi 2.2 protein was localized to the tubular epithelial cells of the cortex, with no staining of the medulla. Spi 2.3 protein localized predominantly to the basolateral portion of the cortical tubular epithelial cells. The divergent tissue localization suggests that, despite extensive sequence homologies, these proteins have unique functions.

90. CHARACTERIZATION OF HUMAN PEDF: OVEREXPRESSION IN EUKARYOTIC CELLS

[¶]Efstratios Stratikos, [§]Elena Alberdi, [¶]Peter G. W. Gettins, and [§]S. Patricia Becerra

[¶]Department of Biochemistry, University of Illinois at Chicago and [§]National Eye Institute, NIH

Pigment epithelium-derived factor (PEDF), a member of the serpin superfamily, is an extracellular neurotrophic factor that induces neuronal differentiation in retinoblastoma cells and neuronal survival in cerebellar granule neuron cells. No proteinase inhibitory activity has yet been demonstrated. Recently bovine PEDF has been purified from the extracellular space between the retinal pigment epithelium and the neural retina, termed the interphotoreceptor matrix, from bovine eyes, with yields of ~1mg of native protein per 750 eyes. No comparable source is available for the human protein. An alternative recombinant source of folded PEDF is therefore highly desirable for biological and biochemical studies. Overexpression of a truncated form of human PEDF (residues 44–418) in *E. coli* resulted in accumulation of the rPEDF in inclusion bodies [Becerra et al. (1993) *J. Biol. Chem. 268*, 23148–23156]. To obtain normally folded and glycosylated human rPEDF in large amount we have stabley transfected baby hamster kidney cells with two types of expression plasmid containing human PEDF cDNA and under the control of either RSV (pBK-PEDF) or SV40 promoters (pMA-PEDF). Stable transfectants were selected by resistance to neomycin and methotrexate. Cells were grown to confluence in the presence of fetal calf serum and then cycled between medium with and without serum. Serum-free cycles were harvested and analyzed for the presence of PEDF by immunoreaction with anti-PEDF in western transfers. Levels of PEDF increased to a plateau by cycle 3, with levels >3-fold higher in pMA-PEDF-transfected cells than in pBK-PEDF-transfected cells. PEDF represented 80% and 50% of total medium protein for pMA- and pBK-transfected cells respectively. rPEDF was purified to homogeneity by ammonium sulfate precipitation, cation exchange chromatography and ultrafiltration, with a 10kDa-cutoff membrane. The yield of purified PEDF was 23mg/L of culture medium. The N-terminal residues of both recombinant and natural human PEDF were identical, Asn-21, indicating that the secreted form of the protein lacks the signal sequence of the first 20 residues. N-glycosidase F treatment increased the mobility on SDS-PAGE indicating that the recombinant protein was glycoylated. Circular dichrosim (CD) spectroscopy indicated that the secreted rPEDF was a normally folded globular protein. Limited proteolysis of rPEDF with chymotrypsin gave a 46kDa fragment on SDS-PAGE, as found with natural PEDF and consistent with cleavage within the reactive center. CD-monitored thermal denaturation of native and chymotrypsin-cleaved rPEDF indicated a reduction in stability following proteolysis, with a change in T_m from 57°C to 52°C, analogous to changes in ovalbumin upon reactive center cleavage. A similar reduction in stability was seen by fluorescence-monitored guanidine hydrochloride-induced unfolding. The mid-points for unfolding were at 1.5 and 1.2M guanidine hydrochloride for native and cleaved rPEDF respectively. The small difference in the CD spectra of native and cleaved rPEDF were similar to what we have found with ovalbumin and are consistent with cleavage in the reactive center without loop insertion into β-sheet A. Although PEDF binds to heparin-agarose, no evidence was found from changes in tryptophan fluorescence for a conformational change upon binding heparin. We have performed neurite-outgrowth assays in Y-79 cells and found that our purified rPEDF had the expected neurotrophic activity. In conclusion, our expression system produces high levels of folded and glycosylated rPEDF which has the same biochemical and biological properties as naturally-occurring human PEDF. The ability to produce large amounts of rPEDF using the expression vector pMA-PEDF will facilitate further biophysical and biochemical studies of PEDF.

Supported by the National Eye Institute (SPB and EA) and HL49234 (PGWG).

91. EXPRESSION OF NORMAL AND VARIANT ANGIOTENSINOGENS FOR FUNCTIONAL ANALYSIS

R. M. A. Streatfeild-James, D. Williamson, P. Coughlin, P. Stein, D. Tewkesbury and R. W. Carrell

University Dept. of Haematology, MRC Centre, Hills Rd., Cambridge, CB2 2QH, U.K.

Angiotensinogen is a non-inhibitory member of the serpin family of proteins and a key component of the renin-angiotensin system which plays a major role in the control of blood volume and blood pressure. Angiotensinogen is cleaved by renin, releasing the N-terminal decapeptide angiotensin I which is further cleaved by angiotensin converting enzyme to form the octapeptide hormone and vasoconstrictor angiotensin II.

Gene linkage and association studies strongly support involvement of the angiotensinogen gene locus in essential hypertension and some studies suggest that specific molecular variants of angiotensinogen may be associated with hypertension and pre-eclampsia. A functional investigation is needed to establish the significance of these molecular variations. In order to do this kinetic studies of the interaction between recombinant normal and variant angiotensinogens and renin will be performed.

We have developed an expression system for the production of normal and site-directed mutants of human angiotensinogen both in E.coli and a mammalian (BHK) cell line. For expression in E.coli, the pelB leader sequence was used to direct the synthesised protein to the bacterial periplasm. Initial expression studies produced non-glycosylated recombinant angiotensinogen which was primarily associated with the insoluble cell lysate fraction. In BHK cell cultures human angiotensinogen has been expressed which is of a comparable molecular weight to the human plasma protein.

92. THE RABBITPOX VIRUS ANTI-APOPTOSIS SERPINS SPI-1 AND SPI-2

Peter C. Turner, Michael A. Brooks, Joanne Macen,[†] R. C. Bleackley,[†] and Richard W. Moyer

Dept. of Molecular Genetics and Microbiology, University of Florida, Gainesville, Florida 32610 and [†] Department of Biochemistry, University of Alberta, Canada

Poxviruses are unique in being the only DNA viruses which encode and express serpin genes. The orthopoxviruses, including the prototype vaccinia virus (VV), the closely related rabbitpox virus (RPV), and cowpox virus (CPV), each contain three serpins: SPI-1, SPI-2 (*crmA*) and SPI-3. The SPI-3 gene (ORF K2L in VV) encodes a glyco-protein which functions to inhibit cell-cell fusion following infection. The SPI-1 and SPI-2 genes located at the right end of the genome probably derive from duplication of a single ancestral host gene; they encode non-glycosylated proteins of 40 kDa and 38 kDa, respectively. Although overall the SPI-1 and SPI-2 proteins are 45% identical, the sequences of the reactive site loops are unrelated, suggesting that the two serpins act against different proteinases. For SPI-2 the P1 residue is Asp; for SPI-1 the predicted P1 residue is Phe. The SPI-2 or *crmA* (cytokine response modifier) gene of CPV has been shown to have anti-apoptosis activity in a variety of contexts, and is an inhibitor of the cysteine proteinase ICE (interleukin-1ß converting enzyme), which cleaves after Asp. Activation of ICE-like enzymes is thought to be a pivotal event in the final stages of apoptosis or programmed cell death.

Our results have shown that the RPV SPI-1 gene is expressed early, prior to viral DNA replication. RPVΔSPI-1, a rabbitpox virus mutant specifically inactivated for the SPI-1 gene, is unable to form plaques on several cell lines including A549 (human lung fibroblast carcinoma), but growth on other lines such as CV-1 (monkey kidney) and RK-13 (rabbit kidney) is normal. The host range restriction observed on non-permissive cells is due to the induction of apoptosis. A549 cells infected with the RPVΔSPI-1 mutant show condensed nuclear morphology, and many free 3' ends are detected by the TUNEL assay, indicating that DNA degradation is occurring. Viral early and late protein synthesis appears normal, as does viral DNA replication and the proteolytic processing of structural proteins that is associated with virion morphogenesis. However, very few infectious mature virus particles are produced. In wild type RPV infection the SPI-1 protein interacts with cellular and/or viral component(s) to prevent the induction of apoptosis.

The features of a poxvirus infection which trigger apoptosis have been investigated by the use of metabolic inhibitors. The protein synthesis inhibitor cycloheximide blocks apoptosis in A549 cells infected with RPVΔSPI-1mutants. Cytosine arabinoside (araC) and hydroxyurea, which prevent viral DNA replication and late protein expression, also prevent apoptosis. These results suggest that although the SPI-1 protein is expressed early, the event initiating apoptosis requires DNA replication and/or late protein synthesis. Rifampicin, a specific inhibitor of viral morphogenesis, did not block apoptosis in A549 cells infected with RPVΔSPI-1, indicating that assembly of mature virions is not required to trigger programmed cell death.

An RPV mutant lacking the SPI-2 gene leads to apoptosis in a pig kidney cell line, but not in A549 cells. This indicates that the host factors that interact with SPI-1 and with SPI-2 are different, and suggests that SPI-1 and SPI-2 act at different points in the apoptotic pathway. An RPV double mutant defective for both SPI-1 and SPI-2 is more attenuated than is either single mutant in mice following intranasal infection and in rabbits following intradermal inoculation, implying that the two genes act synergistically to promote virulence. *In vitro* studies on cytolysis of RPV-infected target cells by CTLs also indicate that *both* SPI-1 and SPI-2 act to inhibit cytotoxic activity toward wtRPV infected target cells.

Stable A549 cell lines have been made in which either the SPI-1 or SPI-2 protein is constitutively expressed. A549 cells containing the SPI-1 gene under the control of either the metallothionein or cytomegalovirus promoter are now able to support productive infection by RPVΔSPI-1 mutants, suggesting that the SPI-1 protein is synthesized and functional. The A549/SPI-1 lines are being evaluated for their resistance to apoptosis triggered by a variety of agents to determine whether SPI-1 is capable of inhibiting apoptosis in the absence of other poxvirus gene products. RPV SPI-1 protein has been overexpressed and purified, and is currently being tested for anti-apoptosis activity *in vitro*, and for inhibition of purified ICE-like enzymes in biochemical assays.

93. UNIQUE REGULATION OF THE SYNTHESIS OF α_1-PROTEINASE INHIBITOR IN THE CORNEA

Sally S. Twining and Goran Boskovic

Department of Biochemistry, Medical College of Wisconsin, Milwaukee, Wisconsin 53226

The cornea is an avascular transparent tissue which synthesizes α_1-proteinase inhibitor (α_1-PI). This protein accounts for 10% of the saline extractable proteins. Organ culture experiments showed unique pattern of regulation of the synthesis of this inhibitor. Synthesis in the cornea was stimulated by IL-1 , retinol and retinaldehyde but not by IL-6 or retinoic acid. In Hep G2, CaCo2 and MCF-7 cells, monocytes and monocytes, synthesis of this inhibitor was stimulated by IL-6 but not by the retinoids. Of these cells, α_1-PI concentrations were increased by IL-1 only monocytes and macrophages. Retinol increased α_1-PI at both the level of protein and mRNA. In addition to previously reported 5′ untranslated region splice variants of the mRNA, a new variant was detected in the cornea and in monocytes using RT-PCR techniques. The PCR product was cloned and sequenced. In this variant both exons 1b and 1c are spliced out producing a mRNA which is composed of the exons Ia–II–III–IV–V. A unique transcription start site was detected for the cornea using a primer extension assay. This site is located 19 nucleotides from the 5′ end of exon Ia which is between the two sites previously observed for monocytes and macrophages. The unique regulation of α_1-PI concentrations in the cornea is probably required to protect the cornea from degradation by elastase released by neutrophils which invade the cornea in response to wounds and bacterial infections.
    Supported by NIH Grant EY06663.

The Organizers wish to thank the participants in the 1996 Serpins Symposium and look forward to seeing them again at the serpins meeting in 1999.

CONTRIBUTORS

Eric H. Ball
Department of Biochemistry
University of Western Ontario
London, N6A 5C1 Canada

S. Patricia Becerra
Laboratory of Retinal and Molecular Biology
National Eye Institute
National Institutes of Health
Building 6, Room 308
6 Center Drive MSC 2740
Bethesda, Maryland 20892-2740

Debra L. Becker
Hamilton Civic Hospitals Research Centre
711 Concession St.
Hamilton, Ontario, Canada L8V 1C3

Anders Bjartell
Department of Urology
Lund University
Malmö University Hospital
S-205 02 Malmö, Sweden

Ingemar Björk
Department of Veterinary Medical Chemistry
Swedish University of Agricultural
  Sciences
Uppsala Biomedical Center
S-751 23 Uppsala, Sweden

Stephen P. Bottomley
Department of Biochemistry and
  Molecular Biology
Monash University
Clayton, Victoria 3168, Australia

Robin Carrell
Department of Haematology
University of Cambridge
MRC Centre, Hills Road,
Cambridge CB2 2QH

Anders Christensson
Department of Vascular and Renal Diseases
Lund University, Malmö University Hospital
S-205 02 Malmö, Sweden

Frank C. Church
Departments of Pathology and Lab
  Medicine and Medicine and Center for
  Thrombosis and Hemostasis
School of Medicine
University of North Carolina at Chapel Hill
Chapel Hill, North Carolina 27599-7035

Scott T. Cooper
Department of Biology and Microbiology
University of Wisconsin — La Crosse
La Crosse, Wisconsin 54601

Dennis D. Cunningham
Department of Microbiology and
  Molecular Genetics
College of Medicine, University of California
Irvine, California 92697-4025

Alvin E. Davis III
Division of Nephrology
Children's Hospital Research Foundation
3333 Burnet Avenue
Cincinnati, Ohio 45229-3039

Yves A. DeClerck
Department of Pediatrics and Department
  of Biochemistry and Molecular Biology
Childrens Hospital Los Angeles
University of Southern California
Los Angeles, California 90027

Frances M. Donovan
Department of Microbiology and
  Molecular Genetics
College of Medicine
University of California
Irvine, California 92697-4025

Daniel T. Eitzman
Howard Hughes Medical Institute
University of Michigan Medical Center
MSRB 1, Room 4520
1150 W. Medical Center Drive
Ann Arbor, Michigan 48109-0650

Jim C. Fredenburgh
Hamilton Civic Hospitals Research Centre
711 Concession St.
Hamilton, Ontario, Canada L8V 1C3

David Ginsburg
Howard Hughes Medical Institute
University of Michigan Medical Center
MSRB 1, Room 4520
1150 W. Medical Center Drive
Ann Arbor, Michigan 48109-0650

Peter J. Hansen
Department of Dairy and Poultry Sciences
University of Florida
Gainesville, Florida 32611-0920

Mary Hendrix
Department of Anatomy and Cell Biology
College of Medicine, University of Iowa
1115 Bowen Science Bldg.
Iowa City, Iowa 52242-1107

Paul C.R. Hopkins
Gladstone Institute of Cardiovascular
  Disease
University of California
San Francisco, California 94141-9100

Suzan Imren
Department of Pediatrics
Childrens Hospital Los Angeles
University of Southern California
Los Angeles, California 90027

N. Jain
Institute of Molecular and Cell Biology
10 Kent Ridge Crescent
Singapore 119260

Haobo Jiang
Department of Biochemistry
Kansas State University
Manhattan, Kansas 66506

Michael R. Kanost
Department of Biochemistry
Kansas State University
Manhattan, Kansas 66506

Walter E. Laug
Department of Pediatrics
Childrens Hospital Los Angeles
University of Southern California
Los Angeles, California 90027

Daniel A. Lawrence
American Red Cross
Jerome H. Holland Laboratory
Department of Biochemistry
15601 Crabbs Branch Way
Rockville, Maryland 20855

Hans Lilja
Department of Clinical Chemistry
Lund University, Malmö University Hospital
S-205 02 Malmö, Sweden

Wen-Jun Liu
Department of Dairy and Poultry Sciences
University of Florida
Gainesville, Florida 32611-0920

David Lomas
Department of Haematology
University of Cambridge
MRC Centre, Hills Road
Cambridge CB2 2QH

Alexandra Lucas
Division of Cardiology
Department of Medicine
University of Alberta
Edmonton, Alberta, Canada T6G 2H7

Edwin L. Madison
The Scripps Research Institute
Department of Vascular Biology
La Jolla, California 92037

Grant McFadden
Department of Microbiology and Immunology
University of Western Ontario
Robarts Research Institute
1400 Western Road
London, Ontario, Canada N6G 2V4

Peter Mikus
Department of Medical Biochemistry and
  Biophysics
Umeå University
S-90187 Umeå, Sweden

Anthony M.P. Montgomery
Department of Immunology
The Scripps Research Institute
La Jolla, California 92037

Barbara M. Mueller
Department of Immunology
The Scripps Research Institute
La Jolla, California 92037

Piers Nash
Department of Biochemistry
University of Alberta
Edmonton, Alberta, Canada T6G 2H7

Tor Ny
Department of Medical Biochemistry and
  Biophysics
Umeå University
S-90187 Umeå, Sweden

Steven T. Olson
Center for Molecular Biology of Oral Diseases
University of Illinois
Chicago, Illinois 60612-7213

Philip A. Pemberton
LXR Biotechnology
1401 Marina Way South
Richmond, California 94804

Ralph A. Reisfeld
Department of Immunology
The Scripps Research Institute
La Jolla, California 92037

Ruth Sager
Department of Cancer Genetics
Dana-Farber Cancer Institute
44 Binney Street
Boston, Massachusetts 02115

Guy S. Salvesen
The Burnham Institute
10901 North Torrey Pines Road
San Diego, California 92037

B. D. Sanwal
Department of Biochemistry
University of Western Ontario
London, Canada N6A 5C1

Shijie Sheng
Department of Cancer Genetics
Dana-Farber Cancer Institute
44 Binney Street
Boston, Massachusetts 02115

Alan R. Stafford
Hamilton Civic Hospitals Research Centre
711 Concession St.
Hamilton, Ontario, Canada L8V 1C3

Penelope Stein
Department of Haematology
University of Cambridge
MRC Centre, Hills Road
Cambridge CB2 2QH

Stuart R. Stone
Department of Biochemistry and
  Molecular Biology
Monash University
Clayton, Victoria 3168 Australia

Douglas M. Tollefsen
Divison of Hematology
Department of Internal Medicine
Washington University School of Medicine
St. Louis, Missouri

James Travis
Department of Biochemistry & Molecular
  Biology
University of Georgia
Athens, Georgia 30602

Jeffrey I. Weitz
Hamilton Civic Hospitals Research Centre
711 Concession St.
Hamilton, Ontario
Canada L8V 1C3

James C. Whisstock
Department of Haematology
University of Cambridge
MRC Centre, Hills Road
Cambridge CB2 2QH U.K.

INDEX

The manufacturer's authorised representative in the EU is Springer
Nature Customer Service Centre GmbH, Europaplatz 3, 69115 Heidelberg,
Germany. If you have any concerns regarding our products, please
contact ProductSafety@springernature.com

Printed and bound by CPI Group (UK) Ltd, Croydon, CR0 4YY
26/04/2026
02097340-0001